The Common Fisheries Policy

The Common Fisheries Policy

The Quest for Sustainability

Ernesto Penas Lado

Directorate General for Fisheries of the European Commission
Brussels
Belgium

WILEY Blackwell

This edition first published 2016 © 2016 by John Wiley & Sons, Ltd

Registered Office
John Wiley & Sons, Ltd, The Atrium, Southern Gate, Chichester, West Sussex, PO19 8SQ, UK

Editorial Offices
9600 Garsington Road, Oxford, OX4 2DQ, UK
The Atrium, Southern Gate, Chichester, West Sussex, PO19 8SQ, UK
111 River Street, Hoboken, NJ 07030–5774, USA

For details of our global editorial offices, for customer services and for information about how to apply for permission to reuse the copyright material in this book please see our website at www.wiley.com/wiley-blackwell.

Library of Congress Cataloging-in-Publication data applied for

ISBN: 9781119085645

A catalogue record for this book is available from the British Library.

Wiley also publishes its books in a variety of electronic formats. Some content that appears in print may not be available in electronic books.

Cover image: Fishing trawler at sea in rough seas, North Sea / The Image Bank © Arnulf Husmo

Set in 9/12.5pt Meridien by SPi Global, Pondicherry, India

1 2016

Contents

Preface

The Common Fisheries Policy (CFP) is one of the most integrated policies of the European Union. Under the Treaty of Lisbon the CFP (in particular its core business, the conservation and management of biological resources) remains in the limited group of its five exclusive competences, together with such fundamental pillars of the EU as trade policy, internal market, monetary policy (for the Member States of the Euro) and the customs union.

This makes the CFP a showcase of EU policy; its success or failure is important in terms of the way public opinion, both in Europe and abroad, sees the added value of the Union, way beyond the economic importance of the fisheries sector in the context of the European economy.

The world of fisheries tends to be conservative and resistant to change. For this reason, a fundamental characteristic of the CFP is that it has been and is still today, considerably shaped up by the weight of its own history. Surprisingly, the CFP today is influenced by political decisions taken 30 years ago and the established *status quo* tends to be extremely difficult to amend or reform.

To understand the policy of today it is essential to know the historical background leading up to its establishment. The Spanish-American writer George Santayana said: 'those who cannot remember the past are condemned to repeat it'. It is important to ensure that the historical experience of those who lived through the birth and development of the CFP can be transmitted to successive generations dealing with this policy in the present and the future. This is the purpose of this book.

I do not try to be systematic or comprehensive. Rather than an academic tutorial on the CFP, I believe it is more useful to offer an account of my own experience over more than 20 years dealing with the CFP in the European Commission and several more dealing with fisheries management outside it. For this reason my account of the CFP is partial and focuses on those policy areas where I have direct experiences allowing me to draw some lessons. The reader will probably miss a chapter on the fish market policy, or fish trade policy, and certainly much more detail about bilateral fisheries agreements, for example. These are the areas where I don't have a substantial direct experience.

This book is addressed to all those who join in the fascinating task of dealing with the Common Fisheries Policy from any angle and I hope this book will be useful for them. I have written it thinking of the kind of book I would have liked to read when I started dealing with this policy many years ago.

I assume that potential readers are reasonably familiar with both fishery management policy and the European Union, its institutions and procedures. In any case, a Glossary has been included to help.

The idea of an official from the European Commission writing about the CFP is not new. Other colleagues have done this before me: John Farnell and James Elles in 1984 and Mike Holden in 1994 have written about the origins of the CFP. Ronán Long and Tony Curran in 2000 have written about controlling the CFP after their experience of working for the Commission services. But these important contributions must be complemented by a more updated view on the developments of the CFP in recent years.

These developments have consistently pursued one goal, which summarises the objectives of this policy in the Treaty: achieving biological, economic and social sustainability, as the basic tenet of the CFP.

The reader will find in the following pages a wholehearted defence of the role of the European Commission in the CFP. Needless to say the Commission's role in running this policy can and will always be questioned. However, I remain convinced that most of the positive developments in the policy have been achieved thanks to the Commission's determination and long term vision.

Brussels, June 2015

Acknowledgements

I am indebted to a number of key contributors to this book.

My former Director General Lowri Evans encouraged me to write these pages and authorised me to publish them. She also provided useful guidance on orientation and content.

Willem Brugge did a very complete revision of an early draft and provided me with very useful comments and suggestions on both content and style.

Ken Patterson gave me the benefit of his in-depth knowledge of a number of aspects of the CFP to enrich the text and get certain facts right.

Enrique López Veiga shared with me his encyclopedic knowledge of fisheries with an invaluable insight into key historical developments of the CFP.

Martin Newman was so kind to spend part of his retirement to do a thorough revision of the text, including a much-appreciated polishing of the English.

Other important contributors to different parts of the text were Eskild Kirkegaard on conservation and scientific matters, Miguel Peña, especially on economic questions, and Adela Rey on international law. Angel Calvo and Tim Lemmens provided some figures for this book.

Nadine Marcos kindly helped me find a number of references. A particularly difficult one could only be found thanks to Fotini Missailidou.

Last but certainly not least, my wife Amalia de Diego used part of her spare time to help me edit these pages, removing bugs and digging out bibliographic references. She also helped me underline the importance of market issues in the CFP.

My sincere gratitude to all of them.

Disclaimer

The opinions expressed in this book are those of the author and not those of the European Commission.

CHAPTER 1

Why the common fisheries policy is important

Europe and fisheries

The CFP as a key European policy

The Common Fisheries Policy (CFP) is, as one of the most integrated policies of the European Union, a showcase of European policy making. As such, its importance in the context of the European integration far exceeds the minor economic importance of the fisheries sector in the wider European economy.

For many years, the CFP has been subject to criticism from almost all walks of life in Europe and, more often than not, it has been a favourite example of bad European policy-making among Eurosceptics. Beyond politics, criticism of the CFP has also come very often from academia (Khalilian et al., 2010). As we will see, the Commission itself has often been very forthcoming in recognising the policy's shortcomings, in an exercise of transparency that has little comparison in other policy areas.

The tradition of CFP-bashing is as old as the policy itself. Yet, the policy has been remarkably resilient over its 30 years of existence. This apparent contradiction of a strongly criticised policy that has, however, ensured a remarkable continuity over the years constitutes a fascinating study.

The need for this policy is obvious. Fishery resources must be managed, like all natural resources. Given their migratory, trans-boundary nature, European fishery resources must be managed jointly among the European countries involved. Thus, there is the need for a Common Fisheries Policy.

The fishing sector in Europe: some facts

The fisheries and aquaculture sector in the European Union is very significant in terms of fishery production in absolute terms. With landings of more than 6 100 000 tonnes *per annum*, the Union of 27 ranked fifth in the world, after China, India, Peru and Indonesia in 2012 in terms of the magnitude of catches (Facts and Figures of the Common Fisheries Policy, 2012, 2014), representing 3.5% of the world total production.

In relative terms, though, the European fishing sector represents a very low percentage of the overall European economy: its contribution to the European Gross Domestic Product (GDP) is only around 0.1%. By way of comparison, this figure jumps to 0.4% in Japan, 0.7% in Norway, close to 1% for Korea and up to 10% for Iceland. In the United States, the percentage goes down to 0.02%.

This percentage is much higher in certain fishery-dependent regions in Europe. For the Highlands and Islands (UK), Galicia (Spain), Ionian islands and Aegean Sea (Greece) the percentage exceeds 2% of GDP and for Peloponesos (Greece), Algarve and Azores (Portugal) and N-E Scotland (UK) it falls between 1 and 2% of total GDP (European Parliament, n.d.). Figure 1.1 provides a glimpse of fisheries-dependent regions in Europe.

The Common Fisheries Policy: The Quest for Sustainability, First Edition. Ernesto Penas Lado.

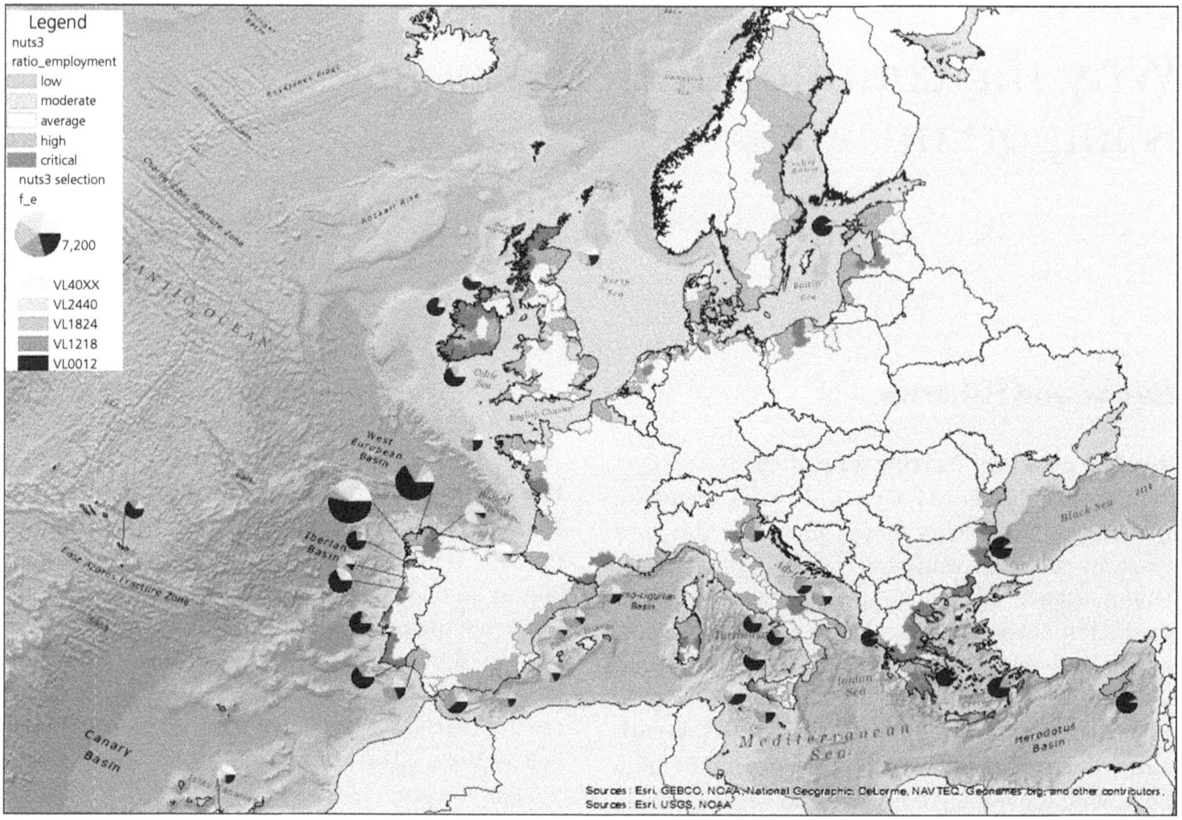

Figure 1.1 Fishery-dependent areas in Europe. © European Union. European Commission. Facts and figures on the Common Fisheries Policy – Basic statistical data – 2014 Edition. Luxembourg: Publications Office of the European Union. 2014 — 44 p. (See colour plate section for the colour version.)

The importance of fishing in certain coastal municipalities can reach values well over 30% of GDP and over 50% in employment, as is the case, for example, in certain coastal municipalities of Galicia (Penas, 2000), and in other European regions. The fishing activity has also a very important multiplying factor in the economy: ancillary activities such as shipbuilding, transport, fish mongering, fish processing and so on represent an important level of activity and employment in fishing areas. The economic importance of fisheries includes all these ancillary activities.

The EU market consumed 12.3 million tonnes of seafood in 2011, worth €52.2 billion. It is the first import market of seafood in the world with 29% of the global exchanges in value. *Per capita* fish consumption reached a plateau in recent years (2008–2011) after years of very dynamic growth. Over time, Europe's own production has fallen increasingly short of the demand for fish products in its domestic market. The EU-27 exported 1 700 000 tonnes of fishery products, worth €2.7 billion in 2012 but imported more than 5 300 000 tonnes, worth €16.5 billion. The EU's self-sufficiency in seafood is today around 45%.[1]

Although the European fishing sector may be very small in terms of its overall economic weight in the European economy it can be much more significant for the regional economy in certain coastal areas of the Union. This helps to explain why, in many instances, the fishing sector and its associated policy, the CFP has become politically far more sensitive than its own economic significance would indicate. One of the reasons for this political importance is the considerable weight of history in its development and the effects that these developments have on the cultures of the countries concerned.

The fisheries sector employs 116 000 people in the catching sector, 115 000 in the processing industry and 33 000 in aquaculture. These figures represent full time equivalents, since the number of people associated part-time to these industries is believed to be higher. The importance of these figures must be seen in a regional context: while this level of employment is relatively small for a population of more than 500 million people in the EU, it is fundamental for a number of European coastal regions, where the fishing and associated sectors are a very significant source of direct and indirect jobs.

In addition, these figures must be complemented by the employment related to ancillary activities referred to previously, as well as by the important level of part-time employment existing in many European areas, from the Spanish *marisqueo* (clam digging) or the French *pêche à pied* to Northern Europe's ice fishing, which represent significant levels of part-time or complementary economic activity with considerable social value.

A historical background

Fishing is a very old economic activity in Europe and its development is closely related to many other historical events. This history has been described with considerable detail by some authors, for European fisheries in general since ancient times (López Veiga, 2000), for the fisheries sector since the Middle Ages (Roberts, 2007), or specifically for tuna fisheries (Doumenge, 1998), among others. This book on the other hand will only attempt to highlight certain examples of historical developments that have contributed to shape attitudes and traditions about fishing in Europe and to illustrate that many, if not most of the problems surrounding fisheries today already existed even several centuries back.

The image of all fishing before the twentieth century being artisanal and focused on local consumption is not real: fishing has been an industrial activity for a very long time. This long history has influenced consumption patterns as well as culture and traditions in the different European regions, helping to establish a particular *ethos* around the fisheries issues that still today conditions social and political behaviour regarding this activity and its management by public authorities.

Let us look at some examples.

Fish consumption

Harvesting and consumption of seafood dates back to the Palaeolithic period, around 40 000 years ago. Archaeologists have traced the consumption of fish in different pre-historical and historical times through the study of fish bones excavated in human settlements.

The Israelites were important consumers of fish, including smoked, salted and dried preparations and ancient Jerusalem had a 'Fish gate' named after a fish market nearby. The Greeks and Romans of antiquity were strong consumers of fish and fish products.[2] In ancient Greece, Bluefin tuna was already an expensive food item. In the Roman Empire, the consumption of many fish species was associated with wealth, as demonstrated by the numerous mosaics showing different Mediterranean fish. Along the North Sea shores, fish consumption was associated mainly with the Roman garrisons.

In Northern Europe, the Scandinavians developed a culture of fishing and consumption of fish due to the climatic conditions that limited their agriculture, while central Europe still relied much more on agriculture and livestock. The activity became an intensive one from the eighth and ninth centuries.

In the British Isles and central Europe, the consumption of marine fish was low, while freshwater fish was intensively exploited and consumed. But this changed dramatically in the eleventh century, when the overexploitation of freshwater fish and the intensive use of water courses for different uses provoked a collapse in these resources that, in turn, allowed for their replacement by marine fish, thus causing the first wave of development of marine fishing (beyond the traditional coastal activity) in central Europe (Roberts, 2007).

Certain fish and fish products were highly valuable food items in ancient times. The best example is that of the *garum* sauce, a delicacy in ancient Greece and in the Roman Empire, made of small fish and fish guts, dried, salted and percolated. The sauce was used to season many dishes of affluent Romans. Its production and marketing became a very lucrative industry.

Fish for trade

The idea of fish as a trade item is also very old. In ancient Israel, merchants imported pickled fish roe from Egypt, among other fish products. The *garum* sauce was the subject of intense trade in Greek and especially Roman times. The market for fish products in ancient Rome was so important that even attracted other products from Northern Europe: oysters (imported in barrels with seawater) and salmon (in snow or ice) were important trade items from Northern Europe to Rome.

The Middle Ages saw a reduction of trade in the Mediterranean basin due to the tension between the Christian west and Islamic east that affected also the fishing industries of the time. However, fishing industry and trade developed in the north of Europe through the expansion of the Hanseatic League, for which the herring became a major trade commodity. Herring from the North Sea could be found as far away as Constantinople. This trade was favoured throughout Western Europe by the limitations imposed by Christianity on the consumption of red meat.

Traditional salted cod (see next) has also been subject to international trade since the sixteenth century. And since the 1960s, the development of freezing has allowed an explosion in the world trade for frozen fish and fish products.

More recently, air transport has also allowed the development of an important and active worldwide market of fresh fish from around the world: in Spain, the main 'fishing port' is the airport of Vitoria, in the landlocked province of Alava.

Long-distance fishing

The search for fish in distant waters is also very old in Europe. According to certain authors, Basque fishermen started catching whales relatively close to shore in the seventh century, and following exchanges with the Vikings on vessel construction, they ventured as far as Icelandic, Norwegian and Faroese waters in search of whales already at the beginning of the eleventh century (Kurlansky, 1997).

The rich cod fishing grounds in North Eastern America were discovered also by Basque fishermen (Kurlansky, 1997). Following the discovery of Newfoundland by Giovanni Cabotto (renamed John Cabot) in 1497, its legendary abundance of codfish allowed French fishermen to start developing a fishing activity of this species, soon followed by the Portuguese. By 1517, around 50 vessels a year brought back preserved (salted) cod from this region. By 1600, England joined in this fishery which, by that time, represented already 150 shiploads of cod per year (Roberts, 2007).

Vessels from Spain and Portugal continue this fishery today but its impact is wider. In France, where this fishery was discontinued in the nineteenth century, it created a whole 'culture' and tradition that impinged upon social habits. To this day the French port of Saint-Malo keeps the memories and traditions of the old cod fishery in Canada as a fundamental cultural asset.

Fish processing

The fish processing industry is also very old. Fish processing facilities have been found by archaeologists as far back as the fifth century B.C. in the Mediterranean and the Black Sea (Bekker-Nielsen, 2004). By Roman times, dried and salted fish were prepared on a large scale and trade in processed fish was very intensive across the empire during its peak.

Likewise, drying and smoking of fish was a well-developed fish processing technique in Nordic countries, already since ancient times. In the middle ages, Vikings were known for their trade in *stockfish* (dried cod) that they exchanged for barley to make beer.

Salting was primarily the way to preserve the cod caught in Newfoundland since the sixteenth century and the tradition of eating this fish is so deeply rooted that has become a commodity in some countries and even a national icon in Portugal. No longer necessary for fish conservation, the salting of cod has become a way to enhance the flavour and increase the added value of the product.

Smoking of fish is also a very old practice. Together with salting, it was the best way to preserve (fatty) fish for consumption and trade. The Dutch fishery for herring, to be smoked and sold in barrels, developed into a very well developed industry in the sixteenth and seventeenth centuries and contributed to the creation of wealth leading to the Dutch *golden age*.

The canning industry is more than 200 years old. It started out in the beginning of the nineteenth century in France, when in 1810 Nicolas Appert won a competition to find ways of providing preserved food for the Napoleonic Army. Very soon, canned sardine became a favourite product. This industry was quickly extended to other European countries as well as to America. And it was there, in California, that the tuna canning industry was born in 1903, as an alternative to the Pacific sardine and has been a favourite food around the world ever since.

Today, the canning industry (for sardine, tuna and many other species) is an iconic one for certain countries (Spain, Portugal, Italy, France) and so is its equivalent (based on sprat) in some Baltic Sea countries, particularly Latvia.

Trawling

There are historical records of fishing with trawl gear as far back as the fourteenth century. A very famous case: the petition to the British Parliament in 1376 to suppress the so-called 'wondyrchoum', a precedent to the beam-trawl used from wooden sailboats, is an indication of the use of trawl fishing already in the fourteenth century, as well as a remainder of the controversy that has often accompanied it.

Many different designs of trawl nets were used for centuries from sailing vessels, particularly in the UK, where different ports acquired historical predominance thanks to this kind of fishing. The port of Brixham acquired notoriety and predominance in the North Sea thanks to its fleet of wooden sailboats practising trawl fishing in the nineteenth century and the Dutch were also known in those years to use the 'dogger', a type of trawler that gave its name to the sandy 'Dogger Bank' in the North Sea.

The full development of the trawl fleets took place in the mid-1800s, notably in the UK, as a result of two parallel developments: the use of ice to chill the catches on board and allow for the handling of very large amounts of fish and the development of railways to distribute quickly the abundant catch to different markets before it could deteriorate (Roberts, 2007).

In the wake of the industrial revolution, the introduction of the steam engine in fishing vessels and the subsequent introduction of this kind of engine in trawlers in the UK and France between the 1870s and the 1880s represented an important leap forward and marked the birth of the modern fishing fleets. Later on, after World War II, trawlers incorporated diesel engines.[2]

From the beginning, the adoption of more sophisticated, efficient and mechanised ways of fishing such as trawling, have been the subject of controversy, notably from those using other fishing gear who suffered from the increased competition from the new methods. Today's controversy over the use of trawling, raised in particular by environmental NGOs, is actually several centuries old.

Overfishing

One may think that overfishing is a problem of modern times, resulting from the excessive development of the modern, mechanised and large scale fishing industry around the world. Yet, overfishing has existed since medieval times. Typically, local fishing exhausted local resources, leading fishers to look for more distant alternative fishing areas. In fact, overfishing is a problem that dates back to the pre-industrial era (Scearce, 2009).

However, the awareness about overfishing is relatively recent in history. Up to the 1800s, fishery exploitation generally consisted of unlimited fishing of available resources. When abundance diminished, other species were sought, or more distant fishing grounds were exploited. It is worth referring to the famous case of the Royal Commission set up in the UK in 1863 to inquire into the complaints against trawling, where the number of vessels had skyrocketed between 1840 and 1860. The Commission concluded by dismissing the case, indicating that management was not necessary and that trawling was a fishing practice that resulted in the production of large amounts of cheap fish, making this food affordable for the poor (Roberts, 2007).

In the late 1800s, however, the idea that marine resources were non-exhaustible started to make headway. In the early 1900s the idea of regulating the harvest started to develop but the lack of a sound scientific basis prevented effective management in practice. The modern notion of overfishing and the need for sustainable management started to develop in earnest only in the 1950s (Lackey, 2005).

The world has seen a number of cases where overfishing (particularly when pursued during unfavourable natural fluctuations) has led to the collapse of resources.[3] Some of these collapses have lasted decades, such as that of the Pacific sardine (*Sardinops sagax*), which supported a very important fishery in the 1920s, to disappear catastrophically in the late 1940s, for reasons that still today are not well understood but probably related to overexploitation and climate change. The fishery came back only in the 1990s. This case, among many others, shows that although fishery resources can collapse they can also bounce back, perhaps many years later. Furthermore, it illustrates that current scares about stock collapse are also nothing new in history.

Social conflict

There are plenty of historical records about the conflict between artisanal, traditional fishing and industrialised fishing. As we mentioned previously, the introduction of new, more efficient gear can provoke social conflict and lead to changes in the dominance of certain ports over their competitors.

A classic example is that of the conflict, in Galicia, between local fishermen and investors from Catalonia (the so-called *fomentadores catalanes*) who settled in Galicia in the eighteenth century to develop a processing industry for sardine to take advantage of Galicia's abundant resources of this species. To ensure a steady, abundant supply of the fish for their industry, they introduced more modern, efficient fishing techniques. This, in turn, resulted in social conflict with the traditional, local fishermen practicing more artisanal methods, who believed the investors from outside Galicia altered the social balance and endangered the resource. The conflict was so difficult to resolve that it lasted more than a century (from 1750 to 1890), in different forms and under different phases (Meijide Pardo, 2002).

Today, the processing industry for sardine in Galicia is an iconic industry defended by everybody regardless of its non-local origin. This illustrates that innovations in fisheries, particularly through foreign investment, can cause opposition and conflict. Over time, however, what was perceived as an intrusion by the local fishermen can end up becoming a widespread practice. It is important that innovations in fishing are looked at from this perspective.

These are just examples to better illustrate that the conflicts in fisheries today, in one way or another are conflicts that have existed in very similar forms for centuries. This also means that learning the lessons from the past remains important to resolving the problems facing today's fishing.

Aquaculture

Aquaculture is a very old activity in the world. There is evidence of well-developed freshwater aquaculture in China 4000 years ago, based on carp. And the first treaty on fish farming, *The Classic of Fish Culture* was written in China in the fifth century B.C. From China this practice extended to India and neighbouring countries and during the Ming dynasty (the fourteenth to the seventeenth century) numerous works described the farming techniques in great detail.[4] Carp farming in ponds continues today in China, four millennia later.

In Europe, the Romans were active farmers of different species of fish, mainly freshwater species and the Roman tradition was developed in the middle ages with pond aquaculture for carp and trout. In central Europe, in the eleventh century, the gradual reduction of wild freshwater fish as a result of overexploitation and urban development along water courses led to the development of intensive farming of fish in ponds, covering thousands of hectares in France, Silesia and Bohemia and managed largely by monks. Many of these, however, disappeared in the following centuries as a result of the availability of more abundant marine fish in the eleventh century (Roberts, 2007).

Marine fish farming developed much later. With the exception of very old aquaculture practices in Japan and other areas, most of the current development of marine aquaculture for fish and molluscs only dates back to the twentieth century, since the previous high abundance and easy access to wild marine fish and molluscs made such developments unnecessary. In any case, freshwater fish farming is a centuries old tradition in many parts of Europe.

The importance of European fishing beyond economics

The history of fishing in Europe is so rich that the fishing sector and, by extension, the Common Fisheries Policy (CFP) enjoys a visibility, for good or bad, that goes far beyond the importance of the fishing activity on the European economy. This is the result of several facts:

Firstly, the fisheries sector is highly concentrated geographically, unlike many other economic sectors that are dispersed all around Europe. The fishing activity is small in global terms but it is essential for certain coastal areas of Europe, where it constitutes the economic and social lifeblood of the local communities. In addition, certain historical activities, as we saw previously, have become over the centuries an icon of the local culture and identity. In these areas, the fishing activity goes beyond pure economics and affects the culture and tradition of the wider society, thus impinging upon national identity issues.

Secondly, the fishing activity still enjoys a somewhat romantic image, as a risky, adventurous activity that distinguishes it from other, normal sectors of the economy. This translates in some Member States into a tradition of paternalism and a sentiment that the sector should not be handled in strict economic terms. In the tradition of many European nations, the primary sector, responsible for the production of food and with traditionally high employment levels, is considered a strategic component whose interests cannot be measured in pure economic terms.

Thirdly, fisheries are somehow the 'last frontier' of national sovereignty in Member States. In a Union where free circulation of goods, citizens and capital are fundamental principles, it is surprising to see how in some cases, the investment of foreign operators in the fishing industry still raises opposition, as if the fishing industry was the last bastion of the sovereignty and even the identity of European nations.

Several times in the history of the CFP, fisheries conflicts have resulted in gunshots being fired and even in a city hall being burned. This led *The Economist* to wonder, in 1994: 'what is about fish that makes democracies send gunboats?' (Schweiger, 2010).

This may be due to a large extent to the fact that international fisheries law was not settled at the time of the birth of the CFP. In fact, the basic rights and obligations of coastal States were still under intense (and conflicting) discussion among different countries in the world when the CFP was born. The declaration of a 200-mile Exclusive Fisheries Zone in 1977 (Council Resolution, 1977) was at the time a unilateral action in a world where such actions were still being taken without international agreement. International fisheries law only consolidated that right through the 1982 United Nations Convention on the Law of the Sea (UNCLOS) which only entered into force in 1994. And still today, some countries such as the United States have not ratified it. The uncertainty of the international legal framework has influenced fisheries management around the world considerably and it continues to do so still today.

Furthermore, given its highly integrated nature the CFP is exposed to considerable public scrutiny. Unlike other policies, where the mixed responsibilities between the Union and Member States can dilute the political responsibilities, the fact that the CFP is traditionally decided 'in Brussels' (though not necessarily 'by' Brussels) puts it under the spotlight and makes it a favourite topic in the political discussion on European integration. As a consequence the CFP has become a showcase of European policy-making and governance.

Last but not least, the nature of fisheries management itself often involves difficult political decisions. A growing interest in the ecology of the oceans and the role of fisheries in this context means that a part of the population with no direct involvement in the sector has very firm views as regards its future development based on the protection of resources rather than the economic sustainability of the industry. The political impact that this body of opinion can exert in some Member States is considerable. On the other hand, restrictions on fishing to ensure the sustainability of the activity are more often than not seen as political 'bad news' by the fisheries constituencies. But for some conservationists they may not go far enough. This results in the Union being regarded by all sides as a favourite scapegoat for all the unpalatable decisions that are made. This has often put the CFP in the centre of the attention of the political debate about European integration.

Fishing and national traditions: the difficult balance

Fishing and the tragedy of the commons

The management of fisheries has always been difficult, since fisheries generally represent the ultimate example of Garret Hardin's 'tragedy of the commons': the shared nature of the resources provides little incentives for individuals to be self-restrained in their exploitation (Hardin, 1968).

As a Malthusian, Garret Hardin disagreed with Adam Smith's theory of the 'invisible hand' that would lead individuals intending their own gain to promote the common good. Hardin believed that when individuals decide on the use of a 'common' (he used the example of herdsmen in a free-access grassland) they expect a gain and a loss but while the gain would be entirely for them, the loss (the reduced productivity through overgrazing) would be shared with others, so the ratio would look beneficial, leading to individual decisions that will eventually ruin all herdsmen through the collapse of the grassland: 'Each man is locked into a system that compels him to increase his herd without limit – in a world that is limited ... Freedom in a common brings ruin to all.'

Hardin further concluded that the 'commons' (that is, the resources under free access) can only be sustained as such under conditions of low population density. The implication is that, to avoid their tragedy, the 'commons' must be managed under a system that replaces freedom of access by 'coercion', understood not as rules imposed by bureaucrats but as 'mutual coercion': decisions mutually agreed by the majority of people affected.

The 'common' of the fishery resource is typically shared but never perfectly assigned to individual 'owners', due to their mobility and to the impossibility of their assignment to a fixed territory. This means that any system of 'ownership' of a part of shared resources is always an imperfect solution, which does not entirely remove the reflex of benefiting from the resource before the neighbour does.

The solution to the tragedy of the 'common' fishery resource has not been subject to scientific studies until relatively recently and particularly not until the CFP was actually established. We have to wait until the 1990s, notably through the seminal work of Nobel Prize winner Elinor Ostrom (1990), for economic studies on how to develop the 'coercion' mechanisms referred to by Hardin.

In the case of the CFP, as well as the difficulties related to the common property of the fishery resources in every country, in conditions of generally high population density in most Member States (thus with considerable social pressure on the exploitation of the common) there is the added complexity of dealing with different countries with different traditions and interests. Since fishery resources under the CFP are managed under the principle of free access, European fishery resources constitute a 'common' shared among the different Member States and thus the potential for 'tragedy' exists if the access to the resource were free.

The world has seen a number of national fisheries management systems that work successfully. However, it is much more difficult to find success stories among the international bodies responsible for managing fishery resources involving different countries. In that sense, the CFP is unique, in that it is the only fully integrated fishery management system in the world regrouping many different countries. The relative merits of the CFP should therefore be looked at from this perspective and simplistic comparisons with the management of fisheries in individual, small and homogeneous countries are meaningless and misleading.

The importance of geography

Attitudes towards fisheries and their management are strongly influenced by geographical factors. Anyone who has flown over the United States or

Japan can easily observe the difference between the seemingly endless surface of prime agricultural land in the former and the hilly and densely populated nature (resulting in a considerable shortage of agricultural land) of the latter. As a result, unlike the US, Japan developed historically a strong dependence on the sea as a source of food. Little wonder that the Japanese look at their seas as a sustainable source of protein and not as a nature sanctuary.

The difficulties of fisheries management are aggravated in cases where the ratio between natural resources and the populations depending upon them is low. Fisheries policy is largely the management of limited natural resources vis-à-vis the social pressure to exploit them. It follows that the higher the population density along the coast and the lower the economic alternatives for that population, the higher the social pressure on the resources and the more difficult and contentious the fisheries management.

The European coast is characterised by a very high population density. Although this is highly variable among Member States and regions, this density is in many areas greater than 500 inhabitants per square kilometre of the 10 km coastal band. This compares with an average of 80 inhabitants for the world as an average and of just 34 for the USA.

Bio-geographical differences are also crucial to explain the differences among fisheries. For example, the rich fishing grounds in the Atlantic have historically produced higher quantities of certain species (cod, herring, sardine), thus allowing for a large-scale approach to fisheries and processing, while the lower productivity and higher diversity of Mediterranean fisheries has traditionally conditioned more dispersed, artisanal activities. These differences have historically determined different approaches to the fishing activity, resulting in different traditions and a different 'culture' around fisheries and their management.

The width of the continental shelf in some cases also determines fishing practices, with countries with narrow continental shelves and a tradition of eating fish developing a historical activity in distant fishing grounds.

Finally, geographical isolation also plays a key role. In remote islands where there is little other economic activity, the importance of fishing achieves an importance that makes this sector the real lifeblood of the local communities.

The importance of economics and culture

Geography does not explain all. In Europe, it is possible to identify land-locked regions with a tradition of eating fish (e.g. Madrid claims a very high level of consumption of fish) and coastal ones that for a long time have largely ignored fish as a significant source of protein (e.g. Ireland until recent times). This is because in addition to geography, other economic and cultural factors have played a role, sometimes for centuries.

Consumption habits among European countries are even today very diverse, both in terms of *per capita* consumption of fish and in terms of the variety of fish appreciated by consumers. The EU-27 has a *per capita* consumption of 23.3 kg of fishery and aquaculture products *per annum*, higher than the world average of 17.8 kg. At the level of Member States, the difference is huge, with Portugal and Spain leading the *per capita* consumption (61.6 and 44.8 kg, respectively) and with Romania, Hungary and Bulgaria being at the opposite end with a *per capita* consumption around only 5 kg *per annum*. Only a few countries outside the EU, such as Japan (56.7 kg) or Norway (51.9 kg) have consumption levels of the same order of magnitude as those of Portugal and Spain (CFP, 2012).

The high consumer demand in certain countries determined the development of fishing activities and the establishment of old historical 'rights' for certain resources, while traditions of low consumption have resulted in a relatively recent development of the activity, focusing mainly on the export to the traditional markets.

Religion has also played a very important role. In southern European Member States with a catholic tradition the prohibition on eating meat on certain days of the year resulted in its substitution by fish protein. This turned these countries (Spain, Portugal, Italy) into significant consumers and importers of the fish that, at the time, was amenable to preservation through salting, drying or smoking. This also led to a centuries-old quest for fishery resources, either in domestic or distant fishing grounds, as an alternative to the forbidden red meat.

Finally, different levels of industrialisation and economic dependency on the fishing sector also influence traditions and approaches. All these differences among Member States and regions in Europe always have something in common: whether or not fisheries are small or large scale, they all consider these activities as a social asset, even a cultural one and not just an economic activity.

All these differences play a major role in shaping up different policy approaches but crucially, they also contribute to create a common policy, in the sense that there is a level of complementarity between Member States having the resources and those having the consumer markets for the fish.

Fisheries and international conflict

Fishing has often been the source of conflict, of different kinds: between European nations, between fleets, between artisanal and large-scale interests and so on.

Conflicts between European nationals about rights to fish are very old. These conflicts have from the beginning been linked to much wider economic interests, particularly trade.

These conflicts were perhaps made clear and formulated in intellectual and legal terms at the beginning of the seventeenth century. In 1609 Hugo Grotius, a jurist of the Dutch Republic, formulated a new principle that the sea was an international area and all nations were free to use it for seafaring trade. In his *Mare Liberum*, Grotius claimed 'free seas', including the right to exploit fishery resources. The ultimate objective of Grotius was not related to fisheries but rather to the need for justification for the Dutch breaking up of various trade monopolies (notably by England, Spain and Portugal) to establish its own.

England, competing with the Dutch for domination of world trade at that time, opposed Grotius' ideas and claimed sovereignty over the waters around the British Isles. In his *Mare Clausum* (1635) John Selden endeavoured to prove that in practice the sea was virtually as capable of appropriation as terrestrial territory.

The need to resolve the growing controversy led maritime States to moderate their demands and base their maritime claims on the principle that their sovereignty extended seawards from land. A workable formula was found by another famous Dutch lawyer, Cornelius Bijnkershoek in his *De Dominio Maris* (1702): restricting the maritime dominion of coastal States to the distance to which they could effectively protect it through cannon fire. This became almost universally accepted and was developed into the 3-mile limit of the territorial waters.

This later evolved in the last half of the twentieth century through gradual extension of the *Mare clausum* well beyond the 3 miles to 6 or 12 miles of the territorial waters and beyond that to the current 200-mile Exclusive Economic Zones (EEZs) under coastal State jurisdiction. The process is not necessarily settled forever: some States such as Chile have actively pursued policies of 'creeping jurisdiction' beyond their EEZs.

In one way or another, the differences between the opposing views of free access vis-à-vis territorialised rights has always produced tension and disputes in fishing, to the point that this conflict of interests strongly influenced the birth of the CFP and some of its features still today. The best-known case of this kind of tension was the so-called 'cod wars' between the United Kingdom and Iceland from the 1940s to the 1970s, which in the end were resolved through pressure from NATO in the context of the cold war (López Veiga, 2000). There have been many more such conflicts in the past. In addition, there have been endless minor conflicts erupting from time to time.

Fisheries and European law

The CFP in the Treaty of Rome

Fisheries policy did not feature in a specific chapter of the Treaty of Rome in 1958. At that time, Europe needed to increase agricultural output to feed a population still recovering from World War II and translated this need into a specific legal basis for an agricultural policy, but fisheries was not considered an area of present or even future action at European level. All six initial Member States exerted their jurisdiction only over the three nautical miles of their territorial sea (this was increased to 12 miles following the European Fisheries Convention of

1964), while a large amount of their fishing activity (particularly in the North Sea and the Atlantic façade) took place outside their jurisdictional waters, in what were at the time only partially regulated international waters.

As a consequence, the CFP was not considered worthy of an explicit legal basis in the Treaty of Rome. Instead, there was a clear basis for the establishment of a Common Agricultural Policy, with specific objectives spelled out in its Article 37. The fisheries sector only featured in this first Treaty through its Article 38(1) stating that the common market shall extend to agriculture and trade in agricultural products. These agricultural products being: 'the products of the soil, of stock farming *and of fisheries* and products of first stage processing directly relating to these products' (EEC Treaty, n.d.).

Since then, it was considered that, since fisheries products were part of the agricultural products, the objectives of the fisheries policy should be considered the same as those of the Common Agricultural Policy.

The legal basis for the Common Agricultural Policy has survived unchanged and these objectives remain identical under Article 39 of the 'Lisbon Treaty' of 2007 (Treaty on the functioning of the European Union (TFEU), 2007). These objectives are:

1. to increase agricultural productivity by promoting technical progress and by ensuring the rational development of agricultural production and optimum utilisation of the factors of production, in particular labour;
2. thus to ensure a fair standard of living for the agricultural community, in particular by increasing the individual earnings of persons engaged in agriculture;
3. to stabilise markets;
4. to assure the viability of supplies;
5. to ensure that supplies reach consumers at reasonable prices.

While these objectives constitute a relatively coherent and complementary set for agricultural and fisheries policy, they are not necessarily so well adapted for certain aspects of fisheries management such as the conservation policy, where these different (and not hierarchical) objectives frequently contradict each other, particularly in the short term. We must bear in mind that, while fisheries are a common pool of resources, agriculture is much more territorialised (even though Garret Hardin based his theory on examples on land, actually the fishery resources represent a much more clear example of a 'common' shared by many users).

For example, in fisheries the objective of increasing productivity can have opposite effects if we look at the short-term or the long-term: unlike agriculture, increased productivity of wild fisheries in the short term will lead to lower productivity in the long term. The same applies to the objective of 'ensuring the viability of supplies' to the European market, that can be seen under a very different perspective depending upon the time scale: many decisions in fisheries management require 'fishing less today to fish more tomorrow'. When deciding on short-term measures, the Treaty does not provide any guidance on whether the short-term and long-term considerations must be combined. As a result, the tradition of the CFP is that the short-term considerations have prevailed.

Furthermore, the economics of fishing and agriculture are fundamentally different. In agriculture, more investment and more labour produce higher yields. In fisheries, more investment and more labour often leads to overfishing and lower yields. This distinction is not reflected in the EU Treaties.

Finding the adequate combination of these objectives and combining their delivery between short, mid and long-term, remains to this day the main conundrum of the CFP. Most decisions in the context of the CFP are related to this, either implicitly or explicitly. In this sense, the CFP is a continuous search for the best way to implement the objectives of Article 39. The resolution of these questions is encapsulated in the context of the discussions on long-term management plans, where the balance between short-term sacrifices and long-term gains must be spelled out.

Fishing in subsequent Treaties

From the Treaty of Rome the legal basis for the CFP has evolved slowly and gradually (like many other EU policies) and, interestingly, it was the very development of the policy that has often created the basis for the successive treaties to recognise.

The Single European Act, signed in 1986, did not modify the legal basis of the CFP. However, its two main contributions (the cohesion policy and the achievement of the internal market in 1992) did have an indirect influence on the CFP and other EU policies. The cohesion policy, with its emphasis on the promotion of economic development of the least developed Member States did result in a special treatment of the Member States concerned (initially, Ireland, Greece, Spain and Portugal) in terms of economic policy.

The Treaty of Maastricht, signed on 7 February 1992, provided for the first time a specific recognition of the CFP by providing for the establishment of a Common Fisheries Policy identified as separate from the Common Agricultural Policy. However, the general objectives of the policy remained unchanged from the Treaty of Rome. This Treaty simply recognised what was, already at the time, a clear reality: a specific policy with its own – important – *acquis*.

The Treaty of Maastricht also introduced the notion of 'subsidiarity' which, among other effects, clearly provided a basis for the Community to act in areas outside exclusive competence: 'only insofar as the proposed action cannot be sufficiently achieved by the Member States and can therefore, by reason of the scale and effects of the proposed action, be better achieved by the Community' (Treaty of Maastricht, 1992).

This was indeed the case for certain parts of the CFP, in particular the control policy where, despite Member State competence, it can be argued that action at Community level is fully justified in monitoring a common policy (Long and Curran, 2008).

The Treaty of Amsterdam, concluded in June 1997, established four new priorities for the Union, none of which has a direct impact on the CFP: new emphasis on employment and citizens' rights, removal of the last barriers to free circulation, the new security policy and the facilitation of the enlargement of the Union. However, the four priorities did, indirectly, have a bearing on the CFP, as for any other policy.

The Treaty of Lisbon

The Treaty of Lisbon, signed on 13 December 2007, which entered into force on 1 December 2009, introduced two major elements on the legal basis of the CFP: the consolidation of the conservation and management of the marine biological resources as one of the five only exclusive competences of the Union and the application of co-decision between Council and the European Parliament as the ordinary legislative procedure for the CFP.

However, the Treaty of Lisbon did not modify the objectives of the CFP, which remained those of the Common Agricultural Policy in the Treaty of Rome in 1958. This is not surprising in the context of such a politically complex exercise. Including changes to the legal basis of the CFP would have created yet another obstacle to the adoption of the Treaty.

Despite the obvious differences between fisheries and agriculture and despite certain requests from the European Parliament in this regard (1995), the Treaties never created a specific set of objectives and instruments for the CFP as a separate policy from the CAP. This was considered unnecessary. Despite the specificity of fisheries management mentioned previously, it was understood that the general objectives of the CFP and the CAP were valid and did not need to be changed.

The Treaty of Lisbon, however, included the core business of this policy, the management of fishery resources, in the select list of exclusive competences of the Union. In particular, it included the CFP as a policy where the ordinary legislative procedure would be co-decision under Article 43(2) TFEU, with the exception of the adoption of fishing opportunities and reference prices under Article 43(3) TFEU, which remain under the sole responsibility of the Council. The change to co-decision established a new way of dealing with the CFP of which the consequences are still not clear today.

In Chapter 13 we will refer to the adaptation of the CFP to the new co-decision process and particularly with regard to the balance of power between the two co-legislators, Council and European Parliament, notably around long-term management plans and their relationship with annual fishing opportunities.

Fisheries policy and European integration

Despite the non-specific treatment of the CFP in the early Treaties, the CFP has made a significant contribution to European integration. This may sound paradoxical, given the conflictive nature of this policy. However, it is precisely because the policy has been very conflictive that there have been many cases taken to the European Court of Justice. As a consequence, there is an abundant jurisprudence from the Court that has consolidated a number of Community principles, thus contributing to reinforce the European integration through case law (Sobrino Heredia and Rey Aneiros, 1997).

This was anticipated by certain authors who, already in 1983, stated that: 'The CFP, though relatively marginal to the Community's economic life, demonstrates many of the possibilities for and limits on the Community action and so has implications that go beyond the sector directly concerned' (Leigh, 1983).

The best example is perhaps the consolidation of the CFP as a field of exclusive competence of the Union, even before such exclusive competence was enshrined in the Treaties. Exclusive competence of the Union regarding fishery conservation measures was in fact consolidated by the Court through several key rulings already in the late 1970s (Schweiger, 2010).

The 'Kramer' case in 1976 clarified the most important aspects of Community competence relating to internal and external policy (Kramer et al., 1976), and the case of the Irish conservation rules in 1978 clearly established two principles that have become basic tenets of the CFP: the Community competence on the conservation of fishery resources and the importance of non-discrimination among Member States (European Court Reports, 1979).

Another example is the Factortame I and II cases, in 1990 and 1991, respectively, on the application of the UK's Merchant Shipping Act (which tried to exclude foreign investors from the UK fishing vessel register through requirements on nationality and residence). Important case law was set, to the effect that in areas of Community competence, the Court ruled that Community law prevails over national law (Factortame I) (The queen vs. Secretary of State for transport, ex parte Factortame Ltd and others, 1990), and consolidated Community rules on the freedom of establishment (Factortame II) (The queen vs. Secretary of State for transport, *ex parte* Factortame Ltd and others, 1991).

In certain situations, the development of European integration has not even required cases in the Court: certain policy developments in the CFP have established precedents that only later on were incorporated into the text of the Treaties. For example, the exclusive competence of the Union for the negotiation of fishery agreements with third parties, today recognised in the Treaty (Article 3 (2) TFEU), was actually developed during the initial stages of the CFP, when the complicated negotiations among Member States, acceding countries and other third States made it preferable for Member States to recognise the advantages of such negotiations being carried out at Community level (López Veiga et al., 1993).

The role of the Court of Justice was by no means easy. In particular, the first of these rulings took place at a time where the basic rules of the CFP had not yet been laid down and in a field of high political sensitivity. The slow progress in agreeing to a CFP between 1976 and 1983 provided a time in which the Court advanced European integration before Member States in Council would agree to do so. In these cases the Court took a rather non-legalistic approach, focusing on the spirit of EU law and not just on the letter and attaching considerable importance to the requirements of Article 5 of the Treaty of Rome on the importance of the loyal cooperation among Member States (Farnell and Elles, 1984).

Who decides what in the CFP?

For readers not familiar with the decision-making process of the European institutions, to understand the following pages requires some basic explanations, particularly on the question of 'who decides what', which is often perceived by outsiders as an arcane system full of jargon that only insiders can comprehend. Very often, the media and stakeholders refer to the decisions under the CFP as being decided by 'Brussels' or even 'the Brussels bureaucrats' in clear derogatory terms.

The CFP is an exclusive competence of the Union. Member States cannot legislate on matters pertaining to the common fishery resources. The legislation to manage fishery resources and accompanying legislative instruments generally take the form of regulations. These regulations were for many years decided by the Council of Ministers (that means, by the Fisheries ministers of the Member States). The historical development of the CFP up to 2008 has been the result of decisions made by this body.

However, after the entry into force of the Treaty of Lisbon, the regulations of the CFP are decided by co-decision, that is, jointly by the Council of Ministers and the European Parliament, through a process of negotiation between them, with the participation of the European Commission as a facilitator. These negotiations are generally referred to as 'trilogue' (a discussion between three parties). Following the adoption of a regulation, its provisions become directly binding upon Member States and individual operators.

All regulations are adopted following a proposal from the EU's executive body, the European Commission. Therefore, the Commission has an important influence in the process by making the proposals. However, the final decisions on the CFP are taken by the Council of Ministers (meeting generally in Brussels or Luxembourg) and, as of 2009, jointly by the Council and the European Parliament. In that process the Commission maintains an important role to facilitate final agreement but the ultimate responsibility for the decisions under the CFP lies on the Council of Ministers and the European Parliament.

There is an important exception to the co-decision in the CFP: the adoption of catch limitations (Total Allowable catches or TACs, divided into national quotas according to relative stability). The adoption of annual catch limitations, together with the allocation of these limits in national quotas is an exclusive prerogative of the Council of Ministers: the European Parliament does not intervene in such decisions.

In this context the responsibility of the Commission is the presentation of proposals, the facilitation of the decisions by the legislators (Council and Parliament) and then certain aspects of the practical implementation. In addition, the Commission can also adopt implementing regulations, pieces of secondary legislation, on technical matters, upon specific mandate of the legislator. When confronted with phrases such as 'Brussels decided…' it is important to bear in mind where the ultimate responsibility lies.

The use of regulations, instead of directives (a directive is an EU legislative instrument that is not directly applicable but has to be transposed into national legislation by Member States) is inseparable from the high level of technical detail that has characterised the CFP from its inception. This is a policy where the deeply rooted tradition is that of very detailed regulations being decided by ministers in the Council and where small technical details can easily become the subject of difficult political negotiations. Only the reform of the policy in 2013 is starting to change that.

Another clarification is the very name of the European Union, which under different Treaties, has evolved from European Economic Community (EEC) to European Community (EC) and to European Union (EU). For ease of reference, the book generally refers to the EU, except in the case of specific historical references where the official denomination of the time is used.

The two co-legislators, Council and Parliament, work very differently. In Council, Member States have a number of votes proportional to the population of the Member State (though the specific formula has changed with the Treaties). In fisheries, the Council decides by the so-called 'qualified majority', which requires a minimum percentage of all the votes. That means that decisions can be adopted by Council with a negative vote from some Member States.

This practice was rare in the early stages of the CFP, when the low number of Member States allowed for the search for unanimity. Furthermore, the use of the 'Luxembourg compromise' (a non-legal political compromise allowing a Member State to refuse a decision that would affect its fundamental national interest) was still common in the 1980s. But as the number of Member States has grown, it has become increasingly difficult to reach agreement by all Ministers and a qualified majority is increasingly used by Council to decide on fisheries matters.

The European Parliament works differently. There, the Fisheries Committee (composed largely of Parliamentarians with a fisheries constituency) makes proposals on draft legislation, as amendments to the proposal from the Commission. The agreed proposed amendments have to be ratified by the Parliament at its plenary session. Often, the Plenary overrules the Fisheries Committee or introduces changes to their proposals, generally on the basis of wider political considerations beyond fisheries. The position adopted by the Plenary constitutes the mandate of the European Parliament to engage in the negotiations with the Council in the trilogue.

Both Council and Parliament may have to modify their negotiating positions to be able to agree on the final legislation. The Commission may also have to modify its initial proposals to facilitate the agreement.

The importance of history

The previous pages illustrate that the CFP is a policy where the weight of tradition is very high. Reconciling the different, centuries-old traditions around fisheries issues into an agreed policy is, for historical reasons, very complex. As a consequence, once an agreement is reached, it tends to be extremely difficult to re-open and renegotiate.

As we will see in subsequent chapters, the policy has been evolving through the establishment of successive 'layers' of policy but while new layers have been established relatively easily, the removal of the old layers proves, still today, extremely challenging. The CFP can only be fully understood if we analyse it from a historical perspective. Table 1.1 summarises the development of the CFP over time, through the most important events that marked its history.

Table 1.1 Timeline of the most decisive events in the development of the CFP.

Year	Event
1958	Treaty of Rome
1967	First document on a future Common Fisheries Policy
1970	First structural and market regulations for the fisheries sector. Principle of free access to Community waters
1973	Enlargement to the United Kingdom, Denmark and Ireland
1977	Declaration of Exclusive Fisheries Zones by Atlantic Member States
1983	Birth of the CFP: TACs and Quotas and technical measures. First Multiannual Guidance Programmes (MAGPs)
1985	Greenland leaves the European Community
1986	Accession of Spain and Portugal
1992	First reform of the CFP
1994	Entry into force of the United Nations Convention on the Law of the Sea (UNCLOS) First Mediterranean regulation
1995	Accession of Sweden, Finland and Austria (Norway votes against accession)
2001	Entry into force of the United Nations Fish Stock Agreement (UNFSA)
2002	Second reform of the CFP
2004	Enlargement to 10 new countries. First EU long-term recovery plan (cod stocks) Establishment of the first Regional Advisory Councils (RACs)
2006	Second Mediterranean regulation
2007	Accession of Romania and Bulgaria: CFP extended to the Black Sea. New financial instrument: EFF (2007–2013)
2009	Treaty of Lisbon: co-decision as the ordinary legislative procedure in the CFP
2013	Third reform of the CFP.
2014	New financial instrument: EMFF (2014–2020)
2015	First discard bans in force

Since its inception in 1983 and through three successive reforms (in 1992, 2002 and 2013) the CFP has preserved a number of constant elements that have proven to be particularly resilient to change. This has produced a characteristic phenomenon, which appears every time there is a CFP reform in sight: while the policy is constantly criticised by everybody, when it comes to changing it many stakeholders, both institutional and private, show a preference for the *status quo*, at least in a number of key areas.

As we will see in Chapter 15, a typical reaction by stakeholders during consultations on CFP reform has been: 'I don't like the CFP but please don't change it'. This statement may look like a caricature but experience shows that it is much less so than it seems. Very often, those who criticise the policy and the Commission's handling of it are the first to complain if the Commission itself presents in public a critical view of the policy, arguing that the 'catastrophic' picture painted by the Commission is unjustified. Indeed, the tradition of the CFP is one of a strong attachment to the *status quo*, even if it is criticised so frequently.

Many of the complex features of the CFP exist to defend the interests of various groupings in a precarious balance. This also makes change difficult. While everybody can agree to dislike the complexity, each wants to keep his own safeguards.

True, the CFP reform in 2013 has witnessed many more statements than ever before about the need for 'radical change' but in general terms these changes have been much more about introducing new policy elements (discard ban, regionalisation) than about questioning the traditional ones (TACs and quotas, relative stability).

As we will see in Chapter 2, the considerable difficulties in the initial steps of setting up the policy have created an image of the CFP as a difficult and contentious policy and this tends to discourage any efforts to modify it substantially. If there are changes, these are generally introduced through new, additional legislative layers but rarely by questioning the past.

It is also true that fisheries management in general is a domain where novelties are scarce and difficult to introduce. In the case of Europe the possible introduction of substantial policy changes is always seen as a way to question the balance of interests established in the past and this results in a considerable resistance to change. This only underlines further the need to know the history of the policy, to know its successes and failures in order to better assess its future.

Notes

1 www.ec.europa.eu/fisheries/market-observatory
2 http://hulltrawler.net/History/History.htm
3 http://pinsky.marine.rutgers.edu/wp-content/uploads/2013/08/Pinsky-et-al.-2011-PNAS-w-supp.pdf
4 www.fao.org

References

Bekker-Nielsen, T. (ed.) (2005). *The Archaeological Evidence for Fish Processing in the Western Mediterranean. Ancient Fishing and Fish Processing in the Black Sea Region*. Aarhus University Press. Gylling, pp. 133–160.

Council Resolution (1976) Council Resolution of 3 November 1976 on certain external aspects of the creation of a 200-mile fishing zone in the Community with effect from 1 January 1977, OJ C 105, 7 May 1981, p. 1.

Doumenge, F. (1998). L'histoire des pêches thonières. Symposium ICCAT sur les thonidés. *Recueil des Documents Scientifiques* L (**2**): 753–804.

Facts and Figures of the Common Fisheries Policy (2012). *Basic Statistical Data*. 2012 Edition.

Facts and Figures of the Common Fisheries Policy (2014). *Basic Statistical Data*. 2014 Edition.

Farnell, J. and J. Elles (1984). *In Search of a Common Fisheries Policy*. Aldershot: Gower Publications.

Hardin, G. (1968). The Tragedy of the Commons. *Science* **162** (3859): 1245–1248.

Judgement of the Court of 14 July 1976. Cornelis Kramer and others. Joined cases 3, 4 and 6–76. European Court Reports 1976, 01279.

Judgement of the Court of 16 February 1978. Commission of the European Communities v Ireland. Case 61/77. European Court Reports 1979, 02345.

Judgement of the Court of 19 June 1990. The queen vs. Secretary of State for transport, ex parte Factortame Ltd and others. Case C-213/89.

Judgement of the Court of 25 July 1991. The Queen vs. Secretary of State for transport, ex parte Factortame Ltd and others. Case C-221/89.

Khalilian, S., R. Froese, A. Proelss, and T. Requate (2010). Designed for failure: a critique of the Common Fisheries Policy of the European Union. *Marine Policy* **34**(2010): 1178–1182.

Kurlansky, M. (1997). *Cod: A Biography of the Fish that Changed the World*. Penguin Books, New York, 294 pp.

Lackey, R.T. (2005). Fisheries. History, science and management. In *Water Encyclopaedia: Surface and Agricultural Water*. J.H. Lehr and J. Keeley (eds) pp. 121–129. John Wiley & Sons, Inc., Hoboken, 781 pp.

Leigh, M. (1983). *European Integration and the Common Fisheries Policy*. London: Croom Helm.

Long, R. and P. Curran, (2008) *Enforcing the Common Fisheries Policy* http://onlinelibrary.wiley.com/doi/10.1002/9780470698815.fmatter/pdf (accessed 5 October 2015).

López Veiga, E. (2000). *Manual de política pesquera*. Tomo II. Xunta de Galicia, 415 pp.

Meijide Pardo, M.L. (2002). *A 'guerra pola sardiña'*. Xunta de Galicia, 167 pp.

Ostrom, E. (1990). *Governing the Commons: The Evolution of Institutions for Collective Action*. Cambridge University Press, Cambridge.

Penas, E. (2000). *Elementos para una ordenación integral de los usos del medio litoral de Galicia*. Xunta de Galicia. 368 pp.

Regional dependence on fisheries (n.d.). European Parliament project IP/B/PECH/ST/IC/2006–198.

Resolution of the European Parliament on the functioning of the European Union of 19 June 1995. OJ C151, 19 Jun 1995.

Roberts, C. (2007). *The Unnatural History of the Sea*. Island Press/Shearwaters Books. 435 pp.

Scearce, C. (2009). *European Fisheries History: Pre-Industrial Origins of Overfishing*. ProQuest Discovery Guides. released August 2009.

Schweiger, L. (2010). The evolution of the Common Fisheries Policy: Governance of a Common-Pool Resource in the Context of the European Integration. Institute for the European Integration. *Austrian Academy of Sciences. Working Paper* No 07/2010. November 2010, 88 p.

Sobrino Heredia, J.M., A. Rey Aneiros (1997). *La jurisprudencia del Tribunal de Justicia de las Comunidades Europeas sobre la política Pesquera Común*. Xunta de Galicia, 335 pp.

Treaty establishing the European Economic Community, EEC Treaty (n.d.) http://europa.eu/legislation_summaries/institutional_affairs/treaties/treaties_eec_en.htm.

Treaty of Amsterdam – OJ C 340 of 10.11.1997 http://europa.eu/legislation_summaries/institutional_affairs/treaties/amsterdam_treaty/index_en.htm.

Treaty of Lisbon – OJ C 306 of 17.12.2007 http://europa.eu/legislation_summaries/institutional_affairs/treaties/lisbon_treaty/index_en.htm.

Treaty of Lisbon amending the Treaty on the European Union and the Treaty establishing the European Community, signed at Lisbon, 13 December 2007. O.J. C 306 vol. 50, 17 December 2007.

Treaty of Maastricht on European Union – OJ C191 of 29 July 1992 http://europa.eu/legislation_summaries/institutional_affairs/treaties/treaties_maastricht_en.htm

CHAPTER 2

The origin of the CFP

Fishing in the 1960s

The context in Europe

Although the Treaty of Rome in 1958 established a legal basis for a Common Agricultural Policy and, with the caveats referred to in the previous chapter, a Common Fisheries Policy also, it is clear that while the former was set up at an early stage of the development of the EEC, the latter took much longer.

Agriculture and fisheries were sectors producing food and ensuring the food supply for the European people after World War II was one of the EEC's primary objectives in the Treaty of Rome. Why is it, then, that while the Common Agricultural Policy was developed very soon in historical terms, the Common Fisheries Policy took considerably longer? There are a number of factors that explain this difference.

One obvious reason is the relatively small proportion of fish in the diet of the six first European Member States. While the development of productive agriculture was identified at an early stage as a strategic objective, the relatively minor contribution of fish to the food supply in these six Member States did not militate in favour of a high priority to develop a more productive fishing sector. As we will see in Chapter 3, the two European countries where fish consumption *per capita* is highest (Portugal and Spain) only became part of the Union in 1986.

From 1958 to 1977, fish stocks were not a real common resource in Europe and national management of fisheries was considered enough: there was no need for a really common European policy. In the 1960s, much of the fishing took place in what were at the time international, partially unregulated waters. Beyond the territorial waters of coastal countries (initially 3 miles, later extended to 12 miles following the London Convention of 1964), fishing was unrestricted and the principle of free access applied. The very incipient management rules set for the North Atlantic by ICNAF[1] and later on by NEAFC[2] did not fundamentally prevent the development of the fishing activity at the time, even though these organisations adopted a number of management measures.

Furthermore, the management of fishery resources was not really seen as a problem for many years, notably on account of the generally good state of fish stocks in the two decades following the end of World War II, together with low general awareness about problems of overfishing and the still limited development of management methodologies.

Although the status of fish stocks was and is still today variable and fluctuating, in general terms there was no strong feeling of overexploitation of European stocks in the 1960s and 1970s, even though at the time there had already been certain events of fish stock depletion, such as the collapse of the North Sea herring stock in the 1970s. This situation in fact lasted for quite a long time and even when the policy was established in 1983, the general feeling was that there were no fundamental

The Common Fisheries Policy: The Quest for Sustainability, First Edition. Ernesto Penas Lado.

problems in the conservation of fishery resources, despite the collapse of that fishery.

Perhaps this impression was the result of the fact that some of the most important resources in European waters were perceived to be healthy at the time. For example, in the 1970s the North Sea witnessed the so-called 'gadoid outburst', with cod abundance well in excess of 200 000 tonnes and haddock reaching as much as 800 000 tonnes. These figures are in sharp contrast with the scientific advice many years later: in 2003 ICES estimated the biomass of North Sea cod at barely 35 000 tonnes.[3]

However, for other stocks the scientific community had already been warning about conservation problems and NEAFC (see later) had been proposing conservation measures for some of them in previous years.

In turn, the STCF[4] in its report of June 1982 did not report any issue of overfishing. Its agenda covered only issues related to mesh sizes and minimum landing sizes for Norway lobster, by-catch levels in the *Pandalus* shrimp fishery, the application of zones where by-catch of the *Crangon* shrimp would not apply and so on. In effect, it just concentrated on technical measures without any indication of overfishing of any of the resources concerned (Commission of the European Communities, 1982).

In the Mediterranean, most fishing took place closer to shore than in the Atlantic. There was little off-shore fishing; most of the activity took place by vessels from the then two Member States in front of their own coasts. Only certain Italian fleets from the Adriatic coast and Sicily had a tradition of fishing in distant fishing grounds.

The activity was still expanding. Fish, just as much as agricultural production, was considered a commodity that Europe needed to feed its population. Concerns about the sustainability were simply absent from the political debate, at least in continental Europe. Fisheries at the time were seen more as a source of food for Europe rather than a matter of management of an exhaustible resource.

Finally, the original six members of the EEC were very 'terrestrial' in that the importance of fishing and the maritime sectors was relatively small, at least as compared to other European countries that became members at a later stage (Denmark, UK, Spain, Portugal) or who never joined (Norway).

All these factors together led to a low priority for the development of a Common Fishery Policy in the first years of existence of the EEC.

The context in the world

Another important factor was that international fisheries law was completely unsettled. Since the first important attempt to extend coastal State jurisdiction way beyond the territorial waters – the Truman declaration of 1945 – the international scene was very agitated. Attempts to agree on common approaches on the jurisdiction of marine waters and their resources in the waters adjacent to the territorial seas failed repeatedly. International law was being established mainly through unilateral actions, such as that taken by South American countries: already in 1947, Chile and Peru declared maritime zones of 200 miles.

These initiatives were followed in the early 1970s by other South and Central American countries and Arab States, through a number of regional conferences where the notion of the rights of coastal States to extend their jurisdiction beyond their territorial waters was gathering momentum, in particular with the Montevideo and Lima declarations of 1970, or the Santo Domingo declaration of 1972.

In Europe, a decision by Iceland, already in 1948, to extend the territorial waters from 3 to 4 nautical miles provoked the first of the 'cod wars' with the UK. Further Icelandic initiatives to extend their jurisdiction followed in subsequent years.

These developments showed clearly that the absence of international agreement to regulate the rights and obligations of coastal and other States for the exploitation of fishery resources was an ideal excuse for coastal States wanting to increase their share of the world's fisheries. This has remained an issue ever since: lack of progress on internationally agreed governance is the best alibi for coastal states to take unilateral action.

As we will see in Chapter 7, this international agreement advanced very slowly. In 1964, following the London Convention, European countries extended their territorial waters to 12 miles but the idea of declaring EEZs beyond these 12 miles was still out of the question. The international community still took 18 years, until 1982, to agree on international rules: the United Nations Convention on the Law of the Sea.

In the 1970s, the Member States of the EEC and the United States were still hesitating about the position to take in the international debate, until they decided to follow this trend and decided in 1976 to declare Exclusive Fisheries Zones jointly and unilaterally too, except in the Mediterranean (see next).

The lack of agreement at an international level had a very important bearing on how the birth of the CFP was perceived both by the existing Member States as well as future ones. The fact that international law was not consolidated gave the CFP a certain image of an *ad hoc* policy, which rather than being based on a clear legal basis, was an arbitrary political deal. This perception of the CFP as the result of political negotiation, rather than the result of universally accepted legal principles, has served to question from the very beginning the legitimacy of the policy by certain of its players. This can be considered as one of the historic factors leading to the negative image that this policy has always suffered from.

Such was the international and European context in which the CFP came into being. This analysis has already been explained by certain authors (Farnell and Elles, 1984; Holden, 1984). I will not duplicate these accounts but present them in a new light: that provided by the long experience of a CFP that has already reached 30 years of age. As we will see, the basic political 'deal' which established the CFP was has over the years proved to be extremely stable politically and some of its elements have never been modified substantially.

The precedents: ICNAF and NEAFC

It is important to refer to the two organisations that, at the time of the establishment of the CFP, were already active in the management of fishery resources in the North Atlantic: the International Northwest Atlantic Fisheries Commission (ICNAF) and the North East Atlantic Fisheries Commission (NEAFC).

ICNAF regulated fisheries in the Northwest Atlantic, including a significant European participation and, although it did not regulate fisheries in the EU waters of the time, it was an important precedent in terms of the development of the governance on international fisheries. ICNAF, created in 1954 following an initiative from the United States, spent its first 15 years as a rather quiet forum for cooperation on the management of fisheries largely conducted in international waters. It was in this framework that, since the 1970s, the strongest voices were heard about the intention of coastal States to change the *status quo*. It was in ICNAF that some of the strongest voices in favour of the declaration of 200-mile EEZs in the Northern hemisphere were heard and the idea started to make headway in that forum (López Veiga, 2000).

Although this organisation dealt with species of interest for a limited number of EU Member States, it contributed fundamentally to shape up modern methodologies of scientific assessment and management in fisheries. During its 20 years of existence, it gathered an outstanding array of fishery scientists and led the way in adopting fishery management measures, in particular through the establishment of overall limits on the exploitation of stocks, adopting TAC regulations, further allocated in national quotas and even established certain precedents of multispecies management through the establishment of 'second tier' TACs (Anderson, 1998).

ICNAF served as a model for later conventions, showing an adaptability that allowed it to be the first international fisheries convention to reform itself to come to terms with the existence of the 200-mile zones, becoming the current NAFO[5] as of January 1979 (Farnell and Elles, 1984).

Perhaps the most direct influence on the CFP was that provided for by the NEAFC. In 1946, the United Kingdom organised an international conference on overfishing that decided the establishment of a Permanent Commission in 1953. This was the forerunner of what gave rise several years later, in 1959, to the establishment of the North East Atlantic Fisheries Commission (NEAFC), which entered into force in 1963. This Permanent Commission started establishing technical measures to protect juvenile fish but gradually moved into the joint management of fish stocks.

In 1969 NEAFC established a ban on salmon fishing in international waters and a closed season for North Sea herring from 1971. NEAFC gradually enlarged its work and in subsequent years established a number of limitations of catches and fishing effort, starting with restrictions on catches for herring in 1974. By 1975, NEAFC was already recommending catch limitations and quota allocations for 15 stocks.

These measures had a relatively limited effect on the development of European fisheries of the time. The lack of an effective system to control the catch limitations reduced considerably the practical value of NEAFC recommendations. For this reason, the NEAFC recommendations had little effect on the status of resources and did not prevent the trend towards an increase in fishing pressure.

However, ICNAF and, particularly NEAFC did have an impact in the early days of the CFP, by establishing a precedent on how fishery resources are evaluated and managed. The original CFP largely drew from the experience from these two organisations, in terms of the type of management instrument used and the type of science-base it sought to establish. When the conservation policy was established, the ICNAF and NEAFC 'culture' was imported to constitute the fundamentals of the CFP that have remained to this day.

The effectiveness of ICNAF and NEAFC in addressing the emerging problems of overexploitation of resources was not impressive. As the measures adopted were a minimum common denominator, coastal States still relied more on other national measures rather than on international cooperation. Nevertheless these two organisations set the tone for the future international efforts to manage fish stocks.

The Commission's first initiatives

The story of the initial stages of the development of the CFP has been described already by several authors. John Farnell and James Elles in 1984, Enrique López Veiga et al. in 1993, Mike Holden in 1994, Christian Lequesne in 2001 and Lukas Schweiger in 2010, among others, have explained these historical events. In this book I will only try to recount the key elements of that story insofar as they contributed to forming the CFP of today.

The 1967 Communication

The origin of the CFP can be traced back to the paper, published in 1967 by the Commission, on 'Basic principles for a common fisheries policy'. Not surprisingly, the emphasis of this paper was on structural, market, trade and social questions. Fisheries conservation was not really an issue at the time in Europe. It has been argued, however, that this was largely due to the absolute lack of fisheries expertise in the Commission services at the time (Holden, 1994).

It may also be the case that the first attempts at managing the sustainable exploitation of fishery resources in Europe took place in the UK and Nordic countries, not EEC members at the time, as continental Europe had relatively little tradition and expertise in this matter at that time. European fisheries science started to develop in the United Kingdom (and particularly in Lowestoft) towards the middle of the 1950s and it took time to reach the continent.

The Commission paper underlined the vision of fisheries at the time, particularly for the Member States where this industry was still not fully modernised. The philosophy was that of the Common Agricultural Policy: a two-fold approach of establishing a common market on the one hand, with support prices and protection against imports at the border and, on the other, an attempt to rationalise production structures by greater coordination and a programme of EEC financial assistance for restructuring of the industry (Farnell and Elles, 1984). This initial orientation was considerably shaped up by French fishermen, who considered that a system similar to that established for the Common Agricultural Policy was a necessary compensation for the acceptance of a common customs tariff for fishery products (Lequesne, 2001).

The structures and market regulations of 1970

This philosophy was clearly reflected in the two first regulations related to European fisheries, adopted in 1970: the first regulation providing for structural aid to the fishing sector (Regulation 2141/70) and the second on the marketing of fishery products

(Regulation 2142/70) (EEC, 1970a,b). These two regulations were adopted in the context of the upcoming accession to the EEC of three countries with important fishing interests and/or important fishing grounds.

The philosophy of the first structural regulation was clearly in line with the approach that characterised the Common Agricultural Policy for many years after the establishment of the Common market in 1958: the emphasis in the increase in production and the financial support to the modernisation and development of the means of production.

In agriculture, more production can usually be obtained by more investment. In mature fisheries, more investment often means lower production, because the wild resources are finite and more overfishing results in less fish being caught. Applying agricultural principles to fisheries has had very long-lasting detrimental effects in the CFP.

This regulation was largely triggered by the upcoming enlargement of the Union in 1972, as we will explain later in this chapter. Beyond the structural actions themselves, it contained a principle that applies to the CFP of today: the principle of freedom of access to Member States' waters. In addition and in order to reassure new Member States that the freedom of access would not entail a high risk of overexploitation of their resources, the regulation established that the Council could adopt any necessary conservation measure. These two seminal elements, in a regulation theoretically dealing with structural policy only, were fundamental in the establishment of the CFP several years after.

The first market regulation (2142/70) was largely inspired by the existing agricultural equivalent. The regulation recognised the importance of fisheries in certain coastal regions and set forth to 'encourage rational marketing of fishery products and to ensure market stability by appropriate measures', thus very much in line with some of the objectives of the agricultural (and fisheries) policy in the Treaty.

This regulation established the basic marketing standards and guide prices to allow for the application of mechanisms of market intervention: private storage aid, export refunds, withdrawals from the market in case of price collapse and so on. Financial aid was also granted to Producer Organisations to prepare and implement marketing plans to adjust supply to demand.

This policy has been remarkably stable, to the point that many of the mechanisms have remained fundamentally unchanged or slightly adapted for more than 40 years, until most were removed in the reform of the CFP in 2013.

The coincidence in time with the first structural regulation was not casual. It was part of a package by the EEC of six to prepare for its first enlargement. As we will see, the importance of the fishing sectors, and the new fishing grounds in the three new Member States of 1972 prompted the adoption of measures to stabilise the industries in the six original members and protect them from the possible effects of enlargement.

However, little in these two regulations anticipated the future shape of the CFP, in particular its conservation angle. What did then determine the birth of the CFP as we know it today?

The birth of the CFP as we know it today was due to two main historical facts:

1. the mentioned accession to the European Economic Community of the time of the United Kingdom, Ireland and Denmark on 1 January 1973, and
2. the declaration of Exclusive Fisheries Zones by Atlantic and North Sea Member States applicable on 1 January 1977.

Let us look at these two developments and to their impact in the birth of the modern CFP.

The enlargement of 1973

Fisheries had not been a significant issue for the first decades of the European Union. The first occasion in which fisheries became an issue was the discussion on the new custom tariff policy during the 1960s. The EEC set in some cases lower tariffs for fish and fish products than were set by certain national governments, thus calling into question the rather protectionist markets that fish had enjoyed so far. In that context, the negotiations for the accession of the United Kingdom, Ireland and Denmark posed a new challenge:

On the one hand, the Member States of the Community of six fished primarily beyond the waters under their national jurisdiction. It has been estimated that these Member States together caught 90% of their fish outside their territorial waters in the 1950s. A very significant proportion of these catches were in fact taken in the waters around the new Member States, particularly the UK and Ireland, whose territorial waters extended to 12 miles, following the Convention of London of 1964.

On the other hand, existing Member States should open their markets, upon enlargement, to the products from new, allegedly very competitive fleets. The six Member States reacted very differently to this challenge: France and Italy preferred to secure a regime of protection to their fleets through structural and market aid, while Germany and the Netherlands preferred simply to secure access to the waters of the future members.

These challenges prompted the first initiatives from the Community of six to prepare for the accession of the three new States.

One initiative was to consider that, in the face of competition from new Member States, certain national fishery sectors would need structural and market aid to renew their fleets and be able to compete with the – supposedly more competitive – newcomers. This gave rise to the birth of the structural policy, through the adoption of Regulations 2141/70 on a common structural policy and to the establishment of a fish market policy through Regulation 2142/70 on the common market organisation in fisheries products, as we saw previously.

The second and more contentious initiative was the regime applicable for the jurisdiction of the fishing areas. Already in the 1960s the question of the jurisdiction over the rich fishing grounds of the continental shelves, beyond the territorial waters, was accelerating. In Europe, the largest continental shelves beyond the North Sea were those around the British Isles, where some Member States of the Community of six and other 'third' countries (and future members) had traditionally been fishing.

While continental Member States wanted to consolidate their access to these waters after accession, the UK and Ireland had the opposite interest and wanted to reserve these waters to their own fleets.

The question was settled in 1970, prior to the negotiations for the 1973 enlargement. The six existing Member States established the principle of free access to each other waters as an *acquis* that new Member States would have to accept. Regulation 2141/70 established that:

> Rules applied by each Member State in respect of fishing in maritime waters coming under its sovereignty or within its jurisdiction shall not lead to differences in treatment of other Member States.
>
> Member States shall ensure in particular equal conditions of access and use of fishing grounds in the waters referred to in the preceding subparagraph for all fishing vessels flying the flag of a Member State and registered in the Community territory.

This text enshrined the principle of equal access in Community law, that has applied to the CFP ever since, underpinned by the Court of Justice through case law (Order of the Court of 22 May 1977).

Regulation 2141/70 entered into force 6 hours after the opening of negotiations with the four candidate countries on 1 February 1971, having been adopted in October 1970. This development left an impression on the acceding Member States of a *fait accompli*. On the contrary, existing Member States considered this a political deal between those countries having the fishing grounds (UK and Ireland) and those having the large, lucrative markets for the fish (France). This idea of the CFP as a basic deal between those having the resources and those having the markets was later reinforced by the accession of Spain in 1986.

Another factor, related to the enlargement that facilitated the agreement was the fact that in their Treaty of Accession the UK, Ireland and Denmark consolidated reserved access to the 12 miles for their own fleets. This consolidation lasted only 10 years, that is, it would expire in 1982. That meant that, in the absence of an agreement on the CFP, the new Member States would lose the derogation of the 12 miles and the principle of free access would guarantee full access of other Member States' vessels to their coastal waters. This factor contributed significantly to create the urge, in the new Member States, to agree on a new CFP based on free access but excluding the 12 miles from it.

As the three countries acceding on 1 January 1973 were North Sea and Atlantic ones, this enlargement did not have effects on Mediterranean fisheries, or indeed Baltic fisheries. This is one of the reasons why the birth of the CFP was fundamentally conditioned and shaped-up by the considerations on North Sea and Atlantic fisheries. The CFP ignored the Mediterranean fisheries for quite a long time.

The declaration of exclusive fisheries zones as of 1977

As we saw previously, during the late 1960s and early 1970s a number of unilateral initiatives were taken by various countries around the world to the effect of reserving access to fishery resources to their own nationals in zones adjacent to their territorial waters.

The international scene had been moving very fast in the 1970s. The development of fisheries beyond the 12 nautical miles of the territorial waters and the lack of effective management of some of these fisheries had been leading a number of coastal States in the world to claim sovereignty over fishery resources well beyond the 12 miles. Examples such as the 'cod wars', a result of the unilateral declaration of sovereignty by Iceland over its rich fishing grounds and the subsequent expulsion of foreign fleets (mainly British, German and French), illustrate this state of affairs.

In Europe, there was much hesitation as to the line to take. The interest to reserve the fishing grounds to their own nationals was compounded by the need to secure access to the distant waters where other countries could also claim the 200 miles. At the time, some of the EEC Member States like the United Kingdom and Germany still had important fishing interests in distant waters and were therefore divided between conflicting interests for their own fleets. European long-distance fishing took place in the Arctic, Northwest Atlantic, Iceland and other distant fishing grounds.

The gains through the declaration of an Exclusive Fishery Zones and the reduction of foreign fishing in the waters of the Europe of nine had to be compounded with the loss of the fishing grounds resulting from similar declarations by third countries. Whether one would compensate the other was a very difficult question to assess, as well as politically delicate. The way in which the loss of fishing opportunities in third country waters (after their declaration of a 200-mile zone) would be compensated by the higher fishing opportunities in the future European waters was an extremely complex question, where all Member States had entirely different interests. In addition, some of them (Ireland) without long-distance fishing, saw the declaration of a 200-mile zone as an unacceptable option unless they would have a right to develop their own fishing fleet, to benefit from the rich waters around their country (Farnell and Elles, 1984).

The position of the United States was crucial in this respect: while the US had an interest to behave as a coastal State in many fisheries, they did not want to accept the legitimacy of the new 200 mile unilateral declaration because this meant the exclusion of their Eastern Pacific tuna fleet from the rich tuna fishing grounds from Central and South American countries.

For a number of years, the unsettled situation gave rise to a number of unilateral initiatives by many countries, without any recognition in international law. In fact, the consolidation of the 200-mile Exclusive Economic Zones in international law took place only at a later stage, when the 1982 United Nations Convention on the Law of the Sea, adopted in 1982, accepted this principle which, subsequently, entered into force in 1994.

The first important decision that started to precipitate the change of position in Europe was the declaration by Iceland of an Economic Exclusive Zone of 200 miles in July 1973. This prompted the last of the 'cod wars' with the United Kingdom, resolved in favour of Iceland in 1976 when they skilfully played their geostrategic position in the context of the Cold War.

The overall balance shifted decisively following the momentum gathered by a number of countries (mostly non-European) at the United Nations Conference on the Law of the Sea, held from 1973 to 1982 (UNCLOS III). There, countries like the United

States, Canada, Norway and the Soviet Union finally undertook to follow the example of other countries around the world to declare a 200-mile EEZ (Long and Curran, 2000). In the case of the US and Canada, the interest to reserve for their fleets the very rich (and allegedly endangered) resources of the Northwest Atlantic (the Grand Banks off Newfoundland, Boston and Georges Bank) seems to have been at the origin of this decision.

After the announcement by these countries, the North Sea and Atlantic Member States of the EEC understood that the balance had shifted and that the 200-mile EEZs were unstoppable. They then decided to follow suit. The Council then agreed on 3 November 1976 that Member States should act in concert to extend their fishing zones to 200 nautical miles with effect from 1 January 1977 along their North Sea and North Atlantic coastlines. This resolution, meaningfully, was only about extending jurisdiction but not about ensuring sustainable use of the resources. The CFP, as we will see, would take years to catch up on that.

The declaration, which gave EEC Member States jurisdiction over the resources in a large space of international waters (Figure 2.1), together with the consideration of the principle of free access by all members of the EEC triggered the birth of the first management measures under the CFP. However, from these preliminary measures until the official birth of the CFP there were 6 years of protracted and often bitter negotiations. Although the Commission presented its proposal for a common fisheries policy already in 1976, in anticipation of the declaration of EFZs in 1977, the CFP as such was only agreed formally in 1983. What made these discussions so difficult was in fact, only one issue: the allocation of the common resources among the Member States.

During this interim period, there were attempts to adopt conservation measures, on the initiative of the Commission, in particular as total allowable catches for key stocks following the example of NEAFC. However, only in one year, 1980, were such measures adopted. This created a difficult situation, where the need to resolve the problem of quota allocation prevented an effective conservation policy. The CFP would need many years to face the challenge of the sound management of the fish stocks concerned.

The declaration of EFZs did not take place in the Mediterranean, for a number of reasons but mainly because of the difficulty of reaching agreement on a common approach between France and Italy on the one hand and the remaining countries of the Mediterranean basin on the other. This was obviously made difficult by both the narrowness of the Mediterranean basin, where no coastal State can declare a 200-mile EFZ without overlapping other States claims (thus requiring very complex agreements on median lines) and by the well-known political difficulties among certain Eastern Mediterranean countries. As a consequence, the Mediterranean was left behind in defining the shape of the CFP.

The 6 years of negotiation for the establishment of the CFP, from the first Commission proposal at the end of 1976 (EEC, 1976) until the adoption of Regulations 170 and 171/83 (see next) encapsulate the difficulties of agreeing on a policy among countries with considerably different interests and traditions.

These diverse interests were complex and included *inter alia* the interest by new Member States to reduce the consequences of the principle of free access adopted in 1970, the consequences of the loss of fishing grounds in third countries (and the different attitude to imports from these countries), the interest to protect certain coastal fishing communities and the interest of Ireland to further develop its own industry.

Although the final agreement of 1983 was due to a series of circumstances, at different times, there are several factors that contributed to the final deal. Some of these are still valid today as political elements to underpin the basic tenets of the CFP:

- The political linkage in the UK between agreement on a CFP and the wider issue of the UK's financial contribution to the Union in 1980 (Farnell and Elles, 1981).
- The adoption of new market rules in 1981, with new arrangements for trade with third countries in fishery products, to handle the effects of the new situation created by the declaration of EFZs (EEC, 1981).

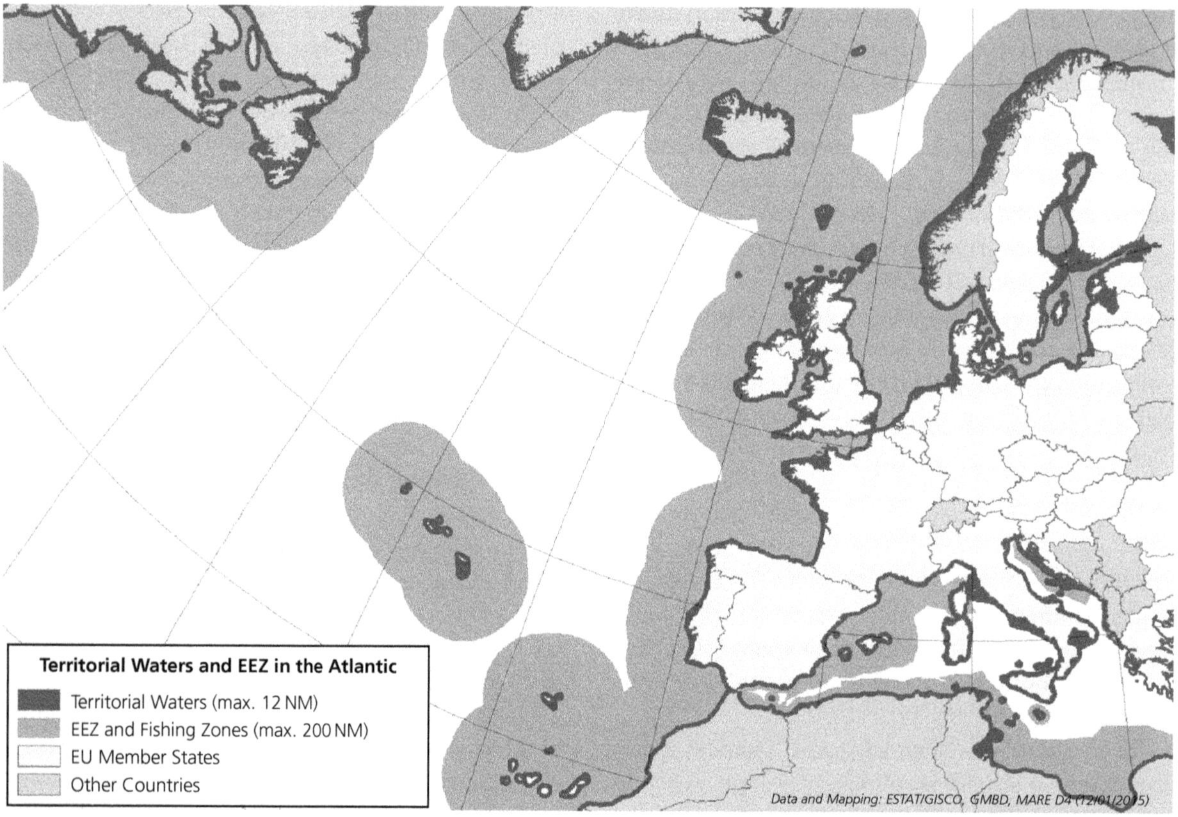

Figure 2.1 The effects of the declaration of 200-mile Exclusive Economic Zones (light blue) as compared with the previous 12-mile territorial seas (dark blue). Source: European Commission Services. (Disclaimer: the EEZs in this figure are illustrative, not authoritative.) (See colour plate section for the colour version.)

- The adoption of a financial package for the scrapping of fleets that lost their traditional fishing grounds in third countries waters.

It has been argued, however, that in the end, it was the perspective of the accession of Spain that convinced Member States that before such accession would take place, it was necessary to establish the basic rules, as happened with the 1970 regulations that were a preparation for the 1973 enlargement (Farnell and Elles, 1984).

Relative stability and The Hague preferences

To agree on the conservation and management of fishery resources, it was paramount for Member States to establish a secure share of the resources of their interest. The establishment of relative stability, that is, the formula to allocate fishing rights among incumbent Member States of the time, was therefore the precondition for the establishment of the CFP. If the policy was born in 1983, 6 years later than the declaration of Exclusive Fishery Zones, it was because of the political difficulties to agree on the specific allocation of the common resources.

Relative stability is the system of allocation of fishing rights among Member States, through fixed percentages per Member State of the catch limitations established in the CFP. These catch limitations, in the form of Total Allowable Catches (TACs) are established by the Council of Ministers of the EU and are automatically divided among Member States according to fixed percentages, corresponding to the relative stability keys.

Relative stability in 1983 was based on the consideration of three elements:

1 historical catches by Member States;
2 losses of fishing opportunities as a consequence of the establishment of 200-mile EEZs by third countries, notably Iceland, and
3 special needs of coastal communities with a strong dependence on the fishing sector.

These three elements were used only for the initial allocation of rights in 1983 among the 10 Member States of the time (or rather, among the seven North Sea and Atlantic Member States concerned). These elements have never again applied when establishing relative stability keys for the accession of new Member States in subsequent years. In these cases, historical catches were the main criterion.

Let us look at the history in more detail.

The basic allocation key

The declaration of EEZs and their status as Community waters led immediately to the question of how to allocate the fishing rights among Member States.

This question of allocation became extremely difficult because it was seen as the only way by the three acceding Member States to secure a significant share of the resources for their industry. During the accession negotiations, the UK and Ireland had been trying to negotiate an access regime that would reserve 50 miles of their waters to their own fishermen. This was not accepted by the Community of six, based on the principle of equal access enshrined in Regulation 2141/70 (EEC, 1970b).

The alternative was for each Member State to secure, by negotiation, permanent access to a significant part of the shared resources. At the time and to take into account the fluctuation of catches, the only way to do this realistically was not to agree on absolute amounts but on percentage shares of allowed catches. All parties agreed that this allocation should be stable, so it would not have to be re-discussed every year. The allocation of percentage shares of the stocks was given the name 'relative stability', which has remained unchanged to this day.

This discussion on how to share the resources took place against the background of a very changing pattern of fishing. By the time this debate took place, the UK fleet had had to withdraw from the waters around Iceland. The loss of the rich cod fishing grounds around Iceland, as a result of the 'cod wars' left the British fleet without access to one of its most traditional fishing areas. To agree on an allocation key for shared resources, this fact should be taken into account.

In addition, the UK also wanted to be compensated for what they considered a threat to their coastal fishery communities (in particular around Scotland) that the principle of equal access would represent.

On the other hand, Ireland had developed very little its own fishing sectors during the times of the free, almost unregulated international waters. In political terms, Ireland could only accept the deal if it involved the right to expand its own fishing industry, particularly for pelagic species such as herring, mackerel and horse mackerel.

This resulted in a system that had two components: (1) a system of basic allocation of rights, based on an objective criterion that would ensure transparency and (2) a system of compensations to the UK and Ireland to take into account these factors. The first system consists of the basic relative stability keys. The second is comprised of the so-called 'Hague Preferences', as they were agreed at a meeting in the Dutch city of The Hague in 1976.

The relative stability keys were not only adopted for the main commercial species at the time but also for a number of stocks of relatively secondary interest. The reason for this was preventive. Although for some of these species there was no scientific advice, and therefore no meaningful conservation policy could be adopted, the idea of 'precautionary TACs' being allocated was to prevent the development of fisheries for 'free resources'; that is, unregulated species. This has given rise to some TAC and quota allocations, in certain areas that did not have a real conservation basis. Certain phenomena such as the 'paper fish' (that is, fish that has quotas in the regulations but that does not occur in these amounts in reality) can be traced back to this time. Although some of these TACs have been removed from the CFP regulations over time, the idea of adopting TACs to prevent unregulated expansion of fisheries that would alter the *status quo* has remained a feature of the CFP.

The basic criterion to calculate the percentage shares was historical catches. As is always the case, the choice of the historical period concerned is not neutral: longer or shorter, more recent or more distant, different reference periods to compare historical catches favour different Member States. Between 1976 and 1983 this discussion proved extremely difficult. In the end, the period 1973–1978 was chosen simply because it allowed a balance among Member States. It also had another advantage: the catches of that period were already consolidated in the statistical bulletin of the EEC and could no longer be altered in Member States' interest.

Overall, the criterion finally chosen corresponded to the maintenance of the recent *status quo*. This set a precedent in the policy, whereby maintaining the *status quo* with respect to other Member States has become a permanent objective: hence the term 'relative stability'.

The Hague Preferences

The Hague Preferences were a system whereby certain Member States (the United Kingdom and Ireland) were guaranteed a minimum quota of certain stocks in case that agreed TAC was very low. This was part of the deal on the relative stability keys. The idea was that the UK and Ireland would accept lower quotas than they wished in exchange for a mechanism to ensure that if total allowable catches were low, they would have a preference to catch them, despite the allocation keys under the basic relative stability keys.

These Hague Preferences also applied to Greenland, still a Community territory at the time and applied until this Danish territory left the EEC in 1985.

The calculation of The Hague Preferences was more complicated. Mike Holden provides a first hand, detailed explanation of how exactly they were calculated (Holden, 1994).

The basic agreement was difficult to reach. Six years to agree on an allocation of fishing shares among Member States for which the fishing sector represented at the time (and still today) less than 1% of their Gross Domestic Product is the most clear demonstration of this phenomenon referred to in Chapter 1. This is also illustrated by the repeated application by Member States of the so-called 'Luxembourg compromise', a non-legal, political agreement of 1966 whereby the Council undertook not to vote a proposal if a Member State considered that it would affect its 'vital national interest'. Applying the 'Luxembourg compromise' for a discussion on quota allocation of fishing rights is the most clear example of how fisheries issues and the CFP, can become issues of national interest far beyond the objective economic importance of the sector.

These considerations only affected the initial allocation keys of relative stability, as a way to establish the basic political deal that allowed the birth of the CFP. As the EU has successively enlarged, the considerations that justified The Hague Preferences have not been applied again.

The consolidation of relative stability

The relative stability of 1983 has been consolidated and extended ever since. This consolidation has been the result of a majority, confirmed political support.

As we saw previously, these considerations did not apply in the Mediterranean, where the lack of identified 'common resources' resulted in the absence of any TAC and consequently the absence of relative stability. It is important to underline, however, that on the occasion of the first TAC (and so far the only one) to be applied in the Mediterranean fishery (the Bluefin tuna TAC adopted by Council in 1998) it was of course understood that the same principles applicable to relative stability should also apply in that case.

Relative stability has proven extremely resilient over time. In fact, only two Member States have been openly critical at some stage of the distribution of fishing rights, although for exactly opposite reasons: Spain and Ireland. Spain has always considered that the relative stability keys it could negotiate upon accession did not reflect their historical weight or their current interest in the fisheries concerned. On the opposite side, Ireland has taken the view that, despite having a large percentage of the EU fish resources in its waters, it is unfair that as a consequence of relative stability it receives a lower percentage of these resources. These two arguments represent opposite points on the spectrum and tend to cancel each other in any possible discussion on relative stability. The Spanish argument is that allocations should be based on historical track records, that is, on the demonstrated activity of their fleets. The Irish argument is in favour of the so-called 'zonal attachment'; that is, that the allocation should be based on the proportion of the fish that occurs in the waters under their jurisdiction.

What makes relative stability so stable is not its legal basis but rather the political difficulty of renegotiating the allocation keys. The best example is the new allocation keys established for new Member States in their Treaty of Accession as they join the Union. In these cases, it is recognised that the allocations agreed in the treaties apply only for the first year. The Council is free to modify these keys in subsequent years. However, in practice, this is hardly ever the case, so it is the preservation of a difficult *status quo* and not a legal basis that makes relative stability so resistant to change.

Over the years, relative stability has been extended but existing allocation keys have fundamentally remained unchanged (with some exceptions, as we will see next). It is remarkable that these allocation keys have remained unchanged despite the evolution of European fleets (see Chapter 5) and changing consumption, regulatory and economic patterns. This illustrates that the allocation keys are more the result of a political deal (extremely difficult to reopen) than the result of an economic reflection on how resources can be best allocated in a rational manner.

The regulations of 1983

It is generally accepted that the birth of the CFP corresponds to the adoption, in 1983, of two regulations: one establishing the general regime for conservation and management of fishery resources (EEC, 1983a) and another on technical conservation measures (EEC, 1983b). These regulations defined the two basic components of the conservation pillar of the CFP, which can be summarised as: (1) do not catch too much fish (catch limitations) and (2) do not catch small fish (technical measures). These two components of the policy have also remained remarkably stable over time.

The adoption of these two basic texts also allowed in parallel the adoption of the first accompanying annual TAC and Quota regulation: 172/83 (EEC, 1983c) that completed the package by establishing catch limitations to a total of 87 stocks of 24 species (including 14 stocks of herring), in European waters of the Atlantic and Baltic Seas as well as NAFO and French Guyana waters.

Regulation 170/83

Regulation 170/83 established the principles for the management of the common resources. As we will see, most of these principles remain today as the cornerstone of the CFP:

- The regulation defined the main types of measures to manage stocks, including the restriction of fishing effort, by limits on catches and by technical measures (mesh sizes, minimum landing sizes and closed areas).
- It enshrined in law the principle of relative stability, including a recital on the need to take into account the special needs of fishery-dependent regions, in a reference generally understood as referring to The Hague Preferences. The principle of relative stability was open to possible adaptations.
- It enshrined in law the establishment of the Commission's advisory body, to complement the advice provided by ICES: the Scientific and Technical Committee for Fisheries (STCF). This committee in fact had already been working but its role as an advisory body to the CFP was formalised in this regulation.
- The regulation also consolidated the access regime within the 12 nautical miles of the territorial sea. In effect, the CFP incorporated the conclusions of the Fisheries Convention of London in 1964. This convention, organised to try to put order into the different approaches by different coastal States, agreed to the principle that the territorial sea would be extended to 12 miles, that the first 6 miles coastal states would have the right to restrict access to their own nationals but that the 6 to 12 miles

they should allow the continuation of the access by vessels from other Member States, on condition that this activity would correspond only to the continuation of existing practice, not to the development of new fishing activity.

- This principle is in fact a derogation to the principle of equal access and initially it applied for 10 years but has worked so well that it has been remarkably stable and has been re-incorporated into the CFP after every reform. The only exception to this is the case of Spain, having historical rights within the Irish 6–12 miles, that in order to facilitate accession on the very difficult fisheries Chapter, gave up these rights through a bilateral protocol signed between the two countries in 1985.
- The delicate balance to agree on quota allocation also required the inclusion of a zone around the Shetland Islands, where a special regime of limitation of the number and size of vessels was established: the so-called 'Shetland Box'.
- The regulation also established a Management Committee for Fisheries Resources composed by Member States to advice the Commission on the preparation of new legal acts under the CFP.
- Finally, the regulation also established the basis for its own partial revision, by requiring the Commission to present a report on its implementation in 1991 with a view to a possible revision of the policy in 1992 on two aspects: the 12-mile regime and the Shetland Box.

It is noteworthy that this regulation did not establish a clear geographical scope and it did not contain a single mention of the Mediterranean, with the exception of a recital whereby the concerted action to declare exclusive economic zones was without prejudice to possible future action in the Mediterranean. This is a very clear indication of the reason why the CFP was set up: not to manage resources (as there were probably as many reasons to manage Mediterranean than Atlantic stocks) but rather to agree on how to share the newly 'common' resources. And these 'common' resources were simply not established in the Mediterranean basin.

As we saw previously, the new basic regulation of the CFP coincided with the first regulation on TACs and Quotas (172/83, see previously), which laid down catch limitations to 87 stocks, largely in the waters of the Community of nine but also with a few species regulated by NAFO. These TACs were already allocated according to relative stability and their very structure has remained practically unchanged to this day.

Regulation 171/83

The other regulation, 171/83, laid down technical measures for the conservation of fishery resources. These measures included minimum mesh sizes, by-catch rates of certain species in certain fisheries, minimum fish landing sizes and closed areas/seasons to protect either juvenile or spawning fish.

This regulation was also difficult to negotiate, due to the large variety of fishing practices in Europe and the need to establish a clear 'level playing field' among them while trying to complement the catch limitations with measures to make fishing more selective.

The scope of this regulation reflected the Atlantic and North Sea traditions that characterised the birth of the CFP. This regulation applied not only to the Atlantic façade of the EEC of nine but also to the French overseas territories of St Pierre et Miquelon, Guyane, Martinique and Réunion but not to the Mediterranean Sea.

From today's perspective, it is hard to understand why the original CFP covered French overseas territories while the Mediterranean was completely missing.

The idea at the time was that, as the CFP and its technical rules were a result of the establishment of 'Community waters', then the policy should cater for all those areas where EEZs were declared or where they would be in the future. Clearly, at the time the idea of declaring EEZs in the Mediterranean was completely out of the question, so the concerned Member States of the time (France and Italy) preferred to keep the *status quo*. In fact, the first conservation measures for the Mediterranean under the CFP still took 11 additional years to be adopted, with Regulation 1626/94.

From the beginning, these measures were extremely detailed. Technical details such as the attachments to nets were governed by technical rules to be developed as implementing regulations (that is, by Regulations adopted by the Commission to provide more detail to the more general rules established by the Council of Ministers). This approach of deciding all the rules, even the most technical ones, through European regulation was both necessary and possible at the time. Necessary because of the need to ensure a 'level playing field' among the Member States that, given the extremely difficult negotiations regarding quota allocation, constituted a way to ensure a fair deal for everyone. It was also possible because a relatively small number of Member States and limited geographical areas were involved. The problem is that, over time, this approach created a precedent, even a 'culture' of law-making in fisheries, that has remained as an integral part of the CFP even when it grew in number of Member States, areas concerned and subsequent complexity.

The complexity of this first technical measures regulation is illustrated by the fact that, by 1984, Regulation 171/83 was amended by the Council as many as five times.

Here too, the historical circumstances of the difficult political negotiation to establish the CFP have contributed to this culture of law making that has survived to this day.

What remains of the 1983 policy?

From its inception, the CFP has had a structural and a market policy, a conservation policy based on TACs and quotas and technical measures, subject to changes over time. In addition, the two basic instruments for fisheries management, TACs and technical conservation measures, have remained also remarkably stable: they have evolved throughout the last 30 years but their structure, philosophy and legislative style have remained practically unchanged.

There are two elements that, although considered as provisional arrangements that facilitated the birth of the CFP, have remained fundamentally unchanged and that are the clearest demonstration of the tendency of the CFP to preserve the results of its most difficult political negotiations: the 12-mile access regime and relative stability.

The 12-mile regime

The regime applicable to the 12 miles of the territorial waters of Member States is a derogation to the principle of equal access to waters. As we saw previously, the basic agreement was to reserve the initial 6 miles to the national fishermen and to allow for the continuation of the existing activity of fishermen from other Member States in the 6–12 mile zone.

This regime is so stable, that the definition of fleets from Member States that can fish in the 6–12 miles remains unchanged from 1983 until the 2013 reform. The regime has only been expanded as a consequence of enlargement but the basic historical rights have not changed. The annex of the basic successive regulations laying down Member States' rights in the 6–12 miles of their neighbours have been identical over time (albeit with new additions as a result of enlargement), despite the evolution in the structure of the fleets in different Member States. In fact, some of the access conditions within the 6–12 miles are not fully utilised by the concerned Member States, showing that the stability of the scheme does not necessarily derive from current fishing activities in all cases.

Relative stability…

The extraordinary resilience of relative stability is probably due to two factors: (1) the need to secure a share of the resource as a pre-condition to consider any sacrifice for management purposes and (2) the fact that this basic undertaking of the policy was so difficult to negotiate (6 years among five countries), that calling into question that basic agreement would risk collapsing the whole policy.

However, relative stability is a living instrument. While the notion itself was established in 1983, its application has been growing steadily over the last 30 years. As the CFP included more stocks and new Member States the number of relative stability keys has increased. Many more species have been added.

The growth of relative stability has not followed any specific pattern, in terms of the criteria used. This means that the original basis used to establish this approach in 1983 has not necessarily been used in subsequent decisions on its application.

Different allocation keys adopted after 1983 reflect entirely different approaches. Although track records of catches are always part of these allocations, the way in which they are taken into account varies. There are many examples to illustrate this.

In some cases, the historical track records are based on a 10-year period, as was the case for the allocation of deep-sea stocks in the first TACs adopted in 2004.

In other cases, the criteria for internal allocation followed the criteria used by the EU in international negotiations to agree on its share of resources in international waters. This has been the case with highly-migratory fish stocks such as Bluefin tuna in the context of ICCAT, where the internal allocation of the EC quota of Bluefin tuna negotiated in ICCAT, adopted by Council in 1998 (EC, 1999), was undertaken according to the same criteria used in ICCAT to allocate the general TAC among its contracting parties of the time in 1994.

It is important to note that one Member State (Italy) took the Council to the Court of Justice, contesting the fairness of this criterion that did not favour Italy since this Member State was not a member of ICCAT at the time and its statistics were poor and incomplete. The Court upheld the allocation key decided by Council, establishing interesting case law for subsequent allocation exercises (Judgment of the Court of 25 October 2001).

In other cases the relative stability key reflects the historical allocation by country prior to accession to the EU. For most Baltic stocks the allocation keys are derived from the allocation agreed by individual countries under the old IBSFC[6] prior to their accession to the EU.

In other cases the allocation keys were entirely *ad hoc* and agreed politically as a precondition to establish an international management regime. This was the case, for example, of the northern albacore stock, where internal differences among Member States prevented the Union from agreeing a management regime in ICCAT. Only after an agreement on the allocation of that stock among EU Member States (on a purely political basis through a negotiation) could the Union successfully propose a management regime for this stock in ICCAT.

The case of the 'cohesion cod' is perhaps the clearest case showing that the establishment of relative stability can be an entirely political agreement, without any specific criteria to support it. This case was originated by Norway's decision, as a consequence of a referendum held in 1994, not to join the EU in 1995. Subsequently, on becoming a member of the European Economic Area, Norway agreed to provide quotas for cod to the EU in Norwegian waters. The agreement was that these quotas would be allocated to the 4 Member States benefiting from the cohesion policy of the time: Spain, Portugal, Ireland and Greece.

Given the difficulty of agreeing to an objective criterion to establish the allocation key, the Commission took an entirely *ad hoc* decision: Spain and Portugal, the two cohesion Member States having a tradition and interest to fish cod in Norway, received 45% of the EU TAC each. The other two cohesion Member States, Greece and Ireland, with no such tradition, received 5% each. Track records of historical catches were simply not taken into account.

... But relative stability can change

While the notion of relative stability remains, specific allocation keys may actually change over time. This was actually the initial intention of the legislators in 1983. Article 4(2) of Regulation 170/83, after establishing the principle of relative stability in paragraph 1 and in connection to an article requiring the preparation of a report in 1991 on the fisheries situation in the Community, established that: 'The Council ... shall enact provisions effecting the adjustments that it may prove necessary to make the distribution of the resources among Member States in consequence of the application of paragraph 1.' This clearly indicated that legislators saw relative stability as a principle, whose application may require possible modifications of specific allocation keys.

The common practice, however, has been to consider that relative stability actually means that allocation keys are never questioned and that the principle itself implies permanent keys.

Despite this general trend, there are some specific cases where allocation keys have been modified, to adapt relative stability to changing circumstances. Here are some examples:

A simple one was the division of the cod stock in the Baltic. The cod stock was considered historically a single stock and was subject to a simple allocation key under IBSFC, and this historical allocation key was transformed into relative stability among Baltic Member States upon accession in 2004.

However, following the scientific advice from ICES that Baltic cod was actually divided in two separate biological stocks, the management system of the EU recognised that two separate catch limitations should be established, each one with its own new relative stability key. This new agreement was achieved through a negotiation where different scenarios were presented by the Commission in a completely neutral way. The final allocation was simply the one that produced consensus among the eight incumbent Member States, regardless of its correspondence with other criteria used before on other areas or stocks.

Probably the best example was that of blue whiting in 2005. That year the Union agreed with its partners in the North Atlantic a new management regime for this stock, which straddles European waters and those of third countries of the Northeast Atlantic. For the sake of compromise, the EU had to accept a limitation of its fishing possibilities.

This represented a new challenge for the EU: how to maximise its catches under that arrangement. It was evident that the relative stability agreed in previous years would not allow an effective exploitation of the EU quota, as the allocation by area did not represent necessarily the areas where the fish was found. In other words, the allocation key agreed under relative stability implied large amounts of 'paper fish'. It was therefore necessary to modify the relative stability by area in a way that the allocations would better correspond with the management areas where the fish was really found and exploitable.

This modification of the allocation key was achieved in the Fisheries Council of December 2005, through a restricted negotiation (under the 'ministers + 1' format, that is, in a very restricted session where ministers are accompanied by just one advisor to facilitate the agreement) at which the mismatch between fishing possibilities and relative stability was made evident. An *ad hoc* arrangement was negotiated at the end, including specific bilateral voluntary transfers amongst Member States to facilitate the agreement. This demonstrated that, when necessary and particularly when forced by external factors, the Union can agree to modify a relative stability key to make it consistent with real fishing patterns.

...And relative stability can be adjusted annually: quota swaps

Once annual TACs are established, national quotas can be exchanged voluntarily among Member States. These exchanges are simply reported to the Commission to allow for proper quota management. This practice, known as 'quota swapping' is as old as the CFP itself: Regulation 170/83 already established the possibility for Member States to exchange their quotas. Indeed, this is a practice that has been maintained ever since.

The number of swaps has been gradually increasing over the years as the number of stocks and Member States has increased. In recent years, the quota swaps among Member States have been in the order of magnitude of 2000 per year. Their nature is very varied: in some cases the swaps are annual and respond to specific needs by Member States to complete their annual quota 'basket', while other swaps are virtually perennial in cases where the allocation keys under relative stability are not close enough to the real catch composition of the vessels on the fishing ground.

Quota swaps are a kind of 'market' of fishing rights' where Member States exchange fishing rights on the basis of their perceived reciprocity value.

The practice of quota swapping provides certain benefits:

- It provides for part of the flexibility necessary to compensate the rigidity of the fixed allocation keys in addressing changing circumstances;
- It helps reduce discarding, by providing quotas to Member States with small national quotas. Much of their catch, without those quotas swapped, would have to be discarded;
- It helps the economic viability of the industry by providing them with the quotas they need to continue fishing in certain cases;

- Last but not least, in the context of the discard ban policy adopted in the CFP reform of 2013 (see Chapter 16), quota swaps will be a crucially important instrument to help Member States build a national basket of fishing rights that will correspond as much as possible with the catch composition of their vessels in the fishing grounds, since the discard ban will require a much better match between fishing rights and actual catches than is provided for by the relative stability keys.

However, as with every instrument, swaps also have certain downsides:

- Very often, the need to have 'currency' for Member States to exchange makes those Member States either claim higher TACs or higher shares of a stock (when new TACs and new relative stability keys are established) without a real economic interest in the stock in question, just to have more unused quota to exchange for the fish of real interest to them. This can contribute to the creation of 'paper fish';
- Certain fisheries are strongly dependent upon swapped quotas (as we saw previously, certain swaps are almost permanent). If, occasionally the regular swaps are not agreed for any reason, certain fleets are in an extremely difficult situation to survive. In other words, economic survival on swapped quotas is highly risky and may lead to strong pressure to increase TACs in order to ensure national quotas of the required level.

In any case, the quota swaps are without doubt an element of flexibility that has probably contributed considerably to the survival of relative stability. Without this flexibility mechanism, the rigidity of the fixed allocation keys would have undermined the value of this key mechanism of the CFP.

The application of The Hague Preferences

The Hague Preferences have been regularly applied as part of the CFP since 1983. But their application, still today, constitutes political practice rather than a clearly-defined policy. For the UK and Ireland, The Hague Preferences constitute an integral part of relative stability. However, France, Germany, The Netherlands and Belgium (the Member States whose national quotas are reduced to allow for higher quotas for the UK and Ireland in the event of low TACs) consider that adequate provision for the protection of coastal communities was provided in the basic allocation keys and the additional amounts are not justified. These positions remain unchanged since 1976. Despite these disagreements, The Hague Preferences are still applied every year through *ad hoc* negotiations. The system works as follows:

Every year, when TACs are discussed, the two beneficiary Member States (the UK and Ireland) introduce a number of requests for the application of The Hague Preferences (usually called 'invocations') among a list of stocks for which they are entitled. These requests are admissible only when the TAC levels of the concerned stocks fall below certain thresholds. In other words, they apply in case of shortage of fishing opportunities.

On that basis, the Council negotiate the acceptance of the invocations, always seeking a reasonable balance between the Member States concerned. This is undertaken by the Council Presidency, with the Commission always remaining neutral.

In many cases, Hague Preferences are only applied at 50%, that is, the quotas of the UK and Ireland are calculated as the average between the quota according to the basic allocation key and the quota corresponding to the full application of the Hague Preference.

The persistence of The Hague Preferences is always subject to discussion every time the policy is revised. Despite the contested legal basis for it in the basic regulations, the practice continues with remarkable continuity.

In the future this practice will continue as part of relative stability. However, as the status of the stocks will improve, it is to be expected that more and more TACs will be in excess of the threshold levels that trigger the mechanism. As a consequence, there will be less and less of a need to invoke The Hague Preferences in the first place.

Notes

1 International Commission for the North Atlantic Fisheries, established in 1954, later on replaced by NAFO in 1979.
2 North East Atlantic Fisheries Commission, established in 1963.

3 http://ices.dk/sites/pub/Publication Reports/ICES Annual Report/2003AnnualReport.pdf.
4 Scientific, and Technical Committee for Fisheries, established in the early 1980s as the scientific advisory body for the Commission on fisheries.
5 Northwest Atlantic Fisheries Organisation.
6 International Baltic Sea Fisheries Commission. Multilaterals fisheries management organisation for the Baltic Sea, dismantled after the enlargement of 2004 and replaced by a bilateral agreement between the EU and the Russian Federation.

References

Anderson, E.D. (1998). The history of fisheries management and scientific advice – the ICNAF/NAFO history from the end of World War II to the present. *J. Northw. Atl. Fish. Sci.* **23**: 75–94.

Commission of the European Communities (1982). Fourth report of the Scientific and Technical committee for Fisheries. Brussels, 13 September 1982. *SEC* (82) 1442.

EC (1999). Council Regulation No 49/1999 of 18 December 1998 fixing, for certain stocks of highly migratory fish, the total allowable catches for 1999, their distribution in quotas to Member States and certain conditions under which they may be fished. *Official Journal* L 013, 18 January 1999: p. 54–58.

EEC (1970a). Regulation No 2141:70 OF THE COUNCIL of 20 October 1970 laying down a common structural policy for the fishing industry. *OJ.* No L 236 of 27 October 1970, p. 1–4.

EEC (1970b). Council Regulation No 2142/70 on the common organisation of the market in fishery products. *OJ* L236, 27,10,70, p. 5.

EEC (1976). Proposal for a Council Regulation establishing a Community system for conservation and management of fisheries resources. *OJ* No. C255, 28 October 1976.

EEC (1981). Council Regulation No. 3796/81 of 29 December 1981 on the common organisation of the market in fishery products. *OJ* no L379, 31 December 1981.

EEC (1983a). Council Regulation No 170/83 of 25 January 1983 establishing a Community system for the conservation and management of fishery resources. *OJ.* No L 24. 27 January 1983.

EEC (1983b). Council Regulation No 171/83 of 25 January 1983 laying down certain technical conservation measures for the conservation of fishery resources. *OJ.* No L 24. 27 January 1983.

EEC (1983c). Council Regulation No 172/83 of 25 January 1983 fixing for certain fish stocks and groups of fish stocks occurring in the Community' s fishing zone, total allowable catches for 1982, the share of these catches available to the Community, the allocation of that share between the Member States and the conditions under which the total allowable catches may be fished. *OJ* L 024, 27 January 1983 p. 30–67.

Farnell, J. and J. Elles (1984). *In Search of a Common Fisheries Policy*. Gower Pub, Aldershot. 213 pp.

Holden, M. (1994). *The Common Fisheries Policy. Origin, Evaluation and Future*. Fishing News Books. Blackwell Scientific Publications Ltd., Oxford. 274 p.

Judgment of the Court (Fifth Chamber) of 25 October 2001 in Case C-120/99: Italian Republic v Council of the European Union. *OJ* C 369, 22 December 2001, p. 2–2.

Lequesne, C. (2001). *L'Europe Bleue: A quoi sert une politique communautaire de la pêche?* Presses de la Fondation Nationale des Sciences Politiques, 239 p.

Long, R. and P. Curran (2000). *Enforcing the Common Fisheries Policy*. Fishing News Books. Blackwell Science, Oxford, 379 p.

López Veiga, E. (2000). *Manual de política pesquera*. Tomo II. Xunta de Galicia, 415 pp.

López Veiga, E., D. Carballeira, J. Caamaño, E. Penas, et al. (1993). *La política pesquera de Galicia en el contexto comunitario*. Xunta de Galica, 541 pp.

Order of the Court of 22 May 1977 – Commission of the European Communities v Ireland – Case 61/77 R. http://eur-lex.europa.eu/LexUriServ/LexUriServ.do?uri=CELEX:61977CO0061:EN:PDF

Schweiger, L. (2010). The evolution of the Common Fisheries Policy: Governance of a Common-Pool Resource in the Context of the European Integration. Institute for the European Integration. *Austrian Academy of Sciences. Working Paper* No 07/2010. November 2010, 88 p.

CHAPTER 3

EU enlargement and the CFP

The first enlargements

The CFP has evolved as the EU has been enlarged. New Member States always means new elements, new challenges and new circumstances that the CFP has had to cope with and the CFP itself had to evolve as a result. The birth of the CFP was to a considerable extent the result of the preparation for the first enlargement of the Union in 1973 and subsequent enlargements have fundamentally contributed to bring about an evolution in the policy.

The second enlargement: Greece in 1981

The events surrounding the first enlargement were limited to the North Sea and Atlantic façade of the Community. The debate on EEZs simply did not apply to the Mediterranean at the time. Little wonder that the second enlargement, the accession of Greece in 1981, had no significant effect on the CFP. In fact, the Treaty of Accession of Greece, adopted in 1979, contained only a small Chapter on fisheries, with two articles (110 and 111) to regulate the access rights within territorial waters between Greece and Italy.[1]

The addition of the fishing activity by Greece to the fishing activity of the EEC at the time did not involve a substantial change of the *status quo* in Mediterranean fisheries. Unlike in the Atlantic, where a considerable fishing activity took place by national fleets in the waters of other Member States, in the Mediterranean most fisheries took place close ashore, with relatively few interactions among Member States' fleets.

This made the Greek accession a very simple and non-contentious one as regards the fisheries chapter. Even after the Community incorporated its third Mediterranean Member State, the CFP established 2 years later ignored the Mediterranean altogether.

To see the real effects of new enlargements on the CFP, we have to wait until 1986, when Spain and Portugal, two traditional fishing powers in the Atlantic and other waters, joined the EEC.

Spain and Portugal in 1986: a quantum leap for the CFP

Although different enlargements have always had an influence in the CFP, the third enlargement to Spain and Portugal has influenced the CFP more than any other one, to the point that its effects are still visible in today's CFP. This is due to a number of reasons.

Firstly, the accession took place shortly after the birth of the CFP. When the policy was agreed in 1983 Spain and Portugal were already candidate countries and their accession was taken for granted as just a question of time. Consequently, already before accession the future membership of Portugal and, particularly, Spain had an influence in shaping the CFP.

Secondly, the accession substantially increased the size of the fishing sector in the Community by 65% and the production of fish and shellfish by 45%. This made the EEC a much larger fishing power on a global scale.

The Common Fisheries Policy: The Quest for Sustainability, First Edition. Ernesto Penas Lado.

This accession brought to the Community 50 million new keen consumers of high-quality fish. The consumption of fish *per capita* in the two countries is the highest in Europe (61.6 kg per person and year in Portugal and 44.8 kg in Spain) and among the highest in the world, together with Iceland, Norway and Japan. A whole new market opened up for the European fishing industry, which was able to absorb fish species of hitherto low value in the markets of northern Europe, such as hake in Spain and horse-mackerel in Portugal.

This accession also extended the CFP to long-distant fishing in western Africa, in Canada and Norway (for cod) and other areas and brought the importance of fisheries management in international waters to a new dimension. Spain had before accession a very successful network of agreements on access to third countries' waters through the principle 'access to resources/ access to markets', capitalising on the very lucrative Spanish market for fish. Upon accession, this national policy was no longer possible as access to markets was already an exclusive competence of the EC. This led to a new emphasis on the bilateral agreement policy with a view to ensuring the continuity of the activity of the large external fisheries sector.

This accession also meant the incorporation of new territories with special characteristics: Azores, Madeira and the Canaries, which were eligible for special measures under the Treaty (today's Article 349 TFEU) as outermost regions. Although the initial CFP actually established some conservation measures for certain outermost regions, such as the French *Départements d'Outre-Mer* (e.g. technical measures for the French Guyana or TACs for Saint Pierre et Miquelon) these measures have remained rather marginal or have even disappeared (such as those applied to Saint Pierre et Miquelon, which later became a *Territoire d'Outre-Mer*, thus outside the EU). In contrast, the Portuguese and Spanish outermost regions, by virtue of their closer geographical situation and shared fishery resources with continental Europe have been much more central to the CFP than any other outermost region.

Of the two Iberian Member States, it was the accession of Spain that posed a particular challenge, due to the traditional fishing activity of the Spanish vessels in the waters of the Community of 10. When the CFP was established, the future Spanish accession was already an issue. After the difficult negotiations to set up the policy between 1977 and 1983, the question of how to cater for the Spanish fleet was quite a challenge.

The Spanish activity in the waters of the Community of 10 was quite old. This activity dates back to 1925, when the pair trawlers 'Chimbote' started to fish in the -then- international waters of the Grand Sole (López Veiga, 2000). Soon, many other vessels followed, to supply the high demand of fish in the Spanish market that could not be satisfied from the narrow continental shelf of the Spanish coast. After the interruption of World War II, the activity picked up again and reached its maximum level in the 1970s, when more than 500 Spanish vessels were fishing in the international waters around the British Isles and France.

When the Atlantic Member States declared the EEZ in 1977, the Spanish vessels could have been expelled from these fishing zones. However, the same year Spain filed its application for EEC membership. As a consequence, the Spanish fleet was not completely expelled: fewer Spanish vessels were allowed to continue fishing, subject to a system of effort limitation through a restricted and decreasing number of licenses issued by the EEC. This gradual reduction corresponded to the objective of the existing members of the EEC of 10 to 'cut the Spanish fleet down to size' to allow for accession.

The question of the incorporation of the Spanish fleet also gave rise to the establishment of a special control regime applicable to the Iberian fleets, through additional rules to the existing *acquis* (EEC, 1985).

At the same time, the Spanish fish market was open to the existing EEC countries and exports of fish from EUR-10 countries to Spain sky-rocketed from 400 tonnes in 1976 to 18,000 tonnes in 1981 (López Veiga et al., 1993).

The Portuguese case was different because Portugal did not have a fleet operating in the waters of the Community of 10 prior to accession. Therefore the accession of Portugal was not considered by the Community as big a challenge as the Spanish one.

In addition to the key question of access to the fishing grounds, the Spanish and Portuguese accession also posed significant commercial questions. The Iberian sector of canned sardine (an iconic industry in both countries) was subject to a very long transitional period before full integration. The Portuguese canned sardines could only enter the EUR-10 market tariff-free after 5 years and Spanish canned sardines would have to wait for 10 years.[2] Certain third countries like Morocco enjoyed better conditions of access to the EU-10 market during that time.

The challenge of the accession of the two Iberian Member States, for which the fishing sector was much more important than for any of the existing members of the time, gave rise to the longest transitional period ever in the context of the CFP: 17 years from 1986 to 2002.

This accession led to a number of changes in the original policy of 1983. With Spain and Portugal as members, a number of the original CFP regulations were revamped. The structural policy adopted a new instrument after accession: Regulation 4028/86 (EEC, 1986a). A new set of technical measures were also adopted: Regulation 3094/86 (EEC, 1986b) and fleet policy took a new orientation with the first attempt to adapt fleet capacity to available resources (see Chapter 5). No other enlargement has ever brought about so many changes in the CFP.

The Treaty of Accession

In the Treaty of Accession, the fisheries Chapter was, for Spain, the last to be negotiated, which also indicates the difficulty of the negotiation. In the end, the Spanish fleet operating in Community waters was subject to a triple system. In addition to the normal limitations through TACs and quotas, the activity of Spanish vessels was restricted to a closed list of 300 vessels (the so-called 'basic list'), down from the approximately 400 vessels still existing prior to accession, subject to a specific effort management scheme allowing them to fish with a maximum of 150 standard vessels simultaneously, of which 23 in ICES divisions Vb and VI (West of Scotland), 70 in ICES division VII (Ireland) and 57 in ICES divisions VIIIa, b and d (Bay of Biscay).

Furthermore, a special limitation was established in the form of the so-called 'Irish Box'. This box was a key part of the negotiation with Spain. It consisted of the exclusion of Spanish vessels from the zone around Ireland situated to the south of latitude 56°30′ N, to the east of longitude 12°W and to the north of latitude 50°30′ N, until 31 December 1995. The box excluded only Spanish and Portuguese vessels from the area concerned. However, in practice it only affected Spain because Portugal had no vessels fishing in the area.

Another delicate issue was that of the mutual access of Spain and Portugal to each other's EEZs upon accession. This posed a problem particularly for Portugal, which feared an increased activity of Spanish operators in Portuguese waters, since the CFP included relatively few Iberian stocks in the initial list regulated under relative stability.

The issue was difficult in relation with the access agreements to outermost regions and, particularly to the waters of the Azores. This gave rise to specific provisional arrangements to limit free access to these waters until 1995.

Finally, the Treaty of Accession dealt with the question of the allocation of fishing rights to the new Member States, that is, the adaptation of relative stability. Here, the negotiation in the case of Spain was influenced by the different interests in different fisheries by the countries concerned. Spain had an interest in the species more keenly sought in its markets, particularly hake and megrim. While UK, Ireland and France wanted to reserve for themselves their main species of interest such as cod, haddock or sole.

As a result, the relative stability for Spain was very different from that corresponding to the real catches of its fleets, because it could not obtain quotas for species such as cod and haddock and only managed a very low one for monkfish. At the time, this was not necessarily a problem, since the non-quota fish caught could be discarded. Many years later, on the occasion of the discussion on the future implementation of the landing obligation in the 2013 CFP reform, this would become a very important issue.

The transitional period and the 'Western Waters' Regulation

Although the transitional period for Spain and Portugal was foreseen until 2002, some of the transitional conditions actually finalised in 1995. So the full integration of the two Member States into the CFP took two steps: first the adaptation of certain conditions in 1995 and second, the conditions related to the full integration after 2002.

The first adaptation in 1995

In 1995, 9 years after accession, although the political difficulties of ensuring full access between Spain, Portugal and the rest of the Community were still evident, the Treaty of Accession required that the transitional regime be re-evaluated.

The Council adopted Regulation 1275/94 for that purpose (EC, 1994). This regulation laid down the continuation of all access arrangements (Articles 156 to 166 and 347 to 353 of the Treaty of Accession) but in a way that these arrangements would be fully incorporated into the mainstream legislation of the CFP.

This included the continuation of the Irish Box, the quantitative restrictions to the number of Spanish vessels in western waters and a regime to continue, *de facto*, the exclusion of certain fishing techniques from the waters of the Azores.

A special emphasis was placed on the need to ensure that no major changes of fishing effort by area would take place. This led to the adoption, a year later, of a specific regulation (Regulation 685/95: EC, 1995a) to manage the overall effort exerted in different areas of the Atlantic façade. The objective of this regulation was mainly: 'to ensure that there is no overall increase in the overall levels of fishing effort within the areas and stocks covered in the Act of Accession of Spain and Portugal.' This objective clearly reveals that the objective of this regulation was not to bring effort management as a specific instrument for the management of fisheries but rather to ensure the continuation of the *status quo* negotiated upon accession.

This new scheme was implemented through a subsequent regulation of 1995 establishing the specific effort limits by area applicable in the waters of the western façade (EC, 1995b).

The end of the transitional period in 2002

On the occasion of the CFP reform of 2002, the end of the transitional period was considered in the light of the latest enlargement of the Union in 1995, where the new coastal Member States (Sweden, Finland) enjoyed free access to waters (subject to relative stability). It was then politically unacceptable that Spain and Portugal would still have transitional conditions while the new Member States would not. This made it necessary to revise the regulations laying down new transitional measures regulating access to the waters between Spain and Portugal and the rest of the Community.

For this purpose, the Commission presented a proposal for a new regime (the so-called Western Waters Regulation) where the specific treatment of Spain and Portugal was removed. This affected in particular the 'Irish Box', which limited the number of Spanish vessels only to the ICES area VII around Ireland and was therefore considered discriminatory. However, this proposal maintained some of the elements of the earlier scheme, in particular the notion of general limitations of fishing effort in western areas, on a non-discriminatory basis, to avoid major displacements of fishing effort among different fishing areas and to remove the fears that the end of the transitional period would result in a major change in the fishing patterns in the area concerned.

The discussion of this proposal became a highly political issue, showing once more that the principle of free access was still disputed by a number of Member States in practice. This political discussion led to the adoption of Regulation 1954/2003, the Western Waters Regulation, which laid down the new regime of access and limitation of fishing effort after the expiry of the transitional period for Spain and Portugal (EC, 2003).

The very existence of this regulation, still in force today, shows that the transitional period for Spain and Portugal has, if not *de jure*, at least *de facto* been partially continued. This regulation was a mixture of different provisions, including rules of access to the waters of outermost regions, a scheme to limit fishing effort with specific control provisions and the establishment of a zone of special protection.

The effort management scheme of the western waters

Regulation 1954/2003 established a very general regime to control fishing effort over large areas, with the objective of preventing massive displacements of fishing effort after the expiry of the transitional period. Since the possibility of a specific regime applicable only to Spain and Portugal was out of the question, the solution was to establish an effort regime that would include the effort ceilings for all Member States fishing in the western waters. To ensure that the regime was not discriminatory, the effort system had to apply to all Member States fishing in the areas concerned.

The intention was not to use fishing effort as a real management instrument but rather to prevent massive displacements of fishing effort among major fishing areas. In fact, the limitations in fishing effort did not represent a reduction (as in fact was the case in the first multiannual plans being discussed in parallel) but the fixing of a ceiling, based on the average effort deployed in the respective areas during the period 1998 to 2002.

The 2003 regulation also established an effort regime applicable to certain non-quota stocks (deep-sea species, edible crab (*Cancer pagurus*), scallops (*Pecten jacobeus*)). This was not related to the Spanish and Portuguese accession but to the existence of conflict around fisheries for these species, particularly in the English Channel. As the possibility of regulating these species by TACs and quotas was considered too difficult (agreeing on an allocation key would have been extremely difficult), the solution was to establish an effort management that would simply guarantee, once more, the preservation of the *status quo*, thus preventing a significant increase in fishing activity by any of the Member States involved in the absence of catch limitations.

To translate the objectives of Regulation 1954/2003 into specific effort limits, the Council adopted Regulation 1415/2004 (EC, 2004a). The calculations of the effort deployed in the period 1998–2002 was difficult because of the lack of reliable data for all Member States, since only Spain and Portugal had an obligation to limit and report on effort exerted, so the data on effort deployed in the reference period by the other Member States was largely based on estimations. As a result, the effort ceilings established would not constrain the fishing activity in subsequent years.

The biologically sensitive zone

A politically delicate question in the context of the new regime after 2002 was the continuity of the Irish Box. The proposal by the Commission for the 'Western Waters' Regulation did not include the Irish Box as this was considered discriminatory. However, in the process of negotiation of this proposal in Council, the idea of replacing the Irish Box by a non-discriminatory instrument that would achieve a similar goal made headway. Ireland made of this issue a matter of State and successfully managed political support in Council for the continuation of a new Irish Box, albeit on a new and non-discriminatory basis.

The new Irish Box was based on two principles: (1) it should be non-discriminatory and therefore be applicable to all Member States concerned, not just to Spain and (2) it should have a conservation purpose, and therefore be based on science. All these elements also required a new name, putting the emphasis on the conservation value of the new area, to clearly dissociate it politically from the initial Irish Box.

On these two conditions, the Western Waters Regulation finally adopted by Council included the so-called 'biologically sensitive zone', which was characterised by:

- An area corresponding largely to the main spawning grounds of hake in Irish waters. This area was based on the pre-existing area where specific technical measures had been previously adopted to protect juvenile hake. These technical measures had been established in 2002, in the wake of the cod recovery plan and the need to tackle the management of northern hake (EC, 2002).
- Effort ceilings applicable to all Member States having a fishing activity in this area and not only to Spain.

The two conditions were thus formally fulfilled: the area was applicable to all Member States and was based on conservation criteria. Figure 3.1 shows the shape and extension of this biologically sensitive zone, as compared to the original Irish Box.

The effort levels per Member State in this zone were implemented through Regulation 1415/2004, where specific fishing effort ceilings by Member State were established based on recent effort deployed

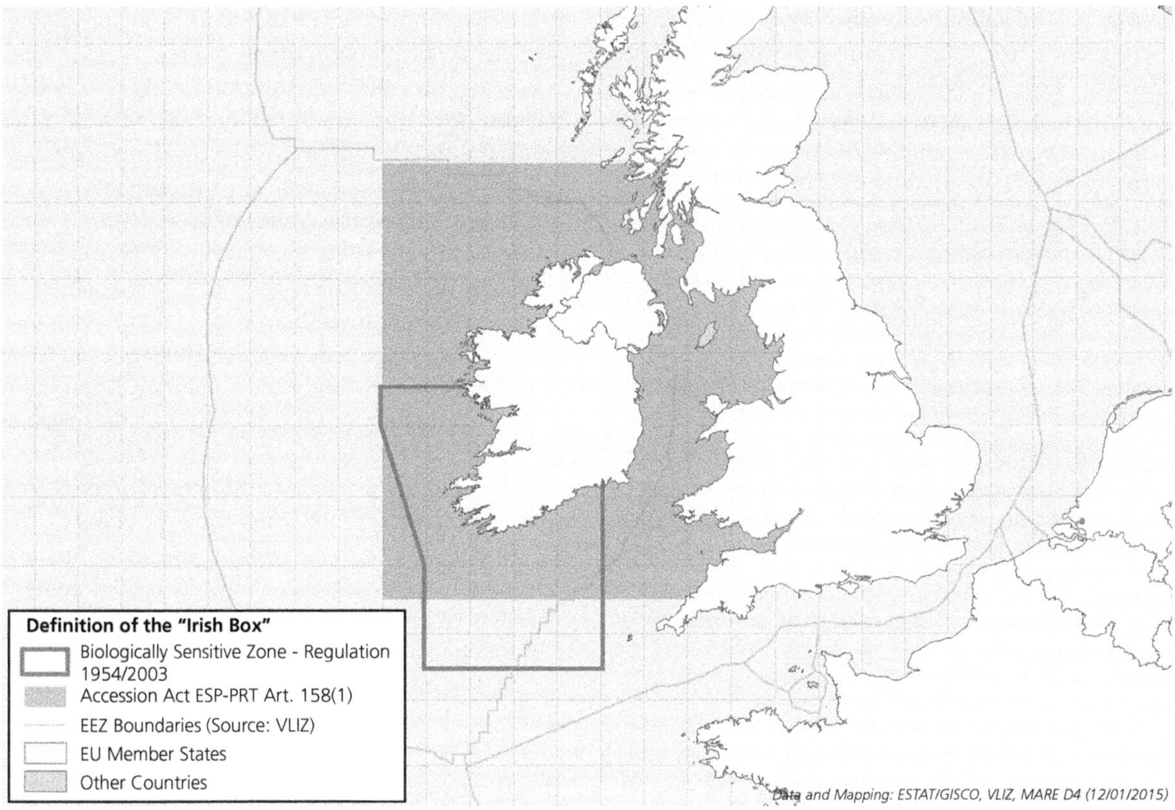

Figure 3.1 The biologically sensitive zone adopted in 2003 and the Irish Box of 1986. Source: European Commission Services. EEZ limits are indicative, not authoritative. (See colour plate section for the colour version.)

(EC, 2004b). In the absence of an obligation to limit effort and report on it, this calculation for all Member States except Spain was made eventually on estimations of fishing levels in recent years, thus representing the *status quo*.

For Spain, on the contrary, this posed a problem: the *status quo* being zero vessels, the maximum number allowed would have to be established *ex novo*. In fact this had to be done without historical references and the level agreed in Regulation 1415/2004 was a number of kW/days corresponding to the operation of 40 vessels.

Beyond the allowance for a restricted number of Spanish vessels, these limitations did not impose actual reductions of fishing effort to any other Member States and have remained constant ever since without any further modification. They remain as a general ceiling that have not even been revised following successive long-term management plans.

Upon adoption of Regulation 1954/2003, Spain took the Council to the Court of Justice, on grounds that the biologically sensitive zone was not a real conservation measure but the continuation of a situation of *de facto* discrimination. The Court, however, dismissed this case on the basis that Spain could not demonstrate that the said zone corresponded to anything more than a scheme to protect juvenile hake (Judgement of the Court (Second Chamber), 15 May 2008).

The regime applicable to the biological sensitive area was subject to a report by the Commission by 31 December 2008.

The question of access

The western waters proposal also established a specific access regime for the waters of the Macaronesian archipelago (Azores, Madeira, Canary Islands) that had enjoyed a specific regime since accession. The question in this case was whether these islands would be granted a derogation of the principle of free access beyond the 12 miles, based on the article in the Treaty (today Article 349 TFEU) that grants special conditions to outermost regions.

This proved to be an extremely difficult issue. The outermost regions, led by a very dynamic Azorean regional government, championed the cause of reserving the 200 miles to the local fishermen, while other Member States defended the idea of the free access, either because of direct interest to fish in these areas, or in order to prevent precedents for other areas.

The specific Azorean interest was understandable, since this archipelago is surrounded by a number of very productive seamounts both within and even outside the 200 miles around it. For Madeira and the Canaries, the 100 miles were a good solution, since they do not have the same kind of extended seamounts as the Azores.

The issue was difficult to resolve, not only because of different views among Member States but also because of different views within the key Member State: in Portugal, while the Azores wanted the 200 miles reserved for their regional fleets, Lisbon was divided between the support to this claim and that to the fleets from continental Portugal wishing to fish in Azorean waters. The latter would be excluded from the Azorean 200 miles, as the special arrangement should apply to outermost regions but not to the metropolitan territory of their Member State.

The final agreement to limit access to local fishermen to the first 100 miles of the EEZ represented a compromise that proved acceptable to everybody in Council. However, for the Azores the solution was unacceptable and the Regional Government took the case to the European Court of Justice. The Court, however, dismissed the case as inadmissible (Judgement of the Court of First Instance (Third Chamber) 1 July 2008).

At that time, these limitations did not apply to the French outermost regions, as France did not present a request for it in that context. Although the political agreement on access included a declaration to the effect that the 100-mile regime would be considered in the future also for the French overseas territories, in practice France never insisted on this possibility, until the next reform in 2013.

Beyond the difficult question of the outermost regions, the regulation had to deal with another political difficulty: the mutual access between Spain and Portugal to their respective continental waters that were subject to transitional provisions since their accession in 1986. Spain wanted free, unlimited access, while Portugal was concerned about a complete liberalisation of access.

This question proved extremely difficult to address, and was resolved only after a bilateral negotiation during an Iberian summit in 2003, at the level of prime ministers. The two Member States agreed to establish limitations in the number of vessels that would have access to each other's waters. The agreement was summarised by the then Spanish Prime Minister as follows: 'We had a legal argument (freedom of access) that Portugal had to accept. And Portugal had a sensitivity that we had to listen to'.

That agreement has worked to this day without significant friction.

The other side effect of accession: the 'quota hoppers'

Following the declaration of EEZs in 1977 by Atlantic Member States, many of the Spanish vessels traditionally fishing in UK waters did not have licenses to fish, as Spain obtained a decreasing number of licenses to continue fishing in the newly-declared EEC waters. One of the effects of this was that a number of Spanish operators took advantage of a very loose legislation in the UK and registered their vessels under the UK registry, thus obtaining access not only to UK waters but also to UK quotas. These were soon known in the British press as 'quota hoppers' or 'flag vessels'. A number of Dutch operators did the same in the North Sea.

As some authors have pointed out, this case is a clear demonstration of the contradiction between an open market and a policy based on protective national policies through quota allocation (Lequesne, 2000).

The UK reacted to this practice by passing a law (the Merchant Shipping Act) in 1988 whereby holders of UK-registered fishing vessels should comply with a number of requirements, including on nationality. At that time, the Spanish operators were already citizens of a Member State of the EEC. This crucial fact conditioned the outcome of this case.

The vessel owners concerned took the British Government to the European Court of Justice which, in a series of famous rulings, established the case law that applies today in terms of rights and obligations applicable to foreign (Community) operators within the CFP.

In addition to the Factortame cases, already cited in Chapter 1, the requirement for nationality was considered by the European Court of Justice which, in two famous rulings, established case law on this question. In the Agegate case, the Court ruled that: 'Community law precludes a Member State from requiring, as a condition for authorising one of its vessels to fish against its quotas, that 75% of the crew of the vessel in question must reside ashore in that Member State' (Judgement of the Court of 14 December 1989a).

In another parallel ruling, the Jaderow case, the Court admitted that Member States could establish conditions to ensure that registered vessels have a 'real economic link' with its flag State (Judgement of the Court of 14 December 1989b). This link could take the form of a proportion of landings of the vessel taking place in the ports of the flag State.

These two rulings have established the basis for what is today a very common practice within the CFP: vessels flagged in one Member State, fishing quotas of that Member State, being owned (and often crewed) by nationals from another Member State. The rulings determine that this practice must be allowed (on grounds of non-discrimination on the basis of nationality) but must ensure an economic link with the flag Member State.

The case law established by these two rulings has always been present when discussions took place about the Member States' rights and obligations regarding national quotas. The impact of these rulings is still felt today. When discussing the Commission proposal on Transferable Fishing Concessions (TFCs, see Chapter 16) some Member States still believed that, on the basis of the Agegate ruling, individual rights limited to Member States' flagged vessels could easily be overruled by the Court on the basis of that case law. This was one of the arguments (though not the only one) to oppose the use of TFCs in the 2013 reform even at national level.

Implementation of the Western Waters Regulation

The implementation of the Western Waters Regulation has been analysed by the Commission, in 2010. In its report (COM, 2010), the Commission stated that the western waters regime was:

> ...a safeguard against effort dislocation from one area to another and against exploring new fisheries without scientific advice... Due to its static nature, the 2003 effort regime cannot be used for stock management or management of multi-species fisheries; it was not conceived for that purpose.

From that point of view, the regime had little impact on the management of key stocks in European waters. In addition, the effort ceilings were set so high that they were hardly limiting in practice.

The report also analysed the biologically sensitive zone. Although the scientific advice on the effects of the zone on the hake stock was unclear, scientists considered that the zone, together with other technical measures, was conductive to the improved status of some important stocks. On this basis, the Commission suggested its continuation, albeit with a clearer link to resource management objectives.

This regime is still a leftover form the Spanish and Portuguese accession, in yet another demonstration of the tendency of the CFP to preserve old instruments whose main objective is to preserve a historic political deal and a perceived stability in fishermen's share of an exploited resource, regardless of their real practical use today.

Other enlargements

Sweden, Finland and Austria in 1995 and the failed accession of Norway

The accession of Sweden, Finland and Austria was very smooth and uncontroversial. The two new coastal Member States had relatively small fishing sectors. Their incorporation into the *acquis communautaire* (the existing set of EU rules that new Member States should accept to implement upon accession) of the CFP was not particularly contentious. The two Member States fished primarily (Finland, exclusively) in the Baltic Sea, an area not initially covered by the CFP and still at the time managed multilaterally under the International Baltic Sea Fisheries Commission (IBSFC). This meant that this enlargement did not fundamentally change the political balance of 1983 concerning fisheries.

Special arrangements were necessary to accommodate the traditional access by Norwegian vessels to Swedish waters, regulated through an old agreement that had to be 'communitarised' upon accession, that is, managed as a bilateral EU/Norway issue.

The two new Member States soon became strong advocates of the enlargement of the EU to other Baltic States (Poland, Estonia, Latvia and Lithuania). That enlargement would be much more far reaching.

The accession of Sweden had a significant impact in a number of discussions in Council. Sweden's very conservationist stance (including in areas beyond their national interest) soon turned this Member State into the Commission's best (and often only) supporter on a number of initiatives concerning the conservation policy. In some cases, only Sweden supported certain principles and although this was often insufficient to influence the final outcome, its positions were in many cases a critical conscience of the Council when deciding on conservation issues.

In that regard, Sweden was sometimes followed by another new Member State: Austria. Although a land-locked country with no direct fisheries interest, Austria was, however, often active in voicing their support to initiatives on fisheries management, contributing to breaking the traditional tendency of the CFP according to which only directly concerned Member States were to have a say in the decision-making process. Sweden in areas beyond their national interest (e.g. Mediterranean policy or NAFO) and Austria as a land-locked country contributed to shape positions in Council, establishing a precedent that has been steadily growing over the years: that Member States not directly concerned by fisheries decisions do take an active part in the decision-making process of the CFP, particularly on issues related to the respect of internationally-agreed principles.

The non-accession of Norway

Perhaps the most interesting question about this enlargement was what did not happen: the accession of Norway was refused by referendum in 1994 on the draft Norwegian Accession Treaty.

The negotiation of the fisheries Chapter of the Norwegian Accession Treaty was one of the most difficult ones. Most of the Norwegian fishery stakeholders were against accession. This played against the background of a strong political interest in the Community of 12 to bring in a rich country like Norway that would immediately become a net contributor to the Community budget. This led to a rather generous negotiation with Norway on the fisheries chapter that even included the promise of the future appointment of a Norwegian Fisheries Commissioner, as a way to reduce the reticence of the Norwegian industry to accession. This, in turn, provoked uneasiness in some existing Member States, which considered that Norway had been offered a much better deal than they had obtained in their own accession negotiations.

One can only speculate about the influence of Norway as a Member State into the CFP, had the Norwegian people decided to vote for accession in 1994. This author would argue that this would have been limited, more limited than the one Norway exerted then and still today, from outside.

Firstly, an accession in 1995 would have meant the acceptance by Norway of the *acquis communautaire* defined in 1983 and consolidated in 1992. In fact, the terms of the draft Accession Treaty did not modify such *acquis*.

Secondly, Norway would have had a small number of votes in Council (by virtue of its population) regardless of the importance of its fishing industry. As the decisions in the CFP moved from *de facto* unanimity (at the times of the Luxembourg compromise, see Chapter 1) to qualified majority, Norway would have had a small weight in the decision-making process, regardless of the amount of 'moral leadership' Norway may have been able to provide. Time and again, the example of certain Member States in implementing modern policies, with less paternalism and more responsibility by the industry, has generally failed to impress a majority of Member States in Council.

It would have been different if Norway had voted for accession in 1972. Then they would have been part of the Member States establishing the first CFP in 1983. Indeed, Norway participated very actively in the negotiations leading up to the establishment of the CFP between 1970 and 1972, where it strongly defended the coastal fisheries against the principle of free access. In fact, Norway signed with the EEC in January 1972 a non-binding protocol to the effect of revising the question of access after a period of 10 years (López Veiga et al., 1993). Had Norway voted for accession, we can only guess what such revision may have led to.

The 2004 enlargement: new Baltic, Mediterranean and land-locked Member States

This accession comprised 10 new Member States from three blocks: four Baltic, three Mediterranean and three land-locked Member States and had very substantial implications. The main ones were: (1) the change of regime of the Baltic Sea, (2) the increase in the number of land-locked Member States and the establishment of a 'landlocked perspective' of the CFP and (3) the sheer number of new Member States (10) that modified substantially the way negotiations were conducted in the Fisheries Councils.

The Baltic

Before 2004, fisheries in the Baltic Sea were still subject to multilateral fisheries management in the framework of the International Baltic Sea Fisheries Commission (IBSFC). Upon accession of Poland, Estonia, Latvia and Lithuania, the vast majority of Baltic Sea waters became EU waters, with only a small fraction of the Baltic Sea corresponding to the coastal waters of the two riparian territories of the Russian Federation: Saint Petersburg and Kaliningrad. This had obvious consequences.

Firstly, several years before accession, the Community of 15 negotiated mutual access agreements with the four Baltic accession countries. These agreements contributed to prepare the ground for accession by anticipating reciprocal access and the implementation of the common rules decided by the IBSFC.

Just as an example, as early as 1993 the Community signed a fisheries agreement with Estonia, to incorporate in EU law the previously existing fisheries agreements of Sweden and Finland with Estonia, in view of the imminent accession of the former (EC, 1996a). Later on, in 1996, the Community of 15 signed a new agreement with Estonia, extending the cooperation between Estonia and the Community beyond the framework of the old agreements with Sweden and Finland (EC, 1996b). On the basis of that agreement, quota exchanges between Estonia and the Community were regularly agreed up to 2004, just prior to accession (EC, 2003). This means that by the time of the Estonian accession in 2004, the Community had established fisheries cooperation and quota exchanges with Estonia for 10 years. Similar arrangements were agreed with other Baltic States prior to their accession.

Secondly, the tradition of cooperation across Member States in the Baltic Sea was very strong, as Baltic fisheries are restricted to coastal States of the Baltic Sea.

Finally, the only non-EU State concerned, the Russian Federation, has only a small part of the surface and its resources. Consequently, the cooperation with Russia, which was negotiated in 2009 following the dismantling of the IBSFC, has not been a leading factor in the development of the CFP in this area.

The accession of the Baltic States did change the participation of the Community in certain regional fishery organisations, such as NAFO, where these countries had a tradition of fishing, either as part of the old Soviet Union, or on their own, as Poland. Their respective historical fishing rights in this area had to be renegotiated and allocated to them in the new context.

What the Baltic *mare nostrum* represents today is an area that, by virtue of its excellent tradition of cooperation is a privileged area for the development of new methods of fisheries management, based on cooperative, regionally-based approaches.

Land-locked Member States

Outside the Baltic area, the most significant change in the CFP following the 2004 accession was the increase in the number and profile of the land-locked Member States. Hungary, Czech Republic and Slovakia represented a net increase in the number of land-locked Member States, whose influence in the CFP is also interesting to analyse.

These Member States, together with the existing land-locked MS such as Luxembourg and Austria, increase the number of Member States which are not directly concerned by the CFP. However, this does not mean they are indifferent to it. At least on two questions they have had a significant influence in the way the CFP is implemented.

The land-locked Member States have brought to the CFP a stronger emphasis on aquaculture in general and fresh water aquaculture in particular. This was also aided by the fact that two of the new maritime Member States, Poland and Slovenia have a relatively small coast compared to the overall size of the country. This implies that they also bring to the EU much of the interest of land-locked Member States on fresh water activity.

Furthermore, even if they are not directly concerned by marine fisheries, they do vote in Council and have the possibility to influence the negotiations by establishing linkages between marine fisheries and their inland aquaculture interests. When dealing with qualified majorities in Council the votes of land-locked States become significant and Member States and the Commission must always think of how to get their support, on issues where they have no national interest.

New Mediterranean Member States

The enlargement to three new Mediterranean Member States (Malta, Cyprus and Slovenia) had a limited impact. The three Member States had small fishing sectors in absolute terms, so their accession did not represent a major change in the Mediterranean policy. However, their accession coincided in time with the discussion on the new Mediterranean regulation finally adopted in 2006 (see Chapter 8) and the three new States brought an extra layer of complexity to the already difficult decision on the first operational regulation on the management of Mediterranean stocks.

Furthermore, the accession of these countries had the effect of extending the relative weight and influence of the EU in the context of multilateral management of resources in the Mediterranean basin, thus creating a higher leverage for the EU in the General Fisheries Commission of the Mediterranean (GFCM, see Chapter 8). The accession of these three countries has contributed to the increasing success of GFCM in adopting multilateral fisheries management measures, hitherto very difficult to agree.

10 new Member States

Beyond the different contributions of the three new blocks of Member States, it was their sheer number that made a difference. This accession took the number of Member States in Council from 15 to 25, the largest enlargement ever. This produced a change in the way Council meetings are conducted. With many more Member States, there was less time to listen to everybody and negotiate deals that would be agreed by all parties. The first result of this was the tendency to rely, more than ever before, on qualified majority instead of consensus to try to make decisions.

While a Council with 15 Member States could still try to negotiate with everyone to reach consensus, the higher number of delegations made such consensus-seeking much more difficult and time consuming. This changed forever the negotiating style of Council, in which the possibility for one single Member State to find solutions to their specific problems was considerably reduced. Further enlargements obviously accelerate this trend.

The enlargement of 2007: the Black Sea

The accession of Rumania and Bulgaria in 2007 has brought to the Union an entirely new issue: the management of fisheries in the Black Sea. We have seen that in all other areas of shared resources with third parties, the Union is among the main players, if not the main player. This allows the Union to play a leading role in the initiatives leading to fisheries management.

In the Black Sea, in contrast, the part of EU waters and specially the contribution of the EU to the total fishing activity in this sea basin are relatively small. The fisheries in this area are dominated by Turkey, Russia and Ukraine. The challenge of managing European fisheries in a context and where there is no common management mechanism represents a new challenge for the Union.

Still today, there is no multilateral cooperation on fisheries management in the Black Sea. Theoretically, the appropriate multilateral body to manage these resources is the General Fisheries Commission for the Mediterranean (GFCM), which includes the Black Sea in its Convention Area. However, in practice GFCM cannot perform this task because some key participants in the Black Sea fishery are still not members of GFCM.

The Commission has taken different initiatives to set up a specific Black sea multilateral management system for this sea basin but results so far have been very limited. The possible role of Turkey in the context of accession negotiations is of obvious importance in this case.

The incorporation of the Black Sea fisheries into the general system of the CFP has been much simpler than in the Mediterranean. The low diversity of this sea and the small number of commercial species in the catch has made the management of Black sea fisheries more amenable to the TAC and quota system than in the Mediterranean. Today, only two commercial species are covered by the quota system and allocated to Romania and Bulgaria: sprat and turbot.

Croatia in 2013

The accession of Croatia in 2013 has been also relatively non-contentious, with the fisheries Chapter being complicated only by the question of the delimitation of areas of national jurisdiction between Croatia and Slovenia. As the two countries had border disputes (both with respect to their zones of maritime jurisdiction as well as on land borders) a resolution to the dispute was a pre-condition for accession. To this effect, both countries agreed on 4 November 2009 to submit the dispute to the Permanent Court of Arbitration in The Hague.[3]

At this stage, it is possible to identify some effects of this accession and to consider its potential to develop fisheries management in the Mediterranean basin.

The accession of Croatia meant that all the Northern Adriatic Sea became an EU-dominated sea basin. This should imply a much better cooperation among coastal States for the purpose of fisheries management. Indeed, the Adriatic-Ionian Sea is already an area where sub-regional cooperation is leading the way in establishing concerted approaches on maritime policy.[4] In the context of the implementation of the CFP in the Mediterranean, the Adriatic claims the first plan for the joint management of shared resources: that on the small pelagic fish that has been adopted by the General Council of Fisheries in the Mediterranean.[5]

In a way the accession of Croatia gives the Adriatic (particularly its northern part) the possibility to become a leading example of regional cooperation in fisheries management similar to what the Baltic Sea represents today.

Notes

1 Act concerning the conditions of accession of the Hellenic Republic and the adjustments to the Treaties. *OJ* L291 of 19 November 1979.

2 Treaty (signed on 12 June 1985) between the Kingdom of Belgium, the Kingdom of Denmark, the Federal Republic of Germany, the Hellenic Republic, the French Republic, Ireland, the Italian Republic, the Grand Duchy of Luxembourg, the Kingdom of the Netherlands, the United Kingdom of Great Britain and Northern Ireland (Member States of the European Communities) and the Kingdom of Spain and the Portuguese Republic concerning the accession of the Kingdom of Spain and the Portuguese Republic to the European Community and to the European Atomic Energy Community – *OJ* L302 of 15 November 1985.

3 www.pca-cpa.org/showpage.asp?pag_id=1443.

4 Maritime strategy for the Adriatic and Ionian Sea Basins: http://ec.europa.eu/smart-regulation/impact/planned_ia/docs/2011_mare_039_maritime_strategy_ionian_and_adriatic_sea_en.pdf.

5 European Union proposal for a GFCM recommendation on a multiannual management plan for fisheries on small pelagic stocks in the GFCM-GSA 17 (Northern Adriatic Sea) and on transitional conservation measures for fisheries on small pelagic stocks in GS 18 (Southern Adriatic Sea).

References

COM (2010). Communication from the Commission to the European Parliament and the Council. Review of the fishing effort management in western waters. Brussels, 11 November 2010. COM (2010) 661 final.

EC (2002). Commission Regulation No 494/2002 of 19 March 2002 establishing additional technical measures for the recovery of the stock of hake in ICES sub-areas III, IV, V, VI and VII and ICES divisions VIIIa, b, d, e. *OJ*. No L77 of 20 March 2002, pp. 8–10.

EC (1994). Council Regulation No. 1275/94 of 30 May 1994 on adjustments to the arrangements in the fisheries chapters of the Act of Accession of Spain and Portugal. *OJEU* No L 140, of 3 June 1994, pp. 1–2.

EC (1995a). Council Regulation No 685/95 of 27 March 1995 on the management of the fishing effort relating to certain Community fishing areas and resources. OJEU No L 71 of 31 March 1995 pp. 5–14.

EC (1995b). Council Regulation No 2027/95 of 15 June 1995 establishing a system for the management of fishing effort relating to certain Community fishing areas and resources. *O.J.E.C.* No L 199 of 24 August 1995 pp. 1–8.

EC (1996b). Council Regulation No 2396/96 of 2 December 1996 on the conclusion of an Agreement on fisheries relations between the European Community and the Republic of Estonia. *OJ*. No L332 of 20 December 1996, p. 16.

EC (2003). Council Regulation No 2287/2003 of 19 December 2003 fixing for 2004 the fishing opportunities and associated conditions for certain fish stocks and groups of fish stocks, applicable in Community waters and, for Community vessels, in waters where catch limitations are required. *OJ*. No 344 of 31 December 2003, p. 1.

EC (2003). Council Regulation No. 1954/2003 of 4 November 2003 on the management of the fishing effort relating to certain Community fishing areas and resources and modifying Regulation (EC) No 2847/93 and repealing Regulations (EC) No 685/95 and (EC) No 2027/95. *OJEU* L 289 of 7 November 2003, pp. 1–7.

EC (2004a). Council Regulation No 1415/2003 of 1 July 2004 fixing the maximum annual fishing effort for certain fishing areas and fisheries. *OJEU* L 258 of 5 August 2004, p. 1–5.

EC (2004b). Council Regulation No 1415/2004 of 19 July 2004 fixing the maximum annual fishing effort for certain fishing areas and fisheries. *OJ* L 258, 5 August 2004, p. 1–5.

EEC (1985). Commission Regulation No 3531/85 of 12 December 1985 laying down certain technical and control measures relating to the fishing activities of vessels flying the flag of Spain in the waters of the other Member States, except Portugal. *OJ* L 336, 14.12.1985, pp. 20–26.

EEC (1986a). Council Regulation 4028/86 of 18 December 186 on Community measures to improve and adapt structures in the fisheries and aquaculture sector. *OJ* L376, 31 December 1986.

EEC (1986b). Council Regulation 3094/86 – of 7 October 1986 laying down certain technical measures for the conservation of fishery resources - OJ L288, 11 October 1986.

European Community (1996a). Agreement on fisheries relations between the European Community and the republic of Estonia. *OJ* No L 332. 20 December 1996, p. 17.

Judgement of the Court (Second Chamber) of 15 May 2008. Kingdom of Spain v Council of the European Union. Fisheries – Regulation (EC) No 1954/2003 – Regulation (EC) No 1415/2004 – Management of the fishing effort - Fixing of the maximum annual fishing effort - Reference period – Community fishing areas and resources – Biologically sensitive areas – Act concerning the conditions of accession of the Kingdom of Spain and the Portuguese Republic and the adjustments to the Treaties – Plea of illegality – Admissibility – Principle of non-discrimination – Misuse of powers. Case C-442/04.

Judgement of the Court of 14 December 1989a in Case C-3/87 (reference for a preliminary ruling from the High court of Justice of England and wales): The Queen vs. Ministry of Agriculture, fisheries and Food, ex parte Agegate Limited. (90/C 16/08). *OJEC* No C 16 of 23 January 1990, p. 6.

Judgement of the Court of 14 December 1989b in Case C-216/87 (reference for a preliminary ruling from the High court of England and Wales): The Queen vs. Ministry of Agriculture, fisheries and food, ex parte Jaderow Ltd and others. (90/C 16/09). *OJEC* No C 16 of 23 January 1990, p. 6.

Judgement of the Court of First Instance (Third Chamber) of 1 July 2008 – Regiao autónoma dos Açores v Council. (Case T-37/04). *OJEU* C 209 of 15.8.2008, pp. 42.

Lequesne, C. (2000). Quota hopping: the CFP between States and markets. *Journal of Common Market Studies*. 38(5): 779–793.

López Veiga, E. (2000). *Manual de política pesquera*. Tomo II. Xunta de Galicia, 415 pp.

López Veiga, E., D. Carballeira, J. Caamaño, E. Penas, et al. (1993). *La política pesquera de Galicia en el contexto comunitario*. Xunta de Galica, 541 pp.

CHAPTER 4

The conservation policy

Conservation: the core business of the CFP

The generally accepted date of birth of the CFP is 1983, when the two first regulations on fishery conservation were adopted and not 1970, when the first structural and market regulations were established. As we saw in Chapter 2, the only exclusive competence of the Union in fisheries is the management of biological resources. This policy is the core business of the CFP.

The objectives of the CFP in the Treaty are very varied and have no order of priority. As a consequence, the combination of biological, economic and social objectives has always been difficult to achieve. All too often, the short-term economic and social objectives have prevailed over the long-term biological ones and this has contributed to the long-term economic and social goals not being achieved.

The conservation policy of 1983 was based on a basic paradigm with two principles: (1) do not catch too much fish (through the instrument of the Total Allowable Catches) and (2) do not catch small fish (through the instrument of technical conservation measures). Over the last 30 years, these two principles have remained stable as the backbone of the conservation policy. While other instruments have been added to the initial two (effort management, long-term plans) the philosophy and structure of these initial instruments has remained fundamentally unchanged.

Whether or not these instruments are ideal ones, their remarkable stability is largely due to the fact that they were adopted after a very long and difficult negotiation that extended over almost 7 years. While it is much simpler to add new layers to the management system, it is extremely difficult to remove these two original ones, since they are still perceived today as an essential part of the CFP. These principles, applicable initially to the Member States of the Atlantic façade and the North Sea in 1983, have been extended to other areas (such as the Baltic Sea and the Black Sea) ever since. The Mediterranean is the only area that does not use catch limitations as a management instrument.

At the time of the declaration of Exclusive Fisheries Zones by Atlantic Member States, although fishery resources were not necessarily in a very healthy state in all cases there was little awareness of the need to ensure their long-term sustainability. The restrictions to the activity for the purpose of ensuring sustainability were not a political priority. For this reason, the first measures under the CFP had a strong emphasis on allocating fishing rights, not so much on reducing catches. The fundamental management instrument already implemented at the time was the system of catch limitations allocated among Member

The Common Fisheries Policy: The Quest for Sustainability, First Edition. Ernesto Penas Lado.

States according to the principle of relative stability: the TACs and quotas.

This has been summarised by early historians of the CFP:

> Because of the minor economic importance of the fishing industry in comparison with its relative political prominence, it has been judged cheaper to buy off protest either by financial assistance towards fishermen's operating costs, by laissez-faire attitude towards conservation measures, or by pressure for a protected Community market. (Farnell and Elles, 1984)

Other authors have underlined that the early decisions on the conservation policy were inspired in the priority to minimise the risk of conflict with fishing communities in the short term, rather than in long-term conservation and management objectives (Boncoeur and Mesnil, 2000). It has taken many years for the CFP to catch up on that.

The resources deteriorated markedly in the first years of implementation of the conservation policy, after 1983. This was not only a consequence of the conservation policy itself; much of this is probably also due to the failure of the early CFP to realise and address the problem of fleet overcapacity and weak enforcement. As we will see in Chapter 5, in these first years there was an increase in the capacity of European fleets, largely funded by the structural policy established since 1970.

After the first years of implementation of the CFP, it became obvious that Europe was not winning the battle for the sustainability of its fishery resources. The conservation policy of the CFP became a long and difficult struggle to manage the resources in a way that would ensure that the objectives of the Treaty could be met. This struggle has been long and difficult and it is by no means finished.

This struggle is also that of developing understanding of the conservation policy not as a stand-alone objective but as one that is the essential element to deliver the combination of objectives of Article 39 of the Treaty. The purpose of the conservation policy is also that of demonstrating that it is not just a policy 'for the fish' but also, or fundamentally, 'for the fishermen'. We will analyse next how the different conservation instruments have developed in these years.

TACs and quotas: the main conservation instrument of the CFP

As we saw in Chapter 2, the TAC and quota system adopted in 1983 was the basic conservation instrument of the CFP. We also saw how the initial system had as a primary objective the allocation of fishing rights among Member States and not the sustainable use of the concerned resources. This has been summarised by some of the early historians of the CFP: 'the distributive nature of the issue ultimately fostered national rather than communautaire attitudes' (Schweiger, 2010). In other words, the initial TAC and quota system focused on how much every Member State would get in terms of fishing rights, not on how much fishing the resource could support sustainably. This initial focus is understandable since it was difficult for fishermen to agree to make sacrifices for the purpose of conservation if their rights were not agreed.

The difficulties of agreeing allocation keys resulted in a strong emphasis by Council on facilitating such agreement over considerations of conservation. Clearly, reducing TACs did not facilitate the agreement, since Member States inevitably translated percentages into absolute amounts of fish. As Schweiger puts it, referring to the 1983 TAC decisions:

> But despite the depletion of stocks in the North Sea and North Atlantic, total TACs for these reference species were only reduced by 8.6%, which indicates that upholding the principle of relative stability was prioritised over the biologically necessary reduction of TACs, setting precedence for inflating TACs. (Schweiger, 2010)

Such was the origin of the TACs. Under these circumstances, it is hardly surprising that in the 1980s, the first TAC decisions (and their practical implementation) led to a clear deterioration of the fishery resources under the CFP. Indeed, during the 1980s and a large part of the 1990s, the TACs were demonstrably inefficient as instruments to control fishing mortality and ensure sustainable fishing in Europe: the Spawning Stock Biomass of all reference species, except whiting, had reached a historical low in the early 1990s (CES, 1992).

To a large extent, the nature of the TACs as instruments where national interest takes precedence over the common European good has remained to this day.

The history of TACs and quotas in the CFP is the history of the evolution from a distributive instrument for the allocation of fishing rights, to a Community instrument to ensure sustainable fisheries for the common good. This has required addressing a number of challenges:

1 Their scope. Which species are subject to TACs and why;
2 Aligning TAC levels to the scientific advice;
3 Turning annual decisions into long-term approaches;
4 Addressing multi-species relationships with mono-specific TACs;
5 Increasing the transparency of the decisions;
6 Rationalising and streamlining the decision-making process;
7 Ensuring adequate enforcement of the catch limitations.

These challenges are permanent ones and, despite gradual but important progress over the years, they still remain important challenges for the CFP.

Although the 2002 reform of the CFP changed the name of this instrument into a the larger concept of 'fishing opportunities' I will refer to TACs and quotas to distinguish them from the effort limitations, also included under the general term of fishing opportunities, that represent a very different management instrument.

As the most iconic instrument of the CFP, the TACs and quotas have always been looked at as a political indicator of performance of the CFP and a battleground for national interests more than for long-term sustainability for all. So it is important to analyse how this instrument has performed ever since the beginning of the CFP.

The scope of TACs and quotas

Since the TACs were primarily an instrument to allocate fishing rights, they were initially set for a limited number of fish stocks. The stocks concerned were not necessarily those in most need of conservation measures but those of a higher commercial value and therefore considered most important for the distribution of rights. As new TACs were established, notably following enlargements, the tendency has continued. In this regard, a first relevant question is: why does the CFP establish TACs for certain stocks and not for others? The answer to this question is not straightforward; for a number of reasons:

- The stocks that made up for the initial allocation in 1983 were fundamentally the main target species for the fisheries of the Community of 10. In fact, the choice of what species would be subject to TACs was not done on the basis of the need to protect the endangered stocks but to ensure the allocation of the most important commercial ones for the Member States of the time.
- The successive enlargements, which brought into the system new stocks in the Iberian Peninsula (1986), the Baltic Sea (2004), the Black Sea (2007) and so on. It is worth noting that the successive additional layers of TAC management upon different accessions followed fundamentally the same approach as in 1983: emphasis was on adding the fishing rights for the most economically important species, not necessarily for those requiring management measures on account of their conservation status.
- Certain TACs were also added following enlargements in order to ensure that the principle of 'access to waters' applicable to new members would not result in new fisheries for hitherto unregulated species. In particular, following the accession of Spain and Portugal who gained access to the waters of the North Sea, TACs were established for certain secondary species (and allocated to old members) to avoid the development of new fisheries for such species by the new members, in another example of the emphasis on maintaining the *status quo*;
- Many TACs were introduced gradually over the years as a result of the recommendations by the regional fishery organisations to which the Union was a contracting party: NAFO, ICCAT, CCAMLR, NEAFC and so on. In many cases this was done on the initiative of the Union.
- The need to bring protection to fragile stocks, once the scientific evidence of their status is known, has been a relatively late evolution of the CFP: in fact, this kind of approach were TACs are used as a real

measure to protect stocks, was only introduced in the 2000s. Several examples come to mind but the establishment of biannual TACs for deep-sea stocks as of 2004 is an example, although in this case the allocation of fishing rights was an important additional motivation.

- In certain cases, the TACs have in fact been established not to allocate fishing rights but to prevent them: for certain endangered species, the TACs were established at zero to prevent the development of fisheries on these fragile species. The example of basking shark (*Cetorhynus maximus*), subject to a permanent zero TAC as of the 2000s is a good example.

As a result of all these factors, the coverage of stocks under the TAC and quota system has been evolving over the last 30 years, in general terms in the sense of an ever-increasing number of stocks under the system.

The TAC decision for 1983 concerned 88 stocks in two main areas: 77 in Community waters and 11 in NAFO waters (EEC, 1983). The 77 stocks in internal waters concerned 22 species, which means that for some of them, many different management areas, with different allocation keys among Member States were established. For example, there were 11 TACs for different management units for herring, 10 for plaice and 10 for sole, all of them with different allocation keys. These management areas did not always coincide with biological divisions in stocks, thus showing how the TACs were from the beginning an instrument to allocate fishing rights, sometimes including very small management units.

In comparison, 267 TACs were agreed for 2014 for 108 species, distributed in several regulations: The general TAC and quota regulation (EU, 2014a) includes 232 TACs for 94 species, in five broad categories: EU waters (146 and 37), NE Atlantic (34 and 18), NAFO (19 and 12), and highly migratory fish (8 and 7), Antarctic (18 and 12) and others (7 and 6). The regulation for deep-sea stocks, establishing biannual TACs for 2013 and 2014 contains 22 TACs for eight species or groups of species (EU, 2012a). The regulation on Baltic TACs contains 10 TACs for five species (EU, 2013a), that for the Black Sea, two TACs for two species (EU, 2012b), and a specific regulation establishes a TAC for anchovy for the season 2013/2014 (EU, 2013b).

It is more difficult to explain why certain stocks that are shared and are significant both biologically and economically are still today not subject to TACs. Two cases in hand are the Iberian sardine and the Atlantic sea bass.

Iberian sardine is an extremely important stock for the two Member States that share it. Yet, no TAC has ever been fixed. The reasons behind this decision are probably related to the traditional opposition by the two Member States concerned, notably on grounds of the difficulty in dividing up the stock in quota shares. The stock is subject to national measures by the two Member States concerned, which appears to have worked reasonably well until recently.

Atlantic sea bass is a more complex case. While scientists have long been warning about the need to prevent the collapse of this stock, Member States in Council are still divided on the need and opportunity to establish such a measure. Not surprisingly, the difficulties associated to the necessary allocation of a future TAC among Member States is at the origin of the lack of progress: certain Member States want a TAC to consolidate their recent share of catches. Others oppose it exactly for that reason: if they have prevented excessive uncontrolled catches by their fishermen, they want to avoid recent catch levels to be consolidated under a newly-established relative stability. They want credit for their past and current conservation efforts. In any case, it is the difficulty of establishing a new relative stability key for such a valuable and widely-distributed stock what in fact prevents the establishment of a TAC.

The lack of progress on a conservation regime for this stock led the Commission to take emergency measures in early 2015 (COM, 2015).

These two cases reveal that when considering TACs and quotas, it is the allocation among Member States that tends to determine positions. Obviously, all these considerations apply to areas other than the Mediterranean, where it has always been the case that TACs and quotas do not apply, except for Bluefin tuna. It is interesting, however, to speculate about what would happen if Mediterranean fisheries were subject to TACs that subsequently would need to be allocated: the only case, that of Bluefin tuna (see Chapter 8) also shows the extreme difficulties in agreeing to an allocation key in this area.

Overall, the scope of stocks covered with the TAC and quota system responds in general terms to those species that either are either economically important commercial species or species requiring conservation measures because of their conservation status. However, this general philosophy has still too many exceptions:

- Extremely important commercial species, such as Iberian sardine are still not covered by TACs at EU level;
- Certain fragile stocks such as certain pelagic sharks are not covered either;
- Certain secondary species, such as for example, dab and flounder in the North Sea are subject to TACs, despite being secondary species for the flatfish fishery, whose catch limitations are meaningless in the context of the management of the main target species (sole and plaice);
- Certain local fisheries, particularly inside the 12-mile limit, are typically not managed under EU TACs but under local arrangements. This includes valuable fisheries for scallops which may extend outside the 12 miles;
- Certain species that are in recent years becoming abundant in areas where they used to be scarce (for example, red mullet in the North Sea) are not regulated either, since their regulation would risk re-opening difficult discussions on the relative stability of the more traditional stocks in the areas concerned.

These few examples illustrate the point that in terms of their value for fisheries management, the scope of TACs and quotas could still be considerably revised in terms of the need and relevance of the inclusion of a number of species and fisheries. Likewise, the need to have certain stocks (of very secondary nature that do not determine fishermen's behaviour) under the TAC system can also be questioned, in terms of the cost/efficiency of the TAC and quota system. However, any such changes would be considerably difficult as they will always be seen as a questioning of relative stability. Such changes could, however, take place as a refinement, or an updating of relative stability in the light of the new developments of the policy, as we will see in Chapter 17.

TACs and scientific advice: a conflicting relationship

Traditionally, TACs for the EU and nearby waters are adopted by the Council of Ministers taking into account the scientific advice provided by two scientific bodies: ICES and STECF (see Chapter 10). This advice, until recently, has been only biological. And also until relatively recently, it focused on annual considerations.

Annual TACs and annual advice

Since the instrument was traditionally annual (the Council decided on the level of catches for the following year, without any consideration at all to long-term effects) the scientific advice was geared to this annual decision. Consequently, ICES traditionally provided advice on how to achieve sustainable levels in 1 year. This led to a number of recommendations for extremely hard reductions in TAC levels, or even for complete closures of fisheries.

Obviously, the recommendation for a zero-TAC, particularly for a stock which is associated to many others in a healthier state, was economically unacceptable for the industry and politically unacceptable for Ministers in Council. This gave rise to a vicious circle: a 'tradition' of ignoring the advice on zero or very low TACs as something not realistic and to the adoption of annual TAC levels very far from the levels recommended by science. In this situation, decision-makers considered scientists as not realistic. Scientists considered decision-makers as irresponsible for ignoring their advice. And the whole policy was discredited as one where decisions are taken against scientific advice.

It is also the case that the advice to reduce or to increase catches has always been taken differently: when ICES' advice allowed for an increase in TACs, Council tended to agree to a full increase of TACs to the recommended levels. However, when the advice was for a reduction of TACs, Council decided on lower TAC reductions than recommended in order to avoid too much short-term economic hardship for the industry.

The tradition of taking the scientific advice as a basis for further negotiation but not as a real policy guidance has characterised the CFP for many years.

In particular, by considering the scientific advice to be unrealistic, this has resulted in the discredit of that advice: on many occasions, while the advice for annual TAC reductions was understandably difficult to take on board, it still conveyed a strong message on the need to take measures to prevent stock collapse. By creating a culture of disregarding the advice, the warnings of the scientific community have been often ignored and certain stock collapses have not been prevented.

A clear case in this regard is that of anchovy (*Engraulis encrasicolus*) in the Bay of Biscay. This highly-fluctuating stock started to show signs of depletion in 2003,[1] when ICES recommended a closure of the fishery for the first time. As a closure was unacceptable for the Member States involved, Council adopted for 2004 a TAC at the same level as in previous years: 33,000 tonnes (EC, 2003a). The story was repeated again in December 2004: ICES recommended again the closure of the fishery,[2] but Council ignored the advice and established a TAC of 30 000 tonnes (EC, 2005a), since it considered the closure economically and politically unacceptable. At the beginning of the fishing season in the spring 2005, the collapse of the fishery became so obvious that the Commission had to close the fishery to protect what was left of the resource through an emergency measure (EC, 2005b). Subsequently, the stock took 5 years to recover and the fishery was reopened only in 2010 (EU, 2010). This is a clear case of how ignoring the scientific advice can lead to stock collapse, and is also an excellent example of what the Commission can achieve through a timely and well-justified use of its powers to adopt emergency measures.

Another clear case is that of the cod stock in the west of Scotland, where ICES has been repeatedly warning about the collapse of the stock. Failure to respond to these warnings led to a long-term collapse of this fishery, which still today shows no sign of recovery.

There is an obvious lesson to learn from this experience: formulating the advice in a way that makes such advice impossible to accept economically and politically may lead Council to ignore such advice altogether, thus leading to stock collapse. Of course, this is no criticism to the scientific community: if they recommend dramatic measures such as closures of fisheries, it is fundamentally because the whole management system was designed to cater for annual decisions. This obliged scientists to provide advice on how to save stocks in one year.

Resolving this problem was only possible through the development of long-term strategies, where the Commission requested ICES to provide long-term advice, thus allowing scientists to make recommendations about gradual approaches to reach sustainable levels.

As the Commission insisted on promoting long-term solutions, where the large reductions in fishing mortality would be achieved gradually over a number of years, the advice started to change: scientists started recommending in the 2000s that a long term approach be adopted. For example, in 2006, ICES recommended for the anglerfish in the Iberian Peninsula that:

> Fishing mortality equal to zero in 2007 is required to bring SSB back to B_{msy} in the medium term. If this is not possible then a recovery plan should be established that will ensure rapid and safe recovery of the SSB towards B_{msy}. (ICES Advisory Committee for Fishery Management, 2006)

This kind of advice marks the change of focus from the one-year recommendation to the multi-annual one. This change has taken years to materialise but as the 'client' (the CFP) has demanded it, the scientific community has gradually provided it.

Single-species TACs and single-species advice

Since 1983, TACs have been fixed for individual stocks, even when these stocks are caught closely together in the same fisheries. This kind of advice was clearly not fit for mixed fisheries.

Later on, scientists sometimes proposed closures of fisheries without any regard for the effect of that closure on the fishery for other, associated species. For example, when ICES proposed to close the fishery for a number of cod stocks in the late 1990s and early 2000s, it did so while recommending certain catch levels for a number of stocks (such as haddock or whiting) that, in certain fisheries, were closely associated to

Table 4.1 Comparison between scientific advice and TAC levels adopted. © European Union. Communication from the Commission to the European Parliament and the Council. Concerning a consultation on Fishing Opportunities for 2015 under the Common Fisheries Policy. Brussels, 26.6.2014 COM (2014) 388 final.

	Percentage excess of TAC compared advice											
Advice for year:	**2003**	**2004**	**2005**	**2006**	**2007**	**2008**	**2009**	**2010**	**2011**	**2012**	**2013**	**2014***
Excess of TAC over sustainable catch (%)	46%	49%	59%	47%	45%	51%	48%	34%	23%	11%	29%	35%

cod, thus making it practically impossible to close the cod fishery without preventing the catch of the other associated species. Obviously this was economically impossible to accept for the fleets concerned and, consequently, the cod fishery was never closed following this kind of advice.

Indeed, the scientific community was aware of this problem. For this reason, they introduced the notion of 'lowest possible level' of fishing, instead of fishery closure. The idea was that, in recognition of the fact that it would not be possible to close all fisheries catching cod, at least their level of catches should be the lowest possible (taking into account the other resources available). This formulation was introduced, for example, in the case of the cod stocks in the early 2000s (see the following).

Again, this is no criticism of the scientific community but rather an inevitable consequence of the annual, single-species approach established in the CFP under relative stability. They had to recommend measures to protect the cod but in a multispecies context were the level of fishing for associated species was not just a biological question but rather an economic and political one.

It is also true that in certain occasions, when the scientific community did make recommendations on mixed species considerations, the fishing industry often reacted by arguing that they could fish in a way that they would catch one stock without catching the associated one and these arguments were used by Council to ignore the recommendations on associated species.

The relationship between the advice and the TACs

Overall, the question of the relationship between TAC and scientific advice goes in two directions: Council very often refuses to follow the scientific advice. But in turn a scientific advice based on annual, single-stock considerations gives rise to economic implications that are often impossible to follow. It is clear therefore that to get TACs closer to scientific advice the management system must integrate the long-term and multi-species considerations. The scientific advice itself must therefore also be based on these principles. Only if this happens will we see a solution to this problem. In fact, this has been happening over the last few years, as shown in Table 4.1, where TAC levels are compared with the scientific advice.

As we can see, the TAC levels are getting ever closer and closer to the scientific advice. This is largely the result of the development of recovery plans, which by establishing recovery strategies for the long-term, ensure that the annual advice provided correspond to a gradual improvement in the stocks, and not to strategies to ensure sustainability in just 1 year. It should also be noted that there is an increasing awareness among ministers that departing too much from scientific advice is negative for the long-term prospects of the industry and increasingly for their own public opinion.

Improving TAC setting: from *ad hoc* discussions to the Policy Statement

For years, the discussion on TACs and quotas was rather unstructured. Political discussions at the December Council would be based on the need to bring Member States to a qualified majority or, in the old times, to a *de facto* unanimity. As a result, concessions on TAC levels were the result of the need to bring Member States into a compromise solution, thus producing results in terms of TAC levels and their relationship with the scientific advice, that often lacked consistency and gave rise to complaints about different treatment or different standards: two stocks with the same conservation status often ended up with very different TAC decisions depending upon whether Member States considered those TAC levels politically important in the final deal.

As we will see in Chapter 13, this kind of on-the-spot negotiation has contributed to create the bad image of 'horse trading' (political bargaining for the highest possible national quotas) that has for many years discredited the policy. This situation was due to several factors, notably by the annual nature of the discussions but, also crucially, because of a problem of timing.

Traditionally, TACs were decided every year in the December Fisheries Council for the following year. This means that the Commission proposal had to be launched at the end of October. But the scientific advice for most stocks was ready only by mid-October, because many stocks were surveyed in August.

This tight timing was also aggravated by the late availability of certain TACs: in particular those negotiated in bilateral or multilateral agreements whose negotiations finalised relatively late in the year. As a consequence, there was literally no time for consultation or for a debate on the principles and standards for TAC setting.

This problem was at least partially addressed through the development of the Commission's 'Consultation on Fishing Opportunities', also known as 'Policy Statement'.

This document was drawn for the first time in 2006 and had several purposes, including in relation to the question of decision-making (see Chapter 12):

Firstly, at a time when the Regional Advisory Councils (RACs) had started providing advice on the CFP it became obvious that the timing of the preparation of TAC proposals left no time for consultations with stakeholders. At the time, the scientific advice provided by ICES was only available in the middle of October and the Commission services only had a few extremely busy days to produce the TAC proposals for the December Council.

In these circumstances, consultation with stakeholders on such a short time scale was not possible. It was necessary to create a different approach to include stakeholder consultation in the TAC and quota workflow. The idea was to publish in the first half of the year a document laying down the approaches to take according to the conservation status of the stocks, as defined in a series of categories related to their respective level of sustainability and data availability. This would allow consultation with stakeholders on principles for TAC setting, even before the annual scientific advice was available.

Secondly and most importantly, the Policy statement put the discussions on TACs on a systematised basis, where all stocks of a similar conservation status would be treated in the same way. The aim of this approach was to increase transparency and fairness in the Commission proposal, and thereby avoiding too many inconsistent decisions by the Council.

Thirdly, the Policy statement also served to ensure transparency on the status of the stocks, through a reporting on their situation in a way that would allow for trends to be appreciated. The purpose of this approach was to provide the TAC decision-making with a more strategic, long-term perspective, which would take less notice of pure short-term considerations. This transparency contributed to bringing the issue of the conservation status of fishery resources to the attention of the public at large, thus helping to raise awareness about this question. This was a key factor in the context of an increasing public consciousness about fisheries management issues in recent years.

The first Policy Statement dates back to 2006 (COM, 2006), and has been published every year ever since with gradual changes.

In this first Policy Statement, the Commission already classified stocks according to categories, differentiating TAC approaches for such categories. This document, in particular, already proposed specific approaches for the following categories of stocks:

- Stocks exploited consistently with maximum sustainable yield;
- Stocks overexploited with respect to maximum sustainable yield but inside safe biological limits;
- Stocks outside safe biological limits;
- Stocks subject to long-term plans;
- Naturally short-lived species;
- Stocks whose status is unknown but which are not at high biological risk.

Table 4.2 Categories of stocks and TAC approaches proposed in the Policy statement of 2008. © European Union. Communication from the Commission. Fishing opportunities for 2009. Policy statement from the European Commission. Brussels, 30.5.2008. COM (2008) 331 final.

Scientific advice	Action to take in setting the TAC
Stock exploited at the maximum sustainable yield rate.	Aim to set the TAC to the forecast catch corresponding to the fishing mortality that will deliver the highest yield in the long term **but** do not change the TAC by more than 25%.
Stock overexploited compared to maximum sustainable yield but inside safe biological limits.	Aim to set the TAC to the higher value of (a) to the forecast catch corresponding to taking the highest yield in the long term, or (b) fishing at an unchanged mortality rate but do not change the TAC by more than 15%.
Stock outside safe biological limits	Aim to set the TAC to the forecast catch that will result in a 30% reduction in fishing mortality rate but do not decrease the fishing mortality so far as to prejudice long-term yields ***and*** do not reduce the TAC by more than 20%.
Stock is subject to long-term plan and scientists advise on the catch that corresponds to the plan.	The TAC must be set by following the relevant plan.
Stock is short-lived and a one-year forecast cannot be provided.	A provisional TAC is set and will be changed when new information is available during the year.
State of the stock not known precisely and STECF advises on an appropriate catch level.	Aim to set the TAC according to STECF advice ***but*** do not change the TAC by more than 15%.
State of the stock not known precisely and STECF advises to reduce fishing effort.	The TAC should be reduced by up to 15% and STECF should be asked to advise on the appropriate level of effort.
State of the stock not known precisely and STECF advises the stock is increasing.	The TAC should be increased by up to 15%.
State of the stock not known precisely and STECF advises the stock is decreasing.	The TAC should be decreased by up to 15%.
STECF advises a zero catch, a reduction to the lowest possible level or similar advice.	The TAC should be reduced by at least 25%. Recovery measures should be implemented including effort reductions and introduction of more selective fishing gear.
There is no STECF advice.	TACs should be adjusted towards recent real catch levels but should not be changed by more than 15% per year ***or*** Member States should develop an implementation plan to provide advice within a short time.

For these six categories the Commission proposed specific approaches in TAC setting, thus providing a transparent and consistent line to take in TAC setting. In subsequent years this approach was further refined, and in 2008 the Policy statement introduced 11 categories of stocks with the corresponding methodology for setting TACs, as shown in Table 4.2.

In the same year the Commission introduced in the Policy Statement new elements such as an analysis of the trends in the conservation status of stocks, to provide a transparent and simple instrument to judge the level of progress in the policy (COM, 2008). This meant an unprecedented level of transparency, by showing to the general public the trends and therefore the level of success or failure of the EU's own policy, which hitherto was reserved for the scientific community.

In recent years, this document has been supplemented by a public presentation to stakeholders and journalists on fish stock status. This presentation, made by scientists, has become a showcase for the performance of the policy and has contributed to extend the fisheries management debate to the public at large.

The importance of the Policy Statement cannot be underestimated. Beyond its value as a consultation document with stakeholders prior to the specific TAC proposals, the policy statement has provided two extraordinary contributions to the policy:

- The transparency in explaining the philosophy under the TAC proposals, ensuring also a guarantee of 'level playing field' in addressing the conservation policy, and
- The transparency in reporting on stock status and thus on policy progress. This has been key to ensure that the proposals from the Commission and the decisions by Council can be scrutinised by public opinion, instead of constituting an arcane political decision.

After the 2013 CFP reform (see Chapter 16) these objectives will become even more important, since the new policy has since then established more clear objectives with specific deadlines, as well as increased transparency and stakeholder participation.

New types of TACs

Although the initial approach of the TAC system, based on annual TACs decided in December by the Council of Ministers for the following year has remained remarkably stable, the need to cater for more specific cases where such model was not necessarily the most appropriate has led to the gradual setting of new types of TACs that do not correspond entirely to the initial model. Let's see some examples.

Real-time TACs: sandeel and anchovy

The TAC policy has also evolved significantly in certain cases where the traditional structure of an annual TAC decided in December for the following year does not work properly. This is notably the case of certain short-lived species where the annual abundance of the stock (and thus the recommendable TAC levels) are a function of the strength of the incoming annual class (that is, fish abundance is largely the consequence of the size of the 'annual crop') and the abundance of this class is not necessarily well evaluated in December, at the moment of taking the decision. In these cases, a different approach was necessary. Two examples are worth considering: the sandeel (*Ammodytes tobianus*) in the North Sea and the anchovy (*Engraulis encrasicolus*) in the Bay of Biscay.

It is remarkable that in these cases the need to decide within a very narrow time scale makes it necessary to adopt relatively mechanical systems for TAC setting, without time for short-term negotiation. The TACs fixed on these species are today the best example of fixing TACs according to pre-established criteria and applied to the annual scientific advice, without any possible political discussion on short-term considerations.

Sandeel is a short-lived species where in normal circumstances the state of the stocks depends essentially on the abundance of the incoming year class. In addition, this year class cannot be easily estimated until the beginning of the commercial fishery. This means that only when the fishery starts in the spring can the annual biomass of the stock be evaluated. As a consequence, the TACs fixed in the December Council were meaningless.

The solution introduced in the mid-2000s was to establish a mechanism in the TAC regulation whereby a provisional catch level was set but the definitive TAC would only be established once the evaluation was completed based on the initial data from the commercial fishery. The time constraints of this procedure (sandeel fishery starts in April and finishes in June) indeed justified an empowerment for the Commission to fix the definitive TAC during the fishing season through an unprecedented fast-track procedure.

The stock has been successfully managed in this way for the last 10 years, even if the formula for setting the annual TAC has been refined several times in the light of experience, and is still subject to continuous refinement. It is also true that this success was certainly facilitated by the fact that the fishery is largely dominated by one Member State, Denmark, and constitutes an example of how the CFP, for all its heavy decision-making process, can also be amenable to creative, fast-track solutions when necessary.

The case of anchovy is similar, in the sense that the stock abundance is primarily dependent on the strength of the incoming year class. And as for sandeel, that strength is very difficult to evaluate reliably before the commercial fishing starts in the spring. In addition, the collapse of this fishery in 2005 and the tradition of conflict between the fleets of the two Member States concerned make the sound management of this stock particularly crucial.

In this case, the TAC is also established in the spring, after the data at the beginning of the fishery allow a better evaluation of the abundance of the incoming year class. In addition, since the season for the two fleets involved is different, it is also advisable that the TAC does not reflect the natural year but the annual seasons. This has led to the establishment of a specific TAC for the fishing seasons, which straddles two calendar years (EU, 2013b). As in the case of sandeel, these special management arrangements are evolving, as scientists improve their forecasts, and recently ICES has been able to provide advice in the autumn to cover the following natural year.

Bi-annual TACs: deep-sea stocks

The incorporation of TACs for deep-sea stocks (for species such as orange roughy, *Hoplostetus atlanticus* or black scabbardfish, *Aphanopus carbo* among many others, occurring only in deep-water habitats) took place relatively recently in the history of the CFP, since the development of fisheries for these stocks took place in the 1990s, and therefore their management was not part of the original political deal of 1983.

As these fisheries developed, the awareness about their possible high exposure to overexploitation was underlined by the scientific community, based on the very slow growth rates of these species that would make recovery from overexploitation extremely difficult. This led the Council to take the first catch limitations for these species in 2002. However, the TACs established were different from the previous ones: the low frequency of the scientific evaluation of the conservation status of these species (a combination of the high stability of the deep-sea ecosystem and the long life span of the species, together with the shortage of resources by marine institutes to invest in the research on these species) led to the first bi-annual TACs of the CFP. The first regulation (EC, 2002) established TACs for years 2003 and 2004 and these TACs have remained bi-annual ever since.

The discussions leading up to these catch limitations constituted an important questioning of the traditional TAC system as the best management instrument for the conservation of the species, to the point that these TACs were accompanied by a duplicate system of control of fishing effort by the fleets concerned. This questioning, however, was based on considerations of national interest (how to consolidate Member States' rights for these stocks) in yet another demonstration of the distributive nature of TACs. We will refer to this case in Chapter 13 as an example of the duplication of instruments in the CFP.

TACs to protect endangered species

In recent times, the TAC system has been used to protect a growing number of very vulnerable fish species. In doing so, the TACs go beyond their traditional role of managing and allocating fishery resource and turn into a mechanism to protect biodiversity. Beyond the case of basking shark, already mentioned, the TAC regulation for 2014 contained a list of eight species (all selaceans, sharks, skates and rays) for which catches are forbidden, in all areas (EU, 2014a).

TACs delegated to Member States

In recent years this new development has emerged: for stocks that are not shared by a minimum of Member States, the TACs can be delegated to the Member States concerned, under certain conditions and standards. A good example is the case of certain stocks of horse mackerel, traditionally managed by TACs assigned to Spain and Portugal in their waters around the Macaronesian archipelago. These TACs, applicable to the two Member States individually (without any concern about relative stability) have been delegated to the two Member States.

The elusive multi-species TACs

As we saw previously, the traditional TACs are adopted on the basis of single-species considerations. This means that the technological associations (two or more species being caught simultaneously by the same gear) or biological ones (predator-prey or competition relationships among species) are not taken into account. This fact, in turn, has a number of negative consequences on management:

- TAC levels for two or more associated stocks can correspond to very different levels of effort of the fleets catching them. As a result, the TAC of one stock can be exhausted much earlier than the others. Once the fishery for one stock is closed, fishing continues on the other stocks. This means that the fishery continues to produce fishing mortality on the first stock whose TAC is exhausted. Traditionally, this has resulted in higher fishing mortality than recommended for the exhausted stock.

- In turn, this has produced massive amounts of discarding. Under the new CFP, this will no longer be possible under the landing obligation. The concerned stocks will become 'choke species', that is, their quotas will be exhausted very early in the year, even when the quotas of the main target species are still largely available. This will require specific solutions (see Chapter 17).
- The allocation of the stocks under relative stability is not uniform among Member States. Therefore, these effects are exacerbated at the level of national quotas: not only TACs but national quotas are subject to these effects: quotas for some stocks being exhausted much earlier than others in the same fishery, leading traditionally to high discarding and, in the context of future policy, to possible early stoppage of the fishery, once the national quotas for the 'choke species', together with the flexibility mechanisms of the landing obligation, have been used to the full.

These reasons led the Union to agree on the need to develop multi-species approaches relatively early in time: already the basic regulation of the CFP adopted in 1992 foresaw the development of multi-species approaches (Article 8.3.i of Regulation 3760/92). Subsequently, the development of these approaches has always been an objective of the CFP. However, in practice such approaches have never fully materialised.

By multi-species approach we can understand different things. In particular there are two fundamental types of approach:

- The models based on technological relationships, that is, the models dealing with the different commercial fish stocks caught together in the same fishery.
- The models based on trophic relationships, that is, based on predator/prey relationships among the commercial species.

The models based on technological relationships are simpler and ICES has in recent years been able to produce recommendations on the relationship of different exploitation levels for species associated in the North Sea (ICES Advice, 2013). The advice establishes catch scenarios for the main nine stocks in the area taking into account the fishing pattern and catchability of the various fleets. This advice allows for an estimation to be made of the catch levels that can be obtained from different stocks as the result of limitations established in one or several of them.

ICES, however, does not make specific recommendations, because the balance between underexploiting and overexploiting certain stocks as a result of the species interactions is a management decision that should be taken by fisheries managers. On that basis, multispecies considerations will gradually be introduced, notably in the context of the multispecies management plans foreseen in the 2013 CFP reform.

Trophic models are more difficult to develop. More commercial fish prey on others and for many species the natural mortality related to that is higher than the mortality produced by fishing. In many cases fish that feed on plankton also prey on the eggs and larvae of their own predators. The predator/prey relationships therefore often work both ways and their overall effect is not straightforward to model. There are a few exceptions, such as that of capelin (*Mallotus villosus*) in the Barents Sea. This fish feeds on zooplankton and krill and is preyed upon by cod and marine mammals. The ecological role is well known and this has allowed scientists to produce multispecies assessments for this stock since 1990. Taking into account the multispecies relationships can establish the difference between a catch of several hundred thousand tonnes or a recommendation to a year-long fishery closure.[3]

In most cases, however, the required level of knowledge is largely missing and the development of this kind of advice will take time. As the recent experience shows, even in the – theoretically – simplest case, the Baltic, the scientific community has not been yet in a position to develop a truly multi-species model and the proposal for a multi-annual plan for Baltic stocks, adopted in 2014, is still based on individual advice for each of the concerned stocks (EU, 2014b).

Scientists continue making progress in this area, and predation mortality is already included in the assessment of stocks in the Baltic and North Sea. However, the difficulty to develop multispecies advice is not only scientific and not only determined by relative stability. Multispecies approaches will also require difficult decisions by managers on the trade-offs of different policy objectives: if the TACs of certain stocks are reduced to avoid damaging other associated ones, or if predator-prey relationships are considered and higher exploitation rates of the predators are contemplated in order to obtain higher yields of the prey, there could be significant economic implications depending on the market prices of the different species.

This implies that under trophic models, fishery managers would have to take more difficult decisions, in terms of establishing priorities among species. These decisions should be for fisheries managers, not scientists to take.

Enforcing TACs

We cannot separate the discussion about TACs from that of their effective enforcement. Much has always been discussed about whether the TAC levels adopted by Council are the right ones or not. But an entirely different discussion, often avoided in public, is whether the TACs are actually effective in determining what happens on the fishing grounds.

The difficulty of controlling catch limitations already gave rise to the establishment of control measures (the logbook) prior to the first TACs in 1983 (see Chapter 9) as Member States understood that unless the quotas allocated were correctly enforced, the difficult agreement on the allocation of rights would be meaningless. Unfortunately, the experience in subsequent years has shown that the control of the catch limitations has been less than perfect. And, strikingly, 30 years after their first introduction the poor enforcement of TACs is still considered one of the main weaknesses of the CFP. Why?

TACs are difficult to control. This is due to a number of factors:

1 Fleets can land their fish in many different ports, not all of them covered by inspection services all the time. In fact this has led to the use of the notion of 'designated ports' as an instrument of control for recovery plans of overfished stocks: landings can only take place where there is a reasonable level of control. However, much of the catches in Europe are still landed in ports with little or no control. The number of landing places in Europe is such that complete coverage by inspectors is simply not possible.
2 Many vessels land their catches in foreign ports. Member States must then rely on the transmission of data from the landing State to control the landings of their own national vessels. Traditionally this has not been done in real time, thus resulting in overfishing of national quotas once all annual catch data are put together.
3 Quota overshoots by Member States are difficult to demonstrate. Individual cases of black landings are relatively easily to detect but demonstrating that a Member State, as such, has fished more than its national quota remains difficult to prove while respecting legal guarantees. In recent years, however, some cases of overfishing have been demonstrated, giving rise to the implementation of 'pay-back schemes', such as those for the United Kingdom and Ireland with herring and mackerel (EC, 2007a), Poland with Baltic cod (EC, 2008a) and salmon (EU, 2013c) or mackerel with Spain (EU, 2011). But these cases were applied only after very heavy investments in investigating the concerned cases.
4 There is also a problem of 'burden of proof' in the system. Traditionally, it is for control authorities to demonstrate that fishermen have overshot their quotas, and this is often extremely difficult to prove. If, instead, fishermen would have an obligation to demonstrate that they only fish the quota they are allocated, using modern technologies, this could substantially change the way catch limitations are enforced.

In any case, the existence of quota overshoots is commonplace in the CFP. Scientists often report about it on their scientific assessments. This has always been a problem in the CFP. In Chapter 9 we will summarise the situation.

In fact, as we will see later in this Chapter, the perception by scientists that fishing mortality of certain endangered stocks was not being effectively limited by TACs gave rise to the recommendations leading to the establishment of fishing effort limitations. Thus the control (or lack thereof) of TACs has had a considerable impact both in what happens in the reality of the fishing grounds and on how the policy itself has developed.

The enforcement of TACs, being a permanent priority for the CFP, should benefit considerably from the new set of instruments established in the new control regulation adopted in 2009. This new regulation established a number of powerful tools that, if properly implemented, can effectively improve the control of the consumption of quotas. In particular, the introduction of the so-called 'electronic logbook' (the Electronic Reporting System, or ERS) should dramatically improve the enforcement of quotas.

Quota flexibility

In 1996 a mechanism was established for the purpose of providing more flexibility in the inter-annual management of quotas. Regulation 847/96 was adopted with the objective of eliminating some of the rigidities of the strictly annual management of quotas (EC, 1996). The objective was, on the one hand, to ensure that certain excesses in catches would not be penalised, by allowing that certain amounts of quota could actually be overshot, on condition that the excess catches would be paid back the following year. But specially, Member States wanted to 'save' certain quotas for the following year in case they did not exploit them fully in a given year.

This regulation has been applied annually ever since: every year the Commission adopts one or more Commission regulations adapting annual TACs on the basis of that flexibility, as reported by Member States.

This flexibility, together with the annual quota swaps, provides relative stability with a mechanism to soften the rigidities of the system, in helping to adapt the annual fishing opportunities for individual stocks to what fishermen find in the fishing grounds. While the effects of this flexibility on the conservation of the stocks are less clear, the system provides a degree of flexibility that softens the rigidity of the annual quota/relative stability complex.

From annual TACs to multi-annual plans

As we said before, the main problem of the TAC system lies in its traditional annual nature. In effect, deciding TACs on an annual basis without long-term considerations has created a culture of 'short-termism' in the CFP that is at the origin of many of the traditional failures of the conservation pillar of the policy. In Chapter 13 we will refer to the 'psychology' of taking decisions with an annual perspective only and how this short term perspective undermines the achievement of the long-term goals of the CFP.

In addition, the 'ratcheting up' effect, so typical of annual TAC-setting, further undermines the achievement of the policy goals: while the scientific recommendations to reduce catches are often applied only partially or not at all, recommendations for TAC increases are taken to their full extent. That produces a tendency for the TACs to be consistently above scientific recommendations.

Consequently, a fundamental question to ensure the sustainability of European stocks has been to introduce a change in the paradigm: from annual decisions on short-term arguments, to annual decisions as part of a multiannual strategy to achieve long-term goals. Long-term plans are designed to produce positive results in the long term, while annual management simply tends to avoid a bad result in the short term.

The limitations of a conservation policy based on annual decisions have long been identified by the Commission as one of the main problems of the CFP. In fact, the CFP reform in 1992 already recognised and the basic regulation adopted at the time already included the need for the CFP to ensure that: 'with a view to effective conservation; the rates of exploitation of certain resources should be limited and may be fixed on an annual or, where appropriate, multiannual and/or on a multispecies basis' (EEC, 1992).

However, attempts to introduce long-term considerations in annual TACs in the 1990s met strong opposition in Council and did not materialise. A number of Member States considered that this proposal would remove room for manoeuvre for decisions on annual TACs at the December Council. For a period of almost 10 years, the principle recognised in the basic regulation of the CFP did not translate into policy decisions.

Things started to change in the early 2000s, when two circumstances coincided to bring about the introduction of long-term approaches to fisheries management: the insistence of the Commission on these approaches in the context of the 2002 CFP reform and the collapse of the cod fisheries and the serious decline of the hake stocks in European waters. It became obvious that the recovery of these stocks would require a continued effort during a number of years.

In Chapter 15 we will refer to multiannual plans in the 2002 reform. However, the real birth of the multi-annual plans as a mainstream instrument for fisheries management in the CFP took place as a parallel process, in the framework of the recovery plans for cod and hake. In addition, two important precedents also contributed to this development, both as a result of bilateral and multilateral cooperation: the long-term plan for the recovery of herring in the North Sea, through bilateral cooperation with Norway, in response to the very poor state of this stock in the mid-1990s and the long-term plan for Atlantic swordfish adopted in ICCAT in 1997 (see Chapter 7).

The case of cod

In the late 1990s, ICES started to warn about the very low abundance of most of the cod (*Gadus morhua*) stocks. In the Irish Sea, early evidence of stock depletion led to a series of measures to try to recover the stock, although these first attempts took the form of specific technical measures, such as gear specifications to reduce catches of cod in fisheries targeting other species (EC, 2000).

But it was the cod in North Sea that changed things dramatically. In the late 1990s and particularly in the early 2000s, ICES warned about the collapse of this stock. In 2001, ICES alerted that the spawning stock biomass of this stock had fallen from 250 000 tonnes in the late 1960s and early 1970s to less than 50 000 t in the late 1990s, a historical low representing a high risk of stock collapse. On that basis, they recommended that:

> ICES recommends a recovery plan that will ensure a safe and rapid recovery of SSB to a level in excess of 150 000t. If a recovery plan is not implemented, ICES recommends that fishing mortality on cod should be reduced to the lowest possible level in 2002. (ICES Advice, 2001)

This kind of advice prompted the establishment of a task force to prepare a specific plan to 'save the cod'. The task force was not the first to be set up, an earlier one having been created in 1992 whose recommendations were largely ignored. The seriousness of the situation in the late 1990s caused the new task force to be taken much more seriously.

Crucially, ICES also recommended that TACs were not limiting fishing mortality and that direct control of the fishing effort was necessary. To follow the scientific advice, this had to be taken into account.

The grave situation of cod and the need for very strict measures for its recovery and the consequent political difficulties to get such measures adopted by the Council led the Commission to develop a 'north/south' package, where the – very unpalatable – measures to recover cod by the Member States concerned (particularly the UK) would be accompanied by similar sacrifices to recover other stock that was in trouble at the time (northern hake) of interest to other Member States (particularly Spain). The idea was that by distributing the sacrifices among more Member States, the necessary difficult decisions would be more likely to be taken. In this vein, in 2001 the Commission made a first proposal for a fully-fledged recovery plan for the cod stocks concerned and northern hake (COM, 2001).

Again, the tendency of the CFP to stress the 'distributive' nature of the resources took over other considerations. The problem of that proposal was that, by trying to be more 'balanced' in asking different Member States for sacrifices, it actually had the opposite effect: the evidence that the northern hake stock, though overexploited, was not as close to collapse as the cod stocks made the package approach politically unacceptable. Rather than providing balance, it provided imbalance by asking different Member States to take similar sacrifices for very different situations. The proposal was never adopted. Instead, a new proposal had to be submitted in 2003 for the cod stocks alone (COM, 2003). Precious time was lost in this regard to advance the recovery of cod. The northern hake stock would be handled in a separate regulation at a later stage.

The cod recovery plan, adopted finally in 2004, covered the cod stocks in the North Sea, Skagerrak, Kattegat, West of Scotland and the Irish Sea and introduced two essential new elements:

1 A 'harvest control rule' establishing a long term objective and the annual steps to achieve it. This established a multi-annual discipline for the adoption of annual TACs, so these annual decisions would no longer be based on short-term considerations but on a long term objective.
2 Provisions on the use of direct management of fishing effort for the cod fisheries concerned, initially for the North Sea and Skagerrak to be gradually extended to the other cod stocks.

These elements were for the first time in the CFP enshrined in the cod recovery plan (EC, 2004a), covering the cod stocks of the North Sea (including Skagerrak and the eastern Channel), Kattegat, West of Scotland and Irish Sea.

However, even before the cod plan was formally adopted recovery measures, including effort management were adopted as of December 2002 and implemented in 2003 and 2004 as provisional measures in the annual TAC and quota regulations for these years. For example, in December 2003, before the cod recovery plan was adopted, Council already decided on a system of simple effort limitations in the number of fishing days allowed for different types of gear from 10 days per month for bottom trawlers using 100 mm mesh size (those targeting cod) to 20 days for those using 16 to 31mm mesh (that catch very limited amounts of cod) (EC, 2003a).

The plan established a formula to calculate annual TACs as part of the multi-annual strategy to achieve the long-term goal of rebuilding the biomass of the stocks to their estimated potential. This formula, known as 'harvest control rule' became a fundamental instrument for fisheries management: it established the balance between the biological objectives and the socio-economic considerations for the concerned fisheries.

The establishment of a harvest control rule was a fundamental step in the evolution of the CFP. For the first time, it provided a way to resolve the problem of the combination of objectives of the CFP (Article 39 TFEU, see Chapter 2) and replaced short-term considerations with a long-term objective and a criterion to take annual decisions in that framework.

The harvest control rule finally adopted for the cod stocks deserves an explanation. First, it is important to explain why this plan established a long term objective expressed in biomass: a level of B_{msy} established at 150 000 tonnes. While nowadays it is commonly accepted that long term plans should be based on fishing mortality objectives, rather than biomass objectives, the first cod recovery plan did actually establish a biomass objective. To understand this, we have to consider that at the time the political debate about fisheries management was completely dominated by short-term considerations on catch levels.

The idea of a strategy based on a target F_{msy} was not well understood outside the scientific community. In contacts with stakeholders a number of them were convinced that a gradual reduction of fishing mortality would be accompanied by a parallel reduction of TAC levels. The idea of higher TACs with lower fishing mortalities, as the stock abundance increased was not necessarily understood by everybody. While the short term sacrifices were certain and easy to understand the promise of future higher TACs as a result of stock recovery was considered by many as an elusive, untrustworthy promise. In other words: it was not possible to engage a political debate around a long term objective based solely on fishing mortality.

However, over time and when the notion of fishing mortality started to become more familiar and the cod recovery plan was consolidated, this initial approach was changed through major amendment adopted in 2008 (EC, 2008b), where the objectives were expressed already in terms of fishing mortality corresponding to maximum sustainable yield. This case, among many others, underscores the importance of adequate pedagogy as an essential element of all new approaches on fisheries management.

Today, the idea of basing the multi-annual plans on fishing mortality, as opposed to biomass, is consolidated policy: this was enshrined in the CFP reform of 2013 (see Chapter 16) in recognition of the fact that biomass objectives are very unstable and may not be achievable even after the fisheries are closed.

Other essential elements of the harvest control rule were the annual increase in biomass of 30% and the limitation of the inter-annual variation of annual TACs. In a context of stock collapse, the establishment of a maximum annual variation of the TACs was an essential element in the adoption of the plan. This provision was a guarantee that there would be no annual dramatic reductions in TAC levels. In a context of stock collapse, this provision was instrumental to make the plan politically acceptable. However, that provision was adopted as part of a harvest control rule which established an annual increase of 30% biomass, to ensure a reasonably rapid recovery of the collapsed stock.

The socioeconomic consequences of applying such a dramatic annual reduction of effort were such that during the December 2002 Council, even the Prime Minister of one of the most affected Member States rang the Fisheries Commissioner of the time to complain about such a measure. In fact, this 30% was a compromise between the biological advice (which would actually require an even greater increase) and the socio-economic considerations that, unlike the biological advice, were taken into account without a detailed analysis of the economic consequences in the short and long term. This in turn showed the need for management measures to be gradual and to have economic as well as biological advice.

In fact this case also brought to the table the question of how to measure a 'gradual' approach in long-term strategies. As the annual 30% increase of biomass was considered by some of the Member States concerned as excessive, the question was then, for future plans, what would be an acceptable level of annual variation of biomass that the industry could absorb.

The subsequent setting up of the Regional Advisory Councils (see Chapter 12) addressed this question and a recommendation to the effect that the long-term plans should try to limit inter-annual variations of fishing possibilities to 15% up or down made headway. This has consolidated itself as a kind of standard of what stakeholders consider an acceptable inter-annual variation.

But perhaps the most significant novelty introduced by this plan was the introduction of effort management, which we will examine next.

The results of this management plan are mixed but partially positive. Some of the stocks, in the North Sea for example, have recovered gradually and are exploited at levels of fishing mortality approaching F_{msy}, according to the most recent estimates (in press). In the case of the cod stock in the West of Scotland the situation has not significantly improved as a result of the plan and since 2006 fishing mortalities are above F_{lim} and biomass below B_{lim} (JRC, 2014a).

The reform of 2002 and the consolidation of long-term plans

The case of North Sea cod had a very strong influence on the position of Member States vis-à-vis long-term plans in the context of the reform of the CFP adopted in 2002. It is clear that without the example of the cod stocks and the evidence that other stocks would need to follow a similar route, the long-term approaches would have never been adopted.

That negotiation of the reform, however, was hampered by the role of fishing effort management in these plans, even though effort management was applied before the adoption of the cod plan. While Member States in general accepted the need for the plans and even the need for some kind of effort limitation in the most severe cases, the possibility of using effort management in the case of stocks closer to safe biological limits was strongly contested. There were several reasons for this reticence:

- The possibility that effort could actually prevent the full exploitation of national quotas;
- The possibility that effort limitations would somehow undermine relative stability and;
- The fear that effort limitations would affect the ability of fishermen to fully benefit from abundant quotas of healthy stocks. In other words, the multi-species effects of an instrument applied to a mono-specific case.

Although the proposal from the Commission included only one article on long term plans, with fishing effort as an option, Council went on to establish two articles dealing with long-term plans: Articles 5 and 6 of Regulation 2371/2002. While Article 5 on recovery plans contemplated the use of effort management, Article 6 on stocks within safe biological limits did not refer to this instrument.

The idea was to limit the use of effort to the most extreme cases.

However, this was an example of rigid law-making: while some cases were clearly identifiable with Articles 5 or 6, respectively, this distinction did not cater at all for situations where a single stock would be around the precautionary limits of fishing mortality or biomass and could be inside or outside safe biologically limits in alternative years, thus giving rise to the absurdity of having a legal basis that could change from one year to the next and not having a stable basis to establish effort management for the concerned stock. In addition, the two articles did not cater for the situations where two or more stocks would be managed together in the same plan and one being inside and the other outside safe biological limits.

This absurdity shows a certain tendency to attempt to regulate fisheries as if they were predictable, forgetting the essentially unpredictable nature of natural resources and thus the need for legislation to be flexible and adapt to that. As a result of this, some later plans where the concerned stocks were fluctuating around safe biological limits could not be based on either Articles 5 or 6 and had to have the general legal basis of the CFP. The lesson, in any case, was well learnt in the 2013 CFP reform, where no such distinction was made between two types of long-term plan (see Chapter 16).

Other long term plans

The cod recovery plan was soon followed by others. The combination of evidences of overfished stocks, together with a certain sense of fairness in proposing the same kind of instrument to different fisheries, justified the following plans. These were certainly influenced by the cod plan but they all incorporated specific features, including one particularly appreciated by the fishing industry: the limitation of inter-annual TAC variation to a maximum of 15%, a guarantee against too dramatic reductions of annual TACs, which became a standard for practically all plans.

Northern hake

The plan for the northern hake (*Merluccius merluccius*) stock came soon afterwards (EC, 2004b). This plan had a double justification: on the one hand, the evidence of serious overfishing (although certainly not at the same level as for cod) and, on the other hand, the fact that the political effort required from the Member States affected by the cod plan was more acceptable if similar efforts would be asked to other Member States affected by fisheries for other (overexploited) stocks.

The plan for northern hake introduced a two-step approach. The CFP reform of 2002 established two articles regarding multiannual plans: one for stocks outside biological limits (Article 5 of Regulation 2371/2002) and another for stocks within safe biological limits (Article 6 thereof). As we will see, Article 6 did not establish any specific long-term objective for stocks within safe biological limits. For this reason, the plan was based on Article 5 and established that once the objective was achieved two years in a row, a new plan should be prepared based on Article 6.

However, the plan did not establish any scheme to limit fishing effort. The reasons were at least two:

- Firstly, the evidence that the hake stock was not nearly as endangered as some of the stocks of the cod plan raised the issue of the proportionality and the real need for a 'heavy duty' instrument, such as the effort management measures already in application of the cod stocks.
- Secondly, because by this time the effects of the effort limitations on cod were already known and, in particular, the effects of the plan to protect cod on other fisheries targeting other species had already been identified as the main practical and political difficulty of the cod plan. There was little appetite in Council to open another case of the same kind.

Since its adoption, the stock has recovered and reached the objective, in what constitutes to this day one of the best success stories in long-term management plans.[4]

Iberian hake and Norway lobster

The two previous plans were based on single species considerations but already from the beginning the need to address mixed-fisheries plans had been recognised. In fact, the next plan addressed, for the first time, a mixed-species approach: Hake (*Merluccius*

merluccius) and Norway Lobster (*Nephrops norvegicus*) in the Iberian Peninsula (EC, 2005c).

This plan was driven by the evidence of heavy overfishing on the Iberian hake and also on most of the functional units of Norway lobster in the waters of the Iberian Peninsula (ICES Advice, 2005). The need for a joint approach was also justified by the fact that most of the vessels fishing for the Norway lobster (bottom trawlers) also caught different but significant amounts of hake, so there was a clear technical interaction between the two fisheries.

The strategy could not be one of joint management with similar instruments. The scientific advice available for the two species did not allow for the same kind of approach. Rather, a new line of attack was proposed based on a combination of a long-term strategy based on fishing mortality for the hake and a series of closed areas to protect the Norway lobster, for which scientists could not provide a quantitative assessment. This plan was very innovative and set an example for future multi-species plans after the 2013 reform (see Chapter 16): establishing long-term objectives for the key species and adopting special technical measures to protect associated ones.

The plan established a precise and timed objective: the hake should recover to a biomass target of 35 000 t in a period of 10 years. As for the Norway lobster, the plan established the objective to attain safe biological limits within 10 years. The combination of a fixed biomass target (though open for revision) in a given time frame looks today somewhat inappropriate: the biomass target and the time scale cannot both be fixed because the rate of recovery depends on natural recruitments. However, this kind of regulation was the result of the political need, at the time, to set objectives that would be easy to understand (fish abundance being more 'user-friendly' than fishing mortality) and where Member States and the industry would be reassured of the gradual approach to achieve such objectives and thereby avoiding very large TAC reductions from one year to the next. These elements were the price to pay for the introduction of the long term plans.

To achieve these objectives the plan included a scheme of effort control, based on the previous experience of the cod recovery, with a reduction of effort levels by 10% per year until the target mortality level would be reached.

This effort system, in addition to the difficulties already explained for the cod plan, showed the limits of effort management schemes based on single stock considerations but having an effect on other associated stocks: the hake was caught by many coastal fleets targeting a variety of species and not just hake. In some cases, the vessels affected by the plan were bottom trawlers whose catches of hake were around 10% of their total catch. This illustrates again the need for multi-species approaches.

Overall, and despite a less than impressive implementation of the effort management, the southern hake stock has been recovering, with higher biomass, lower fishing mortality and high recruitments in recent years. Despite relatively strong recruitments since the early 2000s, allowing for TAC increases, the stock is still exploited above F_{msy} levels and the management plan is no longer adapted to the F_{msy} targets.

The latest reports indicate that the hake stock is slowly recovering, thanks to recruitments since 2005 being above the historical average, and the biomass is already above B_{lim}. However, fishing mortality is still higher than the F_{msy} and while biomass is recovering, it is still far from the levels of the 1980s.[5]

Sole and plaice in the North Sea

The sole (*Solea solea*) and plaice (*Pleuronectes platessa*) plan in the North Sea (EC, 2007b) came relatively soon afterwards and it also broke new ground, given its particular characteristics:

The relationship between the two species, particularly in the beam-trawl fishery, is among the closest any two species have in any fishery in Europe. It is practically impossible to fish sole without catching plaice, although the proportion varies in different areas.

This was the first plan developed in an area where another plan already existed: the cod recovery plan. This was, therefore, the first case where the question of the relationship, the overlap and the potential synergies between two recovery plans in the same geographical area had to be addressed.

The plan was ground-breaking in some aspects:

- It was the result of the best-to-date consultation process with stakeholders (within the North Sea Regional Advisory Council) whose opinions were significantly influential in the proposal (although the advice was not followed at 100%).
- It established long-term objectives clearly identified as corresponding to Maximum Sustainable Yield.

A delicate question was the application of effort management notably in relation with that of the cod plan, which partially affected the fleets targeting sole and plaice. Learning from the cod experience, the plan excluded from this effort regime vessels from Member States whose share of the TAC of these species was less than 5%. However, the plan did not clarify the question of the possible cumulative effects, in certain vessels, of the effort reductions for the cod and the flatfish plan.

In any case, this was a precedent of a plan with two species under a 'technological' association, that is, two species caught together by the same gear but not really a 'biological' association, in that the biological relationship between the two species was not considered.

Indeed, the objectives established for both stocks were their individual objectives, following the single species scientific advice. This means that, if in application of the plan, the TAC of one of the species was to be exhausted long before the other, fishing for the latter can continue but at the cost of high discards of the former. This was acceptable under the CFP up to 2012 but the introduction of the discard ban in the 2013 reform renders this kind of plan insufficient to cater for the possible mismatch in the consumption of the two associated stocks. While this plan was a good precedent under the old policy the discard ban of 2013 makes it obsolete.

Since the implementation of the plan, the sole stock has improved and in 2014 ICES classified the stock as having full reproductive capacity and as being harvested sustainably, although the fishing mortality was still above the MSY target.[6]

Sole in the Bay of Biscay

The plan for sole in the Bay of Biscay (EC, 2006a) affected primarily one Member State: France, which holds the lion's share of the total TAC. So the plan was largely adapted to the characteristics of the French fishery.

In particular, the experience of the effort management for the cod plan led to a very specific approach on effort management: a system of ring-fencing the capacity of the fleets targeting the sole stock, through a licensing system and an associated capacity ceiling for the vessels holding such license. The system did not establish a regular, compulsory annual reduction of the effort level, as was the case for the cod and southern hake fisheries.

The objective of the plan of achieving a spawning-stock biomass of 13 000 tonnes is almost achieved, with 12 800 tonnes.[7]

Western Channel sole

The plan for sole in the western Channel also introduced certain novelties (EC, 2007c). Unlike the other sole plan, this one established a specific objective expressed in fishing mortality: 0.27 and for the first time it required a reduction of the fishing mortality by making a stepwise reduction every third year. This feature reflected the preference of the Member States and the stakeholders for more stability in TAC and effort levels, even at the cost of making bigger sacrifices in certain years. This feature, particular to this plan, showed also the importance of adapting the plans to the requests from Member States and stakeholders.

Baltic cod

This plan established, for the two Baltic cod stocks long-term objectives expressed as fixed values of fishing mortality corresponding to F_{msy} levels as estimated at the time (EC, 2007d). The plan also established two closed area/seasons, building on the pre-existing 'summer ban' applicable already in the Baltic to protect spawning cod and established a system of limiting fishing effort (expressed in simple terms as fishing days) in a way that would correspond to the annual changes in fishing mortality.

Contrary to the one for Atlantic cod, with all its complexities dealing with derogations based on better selectivity, the Baltic effort system was a model of simplicity: it was based on a total number of fishing days per vessel and year.

Of course this simplicity was possible because of the relative simplicity of Baltic fisheries as compared to those of other European waters. But it is also true that the experience of the cod plan in the Atlantic already served as a lesson to Baltic Member States (and of course for the Commission) with respect to the need for simplicity in the system. As we will see later on, this simplicity has provided a relative good evaluation of this effort management regime as compared to others.

The Baltic cod plan established a number of additional control measures and standards. This was the first long-term plan where the issue of effective control and enforcement became a priority.

The results of the plan are mixed. Western Baltic cod is still exploited at levels above MSY target. ICES does not consider the plan precautionary and, for 2015, has recommended following (more restrictive) MSY advice (JRC, 2015). For the Eastern Baltic stock, the situation is not well known and precautionary advice is given by ICES.

Herring in the West of Scotland

This plan for herring (*Clupea harengus*) was the first one agreed for a non-demersal stock (EC, 2008c). It was also the first one developed for a fishery that was not under the pressure of a significant overexploitation. It represents the first attempt to rationalise the management of a fishery that was not in a bad conservation status. As the industry's main priority was the stability of catches, the plan developed a sophisticated 'harvest control rule' to ensure stability of catch possibilities through maximum inter-annual variation of TACs, as a function of estimated biomass. Meaningfully, it also established a biomass threshold below which the annual TAC would be fixed at zero, thus turning a difficult decision such as a fishery closure into a mere implementation of a plan according to scientific advice.

Other plans adopted

Beyond the long-term plans developed by the EC in its waters, it is important to refer to other multiannual plans agreed either bilaterally or multilaterally with other partners. Some of these have been important precedent for the CFP. Others have been largely the consequence of the EU's growing experience on long-term approaches. And, in all of them the EU has played a key role in developing long-term approaches to fisheries management in international and shared waters.

Of particular relevance is the long-term plan for the management of Atlantic swordfish in ICCAT in 1998. This plan constituted a very good precedent of what ICCAT itself would develop later for other stocks and also prepared the ground for the internal debate on long-term approaches within the Union. Its success in recovering the Atlantic swordfish stocks has also been a positive example for long-term approaches in ICCAT and elsewhere. In more recent years, a multiannual plan was also adopted for Bluefin tuna, very much drawing on the experience of the old swordfish management approach. This plan has seen a quite spectacular success in recent years (see Chapter 8).

The bilateral agreement with Norway for shared management of fish in the North Sea (cod, haddock, plaice, mackerel etc.) has also developed joint long-term approaches. These plans, however, are a reduced version of the full plans developed by the Union, excluding effort management. While Norway has always refused to accept effort limitations they have agreed to establish joint 'harvest control rules' that determine annual TACs according to a previously-agreed strategy and long-term goal. These plans have actually contributed considerably to ease the annual bilateral negotiations, notably by facilitating the agreement on common TAC levels for certain shared stocks.

Other plans proposed by the Commission

Other plans were also proposed in recent years but never adopted by the co-legislators as they were the first ones to be adopted by co-decision under the Treaty of Lisbon (see Chapter 13 on decision making). These plans have already had an important impact on management, as they have *de facto* been applied in the annual TAC exercise even before they enter in force. They are the following.

Western stock of horse mackerel

This plan (COM, 2009b) was proposed by the Commission following a recommendation from the Regional Advisory Council (RAC) of the North

Western Waters (see Chapter 12 on stakeholder consultation), becoming the first proposal for a long-term plan ever made on the initiative of a RAC. A fundamental reason for this to happen is that when the RAC proposed the plan, the stock was not over-exploited. So the RAC recommendation was done under little pressure to make sacrifices. This is an important lesson too: while most previous plans addressed stocks under different degrees of over-exploitation, requiring important and difficult sacrifices, the horse mackerel plan represented a different approach: agreeing on a long term strategy when it is easier to do so because the stock is healthy.

Anchovy in the Bay of Biscay

This plan also has a very special importance (COM, 2009c). Proposed in 2009, it has been blocked in the co-decision process because of the institutional deadlock on long-term management plans.

The case of anchovy has something in common with that of sandeel: a directed fishery for a short-lived stock where the annual abundance can only be estimated at the beginning of the fishing season. Anchovy is a short-lived stock where, in an average year, up to 80% of the spawning stock biomass is accounted for by the incoming year class. In other words, the stock abundance is primarily a function of the size of the 'annual crop'. And the size of the stock is in any case highly fluctuating.

But beyond that, the situation was more complex: the fishery is shared by two Member States and has been subject to difficulties. In addition, the two fleets concerned fish at different times of the year.

Here, the need for a creative approach to TAC setting was also determined by the collapse of the fishery in 2005, as described previously, prompting the Commission to close the fishery through an emergency measure (EC, 2005b). In this context, the development of a new approach to TAC setting was based on the need to prevent such a kind of event.

These characteristics justify a specific approach to TAC setting: establishing a definitive TAC once the real size of the incoming year class is well evaluated through direct assessment (notably eco-sounding). This kind of management justified also a specific type of harvest control rule: rather than fixing a specific F_{msy} objective, the plan proposes to reserve, when establishing the annual TAC, a given percentage of the estimated annual biomass for reproduction, the TAC being the remaining biomass.

Beyond this question, the anchovy plan represents a success story in terms of the contribution of the SWW RAC. In fact, this fishery had been, in previous years, a very contentious one for the two Member States involved (Spain and France) and their respective industries have had frequent rifts over the modalities of managing this stock. In fact, however, the RAC allowed and promoted a dialogue between the industries of the two Member States that led to an understanding on certain aspects of the management of this stock that have allowed a very smooth operation of this fishery in recent years.

This is perhaps the best example of the value of long-term approaches, even before they become EU law. In fact, in a number of cases like this, the existence of a Commission proposal for a plan, particularly if it carries an important backing by the stakeholders in the corresponding RAC, constitutes an excellent basis for annual TAC setting even before the plans are formally in force. In some cases, when and where the Commission proposal is based on an approach agreed by the RACs, the implementation of such proposal in the annual TAC decision can be just as automatic as that of the plans really in force in strict legal terms.

Baltic salmon

Finally this most recent plan (COM, 2011) was the first one presented already under the Treaty of Lisbon. It was, consequently, the first test of how the new Treaty would be interpreted by the co-legislators in terms of the elements to be adopted under Articles 43(2) and 43(3) TFEU. But it also presents a particularly difficult characteristic: the fact that a crucial part of the life history of the salmon takes place in fresh waters, outside the scope of fisheries management under the CFP.

Due to the changes introduced in the decision making process as a result of the entry into force of the Lisbon Treaty and in connection with the institutional disagreement between Council and Parliament over the use of Articles 43(2) or 43(3) TFEU (see Chapter 13), these proposals were never adopted. They were finally withdrawn by the Commission at the end of 2014.

The introduction of fishing effort as a management instrument

The 1983 CFP contained references to effort: Regulation 170/83 defined conservation measures under the CFP as including: 'The restriction of fishing effort, in particular by limiting catches'. However, the initial reading of this provision was that effort was exclusively, not just 'in particular' applied through catch limitations. Indeed, the first effort management measures were those applicable for Spain and Portugal upon accession in 1986 as we saw in Chapter 3.

The 1992 basic regulation also included references to effort management but in a way that was more related to the maintenance of the effort limitations stemming from the accession of Spain and Portugal than to the use of effort as an instrument for the mainstream management of stocks under the CFP. Likewise, upon the enlargement of 1994, Council adopted a similar approach in the Baltic, to the effect of avoiding major displacements of fishing effort following accession (EC, 1997a).

So for the first two decades of the CFP the effort management was seen as a general system to prevent major displacements of the fishing activity form one area to another, in order to ease the accession conditions for existing Member States under the principle of free access but not really as an instrument to manage specific stocks or fisheries.

However, this started to change due to the crisis of the cod stocks in the North Sea and adjacent waters in the late 1990s. In these years, the scientific community was very critical with the way the individual TAC and quota system was controlling fishing mortality and they strongly advocated the use of effort restrictions to ensure the desired levels of fishing mortality on the cod stocks would be delivered.

For example, in 2001, in the middle of the crisis of the cod stocks in European waters, ICES recommended for North Sea and the West of Scotland cod stocks that: 'fishing mortality on cod should be reduced to the lowest possible level in 2002. ICES has repeatedly stated that for various reasons, TACs alone are not effective in regulating fishing mortality' (ICES, 2001).

For the North Sea, ICES went on to indicate that: 'a reduction of effort by 30% or an increase in mesh size of around 30 mm would yield the same results on the spawning biomass per recruit'. A 30 mm increase in mesh size would have had immediate economic repercussions on the industry: such an increase would have reduced catch rates of cod dramatically and would have practically eliminated certain other species from the catch, such as whiting. As a consequence, applying effort reduction was considered the only viable option to try to ensure the recovery of cod.

This strong message from the scientists, in the context of a collapse of such an iconic stock as cod prompted the Commission to consider seriously the use of direct effort management to ensure the reductions in fishing mortality necessary to recover the cod stocks. Indeed, scientists continued advocating effort management for a number of years afterwards. In 2006, in the context of the advice on Irish Sea cod, ICES recommended:

> ICES would often advise on a precautionary TAC based in recent landings. However, when the landing data are not reliable due to misreporting a relevant number for such a TAC cannot be established and a TAC regime does not regulate fishing mortality. Therefore, ICES has concluded that in such situations a possible route is to change management to focus on effort, which can be controlled through instruments like VMS, in order to reintroduce effective control of the fishery and to restore a reliable future data base for advice and management decisions. (ICES Advice, 2006)

That was the kind of advice upon which the Commission built its case for effort management in the cod fishery first and in other stocks later.

Cod recovery and effort management

The effort management scheme proposed in 2002 for cod recovery already anticipated some of the ideas that became mainstream some years later and notably the need for decentralised management.

Instead of a very prescriptive system of allocation of effort quotas for the fleets targeting or catching cod, the proposal established a system of national effort quotas, expressed in 'kilowatt/days' to be allocated among their own fishermen according to national criteria based on their respective contribution to cod mortality. This approach, dubbed the 'bucket of kilowatt/days' soon became a major difficulty in the political negotiations and ultimately it was never adopted. Instead, Council had to prepare an alternative system almost literally overnight. The system was extremely complicated and prescriptive and, with certain variations, it is still in place today, at least theoretically.

It is important to analyse the reasons for the failure of the 'bucket of kilowatt/days' approach at the time. This was w very innovative system that anticipated some of the features of the new policy adopted in the 2013 reform. But at the time it was completely rejected by Member States. There are at least three reasons that explain this.

Firstly, that the calculations on the real size of the 'bucket' per Member State revealed the discrepancies among Member States in the way they had adapted their national fishing effort to their respective national quotas. As the effort levels were based on recent levels of effort deployed, which had been recently cut, Member States whose effort had been limited saw their total effort allocated as being 'unfairly low'. In other words, some Member States considered that the system actually penalised those Member States that had done a more serious work to adapt the activity of their fleets to the available quotas in the previous years.

The second reason was much less explicit. The 'bucket of days' system actually implied that some of the usually difficult decisions of allocating fishing rights among fleet segments were clearly left to Member States. And, given the traditional 'culture' of blaming Brussels for all unpalatable decisions, some Member States were not comfortable with the idea of having to take difficult decisions without having Brussels as the usual scapegoat.

Finally, there was an important question of image and perception: the annex to the proposal included a formula on how to calculate kilowatt/days per Member State. The formula was actually quite simple in mathematical terms but it seemed complicated at first sight, since it contained for example the Greek character Σ to indicate the sum of certain data. The psychology of it was not positive: it gave the impression to administrations and stakeholders that an important political decision (the number of fishing days per Member State) would be churned out by an arcane and non-transparent mathematical formula. As we will see in Chapter 13, the psychology of decisions within the CFP has often been crucial in their success or failure.

The alternative effort management system worked out subsequently consisted of a gradual reduction in the number of fishing days per year for the fleets affected. The problem was to define 'the fleets affected'. While fleet segments targeting cod would undoubtedly have to qualify for an automatic effort reduction, the situation was much more complex for all the categories of vessels for which cod was not the primary target species but which caught significant amounts of cod. These segments had to be identified, classified and subject to a system whereby the reduction of fishing effort would have to be proportional to their relative contribution to the fishing mortality they exerted on cod.

But that was not all. The following problem was raised immediately: if vessels traditionally caught large amounts of cod but they would adopt measures allowing them to reduce their catches of this species, thus re-orienting their effort to alternative species, then these vessels should not be penalised by a large reduction of effort. This gave rise to a system of bonuses (in terms of number of days per year) that would be proportional to the kind of measures they would take to reduce cod catches.

The resulting system was an extremely complex table of different vessels and types of fishing gear with different degrees of contribution to cod mortality and different number of days per year allocated as a result. For example, already for 2005, the system distinguished six different grouping of vessels and gear with different number of fishing days per month (from 9 to 21) as a function of their contribution to cod mortality (EC, 2005a).

But this soon proved too simple and the following years saw a tendency for new, more refined fleet segments and gear being identified as requiring a specific treatment, giving rise to an increase in the number of vessel/gear combinations with different fishing days. For example, for 2007, the regulation considered a total of 38 fleet categories, with different number of fishing days per year (EC, 2007e). These categories, when applied to the different cod stocks and management areas, resulted in a table with more than 150 entries!

The technical basis for these different effort allocations was also rather doubtful. While scientists tended to base their estimates of cod by-catch rates on past data, the industry tended, often overoptimistically, to expect that they could reduce their cod by-catches to an important extent. The table of days at sea allocations was finally decided by negotiation in Council.

Not surprisingly, these continuous modifications and increased complexity ended up creating a system extremely difficult to understand, let alone to manage and control, constituting one of the cases of 'micromanagement from Brussels' that has discredited the policy, even though it was driven by the efforts by many separate groups of fishermen to obtain concessions particular to their fleets.

The days at sea were also exchangeable for capacity and these motivated significant fleet reductions. This was also a side effect of the effort management plan.

Different effort management plans have been applied in other areas. The effort management of the 'Western Waters' Regulation was referred to in Chapter 3 and the effort management in the Mediterranean, which replaces TACs in that sea basin, is referred to in Chapter 8.

The implementation of effort management

The practical implementation of the different effort management schemes by Member States has been extremely varied. There are a number of reasons for this:

1 In some cases, the Member States concerned have not fully implemented the plans. Part of the problem is that, by the time effort management was introduced, national administrations were saturated with the implementation of existing instruments and had little administrative capacity to handle an entirely new instrument.
2 While in some areas the effort ceilings have indeed resulted in real limitations of the fishing activity, in other cases a number of fleet segments have not fully used their effort quotas and can be considered as not being actually limited by the effort schemes.
3 The initial allocations of effort levels were not always well aligned with the real effort levels necessary to catch available TACs. Very often, when effort levels did actually limit the capacity of fleets to fully utilise available quotas, the effort limitations have been ignored, as being contrary to the TAC and quota system.
4 But perhaps fundamentally, the implementation of effort schemes has been politically difficult. The use of instruments based on single-species considerations but having effects on many other species was never fully accepted. And allegations of effort undermining one way or another relative stability has made this instrument largely unpalatable for many national administrations.
5 A counter-argument can also be made: that the effort managed system became unpopular because – in some cases – it worked. Certain evidence points at effort management being really constraining in some cases: the high prices of the days-at-sea market, the parallel fleet reductions that took place in certain segments and so on.

The effects of effort management are difficult to evaluate. Strongly favoured by the scientific community with regard to the cod recovery, the instrument was extended to other multiannual plans as recommended by scientists in situations where TACs were not limiting fishing mortality. Effort management had other potential advantages: the basic system was simple to control, at least for vessels equipped with Vessel Monitoring Systems (VMS), although the very complicated derogations made such control difficult.

Effort management, however, was always adopted in addition to catch limitations, not instead of them, thus adding to the management measures on the fisheries concerned. It is difficult to assess the real effect of effort limits for a number of reasons:

1 Several improvements to fisheries management were put into place at the same time as effort limitations: lower TACs, more effective control, fleet

reduction and so on. It is difficult to assess the effect of any of these measures separately.

2 Since the catch limitations through TACs uphold relative stability, Member States always have trouble in accepting that effort limitations could interfere with full quota consumption. Every time effort limitations would actually prevent certain fleets to fully utilise their quotas, the whole notion of effort management was called into question as not corresponding to relative stability.

3 The addition, not substitution, of a new, complex management system contributed to increasing the administrative burden of the Member States concerned. In fact, when it came to implementing a complex effort limitation scheme a number of national administrations that were already saturated by traditional quota management were not in a position to invest in the implementation of the new system.

4 The mechanism that most characterised effort management systems is perceived as too rigid. The idea that total effort must continue to be reduced automatically every year even when the stock is improving is very difficult to understand and often politically difficult to accept.

Overall and despite these shortcomings, the implementation of effort schemes in recovery plans has indeed provided for a general reduction of fishing effort, as shown in Figure 4.1 for the cod fishery in the North Sea. However, these general trends are due to the effort reductions of certain schemes, while others have not succeeded in reducing effort significantly.

Furthermore, even in cases where the total effort has been reduced, the fishing mortality of the target stock has not been significantly reduced. This seems to have been the case for cod in the West of Scotland, where the reduction of effort has not been accompanied by a consequent decline of the fishing mortality (see Figure 4.2). This may be due to the fact that, when confronted with low effort levels, fishermen still have a certain capacity to exert that level of effort in a way to maximise the catch of the most valuable species.

The overall results of the effort management schemes have been evaluated by STECF (JRC, 2014b) and can be summarised as follows:

- for stocks that are recovering, the reduction of fishing mortality corresponds to a reduction in fishing effort. However, it is not self-evident than the latter is the cause of the former: other measures taken at the same time (fleet reduction, economic reasons, TACs more adjusted to scientific advice, better control) may have also contributed.

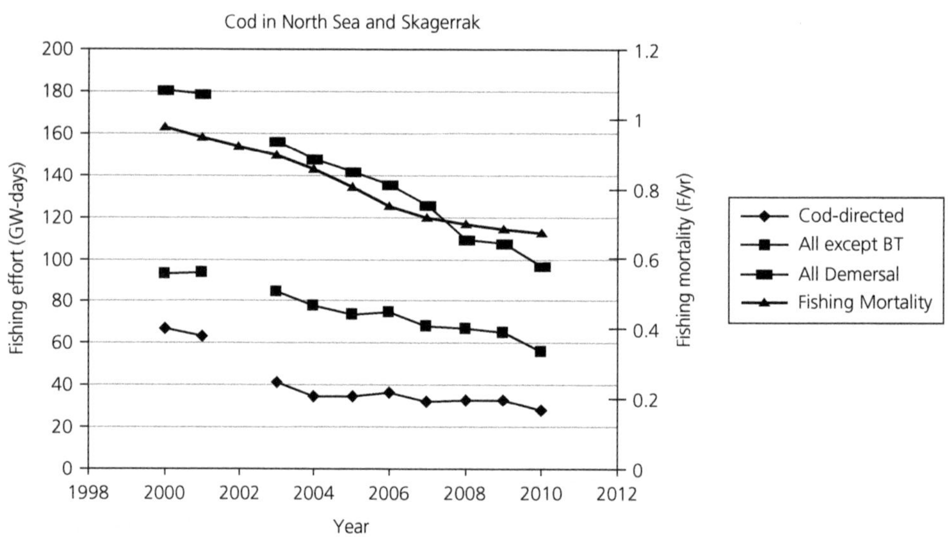

Figure 4.1 Time trends of fishing mortality and fishing effort (as total demersal effort, cod-directed effort or demersal effort excluding beam-trawls). Source: European Commission Services. (See colour plate section for the colour version.)

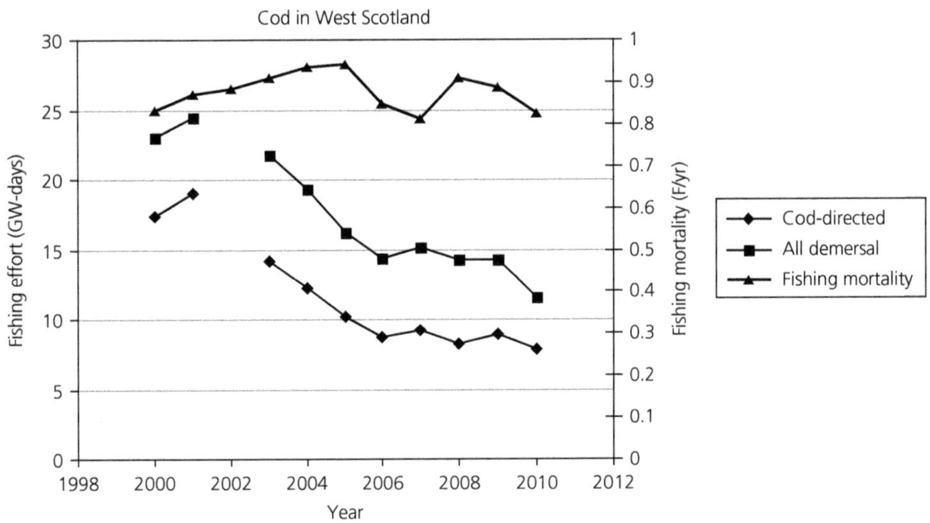

Figure 4.2 Fishing mortality and fishing effort of cod in the West of Scotland. Source: European Commission Services. (See colour plate section for the colour version.)

- The effort management of the cod plan has contributed in some cases to improve selectivity, since the plan gave more fishing days to fleet segments improving their cod selectivity.
- The overall reduction of effort has been more marked after the implementation of the plans than ever before. But as indicated previously, it is not sure that this reduction was due to effort management alone. Indeed in some cases the reduction in effort did not result in a proportional reduction in the fishing mortality of the target species.

This leaves the evaluation of the added value of effort management as part of recovery plans very difficult to ascertain.

One particular difficulty of the effort management regimes, that has always resulted in their potential limitations to the fishing industry being questioned or bypassed, is the fact that, as a 'secondary' instrument to the TACs, both national administrations and the industry consider that the effort limitations should not prevent them to fully catch their quotas. Otherwise, they argue, effort management can be contrary to relative stability. This argument has always undermined the usefulness of effort management: if they can't limit the catch of available TACs, then their value is very limited.

The 2013 CFP reform provides a new element on the potential usefulness of effort management. In the past it has been perfectly possible, until the entry into force of the discard ban, to actually increase effort and fishing mortality while respecting lower quotas, simply through massive discarding and black landings. After the discard ban is in force, this approach will no longer be possible if the enforcement of the new policy is effective.

The future of effort management

Overall, the use of fishing effort has contributed in some cases to reduce fishing mortality, increase selectivity and promote reductions in fleet size but has created an extra administrative burden that has often not demonstrated to have a significant added value as a duplicate management instrument. The future of effort management seems to lie on those cases where such added value can well be demonstrated.

Effort management remains potentially an effective instrument for fisheries management. In particular, for mixed fisheries for non-shoaling species, it can provide a fundamental advantage of facilitating the discard ban: the fishing activity ceases when the effort is exhausted, not when individual quotas are exhausted. However, its relationship with the TACs and quota system is difficult and, if and when effort management is considered as a duplication of the former, it can be perceived as somehow undermining relative stability. This results in its political acceptance and its effectiveness as a management tool being seriously questioned.

There is also a strong case against effort management as a duplicate management instrument in the context of the simplification of the policy. It is true that having a duplicate system of management with TACs and effort makes the policy very complex. This duplication was politically acceptable at the time of the first long-term plans for severely depleted stocks being considered a necessary step towards the recovery of such stocks. But as the simplification becomes a higher priority in the CFP and as – some – of the stocks that justified the introduction of effort management show signs of recovery, the added value of effort management is increasingly questioned.

The new proposal for a multiannual plan for Baltic stocks, presented by the Commission in 2014 (EU, 2014b), did not include the effort management regime applicable in the existing Baltic cod recovery plan. The discredit of effort management as a duplicate system to catch limitations was such that the Commission did not need to explain its suppression, and Member States and stakeholders alike welcomed the elimination of this instrument.

Unless the effort management demonstrates a real added value as an additional management instrument to the TACs and insofar as the management by TACs will remain the preferred management option, its future will remain seriously questioned. If and when effort management could replace TACs rather than duplicate them, could the real potential of this instrument become apparent?

Technical measures

Technical measures: are they just technical?

The importance of technical measures (gear characteristics, mesh sizes, closed areas/seasons and other measures) is illustrated by the fact that of the two founding regulations of the CFP, one was on technical measures (Regulation 171/83).

Technical measures have a special characteristic that make them fundamentally different from TACs in terms of the effects on the fishing industry. Problems related to quotas can be resolved through adaptation of capacities at Member State level, so that the remaining vessels can have enough quotas to ensure economic survival. However, problems with technical measures do affect all vessels concerned regardless of fleet capacity adaptation, because they affect the catch rates of individual vessels and, consequently, their individual economic performance. If technical measures are uneconomic to apply, solutions at Member State level are much more difficult, if not impossible.

In addition, there are a certain number of cases where technical measures have been used to resolve political problems. Many examples come to mind: the current 'biologically sensitive zone' around Ireland in the context of the Western Waters Regulation (EC, 2003b) looks today as a measure to protect the hake spawning ground around this Member State. However, as we saw in the previous Chapter, this is a measure still closely related to the Treaty of Accession of Spain to the Union and this 'technical measure' can only be fully understood in that historical context.

Likewise, the ban on certain kinds of gear (large-scale purse-seines to fish for tuna around the Azores, in force from 1986 to 1995, does not respond to any particular conservation purpose but rather to the agreement to protect the more artisanal tuna fishing of the Azores, whose government decided not to develop purse-seining to protect their more traditional methods. Another example is the ban on fishing for anchovy with pelagic trawls in ICES area VIIIc, in order to protect traditional purse-seining as preferred by Spain.

Furthermore, the current ban on driftnets, adopted in 1998 (EC, 1998a), is after all the end result of a difficult conflict over the use of different gear to catch the same stock and sell it to the same market, so the issue is still today seen as the result of a political battle and not just that of a technical discussion.

Finally, a number of technical measures are not directly related to the management of fisheries, which the fishing industry understands well but to environmental objectives such as the protection of certain species and habitats that the industry often finds more difficult to accept.

For these reasons, contrary to what their name seems to suggest, the adoption of technical measures has always been highly political and still to this day remain the ultimate example of 'micromanagement by Brussels'. In fact, although the industry (and some National Administrations) has always been critical of this top-down, micromanaged legislation any attempt in the past to change this in practice has always failed.

As we will see next, we can divide technical measures in three broad categories of measures:

1 Fishing gear characteristics;
2 Minimum fish sizes and;
3 Closed areas/seasons.

These different types of measure have been combined traditionally in the different technical measures regulations.

Some historical background

We have already referred to the first technical measures regulation (171/83) as one of the initial elements of the CFP from its inception. These measures have been remarkably stable in their nature over the years and their evolution has been slow and difficult.

In 2015, there are four frameworks on technical measures: for the Atlantic and North Sea, direct successors of those adopted in 1983 (Regulation 850/98: EC, 1998b), for the Mediterranean (Regulation 1967/2006: EC, 2006b), for the Baltic Sea (Regulation 2187/2005: EC, 2005d) and for highly migratory fish (Regulation 520/2007: EC, 2007f). The Mediterranean measures will be discussed in Chapter 8 in more detail. The regulation on highly migratory fish is the result of incorporating in EU law the different technical rules adopted at the level of tuna RFMOs.

The initial regulation 171/83 became obsolete soon after its adoption for several reasons:

- it was amended six times in 2 years and it soon started to become confusing;
- It had to remove the technical measures applicable to Greenland that, in 1985, became a non-Community territory;
- It had to find a new way to determine mesh sizes for mixed fisheries. Regulation 171/83 defined as many mesh sizes as target species and this proved extremely confusing for mixed fisheries;
- It had to incorporate new areas as a result of the accession of Spain and Portugal: the whole of the Bay of Biscay, Iberian waters and the waters around the Macaronesian archipelago had to be regulated.

The resulting new regulation (Regulation 3094/86) introduced a new way to legislate on mesh sizes, by associating specific mesh sizes with minimum percentage of target species. It also included new species under minimum landing sizes. These minimum sizes already showed the increasing diversity of waters and fisheries in an enlarged Community. For example, several species (notably the cod, hake, haddock and saithe) already in this regulation had different landing sizes in different areas. These differences were later aggravated when minimum sizes were introduced for the Mediterranean in 1994, some of which were different than those in the Atlantic for the same species.

Regulation 3094/86 was also modified numerous times, giving rise to the need to codify the text to avoid confusion. This resulted in the adoption in 1997 of Regulation 894/97, which did not change the approach of Regulation 3094/86 but simply codified all the previous and often confusing amendments.

At the same time, however, the Commission proposed a new technical measures regulation, notably changing the approach on mesh sizes (see next the issue of 'fishing for'), which resulted in Regulation 850/98 that through different amendments, is still in force today. An additional attempt by the Commission to codify again this regulation in the early 2000s failed to obtain the support of Member States in Council: the complexity of the rules was such that few administrations could actually handle the proposal.

One important lesson from this short historical background is that technical measures are in a permanent process of revision. These constant revisions introduce complexity and confusion. Such permanent amendments were already complex when Council alone decided on them. Now under co-decision, this will become even more difficult. This clearly points to the need to rethink completely the way to legislate on technical measures. The reform of the CFP in 2013 has provided the legal basis to do so.

In the Baltic, the first technical measures were introduced in 1998, very much following the same model as in the Atlantic and the North Sea (EC, 1997b). These were amended following the enlargement of 2004 that brought most of the Baltic into EU waters. In this new regulation (2187/2005 as referred to previously), the ban on driftnets applicable to other European seas since 2002 was also introduced, in this case applicable to salmon driftnets.

Gear characteristics

Mesh sizes

The minimum mesh sizes constitute the essence of technical measures, in that their selectivity is key to achieving one of the two objectives of the conservation policy: avoiding the catch of small fish. Minimum mesh sizes have been a fundamental part of the CFP since the first technical measures regulation in 1983. However, mesh sizes have proven very difficult to regulate in practice. There are a number of reasons for this.

Firstly, the diversity of fisheries, with different catch composition and target species, makes it very difficult to find a mesh size that would represent a fair balance among several factors:

- The mesh must be large enough to allow the escapement of immature fish. This is indeed a very elusive question, since different species caught together by the same nets have very different ages of sexual maturity and very different percentages of retention by a given net.
- The mesh should also be such that the overall catch rates of the vessels allow them a sufficient revenue. For this reason, the room for manoeuvre to increase mesh sizes is always limited by economic factors and strongly contested by the industry.
- In technical terms, it is extremely difficult to separate different species in the catch. While certain separations are easy (e.g. separating Norway lobster from fish) others are very difficult technically (e.g. separating different gadoids, such as cod and haddock).
- And, crucially, when these separations are possible, they are not necessarily economical for fishermen: certain technical solutions are easy but they just mean fishermen would lose a substantial part of their revenue.

The history of the establishment of mesh sizes in the CFP shows how elusive this kind of regulation is. In particular, since different meshes are used for different fisheries with different target species and catch compositions, it is a challenge to define which mesh is the legal one for each fishery.

Regulation 171/83 resolved this by defining a minimum mesh size per species targeted, under the notion of the mesh to be used 'when fishing for species x'. As a result, this regulation established minimum mesh sizes for well over 60 species defined individually. The regulation did not cater for the case where two or more of these species, requiring different mesh sizes, would be caught together. This type of definition of the fishery was retained in the next generation of technical measures (Regulation 3094/86).

The problem with this type of regulation was that in front of a judge, it was impossible to demonstrate that a skipper was 'fishing for' a certain species and not for another. This produced a number of cases where offenders were cleared in courts on account of the impossibility of demonstrating what they were 'fishing for'.

To address this question, the Commission changed its approach in the preparation of Regulation 850/98. Instead of defining the fisheries by the expression 'fishing for' each fishery was defined by its catch composition. This gave rise to a new approach where each segment of mesh size must comply with a certain catch composition, reflected by a list of species that should constitute at least a certain percentage of the total catch.

This approach, despite its complexity, is based on a very sound principle: that mesh sizes should be established for fisheries (with a mixture of species in the catch) and not for individual species, as was the case in the initial technical measures of 1983.

Under this scheme, the different fisheries requiring specific mesh sizes are defined by lists of combinations among a number of species constituting a minimum percentage of the total catches. This meant that the notion of 'fishing for' was given a more exact meaning: if a fisherman 'fishes for cod', for example, he must do so by using a large mesh size that will normally result in his catch composition being dominated in a high percentage by species of relatively large size. If this is not the case, then fishermen must discard to ensure that their catch composition is according to the rules for that mesh size.

This new approach resolved the earlier problem. But it introduced a new one: the additional complexity of the regulation. For example, Regulation 850/98 includes tables on catch composition for different mesh size categories for the main Atlantic areas and the main type of gear. The table corresponding to towed gears in the North Sea considers seven different mesh size ranges with different percentages of catch composition for over 50 species. This table is reproduced in Table 4.3 as an example.

This complexity is aggravated where vessels are authorised to carry on board more than one mesh size, in order to be able to conduct fishing for different target species during the same fishing trip. In that case, Regulation 850/98 requires that a combination of the catch composition of the two groups of target species be calculated: something far from being straightforward and easy to enforce. This complexity is due to the fact that Council refused to accept the notion of the 'one-net rule' that is the obligation to use only one mesh size per fishing trip.

In the Mediterranean, as we will see in Chapter 8, mesh sizes were the first management measures to be adopted, in 1994. Together with minimum landing sizes they have constituted the bulk of the fisheries management instruments in this basin until recently. But as we will see in more detail in Chapter 8, the first regulation of 1994 actually aimed to consolidate existing rules and made no attempt to actually increase selectivity in Mediterranean fisheries. The first occasion where real increases in selectivity were established was in the adoption of the regulation of 2006.

Mesh sizes are always difficult and contentious. The reasons are simple: by modifying selectivity, catch rates can be considerably reduced and, in some cases, certain species can be almost lost. The best-known case is that of the selectivity of the flatfish fishery for sole and plaice with beam trawls in the southern North Sea.

Sole is more streamlined than plaice. The mesh size of 80 mm used in this fishery retains large amounts of plaice below the minimum legal size, forcing large discards. But if the mesh size is increased, then most of the sole (including relatively large ones) would not be retained in the net and, consequently, the fishery would not be economical, since sole is by far the main source of revenue in this fishery. For this reason, a significant increase in the mesh size of this fishery has never been acceptable for the Member States concerned.

This example is repeated in many other fisheries in Europe and that makes mesh sizes (especially for mixed-species trawl fisheries) extremely difficult to change. In addition, the effects of any substantial change in mesh sizes on the economic performance of the individual vessels makes the adoption of these 'technical' rules highly controversial and political, resulting in a tradition of these measures being decided by ministers, thus contributing once again to the image of the CFP as a policy 'micromanaged' from Brussels.

In addition, mesh sizes are a very imperfect way of regulating selectivity. For a given mesh size, selectivity can also be determined by fishermen's behaviour: where and when to fish, and how to rig the net. So the general rules on selectivity are only part of the decisions driving the selectivity of individual vessels.

Mesh sizes in the new CFP

As we will see in Chapter 17, the CFP reform in 2013 should in principle change quite dramatically the way the CFP handles mesh sizes, for two reasons:

1 The introduction of the discard ban, which creates a strong incentive for the selectivity of the fishing activity to be improved, so as to minimise the amount of unwanted fish in the catch and,
2 The regionalisation process that will allow the possibility for results-based approaches with a decisive intervention by stakeholders is now possible under the CFP.

These developments, however, will require a process of confidence building among the different actors, as we will refer to in Chapter 17.

Table 4.3 Tables of mesh sizes for area in Regulation 850/98. © European Union. COUNCIL REGULATION (EC) No 850/98 of 30 March 1998 for the conservation of fishery resources through technical measures for the protection of juveniles of marine organisms. *OJ* L 125, 27.4.1998, p. 1.

Mesh size / **Species**	**1111111**	**40 to 49 mm**	**50 to 59 mm**	**60 to 79 mm**	**80 to 99 mm**	**>100 mm**
Sardine (*Sardina pilchardus*)	X	X	X	X	X	X
Shrimps (*Palaemon* spp.)	X	X	X	X	X	X
Rainbow wrasse (Coris *julis*)	X	X	X	X	X	X
Bogue (*Boops boops*)	X	X	X	X	X	X
Shrimps (*Penaeus* spp.)		X	X	X	X	X
Mantis shrimps (*Squilla mantis*)		X	X	X	X	X
Red mullets (*Mullidae*)		X	X	X	X	X
Wedge sole (*Dicologoglossa cuneata*)		X	X	X	X	X
Wrasses (*Labridae*)		X	X	X	X	X
Horse mackerel (*Trachurus* spp.)			X	X	X	X
Mackerel (*Scomber* spp.)			X	X	X	X
Bib (*Trisopterus luscus*)			X	X	X	X
Cuttlefish (*Sepia officinalis*)			X	X	X	X
Gurnards (*Triglidae*)			X	X	X	X
Sea breams (*Sparidae*)				X	X	X
Rockfish (*Scorpaenidae*)				X	X	X
Eyed sole (*Microchirus acevia*)				X	X	X
Shortfinned squids (*Ommatostrephidae*)				X	X	X
Conger (*Conger conger*)				X	X	X
Forkbeard (*Phycis* spp.)				X	X	X
Brill (*Scophtalmus rhombus*)				X	X	X
Weevers (*Trachinidae*)				X	X	X
Picarels (*Centracanthidae*)				X	X	X
Sea bass (*Dicentrarchus labrax*)					X	X
Whiting (*Merlangius merlangus*)					X	X
Turbot (*Psetta maxima*)					X	X
Pollack (*Pollachius pollachius*)					X	X
Flounders (*Pleuronectidae*)					X	X
Sole (*Solea vulgaris*) (1)						X
Hake (*Merluccius merluccius*) (1)						X
All other marine organisms (2)						X

(1) In ICES Divisions VIIIc and IX the minimum mesh size will be 60 mm. However, with effect from 31 December 1999, the minimum mesh size will be 80 to 90 mm.

(2) Catches of anglerfish (Lophius spp.) and retained on board in excess of 30% of the total catch on board must be taken with minimum mesh size of 220 mm or greater.

Other gear measures

Many other devices have been regulated under the CFP for different purposes. In particular, one of the main problems in fisheries management is the need to treat differently the catches of species that are caught together by the same gear. Over and beyond the sole and plaice case explained previously, many other cases justify such species selectivity measures. For example, fishing for crustaceans such as Norway lobster or shrimp necessitates small meshes in the cod-end. But these small meshes have a low selectivity for the fish that is caught together with the targeted crustaceans. In these cases, it is necessary to ensure that vessels targeting the crustaceans have a way to exclude the fish from the catch. This is commonly done for example through the use of grids, rigged in front of the trawl net, or simply by using a trawl with low headline height.

In some cases, however, the need and opportunity to set up these grids is economic, rather than biological. The best example is that of the so-called 'Swedish grid' which, when rigged in trawl nets fishing for Norway lobster, can prevent the catch of cod with a very high efficiency. Thanks to this characteristic, vessels equipped with these grids were exempt from the effort limitations established under the cod recovery plan.

However, this was possible only because the fishermen concerned could make enough profits by catching and landing only Norway lobster, without accompanying round fish. When the idea of the grid was presented to fishermen from other Member States, the idea was rejected, not on technical grounds but just in purely economic ones: the Swedish grid (or any other similar one) was just fine technically but it would simply remove catches of whitefish, which represented 50% of fishermen's income in the fisheries concerned. In these circumstances, the use of the Swedish grid has remained restricted to very few cases where the economics of a clean Norway lobster fishery are good enough.

A very specific case: the 'pulse trawl'

The 'pulse trawl', developed in the Netherlands, is an alternative to the fuel-guzzling beam trawl fishery. It replaces the traditional system of the beam trawl, which scratches the sandy sea bed with a set of chains, forcing the flatfish to jump out of the sand and be caught by the trawl net. This produces a very strong drag that results in very high fuel consumption.

By replacing the chain mat by an electric pulse, the fish jumps as a result of the electric discharge, so the gear can slide over the bottom and achieve similar catch rates with lower drag and slower towing speed, thus allowing for a very substantial (up to 50%) reduction in the fuel consumption. These advantages led the Netherlands to ask for the authorisation for this type of fishing.

However, the method had a problem: 'electric fishing' is prohibited in the technical measures regulation (850/98). Authorising the pulse trawl implied a derogation from that well-consolidated principle. Here again, the tendency of the CFP to stick to its *status quo* made it extremely difficult to agree to such derogation. In the context of the TAC and quota regulation an initial allowance for 5% of the fleet to use it on an experimental basis was allowed.

Different requests from The Netherlands to increase the number of vessels authorised to use this technique met considerable resistance from many Member States and Parliament, despite the advantages of this kind of fishing: beyond low fuel consumption, the disturbance of the sea bottom is much reduced and the amount of unwanted catches of fish and invertebrates is considerably lower. However, allegations that the pulse can produce spine damage and other negative effects (ulcerations) in the fish have turned the use of this fishing technique very controversial, not only in terms of the unintended mortality they may produce on fish, but also on account of animal welfare considerations.

Scientists have intensively been studying these effects in recent years, and while the advantages are clear (lower fuel consumption, enhanced selectivity, lower seabed disturbance) the controversy on side effects remain. By contrast, the effects of other fishing techniques on fish survival or fish welfare have never attracted so much attention.

Beyond its objective side effects on fish, it seems clear that this debate encapsulates the permanent tension in the CFP between innovation and the preservation of the status quo. The next generation of technical measures will be an opportunity to see what the political choice is of the European co-legislators to resolve this dilemma.

Minimum landing sizes: a difficult compromise

Minimum landing sizes (MLS) are a traditional conservation instrument, not only in the CFP (where they have been an instrument since 1983) but also in practically all fisheries management systems around the world. They have a considerable advantage over other instruments: they are easy to understand, even at the level of non-specialists.

The technical measures always require a very difficult technical (and often political) compromise. Ideally, the minimum landing size should correspond to the size at which the yield-per-recruit curve provides maximum yield. That is, if too many small fish are caught, the proportion that will reach large size will be small and the overall yield in weight will be low. If the size is too large, then a too high proportion of the fish will not be caught and the yield will also be low. By avoiding the catch of the small fish at a certain limit the yield of the stock by each generation of fish can be maximised.

At the same time MLS should also be based on the selectivity of the legal mesh sizes: it would not make sense to have a legal mesh size catching very large amounts of illegal fish but it is not always possible to adapt mesh sizes to biological parameters of the fish due to the very different size selectivity of different gear catching a given species. And the selectivity of the gear is, as we saw previously, the result of a compromise among the different species associated with the catch.

An additional complexity results from the regional differences for the same species in different areas and fisheries in Europe. Given the geographical and technical diversity of species and fishing gear, the MLS always represent a difficult balance between the need to harmonise and the need to take into account regional differences.

As a result, there are many factors coming into play when establishing MLS:

- The life history of the species, particularly the growth pattern. This may be geographically very variable, thus justifying the existence of different MLS for the same species in different areas. Typically, MLS for the Mediterranean are different than those for the Atlantic. For example, hake has a MLS of 27 cm in the Atlantic and 20 cm in the Mediterranean on account of different growth rates and ages of maturity in both areas. Similar cases include horse mackerel (15 cm in the Atlantic, 11 cm in the Mediterranean).
- The same species being caught by very different gear, all having very different selectivity. The same species can be caught with trawls, with low size selectivity and by long-line, with very high selectivity.
- The same species being accompanied in the catch with very different by-catches: cod in a mixed-gadoid fishery and cod in a *Nephrops* fishery.

The MLS have traditionally been accompanied by a policy of compulsory discarding of any undersized fish that may be caught. Although the practical implementation of the MLS has never been perfect (see next) it is true that the complete ban on the catching, transport, consumption and marketing has a considerable advantage: it is understood by fishermen and consumers alike and it creates an incentive to target larger fish.

The MLS are also applicable to imported fish: imports of the species subject to MLS cannot be placed in the EU market. This ban, however, does not apply to other species, no matter how similar they are, that are not strictly subject to EU conservation measures (see next).

The scope of MLS

The species subject to MLS in the CFP have traditionally not coincided with those managed by TACs and quotas. There are several reasons for this discrepancy:

- The Mediterranean is an obvious case: there are no TACs in this area. However, the need to protect juveniles and ensure size-selective fishing there has justified the existence of MLS in this basin since the first regulation of 1994.

- It has been possible to agree on a MLS for certain species for which, in turn, a TAC has not been agreed notably on grounds of relative stability. The examples of sardine (*Sardina pilchardus*) with a MLS in the Atlantic but no TAC is the best example. Another one in exactly the same situation is sea bass (*Dicentrarchus labrax*).
- Even certain inshore or coastal species (largely used also in aquaculture or semi-aquaculture practice) such as bivalve molluscs, also have a MLS even though they are entirely outside the TAC system and are managed locally. The existence of MLS at European level is due to the fact that there is a real European market for these species. If different Member States apply different MLS, then the local or regional efforts to conserve the resources can be seriously undermined by imports of the same shellfish from other regions or Member States where different rules apply.

The MLS are thus established with one common general goal, the protection of juveniles for the improvement of yields but with very different approaches depending upon species, areas and gear.

MLS and the new CFP

In Chapter 17 we will also refer to the important changes introduced in the notion of MLS as a result of the 2013 reform. Briefly:

- Their name changes to Minimum Conservation Reference Sizes, because the 'landing' size will no longer have a meaning: all juvenile fish will have to be landed.
- They will not be allowed to enter the human consumption market but can be sold for other purposes.
- They will have to be counted against quotas. By counting undersized fish against quotas while not allowing their sale for direct human consumption, the policy will deter fishermen who could lose out if they use part of their quotas in catching small, low-value fish.

A fundamental question related to this is that, in a future context of regionalised measures, the possibility of having different minimum sizes in different areas (which already exists) could explode into a myriad of different rules in different corners of the Union, something that would pose serious difficulties for enforcement and even for the overall 'readability' of the policy by fishermen. We will refer to this in Chapter 17.

Closed areas/seasons as technical measures

While mesh sizes have the clear objective of improving the size selectivity of the gear and while minimum landing sizes aim at discouraging fishermen from catching juvenile fish, the objectives of closed areas/seasons are indeed very varied: protection of juveniles, protection of spawners, protection of habitats and even protection of the food source of protected seabirds. However, in a number of cases the areas adopted in the 1980s and 1990s do not have a clearly defined objective, as we will see next.

Closed areas to protect juveniles

This is the most common type of closed area. As a specific example, the need to ensure protection of juvenile hake around the Iberian Peninsula has resulted in the adoption of two closed areas off North western Spain, where trawl fishing is forbidden during several months of the year (Article 28 of Regulation 850/98). This type of measure can be very effective on one condition: that the areas of high concentration of juveniles are well-known and stable. This is not the case of all stocks and similar attempts to find such areas for North Sea cod have failed because the areas of juvenile concentration are indistinct and very variable over time; thus the need for different approaches such as the real-time closures.

A special case is that of the so-called 'plaice box' in the North Sea (Article 29 of Regulation 850/98). Originally intended to protect juvenile plaice in an area off the coast of the Netherlands, Germany and Denmark, this measure evolved into a fully-fledged management system of access depending upon vessels engine power, including complex derogations and with access granted only to a specific list of vessels, and is of the best examples of the 'micro-management' characterising the CFP.

The plaice box was evaluated by the Commission in 2005 (EC, 2005e). The Commission then concluded that while the plaice box should continue to apply, a revision was necessary to ensure that it would be fit for purpose, in terms of the objectives, instruments and its relationship with the management of plaice. Such revision, however, did not take place. As a result, the plaice box has become a prime example of the complexity of the technical measures under the CFP.

Closed areas to protect habitats or endangered species

In Chapter 11 we will examine in some detail the use of closed areas under technical measures to protect certain fragile sea bottom habitats from fishing (particularly those with cold water coral *Lophelia pertusa* and all the accompanying species (Articles 34d to 34f of Regulation 850/98).

In certain, very specific cases, closed areas have been used to achieve other environmental objectives. The best case is that of the sandeel closure in the North Sea with the objective of preserving the local sandeel population that constitutes the main source of food for the kittiwake seagull (Article 29a of Regulation 850/98).

Other closed areas

In a number of cases the closed areas have mixed objectives such as protecting vulnerable species. A good example is the limitation of the use of gillnets in deep waters, with the objective of protecting deep-sea sharks (Article 34b of Regulation 850/98).

The case of real-time closures

A specific type of closed areas is the so-called 'real-time closures'. They consist of the temporary closure of a fishing area when, in the light of evidence of high concentrations of juvenile fish, the closure can protect many of these juveniles from fishing and improve the exploitation pattern of the stock(s) concerned. This type of instrument is used in certain countries (e.g. Norway) and is generally considered a very positive one, notably because it is triggered by fishermen themselves and tends to have a high level of industry buy-in. However, the effect of these closures has never been quantified.

Real-time closures are difficult to apply in the CFP. These measures have to work within the 24 hours of the juvenile aggregations being identified. The institutions of the Union cannot take decisions in such a time scale. The alternative has been to implement them through Member States initiatives. These have been taken notably in the North Sea.

The best example is the Scottish 'Conservation credits' scheme, developed in 2008, whereby fishermen exchanged effort limitations for the avoidance of time/areas of cod abundance. The scheme worked well, as fishermen saw this system based on their daily experience in the fishing grounds. The overall evaluation in terms of conservation, however, remains elusive, since these schemes do not reduce effort but simply displace it and it is not possible to compare the fishery with or without such a scheme (European Parliament, 2010).

Evaluation of closed areas

The effectiveness of the closed areas is not necessarily straightforward. In some cases, the closed areas were established to address certain temporary problems but the tendency of the CFP to add new layers without removing old ones is very clear here: most closed areas, if necessary and justified at the moment of their adoption, may have partially or totally lost their effectiveness over time. A scientific report by the STECF in 2007 was on the whole unable to recognise the real usefulness of most closed areas in the CFP (SEC, 2007). Among other reasons scientists indicated that it was difficult to evaluate their effectiveness when the very objective of the closed areas was not clear in the first place (at least today). In some other cases, scientists noted that whilst the closure of areas may have been necessary at the time they were established but their need and opportunity were not questioned ever since, even though the circumstances may have changed substantially.

Even in the absence of evidence of their effectiveness, STECF recommended that a number of these areas be maintained. In other cases, STECF recommended that they be maintained but that their conditions are adapted. Finally, in other cases, particularly as regards the closures for herring, STECF saw no added value in the closures and recommended that they be removed.

This problem illustrates again two of the shortcomings of the CFP: the tendency to accumulate measures, always adding new layers without questioning the existing ones (always considered part of a delicate 'political balance') and the difficulty of tracing the past, if and when the few officials involved in their establishment retire. This clearly militates in favour of a new approach to these measures, where the justification and scientific underpinning are clearly established and explained in the regulations and where their effectiveness is regularly checked against its objectives.

Closed areas as a mainstream management instrument

As we saw earlier, closed areas/seasons have a variety of objectives. In some cases, closed areas have been proposed (e.g. by certain NGOs) as the main instrument for fisheries management, replacing the existing ones.

In 2005, in view of the difficulties in accepting the dramatic reductions of fishing effort of the cod recovery plan, particularly for those fleets catching cod as a secondary target, the Commission examined the possibility of establishing closed areas in the North Sea where catches of cod would be highest so as to allow for a better targeting of the necessary reductions in the fishing mortality for this species, while not leaving other fisheries as undisturbed as possible.

The Commission developed a number of scenarios for closed areas in the North Sea, corresponding to the statistical rectangles where the highest catches of cod were reported. Closing these areas would have allowed a selective reduction of fishing mortality for cod, while the effort deployed targeting other species would not be significantly hampered by these areas.

The Commission proposed a large closed area which shrank considerably in the negotiations in Council. This small area was in force for a year, until STECF clearly indicated that, given its relatively small size, fishing outside the area compensated its effects within the area concerned, so the measures were useless in limiting the total fishing effort.

The Commission has never again proposed this approach as part of the long-term plans.

In the Mediterranean the picture is different. In the absence of TACs, closed areas of different kinds have also been considered a potential instrument. However, it was not until 2006 that the new Mediterranean regulation (Article 6 of Regulation 1967/2006) enshrined this instrument by requiring Member States to establish:

> ...the Council shall designate, within two years from the adoption of this Regulation, fishing protected areas occurring essentially beyond the territorial seas of Member States, concerning the types of fishing activities banned or authorized in such areas.
>
> The Council may subsequently designate other fishing protected areas, or change their delimitation and management rules established therein, on the basis of new relevant scientific information.

Today this kind of instrument is used only as a supplementary measure for the purpose of protecting juveniles, spawners or marine habitats but not as the only management system. Whether or not this will change, this will depend on the follow up to the decision, in the 2013 CFP reform (EU, 2013d), to require the Union to: 'endeavour to establish protected areas on ground of their biological sensitivity, including areas where there is clear evidence of heavy concentration of fish below minimum conservation size and spawning grounds'. The formulation is not constraining, so it is difficult how many areas and of which size, may be eventually be established and what their contribution as an instrument for the CFP in the future will be.

Amending technical measures

The complexity of the technical measures and their tendency to grow by accumulation has led to these measures being perhaps the most typical case of over-regulation within the CFP. Technical measures (particularly for the Atlantic) have often been used as the ultimate example of Brussels' complex over-regulation that few can actually understand, let alone implement. Every time the Commission took an initiative on administrative simplification, the technical measures featured among the priorities. This situation has been well known for years. Yet, finding a solution has proven extremely difficult.

In the early 2000s, the Commission introduced a proposal to simplify the rules of regulation 850/98 and its numerous amendments into a consolidated, clearer version. This proposal was discussed a number of times in the technical working groups in Council. But it was never adopted. At the time already, a surprisingly low number of national administrations actually had the knowledge and expertise to understand the rules. There was also a fear of disturbing a situation where fishing sectors felt their rights were maintained and the complexity could be blamed on 'Europe'.

Another, different attempt was made a few years later. Conscious of the difficulties of discussing the current texts, the Commission decided to take a new, fresh approach on technical measures based on the new ideas emerging at the time: environmental awareness, de-centralisation, simplification and results-based management (COM, 2004). This new approach was warmly welcomed by Member States.

The proposal that followed was put out to consultation with stakeholders in the relevant RACs, and its principles were generally welcomed and agreed by the relevant RACs. It contained innovative approaches, such as regionalised measures, much simplified rules to establish mesh sizes, a reduction of minimum landing sizes to those really meaningful in conservation terms and a general administrative simplification. But its merit was also its major flaw: it was too innovative. It made proposals for simplification and regionalisation that Member States in Council were not prepared to accept. In particular, the proposal's region-specific rules, with a considerable degree of simplification and with fast-track procedures to adapt them quickly to new developments and/or new proposals coming from the RACs was seen by Member States as a power-grabbing exercise by the Commission. No amount of explanation by the Commission services could dissipate such a false impression.

At that point in time and before the fully-fledged discussion on regionalisation and de-centralised decision-making as part of the 2013 reform, this proved too revolutionary and was not accepted. After several unsuccessful attempts, the Council discussions derailed and the old regulation 850/98 remains to this day fundamentally unchanged.

This example also indicates that, although the complexity of the rules is always subject to complaints by the industry and Member States alike, when the opportunity comes to simplify them, they do not necessarily contribute to do so, either because they wish to preserve certain specific derogations in their own interest, or on inter-institutional grounds (not giving too much power to the Commission).

Another relevant question in this context is that, once the Treaty of Lisbon had entered into force, the Union lost its main instrument to update technical measures quickly. For many years, the Union used fast-track procedure to adopt new technical measures into EU law: the annual regulation on TACs and quotas. This regulation was not supposed to be used to adopt technical measures but it did allow the adoption of measures intrinsically related to the establishment of TACs. The TAC and Quota regulation incorporated new technical measures that were then applied provisionally for a year.

For example, in the TAC and quota regulation for 2008 (EC, 2008d), Annex III contained 17 pages of provisional technical measures (including for example Natura 2000 areas) pending their consolidation in permanent regulations.

Many technical measures were applied year after year with this method. But the Treaty of Lisbon, in its further clarification of what can and what cannot be adopted by Council alone as related to annual TACs (under Article 43(3) TFEU) closed this route. Now, any modification/updating of technical conservation measures must undergo a full co-decision procedure. This obviously hampers any attempt to adapt quickly the technical measures as necessary.

The reform of the CFP in 2013 provided an ideal opportunity to solve this problem, by allowing technical measures to be adopted under regionalisation. However, the co-legislators decided not to include in principle the technical measures as part of the regionalisation process. As the need to have technical measures quickly adopted through fast-track procedures is becoming evident again that future initiatives will have to be taken in this regard.

Implementing technical measures

The implementation of technical measures has been, perhaps like catch limitations, considerably less than perfect. But a distinction can be made between mesh sizes and closed areas, where implementation was generally better, and minimum landing sizes, that have always had particular difficulties for effective implementation.

Mesh sizes, once the notion of 'fishing for' was replaced by catch compositions, have been a favourite of control authorities. On-board inspection is always very effective to control these provisions because, unlike catch limitations, compliance can be clearly established by on-the spot inspection of individual vessels.

With closed areas the problem is somehow similar: it is easy to control the behaviour of individual vessels, at least for these equipped with satellite-tracking systems (vessel monitoring systems, or VMS). There is one caveat, though: since the VMS emits a signal every two hours, the position of the vessel cannot be ascertained between these two signals, thus allowing a vessel to fish in a closed area without being detected by that system.

A different question is, however, the enforcement of the minimum landing sizes.

Minimum landing sizes

Theoretically, enforcing the minimum landing sizes looks straightforward: these are rules that can be unequivocally measured and that have a high degree of understanding by fishermen and by society at large. However, the implementation of these measures continues to be a challenge for the CFP. Why?

There are several reasons for this:

- The existence of traditional and lucrative markets for the small fish in a number of Member States, these markets being deeply-rooted in local or national culture. Anyone visiting southern Europe will be tempted by a delicious platter of small fried fish.
- The difficulty of avoiding catching small fish even with legal gear, particularly for the less selective gear and the larger bodied species. This makes the catching of some amounts of small fish inevitable. Once caught, it is not easy in fishermen's culture to throw back to the sea what is already caught and dead.
- The existence of certain fish that are small in their adult size and can legally be consumed produces confusion in consumers and sometimes inspectors.
- The differences in MLS among different European regions and the difficulty and burden of ensuring adequate traceability of the origin of the fish in the end markets
- In some specific cases, the effects of imported fish, such as for example, the Canadian silver hake.

A very illustrative example is that of horse mackerel. In the early 1990s, this species had a minimum size of 15 cm in the Atlantic and 11 cm in the Mediterranean. When the Regional Government of Galicia, in the early 1990s, enforced the 15 cm in Galician ports, including the seizure of important quantities of undersized horse-mackerel, lorries started to arrive to the Galician markets with horse-mackerel from Mediterranean ports, with sizes of around 12 or 13 com, illegal for Atlantic fisheries but allowed in the Mediterranean. The arrival of this fish in the Galician markets completely undermined all the efforts made to enforce the 15 cm.

This is also aggravated by the fact that in southern Europe and particularly in the Mediterranean, there are many species that are small in their adulthood which can be marketed legally. This only makes it more difficult for the average consumer, who may not know whether the small fish they find in their dish is a legal one or an illegal juvenile. An illustration of this problem is a species such as wedge sole (*Dicologlossa cuneata*) which reaches sexual maturity at a size of 17–19 cm and can be legally caught and marketed as an adult fish at that size, while the common sole (*Solea solea*) reaches maturity at 30 cm and cannot be marketed at sizes lower than 24 cm. Not surprisingly, most consumers cannot tell the difference between a – legal – wedge sole from a very similar – illegal – juvenile common sole when dealing with a fried fish platter containing small sole at their dinner table.

In addition, the minimum landing sizes established by EU law can be applied to imported fish of the same species but not to similar species not subject to technical measures in the EU. An illustrative example is that of the regional government of Galicia that, in the early 1990s, made considerable efforts to stop the marketing of small hake (*Merluccius merluccius*) caught in coastal waters. The success of this policy led to the creation of an empty niche in the market that at the end was rapidly filled in by a similar species, *Merluccius bilinearis* imported from Canada and not subject to any restriction of size since the species is alien in Europe and is not subject to any enforceable minimum landing size.

The imported hake was indistinguishable from the European variety for most consumers, who did not understand how, after so much campaigning against consuming juveniles, they could still find small hake being sold legally with the acquiescence of authorities. Let alone fishermen who could not market their small hake only to see their markets filled with small imported fish. This means, clearly, that in order to have an effective, enforceable policy against the fishing of juveniles, it is necessary that the policy applies also to imported fish.

This is clearly possible for the species subject to MLS in European legislation. However, it would be against the rules of the World Trade Organisation to extend the ban on species that are not subject to MLS, even when they clearly resemble the European species, at least in the eyes of the average consumer.

The future of technical measures

Technical measures will remain a fundamental instrument of the CFP. However, they will have to look very different in the future. In effect, the technical measures have always been the prime example of the 'micromanagement from Brussels' that has characterised the CFP for so many years. Unfortunately, as we saw previously, earlier attempts to simplify this policy pillar have been unsuccessful.

In the future, the regionalised CFP will provide an opportunity to change the approach fundamentally and legislate on technical measures from the perspective of applying the new paradigms of the CFP: regionalisation, simplification and results-based management.

Whether or not this will be possible will depend on one key factor: trust. First and foremost, trust among Member States. Secondly, trust between national administrations and the industry. And last but not least, trust among the European institutions.

Trust among Member States is likely to happen in certain areas like the Baltic but is much more difficult to achieve in certain other areas. In the absence of trust among the actors, the general response will be to try to specify the rules as precisely as possible in order to ensure that nobody abuses the system. In the coming years we will see if the need for simplified rules for fishermen really outweighs the lack of trust on others, in what constitutes possibly the best test of the ability of the CFP to bring about real simplification.

Beyond that question, there is another, perhaps more fundamental challenge to make technical measures effective, in particular in relation with their contribution to the discard ban. Technical measures (particularly gear measures) are relatively easy to circumvent by experienced fishermen. Besides, if properly implemented, they can produce short-term reductions in catch rates that would normally result in lower income levels. In these circumstances, effective enforcement becomes an extremely difficult task, since there is no incentive for fishermen to apply these rules in practice (Enever et al., 2009).

On the other hand, it is well-known that one of the main factors than influences selective fishing is fishermen's behaviour but this is something that, by its own nature, it is practically impossible to legislate. Consequently, a future scheme of technical measures should include the right incentives for fishermen to *want* to fish in a more selective way. These incentives could be of very varied nature, regulatory or not. Certain authors suggest that certification systems that would provide market advantages to those fishing more selectively and reducing discards would constitute an important incentive (Johnsen and Eliasen, 2011). In any case, the search for these incentives should be an essential element in any future scheme on technical measures that will replace the current detailed, prescriptive, top-down approach that the CFP has traditionally applied.

Is the CFP conservation policy a success?

The slow progress of the conservation policy

For many years, it has been a common belief that the conservation pillar of the CFP was a failure. To be fair, the Commission itself has contributed to this

Table 4.4 Summary of stock status over the last 10 years. As reported by the Commission in the 2014 Policy Statement(COM(2014)388 final). © European Union. Communication from the Commission to the European Parliament and the Council. Concerning a consultation on Fishing Opportunities for 2015 under the Common Fisheries Policy. Brussels, 26.6.2014 COM (2014) 388 final.

	Number of fish stocks											
Advice for year:	**2003**	**2004**	**2005**	**2006**	**2007**	**2008**	**2009**	**2010**	**2011**	**2012**	**2013**	**2014**
Stock with MSY assessments			34	23	32	33	35	39	35	38	41	46
Overfished stocks			32	21	30	29	30	28	22	18	16	19
The stock is fished at the maximum sustainable yield rate			2	2	2	4	5	11	13	20	25	27
% of stocks overfished			94%	91%	94%	88%	86%	72%	63%	47%	39%	41%

belief quite substantially, by being openly self-critical about the policy performance in an exercise of transparency that is difficult to find in other fishery management systems. In the context of the evaluations of the policy prior to reform, the Commission has always stressed the insufficient progress of the CFP in achieving its objectives of fully sustainable and productive fish stocks. In fact, as recently as 2009, the Commission indicated, in the context of its Green Paper on the reform of the CFP (COM, 2009a), that:

> ...88% of Community stocks are being fished beyond MSY: this means that these fish populations could increase and generate more economic output if they were left for only a few years under less fishing pressure. 30% of these stocks are outside safe biological limits, which means that they may not be able to replenish.

The perception, however, has been changing. The objective, independently documented improvement of the general state of European stocks and the transparency introduced every year by the Commission through the 'Policy Statement' paints a more optimistic picture.

The Policy Statement published by the Commission in 2014 on annual fishing opportunities for 2015 summarises the evolution of stock status over the last 10 years (COM, 2014). Some figures (reproduced in Table 4.4) are extremely illustrative:

The proportion of overfished stocks has been reduced from 94% in 2005 to 41% in 2014. The number of stocks inside safe biological limits has increased from 29% in 2003 to 55% in 2014. These data clearly show a spectacular improvement in stock status. Of course, these data are fluctuating annually but the trend is now clear enough to indicate a net improvement on the conservation status of European stocks in the last decade. Figure 4.3 shows the growing number of stocks not being considered overexploited in recent years.

What is the reason for this recent change in the state of the stocks? To understand it, we have to go back in time. Figure 4.4 shows the total fishing mortality exerted on the main three types of fish stocks in Atlantic fisheries in recent decades. Beyond the exceptional developments in the pelagics around 1970, the trend is very consistent among the three groups of stocks: an increase in fishing mortality since the 1960s that continued through the 1970s. The short reduction in fishing mortality in the late 1970s can probably be explained as a result of the declaration of EEZs in 1977 and subsequent reduction of foreign fishing in the newly declared EEZs. But such effect was rapidly compensated by another increase in fishing activity by European States in the subsequent years, and this went on through the 1980s and 1990s despite the establishment of the CFP and its conservation and related policies. This trend only started to change at the very end of the 1990s and the beginning of the 2000s.

Little wonder that the CFP has had for a long time a bad image as a policy unable to ensure sustainable

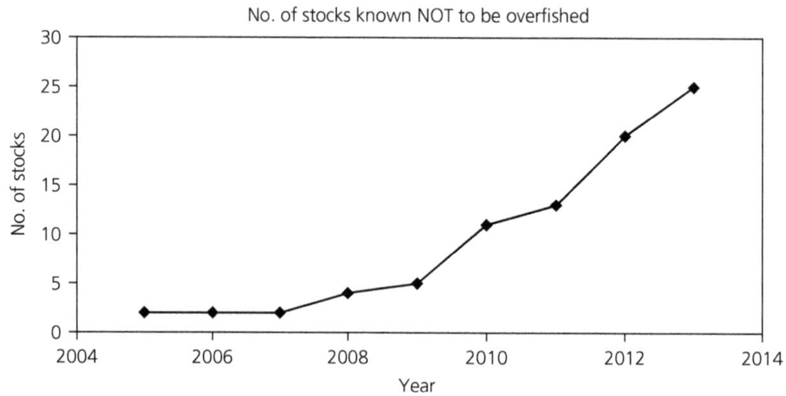

Figure 4.3 Evolution of the number of stocks not overfished in recent years. Source: European Commission Services.

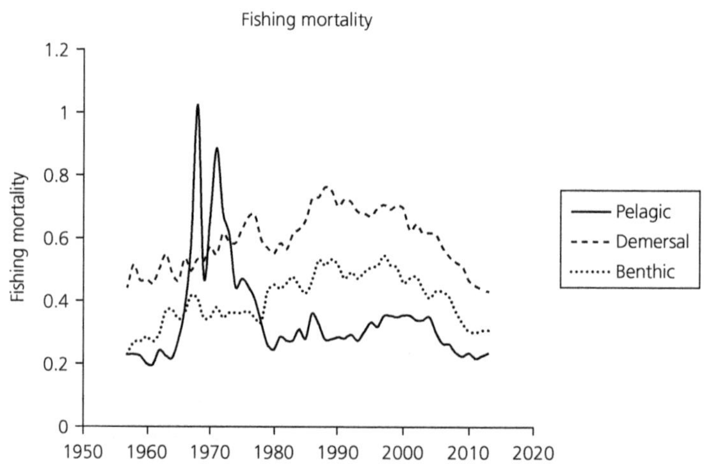

Figure 4.4 Evolution of the fishing mortality in European fisheries in recent decades. ICES (2013). Reproduced with permission. (See colour plate section for the colour version.)

exploitation of European fishery resources. This graph is extremely telling, but it is even better understood if we compare the evolution of the total fishing mortality for all species concerned with the evolution of their abundance, as expressed by their spawning stock biomass. Figure 4.5 shows that comparison, based on a study of 85 fish stocks in Atlantic waters carried out by the International Council for the Exploration of the Sea (ICES, 2013).

This figure shows that, while fishing mortality started being reduced towards the turn of the century, the increase in stock abundance only started to appear clearly around 2007. This figure is extremely important because it encapsulates, better than any other image, what the CFP and fisheries management in general is all about: the efforts to reduce fishing pressure do not give immediate results. A certain time gap is necessary for the resources to show signs of recovery: the number of stocks considered as not overfished only started to increase around 2008, many years later than the start of the reduction in fishing mortality.

This means that long-term planning is absolutely paramount to do fisheries policy. Fisheries management cannot be done by deciding only on short-term considerations, or by expecting immediate

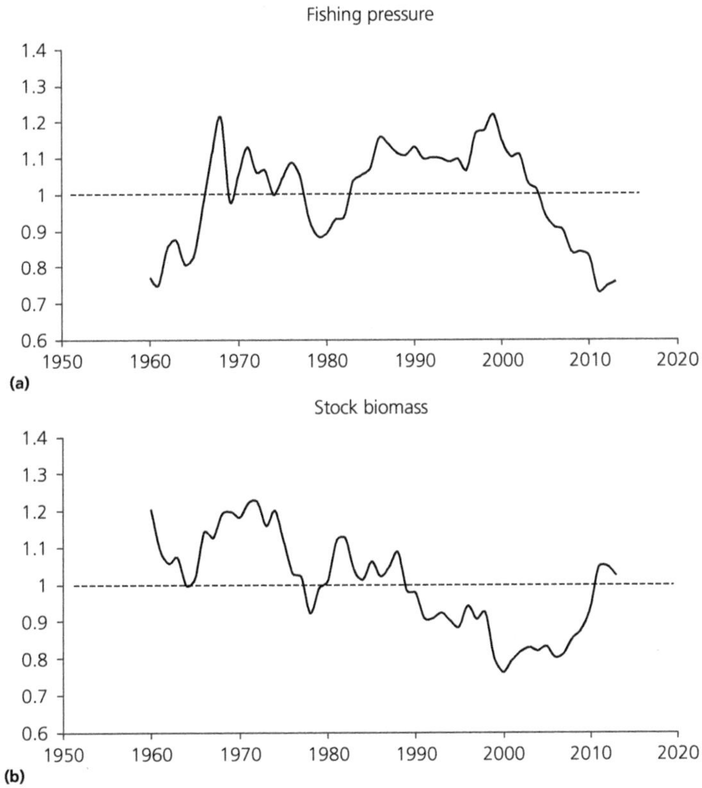

Figure 4.5 (a) Average fishing pressure (measured as fishing mortality) and (b) average stock biomass (right) for 85 major fish stocks in the North East Atlantic, both scaled to the mean over the time period (1960–2013). ICES (2013). Reproduced with permission.

results. Nature takes time to react to good management. Crucially, it also means in political terms that the effort made by fisheries managers may not necessarily show results in time for those efforts to be recognised. While this also happens in other policy areas, perhaps fisheries policy is the best possible example of how policy decisions should be taken with long-term objectives in mind. Trying to obtain short-term rewards does not allow the achievement of the objectives of the policy, as represented by Article 39 TFEU. This is without doubt the most important lesson to be drawn from the implementation of the CFP.

It is also important to underline that these improvements are not just 'good for the fish'. Crucially, they are good for the fishermen. In fact, the economic trends show an improvement in the competitiveness of European fishing in parallel with the improvement of the state of the stocks (see Chapter 6 and European Commission JRC, 2014). More importantly, more specific analysis show what are the economic benefits derived from the recovery of specific stocks under recovery plans. In Chapter 6 we will see some of these trends.

Another question is whether these tendencies are the result of the management alone, or perhaps also the result of economic factors driving the fishing activity itself. Fisheries management may be just one of the factors driving this reduction of fishing mortality. In Figures 4.4 and 4.5 we see the trends in fishing mortality, but it's far from evident that the change of trend operated towards the turn of the century was exclusively (even primarily) the result in changes in the conservation policy. It is even less clear whether these changes were also influenced by other policy pillars of the CFP, such as control, fleet or structural policy, as we will see next.

One can only hypothesise about the reasons for that change towards year 2000. It is true that by then the debate on the collapse of the cod stocks and on the need to recover resources through long-term approaches did create a new awareness about the absolute need for the FFP to start taking the conservation of fishery resources more seriously than ever before. Whether this was the primary reason for that change, it will probably remain a matter for debate.

The Mediterranean

In the Mediterranean the conservation policy has been different, since the main instrument, TACs and quotas, has never applied. As we saw, the 1994 technical measures simply attempted to harmonise the basic rules on selectivity. The 2006 regulation did attempt to constitute a real conservation policy notably through the establishment of management plans based on fishing effort. But their implementation, as we will see in Chapter 8, has been very slow. Those facts, combined with the absence of a systematic evaluation of the conservation status of Mediterranean stocks make the evaluation very difficult.

However, the scientific community has been able in recent years to produce more evaluations of the status of Mediterranean stocks (European Commission JRC, 2013). These evaluations show a worrying picture: the percentage of stocks exploited at MSY level is only 3% for demersal stocks and 0% for pelagic stocks in the western Mediterranean. These percentages are 0 and 25%, respectively, for the central Mediterranean and 0 and 50% for the Eastern Mediterranean.

These data show that while there is a certain proportion of pelagic stocks exploited at MSY levels, the panorama for demersal stocks is that of overexploitation with respect to MSY levels in all demersal stocks except Norway lobster in the western Mediterranean. The slow implementation of the 2006 regulation (see Chapter 8) makes the objective of reaching MSY levels by 2020 at the latest a particularly difficult challenge for demersal fisheries in this area.

The contribution from long-term management plans

Although the reduction of fishing mortality described previously in all likelihood has more than one explanation, it is difficult not to ascribe it largely to the implementation of long-term recovery and management plans.

These have imposed a much better discipline in the setting of annual TACs generally avoiding *ad hoc* decisions on such annual TACs in conflict with the long-term objectives.

The plans have changed the mind-set of both fisheries managers and stakeholders, notably by making them accept more readily the short and mid-term sacrifices necessary to achieve stock recovery.

In any case, there is one very important lesson to learn here. As we saw previously (Figure 4.5), management measures take time to provide results. There are no immediate results of efforts to reduce excessive fishing. This makes it necessary to ensure a long-term framework for decision making. The reductions in fishing mortality as of 1999 or 2000 started showing results towards 2005. This means that fisheries management requires a very long-term perspective by fisheries managers: they have to take difficult political decisions whose benefits are likely to be reaped by their successors. This requires conviction and generosity.

New instruments always need several years to start working properly. During the first years of implementation of long-term plans, the Council often opposed the annual decisions following from these plans, adopted by Council itself. No strategy to recover natural resources can work without a long-term commitment to the long-term objectives and without the necessary patience and determination to overcome short-term scepticism about results.

Linkages of the conservation policy with other policy elements

Although the conservation policy is the 'core business' of the CFP and its only area of exclusive competence, it is certainly not an isolated policy. Much of the development of the conservation policy under the CFP refer to the way in which this policy has related to others.

By the time the conservation policy was established, a structural policy and a market policy existed already and developed further in parallel. However, at this very beginning, there was little in the way of legal or political linkages among the different policy pillars. In fact, there were practically no cross references to the other policy pillars and the development of the CFP consisted, at the beginning, of the parallel but not necessarily consistent development of the different policy strands.

As the CFP developed, the number of linkages among policy pillars has been made increasingly explicit in the legal texts. However, in general terms the mechanisms for ensuring synergies among policy pillars have remained weak or have not been used to their full potential. As we will see in Chapter 15, the reform of 2013 represents a major leap forward in that regard by establishing more clear linkages among policy pillars and by reinforcing the instruments to fully implement such linkages.

Conservation and fleet policy

These two policy pillars are by far the most closely related ones, at least theoretically. In fact, as we will see in Chapter 5, the idea that the fleet policy should aim at establishing fleet capacity levels that are in conformity with available resources was there from the early stages of the CFP. This is in fact the main linkage among policy pillars that has been made explicit in legal text from relatively early in the development of the CFP. Since 1990, with the Gulland report (see Chapter 5) there has been a scientific basis to estimate what would be the basis to ensure that objective.

Another question, however, is how effective fleet policy has been in contributing to the achievement of the conservation objectives. We will discuss this in Chapter 5.

Conservation and structural policy

This is probably one of the weakest links among policy pillars in the CFP. As we saw in Chapter 2, the origin of the structural policy was not related at all to the conservation policy but rather to the need to enhance modernisation and competitiveness of the industry. Much of the initial philosophy of the structural funds was not conducive to achieving conservation objectives, because there were no limits to the development of the factors of production. As we will see in Chapter 6, the application without effective limits of some of the instruments considered traditionally in the structural policy has arguably had adverse effects on conservation.

There has been, however, a potentially powerful instrument to ensure that the structural policy really contributed to achieve conservation goals: the adoption of Member States' Operational Programmes (OPs) for the implementation of the structural policy. The OPs provide an opportunity for the Commission to comment on Member States' programming, to ensure the money is spent in a way that is conducive to the achievement of the conservation objectives.

However, the requirements of the structural legislation did not support a strong stance by the Commission in this regard and the practice has been that the agreements on the OP have been traditionally rather loose in terms of ensuring that Member States used their structural money in a way that would contribute to achieve conservation goals, rather than undermining them. As a consequence, much of the structural money has been used in the past on investments that have undermined conservation goals.

There is perhaps no better example of this lack of linkage than the development of the Bluefin tuna fishery in the Mediterranean. This fishery, already subject to catch limitations since 1994 under ICCAT, saw the increase in the number of purse-seiners built with European structural money in the 1990s. Such development with the aid of structural funds was clearly against the needs of the conservation policy, at the time already calling for restraints in catch levels. Yet, the fleet increased thanks to the use of public funds in a way that contributed to aggravate the problem of the management and control of this fishery. Meaningfully, a number of these vessels had to be scrapped a few years later to ensure proper management of quota levels available. That case was not isolated: similar phenomena occurred in some North Sea fisheries.

Conservation and market policy

This link has traditionally been weak too. A good example is the labelling policy. If labels informed about the status of conservation of fish stocks, consumers could indeed contribute to promote sustainable fisheries. The CFP, however, has not developed certification and labelling systems of this kind, notably in view of the opinion of Member States in Council. As a consequence, private initiatives have taken over.

Today, the Marine Stewardship Council (MSC), as the most developed certification scheme, represents in many markets a fundamental contribution to promoting the consumption of well-managed, sustainable fish. In its annual report 2011/2012, MSC reported 295 fisheries around the world being assessed, of which 147 were certified, 128 were in the assessment process, 10 fisheries withdrew from the assessment process, seven fisheries failed the assessment and three failed part of their wider assessment (MRC, 2012).

The debate about fish certification is extremely important. As public awareness about the need to consume sustainable fish is on the rise (largely due to the very intensive campaigning in that sense by the Commission in the context of the CFP reform of 2013) the debate over the need for an official, public certification scheme of 'sustainability' as opposed to the current system of national or private certification system, becomes even more important than ever and constitutes a key question in the future development of the CFP.

Conservation and external policy

The conservation and the external policy have had a very variable linkage. The bilateral agreement with Norway has been instrumental in developing long-term management approaches that have had an important bearing on the later development of long-term plans at EU level and while certain RFMOs have led the way on sound long-term management of the stocks (e.g. the swordfish and Bluefin tuna plans developed in ICCAT).

However, in other aspects, the external policy has had a very weak link with the conservation policy in EU waters. The main reason is that the external policy (particularly the bilateral agreements) has traditionally privileged the maintenance of the fishing rights and interests of European fishermen rather than the conservation objectives themselves. Moreover, the conservation objectives have been unclear due to the very limited scientific knowledge on the state of resources in many countries. In that context, the determination of the 'surplus' resources that coastal States can allocate to foreign fleets has proved extremely difficult particularly for the resources outside the scientific evaluation of RFMOs.

These surpluses are to be determined by the Coastal State, not by the Union but in practice the Union has been frequently criticised for negotiating fishing possibilities where coastal States were not in a position to evaluate the production potential of their own resources.

This situation, however, has evolved considerably over the years. Gradually, the Union has defended in RFMOs positions increasingly consistent with those promoted in European waters and in the context of bilateral agreements emphasis has been growing on the need to ensure that the agreements are based on the access to 'surplus' resources only, thus requiring better estimates of the productive potential of the waters of the third countries in question.

Leisure fishing

Leisure fishing deserves a special comment. In principle excluded from the CFP, the debate on this kind of fishing has repeatedly surfaced as an issue at different stages.

The general assumption is that the effect of leisure fishing on European fish stocks is secondary. In addition, all Member States have national or regional rules to manage this activity. European legislation has generally been considered both unnecessary and inappropriate.

However, the issue is frequently brought to the attention of decision makers, generally when the industry considers that the level of catches by leisure fishing is such that it has an effect on the overall status of the stocks. This has resulted in certain cases in calls for some kind of limitation to this activity.

The best example of this intervention was the preparation of the Mediterranean regulation of 2006, where the industry denounced a massive use of professional gear by leisure fishermen. This in fact led to the ban of certain kinds of fishing gear (considered 'professional gear') in Regulation 1967/2006 (see Chapter 8).

The issue was again re-discussed on the occasion of the adoption of the control regulation in 2009 (see Chapter 9). Here, the tightening of the control of the professional fishing activity led to the consideration of a mechanism to address possible problems associated to leisure fishing. The mechanism agreed was to consider possible measures under the CFP only in cases where the total catches estimated of leisure fishing would be significant and would have an impact on the conservation status of stocks.

In the international sector, the influence of leisure fishing can be considerable, particularly in the case of ICCAT, given that certain tuna and tuna-like species are highly appreciated as game fish. In particular, the different species of billfishes are the main target of leisure fisheries in North America and the Caribbean and measures to manage them in ICCAT clearly reflect the priorities of leisure fishermen.

Last but not least, leisure fishermen have been recognised a status as stakeholders in fisheries management under the CFP: their associations are eligible as members of the stakeholder consultation bodies established in the CFP reform of 2002: the Regional Advisory Councils (see Chapter 12).

Notes

1 http://info.ices.dk/products/AnnualRep/2003annualreport.pdf.
2 http://info.ices.dk/products/AnnualRep/2003annualreport.pdf.
3 www.ices.dk/sites/pub/Publication%20Reports/Advice/2014/2014/cap-bars.pdf.
4 www.ices.dk/sites/pub/Publication%20Reports/Advice/2013/Special%20requests/EU_northern_hake.pdf.
5 www.ices.dk/sites/pub/Publication%20Reports/Advice/2014/2014/hke-soth.pdf.
6 www.ices.dk/sites/pub/Publication%20Reports/Advice/2014/2014/sol-nsea.pdf.
7 http://ices.dk/sites/pub/publication%20reports/advice/popular%20advice/sol-bisc_popular.pdf.

References

Boncoeur, J. and B. Mesnil (2000). Quelle politique de la pêche ? L'exemple de l'Union Européenne. *Problèmes économiques*, **2650**, 2 février 2000, p. 5–9.

CES (1992). Cooperative research report No 193. Report of the ICES Advisory Committee on Fisheries Management. Copenhague.

COM (2001). Proposal for a Council regulation establishing measures for the recovery of cod and hake stocks. Brussels, 11 December 2001. 724 final.

COM (2003). Proposal for a Council regulation establishing measures for the recovery of cod stocks. Brussels, 6 May 2003. 237 final.

COM (2004). Communication from the Commission to the Council and the European Parliament. Promoting more environmentally-friendly fishing methods: the role of technical conservation measures. Brussels, 21 June 2004. 438 final.

COM (2006). Communication from the Commission to the Council. Fishing opportunities for 2007. Policy Statement from the European Commission. Brussels, 15 September 2006. 499 final.

COM (2008). Communication from the Commission. Fishing opportunities for 2009. Policy statement from the European Commission. Brussels, 30 May 2008. 331 final.

COM (2009a). Green Paper. Reform of the Common fisheries Policy. Brussels, 22 April 2009. 163 final.

COM (2009b). Proposal for a Council Regulation establishing a multi-annual plan for the western stock of horse mackerel and the fisheries exploiting that stock. Brussels, 21 April 2009. COM (2009) 189 final.

COM (2009c). Proposal for a Council Regulation establishing a long-term plan for the anchovy stock in the Bay of Biscay and the fisheries exploiting that stock. Brussels, 29 July 2009. COM (2009) 399 final.

COM (2011). Proposal for a Regulation of the European parliament and the Council establishing a multiannual plan for the Baltic salmon stock and the fisheries exploiting that stock. Brussels, 12 August 2011. 470 final.

COM (2014). Communication from the Commission to the European Parliament and the Council. Concerning a consultation on Fishing Opportunities for 2015 under the Common Fisheries Policy. Brussels, 26 June 2014 388 final.

COM (2015). Proposal for a Council Regulation amending Regulation No 2015/104 as regards certain fishing opportunities. 212 final, 20 May 2015.

EC (1996). Council Regulation No 847/96 of 6 May 1996 introducing additional conditions for year-to-year management of TACs and quotas. *Official Journal* L 115, 9 May 1996 p. 3–5.

EC (1997a). Council Regulation No 779/97 of 24 April 1997 introducing arrangements for the management of fishing effort in the Baltic Sea. *O.J.E.C.* No L 113 of 30 April 1997, p.1–3.

EC (1997b). Council Regulation No 88/98 of 18 December 1997 laying down certain technical measures for the conservation of fishery resources in the waters of the Baltic Sea, the Belts and the Sound. *OJ* L 9, 15 January 1998, p. 1.

EC (1998a). Council Regulation No 1239/98 of 8 June 1998 amending Council Regulation 894/97 laying down certain technical measures for the conservation of fishery resources. *OJ*. 171 of 17 June 1998, p. 1–4.

EC (1998b). Council Regulation No 850/98 of 30 March 1998 for the conservation of fishery resources through technical measures for the protection of juveniles of marine organisms. *OJ* L 125, 27 April 1998, p. 1.

EC (2000) Council Regulation No 2549/2000 establishing additional technical measures for the recovery of the stock of cod in the Irish Sea (ICES Division VIIa). OJ L 292, 21 November 2000, p. 5.

EC (2002). Council Regulation No 2340/2002 of 16 December 2002 fixing for 2003 and 2004 the fishing opportunities for deep-sea fish stocks. *OJ* L 356, 31 December 2002, p. 1.

EC (2003a). Council Regulation No 2287/2003 of 19 December 2003 fixing for 2004 the fishing opportunities and associated conditions for certain fish stocks and groups of fish stocks, applicable in Community waters and, for Community vessels, in waters where catch limitations are required. *OJ* L344, 31 December 2003, p. 1.

EC (2003b). Council Regulation No. 1954/2003 of 4 November 2003 on the management of the fishing effort relating to certain Community fishing areas and resources and modifying Regulation (EC) No 2847/93 and repealing Regulations (EC) No 685/95 and (EC) No 2027/95. *OJ* L 289 of 7 November 2003, pp. 1–7.

EC (2004a). Council Regulation No 423/2004 of 26 February 2004 establishing measures for the recovery of cod stocks. *OJ.* L 70 of 9 March 2004.

EC (2004b). Council Regulation No 811/2004 of 21.4.2004 establishing measures for the recovery of the Northern Hake stock. *OJ.* No 150 of 30 April 2004.

EC (2005a). Council Regulation No 27/2005 of 22 December 2004 fixing for 2005 the fishing opportunities and associated conditions for certain fish stocks and groups of fish stocks, applicable in Community waters and, for Community vessels, in waters where catch limitations are required. *OJ* L 12, 14 January 2005, p. 1.

EC (2005b). Commission Regulation No 1037/2005 of 1 July 2005 establishing emergency measures for the protection and recovery of the anchovy stock in ICES Subarea VIII. *OJ* L 171, 2 July 2005, p. 24.

EC (2005c). Council Regulation No 2166/2005 of 20 December 2005 establishing measures for the recovery of the Southern hake and Norway lobster stocks in the Cantabrian Sea and Western Iberian peninsula and amending Regulation (EC) No 850/98 for the conservation of fishery resources through technical measures for the protection of juveniles of marine organisms. *OJ.* 345 of 28 December 2005.

EC (2005d). Council Regulation No 2187/2005 of 21 December 2005 for the conservation of fishery resources through technical measures in the Baltic Sea, the Belts and the Sound, amending Regulation (EC) No 1434/98 and repealing Regulation (EC) No 88/98. *OJ* L 349, 31 December 2005, p. 1.

EC (2005e). Communication from the Commission to the Council and the European Parliament. Review of certain access restrictions in the Common fisheries Policy (Shetland Box and Plaice Box). Brussels, 13 September 2005.

EC (2006a). Council Regulation No 388/2006 of 23 February 2006 establishing a multiannual plan for the sustainable exploitation of the stock of sole in the bay of Biscay. *OJEU* L 65 of 7 March 2006, pp. 1–4.

EC (2006b). Council Regulation No 1967/2006 of 21 December 2006 concerning management measures for the sustainable exploitation of fishery resources in the Mediterranean Sea, amending EEC (2006) Regulation (EEC) No 2847/93 and repealing Regulation (EC) No 1626/94. *OJ* L 409, 30 December 2006, p. 11.

EC (2007a). Commission Regulation No 147/2007 of 15 February 2007 adapting certain fish quotas from 2007 to 2012 pursuant to Article 23(4) of Council Regulation (EC) No 2371/2002 on the conservation and sustainable exploitation of fishery resources under the Common Fisheries Policy. *OJ.* L 46, 16 February 2007, p. 10–13.

EC (2007b). Council Regulation No 676/2007 of 11 June 2007 establishing a multiannual plan for fisheries exploiting stocks of plaice and sole in the North Sea. *OJ.* No 157 of 19 June 2007.

EC (2007c). Council Regulation No 509/2007 establishing a multi-annual plan for the sustainable exploitation of the stock of sole in the Western Chanel. *OJEU* L 122 of 11 May 2007, pp. 7–10.

EC (2007d). Council Regulation No 1098/2007 of 18 September 2007 establishing a multiannual plan for the cod stocks in the Baltic Sea and the fisheries exploiting those stocks, amending Regulation (EEC) No 2847/93 and repealing Regulation (EC) No 779/97. *OJEU* L 248 of 22 September 2007, pp. 1–10.

EC (2007e). Council Regulation No 41/2007 of 21 December 2006 fixing for 2007 the fishing opportunities and associated conditions for certain fish stocks and groups of fish stocks, applicable in Community waters and, for Community vessels, in waters where catch limitations are required. *OJ* L 15, 20 January 2007, p. 1.

EC (2007f). Council Regulation No 520/2007 of 7 May 2007 laying down technical measures for the conservation of certain stocks of highly migratory species and repealing Regulation (EC) No 973/2001. *OJ* L 123, 12 May 2007, p. 3.

EC (2008a). Council Regulation No 338/2008 of 14 April 2008 providing for the adaptation of cod fishing quotas to be allocated to Poland in the Baltic Sea (Subdivisions 25–32, EC waters) from 2008 to 2011. *OJ.* L 107, 17 April 2008, p. 1–9.

EC (2008b). Council Regulation No 1342/2008 of 18 December 2008 establishing a long-term plan for cod stocks and the fisheries exploiting these stocks and repealing Regulation (EC) No 423/2004. *OJEU* No 348 of 24 December 2008.

EC (2008c). Council Regulation No 1300/2008 of 18 December 2008 establishing a multi-annual plan for the stock of herring distributed to the west of Scotland and the fisheries exploiting that stock. *OJEU* L 344 of 20 December 2008, pp. 6–9.

EC (2008d). Council Regulation No 40/2008 of 16 January 2008 fixing for 2008 the fishing opportunities and associated conditions for certain fish stocks and groups of fish stocks, applicable in Community waters and, for Community vessels, in waters where catch limitations are required. *OJ* L 19, 23.1.2008, p. 1–203.

EEC (1983). Council Regulation No 3624/83 of 20 December 1983 fixing, for certain fish stocks and groups of fish stocks occurring in the Community's fishing zone, total allowable catches for 1983, the share of these catches available to the Community, the allocation of that share between the Member States and the conditions under which the total allowable catches may be fished. *OJ* L 365, 27 December 1983, p. 1–39.

EEC (1992). Council Regulation No 3760/92 of 20 December 1992 establishing a Community system for fisheries and aquaculture. OJEC No L389 of 31 December 1992, pp. 1–14.

Enever, R., Revill, A.S. and A. Grant (2009). Discarding in the North Sea and on the historical efficacy of gear-based technical measures in reducing discards. *Fisheries Research* **95**: 40–46.

EU (2010). Council Regulation No 23/2010 of 14 January 2010 fixing for 2010 the fishing opportunities for certain fish stocks and groups of fish stocks, applicable in EU waters and, for EU vessels, in waters where catch limitations are required and amending Regulations (EC) No 1359/2008, (EC) No 754/2009, (EC) No 1226/2009 and (EC) No 1287/200.

EU (2011). Commission Regulation No 165/2011 of 22 February 2011 providing for deductions from certain mackerel quotas allocated to Spain in 2011 and subsequent years on account of overfishing in 2010. *OJ.* L 294, 24 October 2011, p. 3.

EU (2012a). Council Regulation No 1262/2012 of 20 December 2012 fixing for 2013 and 2014 the fishing opportunities for EU vessels for certain deep-sea fish stocks. *OJ* L 356, 22 December 2012, p. 22–33.

EU (2012b). Council Regulation No 24/2014 of 10 January 2014 fixing for 2014 the fishing opportunities for certain fish stocks and groups of fish stocks in the Black Sea. OJ L9, 14 January 2014, p. 1.

EU (2013a). Council Regulation No 1180/2013 of 19 November 2013 fixing for 2014 the fishing opportunities for certain fish stocks and groups of fish stocks applicable in the Baltic Sea. *OJ* L313, 22 November 2013, p. 4.

EU (2013b). Council Regulation No 713/2013 of 23 July 2013 establishing the fishing opportunities for anchovy in the Bay of Biscay for the 2013/14 fishing season. *OJ* L201, 26 July 2013, p. 8.

EU (2013c). Council Regulation No 1223/2013 of 29 November 2013 providing for the deduction from salmon fishing quota allocated to Poland in 2013 and subsequent years in ICES subdivisions 22–31 on account of overfishing in 2012. *OJ* L 320, 30 November 2013, p. 20–21.

EU (2013d). Regulation No 1380/2013 of the European Parliament and of the Council of 11 December 2013 on the Common Fisheries Policy, amending Council Regulations (EC) No 1954/2003 and (EC) No 1224/2009 and repealing Council Regulations (EC) No 2371/2002 and (EC) No 639/2004 and Council decision 2004/585/EC. O.J. No L354 of 28.12.2013, p. 22–61.

EU (2014a). Council Regulation No 43/2014 of 20 January 2014 fixing for 2014 the fishing opportunities for certain fish stocks and groups of fish stocks, applicable in Union waters and, to Union vessels, in certain non-Union waters. OJ L 24, 28 January 2014, p. 1–145.

EU (2014b) Proposal for a Regulation of the European Parliament and of the Council establishing a multiannual plan for the stocks of cod, herring and sprat in the Baltic Sea and the fisheries exploiting those stocks, amending Council Regulation (EC) No 2187/2005 and repealing Council Regulation (EC) No 1098/2007. COM/2014/0614 final.

European Commission Joint Research Centre (2013) Report of the STECF. 2013 Assessment of Mediterranean stocks. (STECF 13–22) Joint Research centre. Report EUR 26329 EN.

European Commission Joint Research Centre (2014) The 2014 Annual Economic Report on the EU Fishing Fleet (STECF 14–16). JRC Scientific and Policy Reports. EUR 26901 EN. Online, available at https://ec.europa.eu/jrc/sites/default/files/lbaq14001enn_0.pdf (accessed 7 October, 2015).

European Parliament (2010). Real time closures of fisheries. Directorate General for Internal Policies.

Farnell, J. and J. Elles (1984). *In Search of a Common Fisheries Policy*. Gower Publishing, Aldershot, 213 pp.

ICES (2001). *Cooperative Research Report* No 246, 2001, p. 230 and 342.

ICES (2013). *Report of the ICES Advisory Committee*. Book 1 Introduction, Overviews and Special Requests.

ICES Advice (2001). ICES Advice on cod in Sub-area IV (North Sea), Division VIId (Eastern Channel) and Division IIIa (Skagerrak).

ICES Advice (2005). Report of the ICES Advisory Committee on Fisheries Management, Advisory committee on the Marine Environment and Advisory Committee on Ecosystems, volume **7**.

ICES Advice (2006). ICES Advisory Committee for Fishery Management. Advice of the ICES Advisory Committee for Fishery Management, volume **7**, p. 27.

ICES Advice (2013). North Sea mixed-fisheries for Subarea IV and Divisions IIIa North (Skagerrak) and VIId (Eastern Channel). Advice June 2013, Book 6. Online, available at www.ices.dk/sites/pub/Publication%20Reports/Advice/2013/2013/mix-nsea.pdf (accessed 7 October 2015).

Johnsen, J.P. and S. Eliasen (2011). Solving complex fisheries management problems: what the EU can learn from the Nordic experiences of reduction of discards. *Marine Policy*, **35**: 130–139.

JRC (2014a). Review of scientific advice for 2015. Consolidated advice on Fish Stocks of Interest to the European Union (STECF-14–24). *JRC Scientific and Technical Policy Reports*. EUR 26815 EN.

JRC (2014b). Evaluation of Fishing Effort Regimes in European waters – Part 2 (STECF-14–20). Joint Research Centre scientific and policy reports. EUR 27027 – EN.

JRC (2015). Review of Scientific Advice for 2015 – Part 2 (STECF-14–11). EUR 26815 EN.

MRC (2012). Marine Stewardship Council. Annual Report 2011/2012.

Schweiger, L. (2010). The evolution of the Common Fisheries Policy: Governance of a Common-Pool Resource in the Context of the European Integration. Institute for the European Integration. *Austrian Academy of Sciences. Working Paper* No 07/2010. November 88p.

SEC (2007). Evaluation of closed area schemes. STECF Plenary meeting. Ispra, 5–9 November.

CHAPTER 5
Fleet policy

The relationship between fleets and resources

Fleet capacity and fisheries management

The need to ensure that the fleets are in balance with the available biological resources has been, at least in theory, part of the objectives of the CFP from its inception. However, this part of the policy took time to develop. It was in practice absent from the political debate in the first years of the CFP and, since its first stages in the early 1990s, the question of the relationship between fleet capacity and available resources has always been controversial and certainly difficult to quantify and implement. Somehow, the basis of it has been often called into question.

There have always been views that defend the idea that if the conservation policy is well implemented, that is, if catch limitations are respected, then there would be no need for any capacity management: fleets would just adapt their fishing time and catch levels to the available quotas.

This thinking has repeatedly proven too optimistic, notably for three reasons:

1 Firstly, if fleet capacity is clearly in excess of that proportional to the available quotas, then the pressure to overfish these quotas will be too difficult to control. Control can never be perfect and excessive capacity is a clear invitation to excessive catches that probably no amount of control may be able to prevent.
2 Secondly, because of excessive capacity, even if quota uptake is well controlled, it would result in low catch rates per vessel and consequently poor economic performance. And good economic performance is among the fundamental objectives of the CFP (Article 39 TFEU).
3 Thirdly, because excess capacity always leads Member States to push TACs upwards beyond sustainable levels. It has been a common argument that when discussing the conservation policy, some ministers press for higher catch levels than those recommended by science because their fleets 'don't have enough' with the proposed quotas.

These facts have been recognised internationally. OECD has concluded that:

> A regime that does not adequately limit fishing capacity may lead to overexploitation and poor economic performance. The main reason for the poor results is that these regimes do not give the fisherman the incentive to account for the costs of its fishing. (OECD, 1997)

Likewise, FAO has clearly identified this question as a problem. In the motivation for the International Plan of Action against Fleet Overcapacity,[1] FAO concluded that: 'Excessive fishing capacity is a problem that, among others, contributes substantially to overfishing, the degradation of marine fisheries resources, the decline of food production potential and significant economic waste'.

These two examples show why an adequate fleet policy must be an essential element of any package of measures to manage fisheries.

The Common Fisheries Policy: The Quest for Sustainability, First Edition. Ernesto Penas Lado.

Historical background: the first fleet policy

Fleet policy has always been an issue in the CFP from its very inception, at least in theory. As far back as 1970, when the first structural policy was adopted, provision was made for the so-called Multi-Annual Guidance programmes (MAGPs). However, these programmes initially concentrated on the coordination of fleet policy by Member States with a view to ensuring their modernisation but without any real attempt to evaluate or regulate possible overcapacity of the fleets: fleet development and modernisation was the objective of the day.

In 1983, the structural policy adopted very general objectives for fleet management. The need to adjust fleet size to available resources was not necessarily a widely accepted objective. So for the first years of implementation of the CFP, the fleet policy was not conceived as an instrument related to the conservation policy and consequently there was no attempt to adjust fleet capacity to available resources.

As this reasoning was developed and as the consciousness grew about the need to ensure the sustainability of the stocks, the linkage between conservation policy and fleet policy started to develop. A crucial factor in that regard was the accession of Spain, whose fleet was considered by some existing Member States as too large, leading to the more serious consideration of the need to have an instrument to address this kind of question. Gradually, fleet policy was no longer seen as one to develop European fleets but rather to keep these fleets in balance with biological resources.

The first attempt to address this question in fact did not even try to evaluate the magnitude of the problem. But it was clear that the problem was there already. This led to the adoption, already in 1983 and as a complementary part to the birth of the CFP, to the adoption of a Council Directive, 83/515 (EEC, 1983a).

This Directive, short of evaluating the problem and establishing a target reduction of capacity, did provide for the use of three financial mechanisms to tackle the problem: grant aids for scrapping, for temporary cessation of the fishing activity and for the export of vessels to third countries. In other words, short of evaluating the magnitude of the problem, the policy simply provided for financial means to withdraw fleet capacity on a voluntary basis. These measures were theoretically non-targeted, that is, they would apply to anyone wishing to withdraw a fishing vessel for whatever the reason but in practice the ultimate reason was the reduction of the Spanish fleet (see Chapter 3).

The first two measures have survived ever since, in the framework of the successive structural instruments. The subsidies to export vessel to third countries, however, were discontinued in the EFF in 2007, following allegations that these grants reduced fleet capacity in European waters only to increase it in third countries. This was simplistically dubbed 'the Community exporting its overcapacity to third - developing- countries'. In addition, there had also been incidents of vessels that had been exported with subsidies continuing to operate under flag of convenience which engaged in illegal fishing. The damage to the reputation of this measure, whatever its real effects were, (never seriously evaluated) brought about its demise.

In the late 1980s the question was for the first time addressed in a systemic way. The question was, then: is there overcapacity in the European fleets? And if so, how much? Replying to these questions was not straightforward.

The multi-annual guidance programmes

As we saw previously, the question of how to manage the main factor of production, the fishing fleets, was an issue from the inception of the CFP. And from the very early stages of this policy, a specific instrument was established to deal with this: the Multi-Annual Guidance Programmes (MAGPs), better known for many years under their French acronym of POPs (*Programmes d'Orientation Pluriannuels*).

In Regulations 2141/70 and 101/76, provision was made for MAGPs, although at the time they were conceived as guidance documents to better coordinate the structural policy, still based on the philosophy of modernising and developing the factors of production. At that time, there was still no notion of the MAGPs as an instrument to adapt fleet capacity to available resources.

MAGPs for the period 1983–1986

With the establishment of the conservation policy, the MAGPs started to define their role as a complementary instrument to accompany the former. Regulation 2909/83 set out to achieve: 'a satisfactory balance between the fishing capacity to be deployed by the production facilities covered with the programmes and the stocks which are expected to be available during the period of validity of the programme.' (EEC, 1983b).

The principle of using the structural policy to adapt fleets to resources was turned into a political objective of the CFP. These MAGPs provided guidance for the period 1983–1986. However, such guidance was to consolidate the capacity levels of 1982–1983. On that basis, Member States drew up national fleet plans where no reductions of capacity were foreseen. Such plans were fundamentally a framework to carry out the renewal of their fleets using the structural money provided for in Regulation 2909/83.

As an example, the fleet plan for Germany, adopted in 1985, established as its objective the 'maintenance of fishing capacity of the programme at the 1982 level as regards the inshore fleet … and at the end 1983 level as regards the distant fleet' (EEC, 1985).

Clearly, the first MAGPs did not consider a reduction of fleet capacity a necessity. This fact, combined with the effects of fleet renewal in terms of increasing the real fishing power through technical development, most likely led to a *de facto* increase in the real fishing capacity of European fleets at the time.

However, the principle and the instrument were set up, together with other accompanying measures, such as the establishment of a Community fleet register (EC, 2003), or the harmonisation of rules on measurement of capacity, to ensure harmonised approaches by Member States (EEC, 1986).

MAGPs for the period 1987–1991

For the second generation of MAGPs, it was becoming evident, notably through reports by the scientific community, that the scenario of a stable fishing capacity (and a possible *de facto* increase through technical improvement) was not efficient enough to address the emerging evidence on the need to reduce fishing pressure on a number of European stocks. In addition, the accession of Spain and Portugal also had an influence: since 1977, the need to reduce the Spanish fleet fishing in Community waters had been an issue. Clearly, asking only Spain to adapt its fleet could not be politically acceptable and this contributed to the search for a more useful policy in terms of fleet capacity adaptation applicable to all Member States.

On that basis, the Commission tried for the first time to impose a real adaptation of capacities for all Member States. However, the resistance from these was such that only very modest aims could be accepted. The MAGPs for the period 1987–1991 established for the first time specific objectives for fleet adjustment: a generalised reduction of 3% in tonnage and 2% in engine power in global terms, per Member State. Although they were rather modest and even if formally met were likely to be offset by technical progress, the establishment of such targets at least meant the introduction of a new paradigm in the policy: the need to actively pursue the adaptation (reduction) of fleet capacity to available resources. At the time, it was already obvious, at least for scientists, that higher adjustments would inevitably be necessary.

The question of 'technical progress' deserves an explanation. While a vessel keeps the same nominal capacity as expressed in engine power and tonnage, it can, however, increase its ability to catch fish, through the introduction of a number of technical improvements such as, for example, very effective fish-finding equipment (more efficient eco-sounders, new net and gear designs, fish-aggregation devices and many others) that can increase the capacity of the vessels to find the fish and catch it while all other parameters remain unchanged.

It is important to analyse why such objectives were so modest. At this time, it was already clear that the need for capacity reduction was not uniform. On the contrary, it was variable across fleet segments and, crucially, among Member States. However, proving that one Member State would need to reduce its fleet capacity more than its neighbour posed a double challenge: technical and political.

- Technical, because the measurement of fishing capacity was and is still today, very elusive and its quantification is very difficult. The relationship between a vessel's size in GT or engine power in kW and its capacity to catch fish is strongly affected by factors such as the skill of the master, the electronic equipment on board, the knowledge of the fishing grounds and so on.

- Political, because in the absence of an unequivocal quantification of overcapacity, any fleet reduction by a Member States in excess of the reduction of their neighbours would be seen as an unjustified political defeat.

As a consequence, the only way that Council could fix an objective of reducing capacity was to agree on a common reduction to all Member States together, regardless of their situation. For a common reduction rate to be acceptable to all (including those Member States who considered that they had no major problem of overcapacity) it had to be very small.

In any case, the second generation of MAGPs represented an important change: specific objectives for fleet reduction were established. By comparison with its plan of 1985, the new German MAGP for the period 1987–1991 did consider a reduction in both the inshore and long distance fleet of over 3% in tonnage (from 49 900 to 48 200 GRT) and over 2% in engine power (from 137 000 to 132 000 kW: EEC, 1988).

The real effect of this reduction was certainly strongly mitigated by two factors: in some cases the modification of the reference capacity levels of 1982–1983 and, especially the effect of the 'technological creep' that we will refer to next. In addition, vessel construction was still possible under conditions that did not guarantee a real reduction of capacity as a result.

MAGPs for 1992–1996: the Gulland report

It is obvious that the 3% and 2% objectives of the second generation of MAGPs were not the result of an evaluation of the real need to reduce fleet capacity but rather the figures that could politically be agreed upon. For the next period, a different approach was necessary: the objectives of the MAGPs should be established according to scientific advice on the relationship between capacity and resources.

This new approach was not politically palatable and met with considerable resistance among Member States. It would take a considerable determination to get it through. A key factor for this to happen was the arrival at the Commission services of officials with in-depth knowledge of the fishing sector to run the Fleet Division. In line with the reflection by Mike Holden on the importance of expertise in the Commission services (see Chapter 2), this new expertise made all the difference between the traditional approach and a substantially new one. Curiously, at that time the impulse for an ambitious policy on fleet reduction came from the structural side, not so much from the conservation one.

To address these questions the Commission set up in 1988 a group of experts to advise the Commission on the magnitude of the problem. This expert group, constituted by independent scientists from all relevant Member States under the chairmanship of the late John Gulland, prepared a report analysing by fishery the observed level of overcapacity. The report concluded that the overall level of overcapacity of the European fleets was in the order of magnitude of 40%. The report did not recommend that 40% of the European fleet should be scrapped but that 40% was the excess of fishing effort exerted on European resources. Whether this reduction should be achieved through scrapping or effort management, the report did not prejudge (EC, 1990).

The Gulland report, delivered in 1990, was the basis for the discussion of the MAGP for the period 1992–1996.

The corresponding MAGPs did not get anywhere close to the 40% reduction in capacity/effort. However, they included, for the first time, more ambitious objectives for fleet reduction as related to the state of the stocks they fished, through a simplified fleet segmentation. This consisted of dividing national fleets into segments broadly corresponding to the main species they targeted, so as to make it possible, for the first time, to broadly relate fleet categories with the conservation status of the main species they caught. Crucially, for the first time the necessary reductions could be accomplished not only through capacity withdrawal but also by the reduction of the activity of the concerned fleets. The plan, however, required that the reduction of activity be achieved by at least 55% reduction of capacity.

Another important novelty was that, for the first time, Member States had to take into account in their plans the effects of technical progress. Overall, the MAGPs 1992–1996 aimed at a 20% reduction in tonnage and 15% in engine power compared to 1991 levels.

A very good example of these innovations can be seen in the MAGP for France, for the period 1993–1996 (for 1992, a transitional measure was adopted, to make the plans coincide with the new CFP adopted that year). The French MAGP set as objectives: 'an overall reduction in fishing effort, differentiated for the broad group of target species, should be of 20% for demersal species, 15% for benthic species and zero growth for pelagic species'.

In addition, the French plan took account, in establishing those objectives, of an annual increase of 2% in fishing effort linked to technical progress (EEC, 1992).

Overall, the third MAGPs represented a breakthrough in fleet policy, particularly in terms of the introduction of a scientific basis in the fleet policy and the identification of more ambitious fleet/effort reduction schemes as a function of stock status.

The MAGPs 1997–2001: the Lassen report

The fourth generation of MAGPs built on the innovations of the third generation. In particular, the very simplistic segmentation of national fleets under MAGP III and the very generic relationship between fleet segments and resources required a more sophisticated approach in finding more specific fleets segments that could be better related to specific species, since too broad fleet segments catch many species and that makes it difficult to establish clear relationships between these segments and the conservation status of individual stocks.

For the preparation of the following period of the structural policy after 1996, the Commission organised the preparation of the second report on overcapacity, this time chaired by the Danish scientist Hans Lassen. This report (EC, 1996) largely picked up the analysis of the previous report and arrived at very similar conclusions. These conclusions were the basis for the MAGPs for the period 1997–2001.

On the basis of the Lassen report, it was clear that the next generation of fleet policy should be based on a more detailed segmentation of the fleets, so as to allow a more targeted capacity policy focused on the fleet segments depending upon the state of conservation of the main stocks they fished.

The discussion with Member States showed that there was support for the segmented approach, notably to be able to reduce fleet capacity in a more meaningful way but also to be able to find a way to address different fleet reductions for different Member States and segments in an objective, systematic way. However, Member States did not agree that all the necessary reductions in fishing mortality should necessarily have to be accomplished through fleet adjustment. The argument was the following: if we reduce the fleets to address overfishing problems, once the overfishing is solved, then we would have no fleet left to fish the recovered resources. The result of this was the introduction of the management of fishing effort as a complementary or alternative mechanism to fleet capacity management.

The Council decision setting out the objectives for the MAGPs 1997–2001 (EC, 1997) thus included a number of refinements as compared with the previous ones:

1 The fleets would be segmented according to the resources they caught;
2 The reduction of fishing would be proportional to the level of overexploitation of these resources. The reductions would be of 30% for the stocks defined as 'depletion risk' and 20% for those defined as 'overfished'. These reductions were higher than ever before;
3 For each fleet segment, a reduction would be determined based on their dependence upon the resources subject to these categories of stocks;
4 These reductions would be achieved by a combination of fleet adjustment and effort management. As the decision spelled out: 'whereas Member States should be able to choose to reduce fishing mortality by the different means available to them, either by reducing capacity for each fleet segment, or by reducing fishing effort for each fishery; ...'

This solution allowed the establishment of fleet reduction objectives that would not necessarily be identical for all Member States. But the different rates would be established according to their dependence on overfished stocks. This was, at the time, a breakthrough: differential rates of fleet reduction but not based on a national basis, so as to avoid Member States feeling 'named and shamed'. Rather, the higher reductions would be based on objective criteria following scientific advice.

The weakness of the system, however, was that the combination of fleet capacity reduction and effort management was actually an excuse not to reduce fleet capacity when really needed. Crucially, the requirement of the 55% of reduction being achieved through real capacity reduction and not just through activity reduction through effort management, was dropped, due to Member States' interest to have 'more flexibility' to achieve their goals through a free combination of capacity and effort management.

In addition, effort management, in some cases aided by public funds for temporary laying-ups, created in certain cases a different 'culture': that of addressing overcapacity by using public funds to pay for temporary cessation of fishing, thus creating a tendency to preserve overcapacity with public funds. Although this tendency was not generalised, it certainly affected some fleets segments and certain Member States more than others but its effects have remained to this day.

Did the MAGPs work?

In the reform of 2002, the MAGPs were considered an inefficient instrument and were discontinued and replaced by today's policy. However, the formal objectives of these plans were generally met. This is an apparent contradiction that fleet policy has never entirely removed: while the formal objectives tend to be met, the problem of overcapacity is generally considered to remain.

It took, however, some time for this contradiction to come up clearly in the political debate. In the first years of implementation of the MAGPs, it was the formal compliance with the objectives of MAGPs that prevailed. After difficult discussions to establish the objectives and given the difficulty to demonstrate the 'technological creep', the CFP was relatively self-indulgent with its own success in achieving the objectives of its fleet policy. For example, the Commission report on the implementation of the MAGPs in 1994 stated that a real reduction of capacity was taking place, in general terms in accordance with the objectives (COM, 1995). However, it was completely silent as to whether the problem of fleet overcapacity in absolute terms was being addressed.

This situation lasted for a number of years and, in its annual report of 2001 on the implementation of the fourth generation of MAGPs (COM, 2001), the Commission concluded that, overall, its formal objectives had been achieved and even exceeded:

> During the first four years of the MAGP IV, the Community fleet was reduced by 49,983 GT and 459,866 kW, which represents reductions in fleet capacity of approximately 2.5% and 5.9% respectively. At 1 January 2001 the Community fleet was already approximately 17% below the final MAGP IV objectives in terms of tonnage and 12% below the final MAGP IV objectives in terms of power.

This overall picture, however, is very different when looking at different Member States: while some of them achieved and exceeded their objectives, others did not reach them.

This apparent success was instrumental in the decision to discontinue the MAGPs: some Member States considered that their objectives being achieved, they were no longer necessary. However, over and beyond the formal achievement of objectives, the MAGP approach was not considered a success in removing the problem of overcapacity.

There were several reasons why the MAGP approach was discontinued: the modest established objectives, the measurement of capacity, the technological creep and the problem of the image of the policy itself. However, above all, the practical problems of implementation made things difficult. Let us examine these questions in some detail.

The modest objectives and their implementation

The objectives of MAGPs have always been, one way or another, very modest in practice. There are several reasons for this.

Firstly, the Gulland (and likewise the Lassen) report was always called into question by many from national administrations and the industry and the conclusion of a 40% level of overcapacity was strongly opposed. This opposition, however, was rarely based on alternative evaluations but rather focused on the high level of uncertainty in the assumptions of the report. In other words, even if the report was questioned, it is clear that there was no better alternative analysis of the situation.

Secondly, the reports did not attribute the excess overcapacity to specific Member States. This produced a reaction in many capitals: yes, overcapacity is a real problem but in other Member States, not in mine. Consequently, it was extremely difficult to establish beyond any reasonable doubt that some Member States had higher overcapacity than others. In these circumstances, the only solution was to establish objectives for fleet reduction that were identical for all Member States, as the only politically viable solution. But the obvious price to pay for this was that the common objectives of fleet reduction fell short of what was necessary to address the problem effectively.

But even when the objectives were theoretically more ambitious, as was the case under MGPS III and IV, these apparently ambitious objectives were seriously undermined by the fact that effort reduction was as an alternative to capacity reduction. As a result, Member States chose to achieve a substantial part of the necessary reductions of capacity through reductions of effort, thus replacing a permanent, structural solution by a temporary one.

In addition, the implementation of these modest objectives was not ideal. One important reason for this was the lack of consistency of legislation: while the MAGPs established fleet capacity objectives for certain periods, the structural policy established financial measures to address overcapacity in a way that was not necessarily consistent. The periods concerned by the MAGPs and those concerned by the financial instrument were different: the structural instrument of the time, the FIFG[2] applied for the periods 1994–1999 and 2000–2006 while the MAGPs covered the periods 1992–1996 and 1997–2001. This mismatch in timing was an obstacle to ensure that the FIFG actually provided the necessary instruments to achieve the MAGP objectives. In addition, the provisions for the implementation of the fleet measures in FIFG was not clear enough and therefore the FIFGs largely failed to provide the necessary support to the achievement of the MAGPs objectives.

Overall, the rather limited (and highly unpopular) objectives established and the relative weakness of the instruments to enforce them seriously hampered the effectiveness of this instrument.

The measurement of capacity

The measurement of 'capacity' has always been a problem. In particular, the use of two parameters, such as gross tonnage and engine power to define the real capacity, that is, the ability of the fleets to catch fish, has always been extremely elusive, for a number of reasons.

Firstly, the two parameters are static, whereas the ability of fleets to catch fish is dynamic. A given vessel or fleet segment can have exactly the same GT and engine power over many years and yet gradually increase its catching power, for example through the improvement of its fish-finding equipment. Other improvements can concern the storage systems on board, thus allowing vessels to keep on board and market certain species that they could not retain before. Importantly, the skill of the vessel's master makes a huge difference to the effectiveness of the vessel and this often changes and develops over time.

Secondly, the measurement of the engine power has always been very elusive, in particular through the extended practice of engine de-rating. This practice, which in modern engines can be done through a switch-on/switch-off mechanism, means that an engine can be certified as having a low engine power if and when de-rated. When the de-rating is removed, the engines can develop a much higher power than the vessels have been certified for.

As a result, the formal compliance with the capacity limits established in the fleet policy is nearly meaningless. Most Member States complied with their fleet policy ceilings in 2012, even within a comfortable margin. However, such formal compliance does not necessarily translate into a full compliance with the objectives of Article 11 of Regulation 2371/2002, notably the obligation for Member States to adjust the fishing capacity of their fleets in order to achieve a stable and enduring balance between such fishing capacity and their fishing opportunities.

The technological creep

As 'technological creep' we understand the increase in a vessel's ability to catch fish when the basic parameters to measure capacity (tonnage and engine power) stay unchanged. The improvement in the ability to catch fish can be the result of a number of technical improvements:

- The development of new nets and new netting materials as well as new ancillary gear: new types of ropes, the length of the net, number of winch or net drums, and so on all have an effect on catch rates, for the same vessel.
- The growing efficiency of fish-finding equipment (sonars, fish aggregation devices etc.) which make a vessel much more efficient in finding the fish shoals for the same fishing time;
- The improvement in storage capacity, allowing them to store more commercially-valuable fish for the same fishing time.
- The growing skill and experience of the vessel's masters.

While the existence of technological creep is undeniable and has been demonstrated by scientists (Marchal et al., 2007), its quantification is very elusive and certainly very variable among different types of fishing. Certain studies have tackled this problem in historical terms, and have concluded that technological creep is not constant: periods of important increases in productivity alternate with periods of stagnation. In any case, a study on the long-term evolution of the fishing power in the British trawl fleet showed that the real ability of the vessels to catch fish has grown 50-fold for otter trawlers and 100-fold for beam trawlers in the last 120 years (Engelhard, 2009).

The accepted wisdom about technological creep is that it represents an average increase in the catching capacity of vessels somewhere between 2 and 3% *per annum*. This means that if a fleet reduces its capacity by 20% in 10 years, the remaining fleet, with 20% less nominal capacity, has the same ability to cause fishing mortality as the larger fleet 10 years before.

If we compare the evolution of the overall capacity ceilings over the years with the hypothetical evolution assuming a technological creep of 3% *per annum* (as in Figure 5.3), we do not have the impression that the European fleets have dramatically reduced their capacity over the same period. Assessing the real magnitude of the technological creep is and will remain elusive but the recent evaluation methodologies for the levels of overcapacity, established by the Commission in recent years (COM, 2014a), allow refinement of the analysis of the real overcapacity, once the effects of nominal fleet reduction and the level of technological creep have been taken into account.

The unpopular image of the policy

Over and beyond these practical difficulties, the MAGP system had another problem, one of image: the fact that the Union established legally binding obligations to reduce fleets created a very bad image for the policy. This gave rise to easy complaints about 'Brussels' or even worse, 'Brussels bureaucrats' obliging people to destroy their vessels and abandon the profession. This was at some stage a significant reputational issue for the CFP.

This reputational issue is, however, very unfair, as many such issues are. At the time, it was very common when visiting fishing ports around Europe to hear voices from industry leaders and local politicians, bitterly complaining about Brussels obliging them to destroy vessels and reduce fleets. At the same time, however, individual vessel owners were queuing-up to apply for scrapping aid. A contradiction between public positions, the perception of the policy and the practices and needs of private operators that is still today very typical of the CFP.

In any case, this reputational issue had a role to play in the evolution of the policy. The belief that the compulsory (albeit unambitious) reduction of fleet capacity did not work became commonplace, even in the absence of any attempt to update the Gulland and Lassen reports.

Other weaknesses of the MAGPs

In addition to all these shortcomings, the implementation of the MAGPs was subject to a number of problems that undermined their effectiveness. Most of these problems were indeed identified already in 1993 by the Court of Auditors in a special report (OJEC, 1994).

In this report, the Court underlined in particular that the linkage between the fleet policy objectives expressed in the MAGPs and the structural instruments to address capacity were not satisfactory. In fact, the Court already made three fundamental recommendations that have taken quite a long time to develop:

- that there should be a stronger link between structural aid and the conservation policy;
- that capacity development (new constructions) should be limited to fisheries with underexploited resources and
- that the conditions to provide structural assistance should comply with all the regulations and conditions for assistance.

The Court clearly pointed its finger at the fundamental weakness, referred to previously: while the fleet policy established targets in capacity, the fleet measures under the structural policy financed projects of vessel construction and modernisation that were not consistent with the achievement of the fleet objectives.

Overall, the MAGPs achieved their formal objectives. Another question is whether they actually addressed the real problems of overcapacity. In fact, this has been so difficult to do in practice that the official evaluations of fleet policy have, until 2002, limited themselves to the formal, as opposed to the substantial problem.

The last report of the MAGPs was published by the Commission in 2003 and referred to the fourth generation of MAGPs applicable in the period 1997–2002 (COM, 2003).

The report concluded that: 'during the six years of the MAGP IV, the Community fleet (without the vessels registered in the French outermost regions) was reduced by 107,284 GT and 928,973 kW, which represent reductions in fleet capacity of 5.31% and 11.77% respectively'.

The report, again, was silent on the real meaning of these figures. A reduction of almost 12% in engine power is equivalent to a 2% annual reduction, which is of the same order of magnitude as the estimated technological creep. And that, without taking into account the problems of measurement of engine power, as we saw previously. In GT, the reduction was less than 1% per year.

Overall, the MAGPs did achieve reductions in nominal capacity but the real value of these reductions remained unclear. While the objectives of capacity reduction in MAGPs III and IV were significant, much of these reductions took the form of reduction of effort, so these objectives did not imply a reduction of capacity but in some cases just a reduction of activity, at best.

In addition, the MAGP approach, by concentrating on the formal fulfilment of two objectives expressed in static parameters, somehow diverted the attention from the real problem of whether the real capacity (understood as the ability of the vessels to generate fishing mortality) was getting closer to what the biological resources can sustain.

It is very intriguing that, despite the very different objectives of the four generations of MAGPs, the rate of reduction of fleet capacity has remained relatively constant (see Figures 5.1 and 5.2), as if fleet reduction was led more by economics than by policy.

From MAGPs to Member States' responsibility: the policy since 2002

The new policy

All these reasons seriously undermined the credibility of the MAGPs as an instrument to effectively adapt real fleet capacity to the resources. Curiously, the MAGPs were criticised from both ends of the spectrum. For some, they achieved their (formal) objectives, so they were no longer necessary. For others, they did not really address overcapacity in substance. In that context, the reform of the CFP in 2002 brought this MAGP system to an end.

As of 2003, according to Article 11 of Regulation 2371/2002, each MS is autonomous to decide on this policy with two conditions: addressing the overcapacity in a general way and respecting the capacity ceilings (EC, 2002).

The main consequence is that the compulsory capacity/effort reduction objectives were removed.

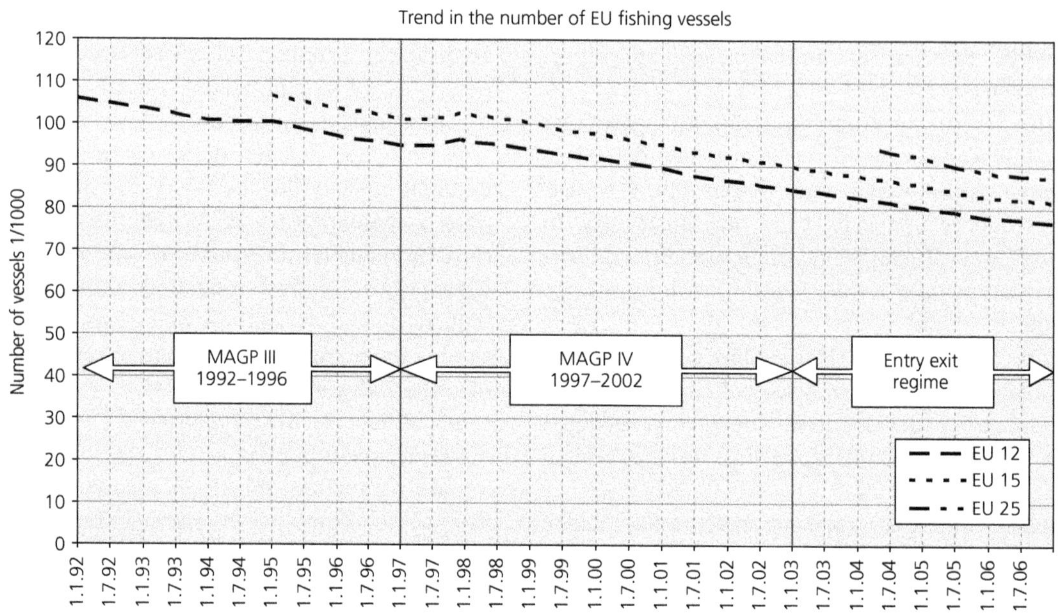

N.B.: The increase in the number of vessels shown in 1998 is due to the inclusion in the Community Fishing Fleet register of the vessels registered in the French Outermost Regions

Figure 5.1 Evolution in the number of vessels in the European fleets in different periods and under different policy objectives. Source: European Commission Services. (See colour plate section for the colour version.)

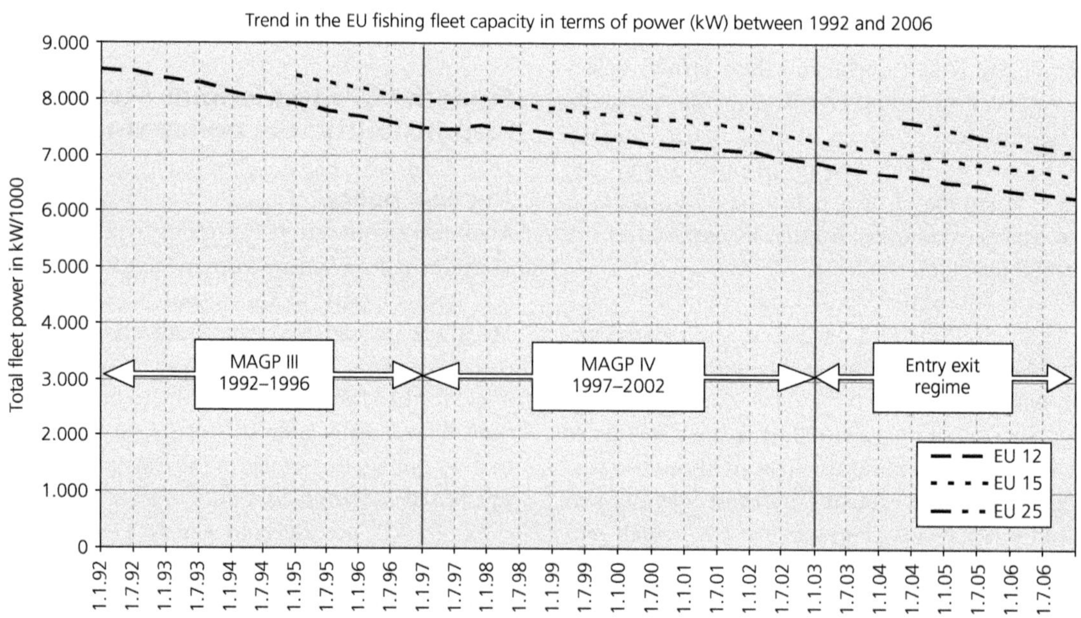

N.B.: The increase in power shown in 1998 is due to the inclusion in the Community Fishing Fleet register of the vessels registered in the French Outermost Regions

Figure 5.2 Evolution of fleet capacity, expressed in engine power (kW) in different periods and under different policy objectives. Source: European Commission Services. (See colour plate section for the colour version.)

And so was the segmentation of the fleets. Instead, Member States have a general obligation to: 'put in place measures to adjust the fishing capacity of their fleets in order to achieve a stable and enduring balance between such fishing capacity and their fishing opportunities'.

The system is complemented by the so-called 'entry-exit regime'. This consists of the obligation for the Member State to revise the national capacity ceiling when a vessel is scrapped with public aid, so that the overall capacity ceiling is reduced by the amount of capacity scrapped through public aid. This means that the capacity ceilings are not fixed forever but are gradually ratcheted down as scrapping with public funds takes place. The obvious intention was to ensure that the capacity withdrawn could not be rebuilt. The new system also contained transitional measures to prevent that vessel construction, still eligible in 2003, would increase capacity.

The new system provided Member States with an unprecedented freedom to conduct their fleet policy without strict objectives, just on the basis of a generic obligation to adjust fleets to resources. Member States were also under an obligation to report annually on their progress. Such reports would be summarised and presented by the Commission annually.

In practice, the very general obligation had the obvious risk of allowing for very varied interpretations on how to estimate levels of overcapacity. This led the Commission to elaborate guidelines on how the overcapacity should be estimated in a harmonised manner (COM, 2014a).

Implementation

Since 2003, the Commission publishes an annual report on the implementation of the fleet policy, based largely on the national reports submitted by Member States. After years of a gradual increase in the completeness and quality of the national reports, the last Commission report, published in 2014 (COM, 2014b), summarises the implementation of the new policy approach:

Firstly, the overall ceilings have in any case been respected:

- In terms of GT, the total average capacity for all Member States combined was 83.55% of the overall ceiling. That means that the EU fleet could grow more than 10% and still respect the overall capacity limits.
- In terms of engine power, the average is 89.57% of the ceilings, which also means that the EU fleet could still increase total engine power by 10%.
- By individual Member State, some data are striking: the Cypriot fleet is at 38% in GT but 95.77% in power of the total national ceiling. For Lithuania, these percentages are 36.97 and 46.80%. For Malta, 53.44 and 80.04%, respectively. Even for the larger fleets, there is room for increase in fleet size while respecting the ceilings: in Spain the percentages are 93.04% for GT and 92.73% in power and in France, 85.52 and 90.35%.

These figures show the limited effect of these restrictions. They do, however, have a certain value: avoiding a possible significant expansion of Member States' fleets, even if a moderate expansion would still be possible.

Perhaps there is no better proof of the meaningless character of the fulfilment of the capacity ceilings than the important demand for scrapping funds during the period of application of the structural policy after 2002 (see Chapter 6). If the capacity ceilings are respected, even with a certain margin, then why should there be demands for reducing further the fleet through scrapping? It seems clear that, irrespective of the achievement of the overall capacity ceilings, there are still fleets segments were overcapacity continues to drive the scrapping demands.

Secondly, the current system requires Member States to adapt these ceilings as a result of the withdrawal of vessel capacity using public money: it is the fundamental feature of the so-called exit-entry scheme. But as much of the vessel withdrawals (around 50% for the Union as a whole) have taken place without scrapping funds, the overall ceilings have been reduced less than the real reduction of capacity. This has made the capacity ceilings even more meaningless.

Thirdly, the capacity ceilings include a number of vessels that are not active. According to Member States' information for the 2015 fleet report, the percentage of non-active vessels is very significant in all Member States and, in some of them, they represent more than 40% of the total fleet. That means that the real capacity being used is lower than the nominal available capacity. The potential use of that inactive capacity is not restricted. If and when such unused capacity would be mobilised (possibly in the context of unregulated species) the system would simply be incapable of avoiding a net increase in the total fishing effort being exerted by European fleets.

The experience of the annual reports is that, in general terms, most Member States have not been in a position to evaluate the overcapacity of their fleets or segment thereof, despite the fact that the Commission has produced guidelines on the preparation of these reports.

- The presentation of Member States' reports has been patchy at best. Some important Member States have even failed to provide their annual report;
- The Commission guidelines have not always been followed;
- Some Member States simply state that their fleets are in balance with biological resources but do not provide any justification or demonstration;
- Others recognise they have certain problems of overcapacity but do not necessarily identify the specific segments where this overcapacity should be addressed.

It is certainly true that there was relatively little incentive for Member States to invest in a very in-depth report. The reason is simple: whether or not the report was serious, there was no obvious policy consequence. So Member States tended to take this obligation as a routine without particular interest.

It is also important to refer to questions that the reports did not mention: the real meaning of the data on engine power. The difficulties in measuring the real engine power, or to avoid increases in the real power delivered by an engine regardless of the certified engine power (see issue of de-rating previously) have been known for quite a long time. Conscious of the need to solve this problem the Commission proposed (and Council adopted in 2010, in the framework of the new control regulation) a system to certify and control the engine power in a more unequivocal manner and obliging Member States to certify and verify the engine power of their vessels. In order to overcome the problem of de-rating mentioned previously, the regulation requires that they: '...shall only certify a propulsion engine as not being capable of exceeding the officially stated power if there is no possibility to increase the performance of the propulsion engine above the certified power'.[3]

This means that Member States cannot certify a de-rated engine power that subsequently can be increased as the de-rating is removed.

The implementation of this new requirement is still incomplete and will certainly be subject to scrutiny and follow-up in the years to come.

Overall, the fleet policy can hardly be considered a success. It has provided a gradual reduction of nominal capacity both expressed in number of vessels (Figure 5.1) or in engine power (Figure 5.2) but this reduction has tended to be constant, regardless of whether there has been a compulsory fleet reduction target or not. This strongly suggests that fleet capacity in Europe is being gradually reduced as an inevitable economic trend, rather than as a result of policy. And, more importantly, if we consider that the nominal reduction of capacity may have been compensated by the so-called 'technological creep' (which is commonly understood to be in the vicinity of 2–3% per year) then we can conclude that fishing capacity in the Union has not been significantly reduced so far.

MAGPs versus Member State responsibility

With the perspective of time, it is relevant to compare the achievements of the two kinds of models of fleet management: the one of the MAGPs, with legally-binding objectives, or the one after 2002, where Member States are free to translate the generic objective into specific fleet reduction targets.

In this sense, Figures 5.1 and 5.2 are very intriguing: in the general evolution of –nominal- fleet capacity over the years, the trend has been very stable: a gradual reduction of capacity around an average of *circa* 2% per year and this, regardless of the MAGP objectives or the fleet policy modalities.

The trends in these figures do not show major disruptions, and the small changes in trends do not seem to be clearly related to the changes in policy. In fact, certain changes of trajectory seem counterintuitive: after 1997, the pace of fleet reduction seemed to slow down slightly, even though the MAGPs as of 1997 had more ambitious objectives than previously.

And, very significantly: once compulsory reductions in capacity were discontinued in 2003, the rate of capacity reduction has continued without substantial change.

The conclusion is inevitably tempting: the evolution of fleet capacity in the CFP may have been as much, if not more, the result of economic and technical factors rather than as a result of the CFP policy measures.

Under the 2002 regime, many Member States, in the context of their annual reports on fleet policy, found it very difficult to identify the level of overcapacity corresponding to each fleet segment. However, these Member States were able to do quantify this in the context of their MAGPs in previous years and more particularly in the fourth generation of MAGPs (1997–2001). This seems to indicate that when confronted with the obligation to specify the level of overcapacity Member States can certainly do so.

All these questions led the Commission to conclude, in its last fleet report prior to the CFP reform of 2013, that: 'The current fleet management policy has failed to bring fleets into balance with the resources they exploit. It is not sufficient to rely on compliance with national capacity ceilings (expressed in vessel size (GT) or power (kW)) only' (COM, 2013).

This conclusion is essential to understand the Commission's position on fleet policy in the context of the CFP reform of 2013, and the proposal for an alternative, rights-based management system, as we will see in Chapter 16.

Fleet and structural policy: have they helped each other?

Among the different pillars and sub-pillars of the structural policy, the most controversial has always been the fleet policy and in particular the three main components: construction and modernisation of vessels, temporary laying-up and permanent cessation. It is important to analyse these three instruments in terms of their contribution to achieve the objectives of the fleet policy within the CFP.

As we saw previously, already in 1993 the Court of Auditors drew attention to the weak links which existed between fleet policy and the structural instruments applicable to the fleets. The Court again insisted in these weak linkages in 2011. Let's look into these linkages.

Vessel construction and modernisation: a contribution to overcapacity?

This debate is as old as the CFP itself. That the initial structural policy did contribute to create overcapacity is beyond doubt. It was estimated that from 1970 when the first structural regulation was adopted, the capacity of the EU fleet almost doubled: from a total of 794 000 Gross Registered Tonnage (GRT) in 1970 it increased to 1 303 000 GRT in 1983 (Holden, 1994). The trend in engine power is much more difficult to quantify but the tendency must surely have been similar.

Up until 2002, the successive structural programmes under the CFP included aid for the construction of fishing vessels, although under conditions that would, theoretically, prevent the increase of fishing capacity as a result.

Grants for the construction of new fishing vessels were available in the different structural programmes up until the Financial Instrument for Fisheries Guidance (FIFG) applicable until 2002.

Although this question has not been studied in depth by fisheries scientists, one can contend that the very modest requirements in terms of withdrawal of capacity for new constructions applicable in a number of Member States (although some did apply higher requirements) are likely to have produced a net increase in capacity in many cases.

In effect, replacing a 30-year old fishing vessel by a new one of the same GT and engine power is not a neutral operation in fleet capacity terms. New vessels with the same nominal capacity as expressed in GTs and kW of engine power are much more capable of producing fishing mortality than their equivalent older vessels.

In addition, the old vessels replaced by new constructions may not always have been fully operational vessels. As a result vessel construction did result in a net increase in fishing capacity.

Finally, the use of this instrument has not been necessarily in a way to ensure balance between fleet and resources. In a number of cases, the new vessels being built and the ones being scrapped as capacity withdrawals did not belong to the same fleet segments. As a result, certain fleet segments have seen a net increase in capacity, while the real capacity reduction has taken place in other segments where perhaps this reduction was not so necessary. In other words, this mechanism has been used in a number of cases without proper targeting to reduce imbalances.

There is an important lesson too from this experience. Many, from Member States and industry have advocated that new constructions, if properly balanced by larger withdrawals of active fleet capacity, could have provided a real reduction of fishing capacity. But failure of many to deliver such mechanism (by refusing to accept a really high percentage of capacity withdrawal for every new construction) ended up discrediting the mechanism itself. The damage to the policy is such that today the new constructions, even under very high percentages of capacity withdrawal, are no longer politically acceptable. Once more, applying a measure under quantitative conditions that nullify their effect is a bad strategy that ends up discrediting the measure itself, perhaps forever.

On this basis, the very use of this structural instrument to modernise fleets while at the same time reduce overall fleet capacity has been traditionally called into question, as a mechanism that de facto increases overcapacity. This led to its suppression as of 2004.

Overall, the lack of consistency between the fleet and the structural policy, that at least until 2004 have been contradictory in relation with construction and modernisation aid, constitutes the best example of how insufficient consistency between the different policy pillars within the CFP has undermined the effectiveness of the whole policy.

Scrapping

Scrapping premiums (subsidies paid to vessel owners to decommission their vessel) have also been part of the structural policy from the beginning and, after the decision on the future European Fisheries and Maritime Fund for the period 2014–2020, they will continue to be used until 2017 (EC, 2006). Unlike new constructions, there is no question about scrapping contributing to reduce capacity. In this case, the question is by how much and with which cost/efficiency ratio.

It is beyond doubt that scrapping has been a very useful instrument, helping the Union to resolve in many cases well-identified overcapacity problems. The necessary reductions of, say, the fleet operating in NAFO, or that of the purse-seine fleet fishing for Bluefin tuna in the Mediterranean are positive examples of how well targeted scrapping can contribute to resolve situations of clear overcapacity as compared with available resources.

However, what is much less clear is the overall contribution of the scrapping premiums to significantly resolve the problem of overcapacity in European fleets as a whole. There are several reasons for this scepticism:

Firstly, the use of scrapping has been in some cases too indiscriminate and not well targeted to the fleet segments that really needed capacity adjustment. In certain circumstances, the criteria to allocate scrapping premiums has been on a 'first come first serve' basis, which implies that the vessels scrapped were those of vessel owners who wished to quit the fishery, for whatever reason, but not necessarily those of fleet segments with the highest overcapacity.

Secondly, in certain cases some of the vessels scrapped were hardly active and operational. There are a number of examples of this. Although this effect is difficult to quantify (there are no systematic studies about this question), abundant anecdotal information from Member States show many cases where the vessels scrapped were hardly active. In these cases, the withdrawal of capacity (at a high cost for the European taxpayer) did not result in an equivalent reduction of the fishing mortality exerted on European fish stocks.

Thirdly, in a number of cases the public money invested in scrapping was reinvested in the modernisation of existing vessels or even on the construction of new, more efficient ones. In these cases it can certainly be argued that the capacity reduced through scrapping was at least partially reintroduced in the form of technical improvement in the remaining vessels of the fleet. As a result, the real reduction of capacity was lower than the nominal reduction resulting from the overall nominal capacity withdrawal. This case has been well documented for Denmark but has probably happened in many other Member States where the question has not been studied.

Scrapping premiums can also be seen as a safety net that encourages vessel owners to overfish the stocks. If stocks collapse, the scrapping premium can be used to exit the fishery without loss. This can contribute to removing the incentive for the industry to support cautious and sustainable fishery policies.

Finally, considerable reductions of fleet capacity have also taken place without scrapping funds, which show that such funds are not the only way to address capacity reductions. According to a recent study, the Financial Instrument for Fisheries Guidance (2000–2006) contributed only to 50% of the capacity reduction observed in that period in the main Member States. For the European Fisheries Fund (2007–2013) this percentage increased to 78% but the percentage of vessels retired without scrapping aid was still very significant (European Commission, 2013).

Overall and despite partial successes, the scrapping aid has been an instrument of a doubtful cost/efficiency ratio. Between 1997 and 2006, Europe has spent over €900 million on scrapping. Despite this huge investment, fleet overcapacity is still a problem in Union fisheries, no matter how difficult it may be to measure it.

It is to be noted that the European Court of Auditors (2011) examined the use of scrapping in the CFP and arrived to very similar conclusions:

- Many investments on board may have increased the ability of the vessels to catch fish, thus contributing to overcapacity while formally respecting the fleet capacity ceilings;
- Scrapping has been to a considerable extent non-targeted, so public money has not necessarily been spent where the reduction of capacity was most needed;
- In addition, scrapping premiums have been decided in a way that does not ensure cost-efficiency of the measure;
- The fleet register is not updated, so the real evolution of the nominal capacity is not well known in all cases.

As previously, the use of scrapping under non-targeted, less than ideal conditions in some cases has resulted in a considerable damage to the reputation of this type of measure, even if in many cases the scrapping was well targeted and resulted in very positive adaptations of fleets to resources. This is another important lesson to retain: its ineffective use of an instrument by some (probably not the majority) can discredit an instrument and ultimately provoke its rejection by policy makers, thus losing its potentially positive effects when properly used.

Temporary laying-ups: an instrument to undermine fleet capacity adjustment?

Another instrument traditionally applied is that of temporary laying-ups. Although in some cases this has been very useful to provide an economic and social cushion to situations of emergency (such as the sudden finalisation of a fisheries agreement, or the unforeseen collapse of a fishery), the extended use of this instrument as a regular instrument of fisheries management can be questioned, since it undermines the necessary adaptation of fleet capacity: by temporarily addressing mismatches between fleet capacity and resource availability, lay-ups can be used as an excuse not to address more structural overcapacity problems.

As with scrapping, we must refer to the success stories, such as the use of this money to pay for the compulsory stoppage of certain fleets that operate exclusively in waters of the third countries, if and when the agreement is temporarily suspended. The necessary stoppage of the anchovy fishery in the Bay of Biscay in 2005 (see Chapter 4) was economically acceptable thanks to the use of this kind of aid. In addition, in case of a catastrophic event such as an oil spill, the availability of these funds has proven crucial to address the necessary (temporal) stoppage of the fishing activity.

Temporary cessation has also been used in certain cases to render some measures more socially acceptable. In these cases, it can certainly be argued that the measure makes politically possible certain necessary restrictions for biological reasons that would otherwise be politically unpalatable and probably not applied for that reason.

But the negative case is when temporary cessation becomes a kind of regular subsidy to a fleet. As that study concludes, for certain specific cases:

> ...public funding has been more useful in rendering the measures politically acceptable than in actually reducing the amount of fishing ... temporary cessation subsidies have become entrenched and incorporated into yearly payments to cover vessel owners' fixed costs during periods where they would be inactive anyway.

Temporary cessation may be biologically or politically useful in a number of circumstances, particularly in the case of unforeseen stoppages of the fishing activity (oil spills, sudden interruption of fisheries agreements etc.). However, it is difficult to find an argument to the effect that this instrument has contributed to address overcapacity on a permanent, structural way.

Has the fleet policy delivered?

As we have seen, the objective of the fleet policy since the inception of the CFP has been to ensure that the capacity of the European fleets is in balance with the available resources. This has been addressed with different instruments and different objectives. The question is: has this policy delivered its objective?

There is no straightforward reply to this question. There seems to be a contradiction between some of the conclusions we can draw from the fleet policy itself and some of the observations on the status of biological resources.

However, the surprisingly stable rate of reduction of capacity over the years, regardless of policy objectives, would seem to indicate that fleet capacity reduction is led by economic or technical factors, not by policy.

While the evolution of the stocks (see Chapter 4) shows that there has been a real reduction in the overall rate of fishing mortality and that this reduction started roughly towards the turn of the century, the overall tendency in total fleet capacity is such that it may have been largely compensated by technical progress (see Figure 5.3). The question is: has this reduction of fishing mortality been caused by a real reduction in fleet capacity?

Figure 5.3 casts serious doubts as to whether the real capacity of European fleets (in terms of the real

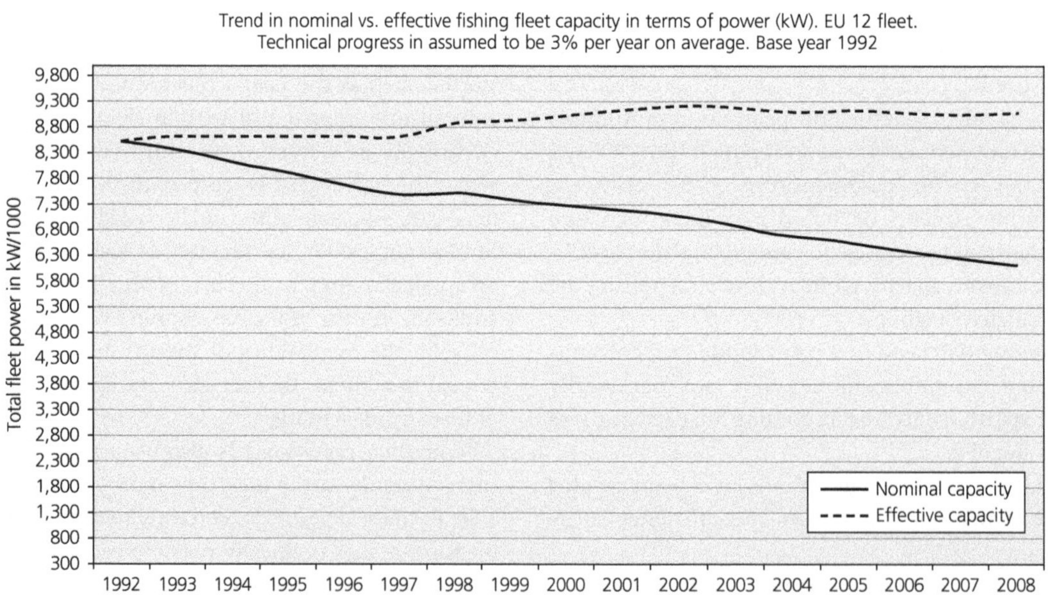

Figure 5.3 Comparison between the nominal reduction in capacity and the effective capacity taking into account technical progress. Source: European Commission Services. (See colour plate section for the colour version.)

capacity to produce fishing mortality) has really been reduced, even though the nominal capacity (expressed in engine power and tonnage) has clearly been substantially cut. Of course this image is the result of an assumption of a 3% annual increase in the technological creep and this may be also questioned.

It is very difficult to establish a clear cause/effect relationship between the trends observed in fleet capacity and the reduction of fishing mortality observed in recent years. At least, the figures on the evolution of fleet capacity (Figures 5.1 and 5.2) show a continuous downward trend over the years. In contrast, the trends in fishing mortality (see Figure 4.5 in Chapter 4) only show a downward trend since the late 1990s or early 2000s. So it is tempting to conclude that, whatever happened with fleet capacity, the reduction of fishing mortality in the last decade may have other causes as well.

There can be several hypotheses around this. Of course these hypotheses are not mutually exclusive and there may have been a combination of them in practice:

- *Technological creep lower than believed.* A first hypothesis is that, despite technological creep, there has been a very real reduction of capacity in a number of fleets. This implies that the technological creep may have been overestimated, so the nominal reduction in capacity is closer to the real reduction than is implied in Figure 5.3.
- *Effort reduction more than capacity reduction.* A second hypothesis is that the reduction of fishing mortality is not necessarily due to a real capacity reduction but to a reduction of the total fishing effort exerted by the fleets. In other words, that fleets have reduced their fishing time overall. In Chapter 4 we have seen that there have been certain effort management schemes that seem to have produced a real reduction of effort.
- *Increase in the number of non-active vessels.* A third hypothesis is that the real reduction of activity has been due to the existence of an increasingly high number of non-active vessels in the European fleets (COM, 2014b). In other words, that even in cases where the capacity reduction has not fully taken place, many vessels stay in port inactive and while they still 'exist' in the overall evaluation of capacity, they do not exert any fishing mortality in practice.

Although it is very difficult to draw clear conclusions, this author would argue that in any case the fleet capacity reduction has made a partial but positive contribution to the reduction of fishing mortality observed. Even if much of the reduction of fishing mortality observed has been the result of a reduction of activity (effort restrictions and inactive vessels) one can certainly argue that without the emphasis on fleet reduction and the European taxpayers' contribution to this end through scrapping funds, the problem would be much worse than it is today.

Is there an alternative to fleet policy? The case for rights-based management

The debate about fleet policy inevitably leads to the consideration of a very well-known but highly controversial system: the rights-based management (RBM) that very often takes the form of the so-called Individual Transferable Quotas (ITQs). These systems consist of allocating fishing rights to individual vessel owners. Such rights can be traded among operators under certain rules and safeguards. On that basis, operators can buy and sell fishing rights and, in this particular context, can decide what kind of fleet capacity they need to exploit their rights in a profitable manner.

A number of countries around the world use these systems to handle fishing rights and, in doing so they do not have any specific fleet policy: private operators decide on such matters. In Europe, the oldest ITQ system has been used in the Netherlands. But many other Member States have introduced such systems in at least part of their fleets. Spain, Denmark, Estonia and Sweden among others, have applied these systems at different stages.

Outside the EU, this kind of system was put into place by New Zealand in 1986. Many other countries in the world have followed. Among them, Canada, United States, Iceland and Norway have all applied RBM-like systems in at least part of their fisheries.

Every RBM model in the world is different. Norway introduced its structural quota system in the 1990s for its offshore fleets. The system was extended to other fleets segments and, since 2007, all vessels longer than 11 m are regulated by that system. According to a study for the Commission (EC, 2009), the system produced, since its inception, a 50% decrease in the number of offshore vessels and around 30% reduction in the coastal fleet segment. Through public scrapping, this reduction would have cost the Norwegian taxpayer around €2 billion. In parallel, the levels of profitability of the Norwegian fleet have increased. Interestingly, the fleet reduction has inevitably led to loss of employment on board but at much lower levels than the reduction in the number of vessels. This is allegedly due to the need for more employment necessary for the remaining vessels, enjoying higher quotas.

In the Union, the case of Denmark is well known. Introduced in 2003 for the herring fishery, it was later extended to industrial fisheries and mackerel and, as of 2007 extended to demersal species. As in Norway, the system has resulted in considerable reductions of capacity: 50% reduction in the pelagic fleet. The profitability levels improved significantly as a result. There has been, on the other hand a clear concentration of the fishing rights in fewer hands and the implementation of the system was aided at the beginning by scrapping with public funds. The study estimated that the reduction of capacity realised through the Danish RBM would have required €100 million of taxpayer's money.

Another case in hand is that of Spain's 'fleet of the 300', a closed list of vessels authorised to fish in the waters of the Community of 10 upon accession in 1986. Given that these vessels could not operate all at the same time (as we saw in Chapter 2) and that Spain never had enough quotas for these 300 vessels to fish year-round, this fleet of 300 vessels represented a clear case of overcapacity (as a result of the conditions for accession).

In order to resolve this situation, an RBM-like system was put into place in Spain allowing the purchase and sale of the days of activity allocated to these vessels, so that vessel owners could buy a vessel of the 300, scrap it and accumulate the fishing days in the remaining vessels of the list. As a result, the fleet could survive economically, albeit with a strong reduction in the number of units: the original 300 vessels have been reduced to less than 100 today.

Are RBM systems a panacea?

Of course RBM systems are no panacea. Like all instruments, RBMs have advantages and disadvantages. These are highly dependent upon the specific design of the scheme and its degree of adaptation to the characteristics of different fisheries or countries and, consequently, one cannot draw one-size-fits-all conclusions on pros and cons of different RBM systems having very different designs and safeguards.

The evaluation of RBM is largely a question of the socioeconomic model favoured by countries for the management of their resources. Where fisheries are considered an industry like any other, RBM tend to be valued positively. Where fisheries are considered as a strategic sector to be treated with government protection, they tend to be rejected. But beyond this classification, the advantages and disadvantages of RBM systems is also intensively debated by scientists and scholars, with strong supporters (Beddington et al., 2007) and strong opponents (Copes, 2000). In any case, it seems clear that RBM systems, on their own, are no guarantee of sustainable management: they must be part of a management system including other kinds of instruments to ensure the sustainability of the biological stocks (Chu, 2009), and to deliver their objectives, they must be very carefully designed (Sumaila, 2010).

In general terms, RBM systems tend to be favoured in economic analyses, given their attractiveness to save costs, to adjust capacity autonomously and to generate economic rent (Sche et al., 2008). But social scientists tend to be more critical, on account of undesired effects such as the accumulation of fishing rights in relatively few hands.

The discussion is endless and sometimes highly ideological. However, a certain number of generic conclusions can be drawn from the experience of the implementation of this type of instrument in a number of countries, including EU Member States:

- In general terms, countries having applied this kind of instrument can claim that they have well adapted the capacity of their fleets to available resources, so they tend to be efficient in achieving capacity reductions when necessary;
- RBM systems address the problem of overcapacity from an economic, not political perspective. In doing so, they remove the contradiction referred to earlier in this Chapter between the political positions and the private individual decisions on fleet reduction;
- Fleets managed under RBM systems tend, in general, to be more profitable that those managed under more traditional systems. However, this is at the cost of fewer vessels and fewer jobs at sea (a trend observed in all industries in any case). Whether the former offsets the latter, it depends of the political objectives of the fishery managers;
- RBM systems do not require heavy investments of taxpayer's money to address overcapacity. On the contrary, they are amenable to cost-recovery in fisheries management. By associating fishing rights with a cost, governments can recover part of the rent generated by the sale of fishing rights to co-finance fisheries management, in particular control or research;
- On the other hand, RBMs handle capacity in a quantitative, not qualitative way. In other words, it is a system whereby the fishing rights end up in the hands of the most dynamic or profitable vessel owners and not necessarily in the hands of those exerting the most ecologically-minded or socially-desirable types of fishing;
- If not properly designed, they can have other effects: the accumulation of fishing rights in a few hands, or 'desertification' of rights in certain regions and segments, etc. Whether these effects are undesirable, or just a normal effect of economic competition, it depends again on the policy objectives of fisheries managers. These effects, in any case, can be addressed through policy design and specific safeguards.

Many of the possible undesirable effects can be corrected or avoided by system design: if the protection of small-scale fishing is an objective, the fishing rights of that segment can be ring-fenced, so that they cannot be sold to larger vessels. This in fact applies in some countries, thus combining the overall efficiency with an objective of social protection.

Overall, there is little doubt that rights-based management systems are effective in reducing fleet capacity, substantially and with no cost for the taxpayer. The question is, however, what are the economic and social consequences of such reduction, whether they are intended or not and whether they can be addressed properly through system design.

The use RBM-like measures for the fleet policy in the EU was proposed by the Commission for the 2013 CFP reform (the so-called transferable fishing concessions, TFCs) but the co-legislators did not support the proposal. In Chapter 16 we will refer to this question in detail.

Notes

1 FAO International Plan of Action for the Management of Fishing Capacity: www.fao.org/docrep/006/x3170e/x3170e04.htm.

2 Financial Instrument for Fisheries Guidance: see Chapter 6.

3 Article 40 of Regulation 1224/2009.

References

Beddington, J.R., D.J. Agnew and W. Clark (2007). Current problems in the management of Marine Fisheries. *Science*, 22 June 2007. **316**(5832), pp. 1713–1716.

Chu, C. (2009). Thirty years later: the global growth of ITQs and their influence on stock status in marine fisheries. *Fish and Fisheries*, **10**: 217–230. doi: 10.1111/j.1467-2979.2008.00313.x.

COM (1995). Annual report to the Council and the European Parliament on the progress of the multi-annual guidance programmes for the fishing fleets at the end of 1994. Brussels, 10.11.1995. COM (95) 463 final.

COM (2001). Annual report from the Commission to the Council and the European parliament on the results of the Multi-annual guidance programmes for the fishing fleets at the end of 2000. COM (2001) 541 final. Brussels, 1.10.2001.

COM (2003). Annual report from the Commission to the Council and the European Parliament on the results of the multiannual guidance programmes for the fishing fleets at the end of 2002. Brussels, 21.08.2003. 508 final.

COM (2013). Report from the Commission to the European Parliament and the Council on Member States' efforts during 2011 to achieve a sustainable balance between fishing capacity and fishing opportunities 85 final.

COM (2014a). Communication from the Commission to the European Parliament and the Council. Guidelines for the analysis of the balance between fishing capacity and fishing opportunities according to Art 22 of Regulation (EU) No 1380/2013 of the European Parliament and the Council on the Common Fisheries Policy. Brussels, 2.9.2014 545 final.

COM (2014b). Report from the Commission to the European Parliament and the Council on Member States' efforts during 2012 to achieve a sustainable balance between fishing capacity and fishing opportunities. Brussels, 29.4.2014 233 final.

Copes, P. (2000). Adverse impacts of individual quota systems on conservation and fisheries harvest productivity. Simon Fraser University. *Institute of Fisheries Analysis*. Discussion paper 00–2. June 2000.

EC (1990). Report of an independent group of experts on guidelines for the preparation of the multiannual guidance programmes in relation to the fishing fleets for the period 1992–1996. SEC(90) 2244 of 30.11.90.

EC (1996). Report of the group of independent experts to advise the European Commission on the 4th generation of multi-annual guidance programmes. 26.4.96. XIV/298/96-EN.

EC (1997). Council Decision of 26 June 1997 concerning the objectives and detailed rules for restructuring the community fisheries sector for the period from 1 January 1997 to 31 December 2001 with a view to achieving a balance on a sustainable basis between resources and their exploitation. (97/ 413/EC) OJEC NO L157 of 3.7.97, pp. 27–32.

EC (2002). Council Regulation (EC) No 2371/2002 of 20 December 2002 on the conservation and sustainable exploitation of fisheries resources under the Common Fisheries Policy, *OJ* L 358, 31.12.2002, p. 59–80.

EC (2003). Commission Regulation (EC) No 26/2004 of 30 December 2003 on the Community fishing fleet register.

EC (2006). Regulation of the European Parliament and of the Council on the European Maritime and Fisheries Fund [repealing Council Regulation (EC) No 1198/2006 and Council Regulation(EC) No 861/2006 and Council Regulation No XXX/2011 on integrated maritime policy.

EC (2009). An analysis of existing Rights Based Management (RBM) instruments in Member States and on setting up best practices in the EU: Study published 01/02/2009. http://ec.europa.eu/fisheries/documentation/studies/rbm/index_en.htm.

EEC (1983a). Council Directive 83/515/EEC concerning certain measures to adjust capacity in the fisheries sector. *OJEC* No L 290 of 22.10.1983, pp. 15–19.

EEC (1983b). Council Regulation No 2909/83 of 4 October 1983 on measures to encourage exploratory fishing and cooperation through joint ventures in the fishing sector, *OJ* No L290 of 22 October 1983.

EEC (1985). Commission Decision of 24 April 1985 on the multiannual guidance programme in respect of the fishing fleet submitted by the federal republic of Germany pursuant to Council Regulation (EEC) No 2909/83. (85/277/EEC). *OJEC* No L 157 of 15.6.85, pp. 1–3.

EEC (1986). Council Regulation (EEC) No 2930/86 of 22 September 1986 defining characteristics for fishing vessels, *OJEC* No L274 of 25 September 1986, p. 1.

EEC (1988). Commission Decision of 11 December 1987 on the multiannual guidance programme for the fishing fleet (1987 to 1991) forwarded by the federal republic of Germany pursuant to Regulation (EEC) No 4028. (88/139/EEC). *OJEC* No L67 of 12.3.88, pp. 14–17.

EEC (1992). Commission Decision of 21 December 1992 on a multiannual guidance programme for the fishing fleet of France for the period 1993 to 1996 pursuant to council regulation 4028/86. (92/588/EEC). *OJEC* No L 401 of 31.12.92, pp. 3–7.

Engelhard, G.H. (2009). One hundered and twenty years of change in fishing power by English North sea trawlers. In *Advances in Fisheries Science: 50 years from Beverton and Holt*. A. Payne, J. Cotter and T. Potter (eds). John Wiley & Sons, Inc., Hoboken. 568 pp.

European Commission (2013). Retrospective evaluation of scrapping and temporary cessation measures in the EFF. Final report, November 2013.

European Court of Auditors (2011). Special report No 12. Have EU measures contributed to adapting the capacity of the fishing fleets to available fishing opportunities?

Holden, M. (1994). *The Common Fisheries Policy. Origin, Evaluation and Future*. Fishing News Books. Blackwell Scientific Publications Ltd., Oxford. 274 pages.

Marchal, P., B. Andersen, B. Caillart, O. Eigaard, O. Gudayer, H. Hovgaard, A. Irioondo, F. Le Fur, J. Sacchi and M. Santurtún (2007). Impact of technological creep on fishing effort and fishing mortality, for a selection of European fleets. *ICES J. Mar. Sci.* **64**(1): 192–209.

OECD (1997). *Towards Sustainable Fisheries: Economic Aspects of the Management of Living marine Resources*. OECD, Paris, France.

OJEC (1994). Court of Auditors Special Report No 3/93 concerning the implementation of the measures for the restructuring, modernization and adaptation of the capacities of fishing fleets in the Community together with the Commission's replies. (94/C 2/01). *OJEC* No C 2 of 4.1.94, pp. 1–64.

Sche, F., H. Eggert, E. Gudmundsson, A. Hoff and S. Pascoe (2008). Fisher's behaviour with individual vessel quotas-overcapacity and potential rent. Five case studies. *Marine Policy* **32**(2008): 920–927.

Sumaila, U.R. (2010). A cautionary note on Individual transferable quotas. *Ecology and Society* **15**(3): 36.

CHAPTER 6
Structural policy

The structural policy: the oldest component of the CFP

The structural policy of the European Union is one of its oldest and fundamental policies. The need to improve the factors of production of the economic sectors through public aid was both an imperative of the economic recovery and development in the post-war Europe and a powerful instrument for European economic integration by redistributing the wealth throughout the European territory. This policy has benefited many sectors, such as infrastructures, energy, communications and so on. But one of its more characteristic uses has been the support to the primary sectors of the European economy: agriculture and fisheries.

Although the official birth date of the CFP is still considered to be 1983, six years after the declaration of EEZs in the Atlantic, in reality the oldest pillars of the CFP can be traced back to 1970: the first regulation on structural aid to the fishing sector, together with the parallel regulation on markets of fishery products were adopted that year. Ever since then, the European fisheries policy has always had a specific structural policy (as well as a market policy) as a fundamental component.

As we saw in Chapter 2, the first attempt by the Commission to set up a Common Fisheries Policy in 1967 (through the document 'Basic Principles for a Common Fisheries Policy') emphasised structural, market, trade and social questions. So the structural side of the CFP was there long before the conservation component was identified as a policy priority. This initial approach was largely inspired by the parallel development of a common agricultural policy focused on the increased production of food and the support to the agricultural sector.

While the initial objective of improving the factors of production of the fishing industry (in a very wide sense of the word, including processing and ancillary industry, infrastructure, aquaculture etc.) has remained remarkably stable over the years, the evolution of the policy has been marked by the increasing relationship between structural policy and the objectives of the policy as a whole. While the first objective has been an impressive success in many respects, the second one has proven much slower in bearing fruit.

This policy has always divided Member States in two camps. Although the composition of these camps has varied over time and also varied as a result of enlargement, the divide is, today, fundamentally the same as in 1970. Certain Member States have seen, from the very beginning, the structural policy as the centrepiece of the CFP, based on the idea that the fishing sector is a strategic one, as a food-producing sector, thus needing public help by definition. Others on the contrary see this policy as an interventionist, paternalistic policy on a sector that should fend for itself, like any other economic sector in the European economy.

In addition, not surprisingly, net contributors to the European budget have always been more critical about the need and opportunity for this policy pillar, which is difficult to justify to taxpayers. This division

The Common Fisheries Policy: The Quest for Sustainability, First Edition. Ernesto Penas Lado.

existed already between the Member States of the Community of six. Today, with 28 Member States it is still alive, although with many more nuances.

These differences of approach are deeply rooted in historical and cultural traditions, as well as in the relative economic weight of the fishing industry in each case. These differences tend to perpetuate themselves in most Member States irrespective of the ruling political party, in a demonstration that these positions are most unlikely to change significantly over time, let alone disappear.

Beyond this question, there is another that remains as a permanent philosophical conundrum: while the intention of the structural policy is to make itself unnecessary (once the structural problems it is supposed to address are resolved) in practice the need for a structural policy has become permanent. This indicates that either the objectives of the structural policy are never fully achieved, or else the real objective of the policy is to have a fishing sector permanently assisted by public aid. This debate always permeates, explicitly or implicitly, the discussions on the structural policy of the CFP, because it represents different socio-economic models and objectives that Member States want for their fisheries sectors, according to their different cultures and traditions.

The structural pillar of the CFP has largely been a stand-alone element for many years, justified on its own goals, regardless of its contribution to the achievement of other CFP objectives that have been developed later. At the time the conservation policy was set up in 1983, the fisheries structural policy had been running for 13 years, creating a 'culture' around it that is extremely difficult to change. The history of the structural policy in fisheries is thus the story of how this policy has gradually been incorporated into the CFP and how it has been adapted to contribute to achieve the fundamental objectives of the policy.

The 43 years of structural policy

Despite its remarkable continuity, the structural policy has also gradually evolved over the years, both in terms of its content and scope, its structure and, crucially, in terms of its linkages with other policy pillars.

The first regulation in 1970

As we saw in Chapter 2, prior to the accession of the UK, Denmark and Ireland in 1972, the Community of 6 adopted Regulation 2141/70 establishing a structural policy for the fisheries sector (EEC, 1970). This was a first attempt to establish some financial support to the fleets of the existing Member States in view of the imminent enlargement. This first regulation aimed at modernising and increasing the competitiveness of the fleets in the six original Member States.

While this objective was agreed, other objectives, such as providing for a social policy in fisheries, defended by some Member States (in line with the initial views of the Commission, expressed in its first communication on the CFP in 1967, see Chapter 2), were not adopted, setting the tone as to what would be the general orientation of the structural pillar within the CFP. Already, in 1971, aid was available under the Guidance section of the European Agricultural Guidance and Guarantee Fund, in particular to encourage the construction and modernisation of inshore and pelagic fishing vessels, as well as processing and marketing of fish.

The measures contemplated by this first regulation had as an objective to: 'promote the rational development of the fishing industry within the framework of economic growth and social progress…'

To this end, eligible measures were aimed at: 'increased productivity through restructuring of fishing fleets and other means of production…'

The regulation excluded, however, any mention to balancing fleet with resources, for example. As we saw in Chapter 5, the idea that the fleet should be reduced to ensure the sustainability of resources took considerable time and effort to make headway. This structural policy was in fact still an extension of the structural policy for agriculture. Its philosophy clearly shows the emphasis on development and production, along the lines of the Common Agricultural Policy of that time.

While in agriculture greater investment means more production, in fisheries more investment leads, at a certain stage, to overfishing and lower productivity. However, the need to introduce restrictions in fishing levels to ensure sustainability was not yet there and as we saw in Chapter 4, the 1960s and

1970s witnessed a significant increase in the levels of fishing activity.

During this period, different programmes contributed to the build-up of fishing capacity in Europe: between 1973 and 1977 a specific aid was given for the reconversion of the cod-fishing fleet in France and Germany through new constructions and modernisation. The inshore fishing fleets of different Member States were also subject to restructuring between 1978 and 1982, with 81.7 million ECU spent on construction and modernisation of vessels between 6 and 24 m. It was not until 1983 that the new structural instrument included aid to the withdrawal of fishing capacity (Hatcher, 2000).

The second instrument in 1976

In 1976, in the context of the imminent declaration of EEZs, leading up to the birth of the CFP, a new approach was agreed with the adoption of a more comprehensive structural instrument for the fishing industry (Regulation 101/76: EEC, 1976a). In parallel a separate administrative body was established to manage it: that year fisheries management was separated from agriculture in the administrative structure of the European Commission and a specific Directorate General for Fisheries (DG XIV) was set up in the Commission's structure. At the same time, a new regulation on the markets for fish products was also adopted (EEC, 1976b).

The new structural regulation established as an objective: 'to promote harmonious and balanced development of this industry within the general economy and to encourage rational use of the biological resources of the sea and inland waters'. Although the objective of this second regulation already introduced language relative to the biological resources and not just about the development of structures, the focus was very similar to that of the first one. In fact, the eligible actions were practically the same as in 1970, notably to increase productivity of fleets and other means of production, to adapt production to marketing conditions and to improve the living conditions of fishermen. No specific measure was foreseen to 'encourage the rational use of biological resources'.

The new Member States of the 1972 enlargement that had considered the 1970 regulation as a protectionist reaction of the Community of six to the imminent enlargement ended up accepting the idea of public subsidies to their fleets. This was due largely to the fact that the instrument was considered by some older Member States as an *acquis* whose discontinuation would be politically unacceptable. This is perhaps the first example of the tendency of the CFP to stick to its own traditions, making reform very difficult.

The funding for the processing industry was in principle not available in the 1970 and 1976 regulations, since the emphasis of these first fishery-specific regulations was to concentrate on the fleets. However, this kind of aid also became available under a different source: Regulation 355/77 to support to food processing industry, both for agricultural and fisheries products (EEC, 1977). This joint support was in force until 1989, when the support to agricultural and fishery processing was split. From then on the support to the processing industry for fishery products has been part of the fisheries structural policy.

Structural policy and the birth of the CFP in 1983

In 1983, coinciding with the 'birth' of the CFP, a new structural instrument came into play through Regulation 2909/83 (EEC, 1983). This regulation sought to explore alternative fishing resources and to export to third countries, through joint ventures, the 'surplus fishing capacity', notably corresponding to vessels over 24 m whose activity would be restricted in their traditional fishing zones.

This was clearly the first sign that, even if the first conservation measures in 1983 were not extremely restrictive, the adoption of a conservation policy inevitably led to the need to address the question of the capacity of the fleet in the new policy context. At the time, it was already perceived that the new TAC and quota policy would not deliver enough quotas to keep all the fleets developed in the 1960s and 1970s economically viable.

However, the policy option was not yet to propose the destruction of the fleets that the policy itself had contributed to build in previous years but rather to try to find alternative opportunities for the redundant vessels, either as new, untapped resources, or as resources outside European waters.

These measures continued for some time as part of the structural policy and a number of joint enterprises with third countries, on that basis, were successful in developing the fishing sector of the third countries concerned. However, the way in which the new policy was formulated would contribute, years later, to its own discontinuation: by recognising that the policy tried to 'export overcapacity' such measures became easy victims of criticism that the EU 'exported its problems to third countries', thus creating a bad image for this type of measure that eventually led to its suppression years later.

This bad image, however, did not do justice to these measures that in some cases contributed to promote healthy fisheries investment in third countries wishing to develop their own industry. We will see some examples in Chapter 7.

The accession of Spain and Portugal and the new regulation in 1986

Following the accession of Spain and Portugal, a new structural regulation was adopted in 1986: Regulation 4028/86 (EEC, 1986). This regulation represented a turning point in the structural policy: from almost exclusive emphasis on developing the factors of production, the new structural policy really started to look into the question of how these factors of production related to the available resources. Although the linkage was still weak, the regulation introduced new elements in that regard. In particular, this regulation set out to: 'facilitate the change in the fisheries sector within the guidelines of the Common Fisheries Policy'.

Two elements are noteworthy in this respect:

1 The more clear linkage between the structural policy and the CFP as such, in other words, the introduction of the notion, albeit in a rather vague manner, that the structural policy should aim to facilitate the achievement of the objectives of the CFP;
2 The more nuanced combination of measures to improve and modernise the factors of production, with the measures to tackle the incipient problem of overcapacity, notably with the inclusion of measures for: 'the adjustment of fishing capacity by the temporary or permanent withdrawal of vessels from fishing activities'.

From the initial objective of modernising the continental fleet to confront the newcomers, the emphasis shifted to adapting everybody's fleets, once the newcomers were in. Once again, the question of the 'digestion' of the large Spanish fleet in the CFP had an important bearing on this policy shift.

This structural policy was to be delivered in the context of the fleet reduction objectives determined in the MAGPs as we saw in Chapter 5. However, whether or not Member States fulfilled the MAGPs objectives, this had no effect on Member States' rights to use the fleet measures of the fund. In other words, the structural measures were applied in a way that was unconditional to the achievement of the objectives. This would continue to constitute a serious shortcoming for the years to come.

It is also worth mentioning that, up to that period, the structural policy was fundamentally managed directly by the Commission services. Individual projects of vessel construction in Member States were handled by the Commission services. This had certain advantages, notably the use of exactly the same standards when deciding on the financing of projects. For example, vessel construction projects were financed, within the financial possibilities, according to the percentage of capacity of withdrawal, that is, the projects with the highest capacity withdrawal (the difference between the capacity of the vessel withdrawn and that of the vessel built) were systematically prioritised, as a way to ensure that new constructions would indeed contribute to reduce fleet capacity as much as possible.

In subsequent years, as we will see, this would change, as the fisheries structural policy was integrated into the mainstream of the structural policy and further structural instruments would be managed under shared management with Member States. This was an inevitable decision, as the enlarged Union would make the old system increasingly difficult to manage for the Commission services. However, this would also come at a cost: under shared management it would be more difficult to ensure that the structural money was spent in as strict a way in terms of contributing to achieve policy goals.

The first FIFG: 1994–1999

The dispersed regulations on the structural policy were first unified into a single instrument in 1994: the so-called Financial Instrument for Fisheries Guidance (FIFG) laid down by Regulation 2080/1993 (EEC, 1993). The first such programme ran from 1994 to 1999.

This new regulation was adopted shortly after the first revision of the CFP in 1992. As we will see in Chapter 15, this revision was very limited and, not surprisingly, the 1993 structural policy did not represent a very substantial change in the structural policy either.

The architecture of the instrument was new but the substance did not change much. For the first time, the structural policy for fisheries was integrated into the general framework of the Structural Funds under Regulation 2052/88 (EEC, 1988). This meant, *inter alia*, the adoption of a delivery system closer to that of the structural funds, based on shared management with Member States instead of direct management by the Commission from Brussels. Individual projects, hitherto managed by Commission officials from Brussels, were transferred to national management, with the Commission taking a new role, that of the overall verification of the work of national administrations.

The adoption of this new structural framework preceded a very critical report of the European Court of Auditors on the way the structural policy was conducted (European Court of Auditors, 1994). This report already signalled a number of the problems that have affected traditionally the structural policy in particular:

- The lack of coordination between the structural policy and the other elements of the CFP.
- The requirement that new constructions were at least fully compensated with the same level of capacity withdrawal was not always achieved, particularly in terms of engine power.
- Modernisation projects frequently led to significant increases in capacity, in particular through increases in engine power.

Council reacted to these findings by trying to tighten up the conditions attached to vessel construction and modernisation but progress was limited by what Member States could accept, as we will see next.

This first FIFG established three main objectives:

1. to contribute to achieving a sustainable balance between resources and their exploitation;
2. to strengthen the competitiveness of the sector; and
3. to improve market supply and the added value of fishery products.

The eligible actions included all the measures previously introduced in earlier regulations. This was a good example of the tendency of the CFP to accumulate new layers of policy but without necessarily re-focusing them on priorities. As the structural policy progressed, it was clear that opening new possibilities was much easier than eliminating existing ones, thus resulting in an increasingly wide range of structural measures available. Theoretically, the inclusion of this wider palette of possible actions would allow for the establishment of clear priorities among the different possibilities available.

From today's perspective, the inclusion of fleet renewal in a regulation aiming at achieving balance between fleet and resources may seem contradictory. We must bear in mind, however, that this instrument was considered under conditions that would ensure that the new constructions would not prevent the achievement of the fleet reduction objectives established in the MAGPs for the period 1992–1996. The implementing regulation for the first FIFG, Regulation 3699/93, in its Article 10, established that: 'Member States may take such measures to promote the construction of fishing vessels as comply with the global annual intermediate objectives and the final objectives by segment under their national multiannual guidance programme within the stated time limits' (EC, 1993a).

This condition was rather vague particularly in a context where compliance with the MAGPs limits was less than perfect. One can only speculate what would have been the effect of vessel construction if it had been made dependent upon a very ambitious objective in direct capacity withdrawal, as indeed was unilaterally applied by some Member States.

In terms of implementation, by and large the fleet measures clearly took most of the budget: appropriations corresponding to 'Renewal and modernisation of the fishing fleet' comprised 26.4% of the total budget, while the 'Adjustment of fishing effort' (scrapping and temporary laying-ups) comprised 22.7%. So nearly 50% of the budget was dedicated to fleet measures while the processing industry absorbed 23% of the budget and aquaculture the 9.4%.

The second FIFG: 2000–2006

The second FIFG, laid down by Regulation 1263/99 (EC, 1999a) and developed through Regulation 2792/99 (EC, 1999b), came about again in a context of policy continuity: this second FIFG, like the first one, was adopted after the CFP reform (or rather policy consolidation) of 1992 but before the following important policy change (the CFP reform of 2002). This means that the second FIFG was adopted in a scenario of little or no policy change as compared to that of the first FIFG. As a consequence, the second FIFG was fundamentally a continuation of the previous one, at least in terms of the eligible actions considered. There were, however, some new elements:

- The inclusion of specific 'socioeconomic measures', fundamentally consisting of the financing of early retirement schemes for fishermen. The social objectives of the CFP clearly re-entered the political debate, notably as a consequence of the emerging emphasis on the need to reduce fleet capacity since the second MAGP (see Chapter 5). As the scrapping premiums benefitted the vessel owner, it was felt that an aid to the crew members was necessary as a complementary measure.
- The inclusion of specific actions regarding small-scale coastal fishing, notably by providing a lump sum to collective projects. This was the result of a problem identified at the time: that when it came to benefit from the opportunities offered by the structural aid, the small-scale sector tended to be squeezed out of the access to the funds by the more proactive bidding for funds by other, more industrial and more organised sectors of the fishing industry.
- Innovative actions. These included training, pilot projects, demonstration projects, experimental fishing and so on.

An important element was the establishment of – theoretically – strict conditions to grant aid to new constructions: the obligation to require the withdrawal of capacity without public aid at least 30% higher than the new capacity being built. However, this condition was applicable only between 2000 and 2001 and 'where the annual objectives of the segments concerned are not yet respected.'[1] This meant that for segments where fleet capacity objectives were met, new constructions could go ahead without capacity reduction. And, given the limited value of such capacity limits, as we saw in the previous chapter, we can easily understand why the instrument of new constructions, under such conditions, gave rise gradually to the bad image of contributing to the aggravation of overcapacity.

The reflection can be made again as to what would have happened if, for example, all new constructions had been subject to a real reduction of capacity of 30% in all cases. Under such conditions, perhaps new constructions may have actually been an instrument to promote the reduction of overcapacity.

In the second FIFG, in comparison with the previous one, there was already an important shift in terms of the percentage of the total budget dedicated to different actions:

Scrapping, modernisation and construction of vessels still took the lion's share of the fund but the percentage went down from nearly 50% to slightly over 30%. Processing and marketing (17.5%) and aquaculture (8.8%) were still important chapters but with a lower percentage of the total budget. In contrast, the chapter dedicated to 'other actions' increased to 33.3%, including notably social aid (departure of fishermen to retirement) and exploratory fishing, among others.

It is worth noting that the measures most related to the conservation policy (fleet measures) were reduced, while social measures increased, in yet another demonstration of the social sensitivity of the policy, as we will see next.

The European Fisheries Fund: 2007–2013

The European Fisheries Fund, applicable for the period 2007–2013, was laid down by Regulation 1198/2006 (EC, 2006). This regulation, unlike the previous FIFG, was a more evident product of the new CFP: that adopted in the reform of 2002.

The EFF established a clearer structure of eligible actions according to priority axis. It laid down four priority axes (the fifth axis being technical assistance):

- Axis 1: measures for the adaptation of the Community fishing fleet.
- Axis 2: aquaculture, inland fishing, processing and marketing of fishery and aquaculture products.
- Axis 3: measures of common interest, including notably port facilities and infrastructure but also including environmental protection, promotion campaigns and so on.
- Axis 4: sustainable development of fisheries areas.

Axis 1 was obviously adapted to the decision, in the 2002 reform, to discontinue aid to new constructions. It did, however, keep certain instruments for fleet management, such as temporary cessation and scrapping.

The new axis also suppressed a type of aid available still under FIFG2: the aid to the constitution of joint ventures by Community operators in the fishing sector of third countries. This kind of aid was very important to promote healthy investment in third countries as an alternative to the traditional bilateral agreements against a financial compensation (in Chapter 7 we will refer to the specific case of Namibia). However, the alleged cases of fraud and abuse of this instrument led to it being discredited and eventually to its eventual suppression.

Axis 2 and 3 were by and large a continuation of previous actions under the two FIFGs.

The new Axis 4 represented an important innovation. Modelled after Agriculture's LEADER initiative, the projects under Axis 4 focused on the problem of rural (in this case coastal) development by promoting socially inclusive economic diversification in fishery-dependent areas. This new axis represented a leap forward in fisheries policy, in that it identifies an entirely new group of beneficiaries: not just fishermen or aquaculture operators but the whole of the social structure of coastal areas.

The total funds amounted to €4.3 billion, of which €3.25 billion were allocated to convergence countries and €1.05 billion to non-convergence countries. As the fund was allocated to Member States under convergence criteria, this meant that the funds were allocated more on the basis of the relative wealth of these Member States, rather than on the size of their fishing sector.

Implementation of the EFF

The implementation of this fund is regularly evaluated by the Commission (COM, 2013). Until May 2013, the EFF money had been committed in five main areas:

- Fish processing (17.41%)
- Permanent cessation (17.25%)
- Aquaculture (14.83%)
- Fishing ports (11.46%)
- Temporary cessation (7.40%)

This represents a further evolution in the expenditure by policy pillar, as compared to FIFG. It is of particular relevance that still 25% of the funds are spent on – not necessarily well targeted – scrapping and temporary cessation.

This is, however, a global picture that hides very different approaches at national level. In fact, while some Member States invested a high amount on fleet measures, others invest more on coastal development under Axis 4, while others concentrate on processing. For example, in the Operational Programme of Ireland (COM, 2014a), 71% of the budget was allocated to fleet measures and only 2% to community-led local development. On the opposite side, Poland (Decyzja Wykonawcza Komisji, 2014) dedicated much more to Axis 4 (32%) than to fleet measures (17%) and Germany (Durchführungsbeschluss Der Kommission, 2012) favoured the collective measures (50%) and the processing industry (29%) to the fleet expenditure (only 2.3%).

This initial picture, however, is mitigated by the adjustments to the initial budget and, especially, by the different rate of consumption of these budgets, indicating that certain measures have a much higher demand from the sector than others.

The EFF implementation showed a limited capacity to measure the contribution of the structural funds to achieve the objectives of the CFP. For example, there has been an improvement in the economic performance of the fleets in recent years. But, is this improvement related in any way to the restructuring of the fleets and other means of production through the structural policy, or is it the result of other factors? This is, still today, extremely difficult to evaluate.

The same applies to the response of the sector to the oil crisis in 2008. At that time, a special amendment to the EFF was adopted to facilitate the adaptation of the fleets to the higher fuel prices (EC, 2008). Over the last few years, certain fuel-guzzling fleets have adapted remarkably well to the evolution of fuel prices. However, these adaptations seem to have been due to modifications in fishing patterns rather than to the implementation of the fuel package.

The implementation of the EFF is still pending a final evaluation by the Commission. This evaluation will be very clear in terms of budget execution, allocation of funds, problems encountered and so on. But it will still be difficult to evaluate the contribution of the EFF to the achievement of policy objectives, since the analysis of the effects of this policy on the economic performance of the fishing and aquaculture sectors is still preliminary (see next).

The new Regulation: EMFF

In early 2014, as part of the 2013 CFP reform package, a new financial instrument was adopted: the European Maritime and Fisheries Fund (EMFF) (EU, 2014). In historical terms, this new instrument was the most innovative one, for a number of reasons:

- It integrated this policy pillar in a more coherent package of structural policies: those represented by the so-called European Structural and Investment Funds, or ESI funds (in addition to the EMFF, the European Regional Development Fund, the European Social Fund, the European Agricultural Fund for Rural development and the Cohesion Fund), so as to allow Member States to prepare coherent packages of structural measures using the different funds to seek integrated solutions for development. In addition, for the first time these five structural funds shared all the basic common rules, through the 'common provisions' regulations for all ESI funds, in an attempt to reduce the complexity and administrative burden (EU, 2013).
- It widened the scope of the fisheries structural fund to include the funding of all policies related to the CFP, such as data collection, control and so on, hitherto funded under a separate cover. Likewise, it included for the first time measures to develop the maritime economy, thus promoting economic development in coastal areas over and beyond the (well-funded) fishing activity itself.
- It was a structural instrument designed, for the first time, to support the fishing industry and national administration in their efforts to deliver the objectives of the CFP as reformed. The links between the fund and CFP policy delivery developed to an unprecedented level.
- The access to public funds by Member States and by individual operators was made more conditional on good behaviour. The introduction of important conditionality mechanisms, whereby Member States must show administrative capacity to manage the CFP in order to have access to the funds, was introduced in the structural policy for the first time.
- The access to funds for fleet measures was restricted both in terms of the total amount available (with specific ceilings in respect of total expenditure) and in terms of the kind of action eligible for funding. On the other hand, a new emphasis was placed on improving control and data collection, as well as on the development of aquaculture and community-led local development.

In Chapter 16 we will refer to this new structural instrument in more detail.

Has the structural policy resolved the structural problems of the CFP?

The philosophy of the structural funds is to address the structural weaknesses of Europe's productive sectors and, ultimately, to make themselves unnecessary once these weaknesses have been addressed. So the question is whether the structural policy being implemented as part of the CFP since 1970 is addressing and resolving the structural weaknesses of the sector? Are those policies making the fishing industry self-sufficient or, on the contrary, have these policies created an industry permanently dependent on public aid? In particular, the question must be asked in connection with the founding fathers of the Union and the earlier Member States, all of which have had access to structural aid for many years (in the case of the founding States, they have benefited from structural aid to the fishing industry for 43 years!).

The evolution of the financial package

The financial package available for the structural policy in fisheries increased continuously. (Figure 6.1). The total expenditure in 1983 (with 10 Member States) was €53.6 million. This amount has been steadily growing to an annual average of €615 million for 27 Member States in the period 2007–2013. The total envelope for this period is €4.3 billion. For the current period, a total package of €5.75 billion was adopted under shared management with 28 Member States, while €647 million were allocated to direct management by the Commission, under the new European Fisheries and Maritime Fund (EU, 2014).

Of course, this increase is related to the number of Member States, also growing in this period. If we compare these figures with the – falling – number of workers in the fishing sector in Europe, one can argue that the structural money available per fishery worker in Europe has probably increased over time. This question certainly deserves a closer analysis.

The results: a mixed picture

This question, in itself, has two sub-questions where the success of the policy is in fact quite different:

1 one question is about whether the policy pillar has contributed to achieving its goals of modernising and improving the factors of production. In general terms this has been a considerable success;
2 another question is about whether the structural policy has effectively contributed to improve other policy goals in the framework of the CFP. This, in contrast, is much more mitigated.

In any case, the results are also very different for different Member States. The structural funds have traditionally offered Member States a wide range of possibilities for funding but the priorities of the expenditure, as set in the Operational Programme of each Member State, can vary very substantially in focus and instruments. But even more importantly, there have been differences in the philosophy applied to the use of the structural funds: while some Member

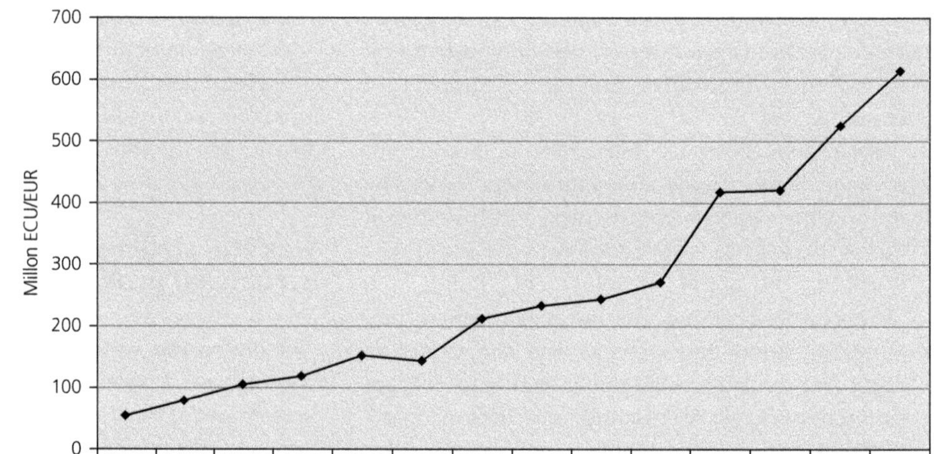

	EUR10 1983	EUR10 1984	EUR10 1985	EUR12 1986	EUR12 1987	EUR12 1988	EUR12 1989	EUR12 1990	EUR12 1991	EUR12 1992	EUR12 1993	EUR15 avg. 1994–99	EUR25 avg. 2000–06	EUR27 avg. 2007–13
Evolution of appropriations for fisheries structural policy - Commitments for the period 1983–2013	53.6	78.4	104.3	117.9	151.3	143.2	210.5	232.7	242.6	271	417	420.7	526.3	615

Figure 6.1 Evolution of the size of the financial package for the structural policy in fisheries, since 1983. Source: European Commission Services.

States have used the structural funds to promote the necessary change, others have tended to use them to preserve the *status quo*. This has obvious consequences on the results observed in different Member States.

The main failures of the structural policy have been those related to its insufficient effect in reinforcing other policy objectives. Although gradual progress has been achieved, this was fundamentally due to a very weak linkage between the implementation of the structural policy and the achievement of the objectives of other policy pillars. We can distinguish three main phases:

1 In a first phase, roughly between the first regulation in 1970 and the first one to design a global package of measures in 1986, there was no actual linkage between the structural policy and the other policies of the CFP. The structural policy had objectives in its own right;
2 In a second phase (roughly 1986–2013) the linkages were formulated to other policies but no effective mechanisms were put into place to ensure that the structural policy really contributed to achieve other CFP objectives. Traditionally, the management of this policy has focused on the formal compliance with the rules but not so much with the achievement of policy targets. In particular, emphasis has been traditionally placed on the eligibility of actions rather than on their real contribution to achieve policy goals;
3 The third phase is very recent; the adoption of stronger policy links and the adoption of specific mechanisms to enforce these links, in particular through the conditionality of structural aid, adopted in the wake of the CFP reform of 2013.

In any case, the structural policy has also been short of indicators that would allow the measurement of the contribution of the policy itself to the broader objectives of the CFP. Not only the linkages have traditionally been weak but our ability to evaluate their effectiveness has been very weak too.

Let us analyse the different cases.

Processing industry

This industry is an essential element of the European fishing sector. According to the last Annual Economic report of the European fishing Industry, the sector involves around 3500 companies (56% micro-enterprises) and more than 110 000 jobs (around 50% of them women) and generates €6 billion in gross value and €1.4 billion in profits. This gives this sub-sector a considerable importance in socio-economic terms, since it represents around 60% of the total value of the production chain and 34% of the employment of the European fishing industry. In addition, these industries are concentrated in coastal areas, where they are a major contributor to economic activity and jobs, particularly for women who represent a majority in this sector.

The sector has remained stable in recent years but has substantially increased its productivity and today it enjoys on average a higher level of profitability (average return on investment of 15% in 2011) than both capture fisheries and aquaculture. A significant component of this success has been the essential contribution from the structural policy: this sector has enjoyed the support of over €500 million between 2007 and 2013, the largest share of any other action under the EFF (JRC, 2013).

This is, beyond doubt, one of the success stories of the structural policy. One clear example is the extraordinary modernisation of fish canning facilities. Another is the development of storage or fish-processing facilities that have contributed across Europe to increase the added value of European fish or, in many cases, to attract landings of raw fish to European facilities where they are given added value and where they contribute to the creation of jobs.

Ancillary industries and infrastructures

The contribution of the structural policy to improve port infrastructures (auction markets, ice production facilities and many others) has also been spectacular over many years and represents another success story for the structural policy. This is by definition difficult to quantify, notably because the result is not just more but fundamentally better infrastructure. Anyone familiar (for a long time) with the fishing sector will agree that the improvement in fishery installations in Europe in the last two or three decades has been spectacular.

There is, however, a caveat to this success story: whether in some cases this development has given rise to some kind of 'infrastructure overcapacity'. In certain areas, the funding has been used, for example, to build or modernise auction markets in small towns relatively close to each other. As a consequence, it is not unusual to see clearly under-utilised auction markets when and where local fishermen prefer to land their catches in nearby larger auctions where buyers concentrate and prices tend to be higher. This is difficult to estimate but at least anecdotal information from certain Member States seems to indicate that such overcapacity may exist in certain European regions.

The development of aquaculture

Aquaculture development has been an important objective of the structural policy from an early stage in the CFP. As the CFP does not regulate aquaculture, the structural policy has been in fact the only instrument of the CFP to promote its development. We will analyse this sector in Chapter 14.

The effects of the structural policy on this sector are also difficult to evaluate but overall we can distinguish two phases in terms of quantitative production:

- The 1980s and 1990s, when aquaculture production in Europe developed quantitatively and qualitatively. Total production increased up to almost 1.4 million tonnes in 2000.
- From 2000 on, production has stagnated and in 2011 it was only 2.28 million tonnes (STECF, 2013), but there has been a process of improvement in competitiveness, particularly in recent years.

The intention to further develop European aquaculture has featured in the successive structural regulations, as we have seen previously. Still in 2002, the Commission had very high hopes for the development of this sector: a Commission strategy for the development of aquaculture set out as objectives a 4% annual increase in aquaculture production and the creation of 8000 to 10 000 full-time jobs over the period 2003–2008 (COM, 2002).

The strategy, however, already indicated that this objective could not be achieved simply by providing public money through the structural policy but that other crucial elements were also necessary, including *inter alia* the development of maritime spatial planning and the implementation of coastal zone management by Member States.

This has been analysed by the Commission in a number of initiatives and clearly shows that the structural policy, on its own, does not suffice to promote the activity in a meaningful and sustainable manner.

In any case, the last report on the economic performance of this sector sheds very interesting light into the effects of the structural policy on its development (JRC, 2014):

The economic performance of the sector has been improving in recent years, after a crisis in 2008 and 2009 that led to mergers and bankruptcies. The sector had, in 2012, a net benefit of €220 million, with a return on investment of 8%. The sector was still largely dominated by micro-enterprises and relied on part-time labour (80 000 workers equivalent to 30 000 full time jobs). Most importantly, the overall production in value has increased in recent years, even if production in weight has not. That means that European aquaculture is improving the average value of its products.

Key elements for the improvement are, as suggested by the study: vertical integration of enterprises ('from cage to table'), multispecies production, organic production and quality labels, investments in the fight against pathogens, improved fish feed conversion and training. All these actions have been financed under the EFF. In Chapter 14 we consider aquaculture in more detail.

The improvements in working conditions on board

This is another spectacular success. To anybody familiar with the fishing industry, the comparison between the fishing vessels one could observe in European ports in the 1970s and 1980s and the vessels one can see today is striking. The improvement in working conditions, hygiene safety on board has been impressive.

This success has been brought about by many of the instruments of the structural policy, including financial aid for vessel construction, which was discontinued in the CFP reform of 2002. This may seem a contradiction in principle. However, if we analyse this question more closely, it is obvious that the improvement of working conditions in fishing vessels is due to a variety of factors, including social policy (national competence), technological developments (e.g. in habitability or in efficient water desalination on board) and, ultimately, in the improvement of the industry's profitability due to a better balance between fleet and resources, which allows for a more capitalised industry capable of building modern vessels without public aid.

In any case, the structural policy has been instrumental in these improvements. Beyond the aid to new construction, the modernisation of vessels, still admissible until recently and especially, the permanent aid to the improvement of working conditions and safety on board (never interrupted) have certainly contributed to this progress.

Community-led local development

This is a relatively new strand in the structural policy. It exists only since 2007, in the context of the European Fisheries Fund for the period 2007–2013 under the Axis 4 of the policy, modelled after the LEADER initiative of the Common Agricultural Policy. Yet, although its implementation is still recent and some Member States are still today in the early stages of its practical application, this new policy strand has scored a number of big successes, in terms of developing activities complementary to the fishing itself and generating new jobs in areas strongly dependent from fisheries.

One of the reasons for a relative slow development of this policy pillar is the existence, in some Member States, of a certain reticence by the fishing industry to accept that what they always saw as 'their money' was now to be shared with local communities beyond fisheries itself. As one official from a northern European Member State once put to this author: 'fishermen think that fleet policy money is for them but that axis 4 money is for their wives'.

However, the potential of this instrument is considerable, for a number of reasons:

- It contributes to structure local communities, by helping to bring together different people around projects. Over and beyond the direct financial aspects, it creates a culture of entrepreneurship in small coastal communities;
- It is very socially inclusive, by providing aid not just to the traditional beneficiaries of the structural policy but also to a very wide range of social interests in coastal areas, including a crucial participation of women.
- It provides a relatively high return, in economic and social benefits, for a very low level of investment.

Although it is difficult at this stage to make a full evaluation of this instrument, there are already many interesting success stories where jobs have been created on activities related to the exploitation of marine resources. Many of these projects, particularly many of the most innovative, have been led and carried out by women.

Overall, however, being a new policy pillar in the fisheries sector, it is taking its time to implement in practice. As we saw previously, some Member States did not believe in this policy pillar and programmed a very low percentage of their total EFF funds for this purpose. In addition, the Axis 4 of the EFF has the lowest percentage of budgetary execution, with only 44.60% of the budget executed by May 2013, compared with 77% for axis 2 or 74% for Axis 1, for example (COM, 2013).

Despite these initial implementing difficulties, the Axis 4 has already shown it important potential: a recent study estimated that *circa* 300 Fisheries Local Action Groups, with budgets between €450 000 and 5.5 million contributed in creating 8000 jobs and to maintain 12 500 and as well as helping to create 220 businesses. The leverage factor was also significant: 1.06. Every Euro of Axis 4 attracted another €1.06 in additional investment from private and other public funding. Crucially, the study estimated that 61% of these businesses will continue after the EFF support stops (Capgemini Consulting, 2014).

Contribution to the achievement of CFP objectives

This is perhaps the most difficult part of the analysis. As we saw in Chapter 5, there is a relationship between fleet capacity and the conservation of resources. But this relationship is not exclusive: other factors (effort management? economics?) may have influenced the evolution of stock status.

However, insofar as the structural policy has not sufficiently helped to achieve the objectives of fleet policy, one can initially conclude that the structural policy has contributed little to help the conservation policy. On the contrary, to a certain extent it can be argued that in some cases the structural policy may have undermined the achievement of the objectives of the CFP. Some authors have argued this very strongly (Holden, 1994).

The best historical example is the use of the old subsidies for vessel construction that, when used without sufficient withdrawal of old active capacity may indeed have contributed in a number of cases to increase the *de facto* fishing capacity and aggravate the conservation status of the stocks.

In any case, this remains a big question mark, fundamentally because the structural policy has suffered, up to this day, from the consideration of being a policy pillar that is self-justified, even beyond its contribution towards CFP objectives. This was the result of two main weaknesses:

- The lack of effective mechanisms to measure this contribution. Traditionally, all indicators of policy performance of the structural policy have referred to the rates of budget consumption, the eligibility of actions funded and other indicators of administrative performance. While these are important in terms of ensuring sound financial management of the European taxpayer's money, they do not indicate at all the real contribution of the structural policy to achieve broader CFP objectives. As a result, the analysis remains to this day completely insufficient to determine cause/effect relationships, or to evaluate cost/benefit ratios.
- In addition to the lack of indicators, the structural policy has lacked the instruments to ensure that Member States spend the structural money in a way that is really conducive to achieving policy results. The structural funds can be suspended if Member States do not spend them properly in terms of financial management or eligibility. But there has been no effective mechanism to withhold or suspend funds to Member States if and when these do not spend the money in a way that effectively contributes to achieving CFP goals.

These two issues have been addressed only recently in the CFP, when the last structural regulation, adopted in 2014 as part of the CFP reform package of 2013 introduced two important elements: the indicators of policy performance and the ex-ante conditionality. We will refer to this question in Chapters 16 and 17.

Fisheries and subsidies

The fishing sector: a highly subsidised industry?

The discussion on the contribution of the structural policy in addressing the problems of the fishing industry is more meaningful if it takes place in a wider context: over and beyond the contribution of the structural policy of the CFP, it is also important to examine other sources of public subsidies to the fishing sector, at national and European level. These are very varied: from the direct subsidies through State aid to the tax exemptions (e.g. for marine fuel) and, last but not least, the general lack of cost-recovery mechanisms in European fisheries: the fact that the fishing industry does not pay (at least directly) for certain services associated to fishery management, such as fisheries control or fisheries research.

That the fishing industry in Europe is a highly subsidised economic sector is another commonplace perception of the CFP. It is therefore important to analyse this question also in the perspective of the public image of the policy and in terms of the international relations of the Union, notably in the context of the World Trade Organisation, where the subsidies of the EU to its fleets have frequently been criticised.

It is obvious that the fishing sector is not the only one being subsidised. Likewise, Europe is certainly not alone in subsidising its fishing industry: many countries around the world provide direct and/or indirect subsidies to their fleets. The purpose of this section is not to single out the question of the European fishery subsidies but to analyse the situation in Europe in terms of the contribution of these subsidies in terms of achieving the objectives of the CFP.

As 'subsidies' we can include several different elements, beyond the structural aid referred to previously: the other direct subsidies in terms of state aid in fisheries and the so-called 'indirect subsidies' understood as tax exemptions or services provided by governments to support the activity.

The most widely accepted definition of subsidy is that provided by the World Trade Organisation, which defines subsidies as:

- A financial contribution by a Government or any public body, where a financial contribution can involve a direct transfer of funds, a potential direct transfer, foregone government revenue, government provision of goods and services other than general infrastructure and government purchases of goods,
- By a government or any public body within the territory of a Member and
- Which confers a benefit.

The real level of subsidies in EU fisheries has been estimated by different sources. While methodologies differ considerably as to what can be considered a 'subsidy', thus making it difficult to compare different countries, it can be safely said that the fishing sector in the EU is still partially subsidised. In any case, different types of fisheries subsidies also exist in many other places in the world. The World Bank's study on fisheries subsidies has estimated the level of subsidies in world capture fisheries as being around US$50 billion per annum and the estimated loss to the global economy over the last three decades around US$2 trillion (The World Bank and the FAO, 2008).

Naturally, not all subsides have similar effects. The distinction, in the World Trade Organisation, between 'good' and 'bad' subsidies, depending upon their contribution to make more or less sustainable, is of relevance in that context. And so is the existence, in many countries of the world, of explicit or implicit protectionist measures. But the main interest in the context of competitiveness is, beyond their good or bad effect on sustainability, whether the sector can be economically self-sustainable without them.

Fisheries subsidies are a global issue. They exist practically everywhere and certainly not only in the European Union. The estimation of their real levels and their comparison among countries and/or regions of the world are very difficult due to the difficulties to define subsidies and difficulties with data availability and comparability but a certain number of conclusions can be drawn.

A study estimates that the global overfishing subsidies total US$16 million per annum, equivalent to roughly 25% of the value of the world fish catch. The study recognises the existence of 'good' subsidies and 'ambiguous' ones. But the figure of US$16 million corresponds to the ones considered negative, as 'capacity enhancing'. Interestingly, these subsidies exist in all regions of the world but are particularly high in Asia (Rashid Sumaila et al., 2010).

The effects of subsidies have been studied by many authors. From an economic perspective the general wisdom is that subsidies fuel the race to fish and encourage inefficient and unsustainable fishing, as well as a misallocation of public funds (Markus, 2010). However, from a political point of view this is not necessarily a problem. As we saw in Chapter 1, different European countries have different traditions on how to manage the fishing sectors. This economic consideration can be seen under very different political lights, between those who want the fishing sector to behave like any other economic sector and those whose priority is to maintain the social *status quo* in the sector, even at the cost of economic efficiency.

State aid

The fishery and aquaculture sector is subject to competition rules (including State aid) to the extent determined by the legislator, in line with Article 42 TFEU.

State aid in fisheries is important to consider, notably in connection with the structural policy. If and when actions not allowed under the structural policy of the CFP were to be financed by Member States as state aid under their exclusive financial means, this would certainly represent a distortion to the structural policy. The relationship between State aid policy and structural policy can be summarised as follows (as laid down in Article 8 of the new EMFF): aids provided in the framework of EMFF are generally not State aid and are exempt from the EU State aid rules. However, aids provided for Member States that represent a top-up to the maximum amounts of aid established by EMFF, as well as aids beyond the eligibility criteria of EMFF, are considered as State aid. However, certain EMFF measures are not clearly outside the scope of State aid, in particular aid to integrated maritime policy (not related to fisheries) and must be subject to the control instruments established, under three instruments:

- The *de minimis* regulation, allowing Member States to provide state aid on condition that its total amount would be lower than a certain threshold level, so that such aid would not have a significant impact on competition. As of 2013, this level was fixed at €30 000 per action (EC, 2007).
- The 'block exemption', which exempts from notification to the Union of certain types of state aid, considered not to distort competition.
- The guidelines for the examination of State aid in the fisheries and aquaculture sector (European Commission, 2008).

In practical terms, the evolution of State aid for the fisheries and aquaculture sector in Europe shows a clear downward trend in recent years. According to the last Impact Assessment Report for the *de minimis* aid, the total aid provided by Member States gradually fell from €418 million in 2007 to €111 million in 2012 (SWD, 2014). Interestingly, this aid was not fundamentally directed to the fishing fleets.

What is important about these figures is their comparison with the total value of fishing in Europe. The last figure represents less than 1% of the total value of landings in European fisheries. If we consider that this state aid has not benefitted the fleets substantially, but rather certain aquaculture activities suffering from zoosanitary problems, one can conclude that the state aid to fisheries in Europe is certainly not a significant distortion to free competition in the fisheries sector of the EU.

Indirect subsidies

Indirect subsidies are the costs associated with an economic activity that are borne by governments, instead of the private sector. The most typical indirect subsidies are the tax-exemption applicable to marine fuels (also enjoyed by certain other economic sectors such as maritime transport) and the costs associated with research and control of the fishing activity, which in Europe are entirely borne by governments.

These two kinds of subsidies have different origins and justifications. The -fuel tax exemption is explained by the high percentage that fuel costs represent in the overall running costs of most fishing activities. This allows the continuation of many fishing activities that could otherwise be uneconomical in certain circumstances. This is, however, a type of indirect subsidy that is highly extended around the world: many other countries have this kind of exemption. It is also a kind of direct subsidy that is very difficult to call into question: long-distance vessels refuel wherever they can. If their flag State does require the payment of fuel taxes, they can always refuel through bunkering in other countries where such taxes do not exist.

The other indirect subsidies (lack of sharing out of scientific and control expenses) are the result of the European tradition (though not in all Member States) of paternalism with the industry. In certain countries around the world, the industry participates in the financing of fisheries science, under the so-called 'cost recovery' schemes.

In any case, the different methodologies also raise an interesting question: can we consider as a 'subsidy' the fact that certain investments related to the activity (such as the control or the scientific advice) are entirely paid for entirely by the fisheries administrations? In fact, certain countries in the world have already introduced certain schemes of co-financing.

The Commission suggested such ideas in the framework of Green Paper on reform published in 2009. However, such debate never made headway: the idea of making the industry to pay for control or research is simply off the agenda in Europe.

A third type of intervention can also be considered an indirect subsidy: the payment of license fees to have access to the waters of third countries, in the framework of bilateral fisheries agreements. This question has been raised by certain members in the World Trade Organisation that presented this kind of payment as an indirect subsidy. While a number of countries around the world also have this kind of aid, the recent reform of the CFP in 2013, with the new emphasis on the need for vessel owners to pay an increasing percentage of their access fees, is an adequate response to these allegations. In addition, a number of European fisheries interests obtain access to third countries' resources through private agreements.

Indirect subsidies are very common in many countries of the world, perhaps with the exception of countries that have advanced more in the development of rights-based management with associated cost-recovery schemes, whereby fishermen pay fees for the access to fish and the money from these fees is partially used to finance the management of the fisheries concerned.

The overall costs of managing fisheries are generally considered a small percentage of the value of fisheries production. However, this does not necessarily imply that resource rents are being generated. In 1999 these costs were estimated as around US$600 million in the EU and USA and US$466 million in Japan (OECD, 2003).

The effects: subsidies and competitiveness

As we saw previously, one of the fundamental objectives of the structural policy has been to 'promote competitiveness' of the European fishing sector. Although somehow diluted in more recent structural regulations, this objective has always featured as an essential one for this policy pillar. So the question is: after more than 40 years of structural policy in fisheries, is the European fisheries sector competitive?

Analysing the effects of direct subsidies and state aid is not easy: data on the competitiveness of the European industry have been available only relatively recently in the history of the CFP. In fact, the competitiveness of the European fishing industry has only been measured on a systematic basis since 2007. In addition, any historical development in the industry's competitiveness over the last decades would in any case be the result of a number of factors (market liberalisation, competition with aquaculture products, oil prices, etc.), subsidies being just one of them. We can, however, describe the general competitiveness of the European fishing sector in recent years as reflected in the STECF economic report on the European fishing fleets (JRC, 2014b):

- In 2012, the EU fleet generated €6.9 billion in revenue, amounting to €3.3 billion in Gross Value Added (GVA), which measures the direct contribution of the fishing fleet to the economy. After deducting for operating costs the fleet generated €1.3 billion of gross profit and €458 million in net profits after accounting for capital costs.
- GVA as a proportion of revenue was estimated at 48%, which means that the fishing fleet turns roughly half its revenue into salaries and profits for the fishing communities. The estimated net profit was estimated at 6.6% of total revenue.
- These data represent an improvement when compared with previous years in terms of profitability, mainly as a result of the reduction of costs, including labour costs (6% reduction as compared with 2011) and depreciation.
- However, the overall trend did not apply to all Member States and sectors. Five national fleets made overall net losses and the performance of different fleet segments also varied: while distant-water fleet increased their profitability, that of small-scale fishing tended to deteriorate.
- In terms of fishing gear, the static gear vessels were generally more profitable than the mobile gear ones. The former generated net profits in 2008–2011 while the mobile gear segments showed net losses in the same period.

This is just a glimpse of the situation in terms of competitiveness, with very limited value in terms of trend analysis, given the short historical data series. It seems obvious, however, that the differences in profitability among segments and gear can probably be attributed to a considerable extent to the influence of fuel costs, which is on average represents 23% of total costs but which can reach significantly higher percentages in long distance and trawl fleets. The differences between Member States, however, are much more difficult to analyse.

In any case, the drivers of the general improvement in profitability seem to have been:

- The higher average price of certain species;
- The recovery of certain stocks, largely related to long-term plans;

- Research and innovation; and
- Capacity reduction (higher quotas for fewer vessels).

The last report also introduces, as an innovation, a long-term analysis of the evolution of the competitiveness of the EU fleets, by rebuilding historical data series before the data collection information was available. This analysis indicates that since 2002 the economic performance of the EU fleets has improved gradually, with short-term deteriorations related to the oil price rises of 2003 and 2008 (see Figure 6.2). Labour productivity also increased in that period. Of course, this trend can be due to different factors and it is tempting to conclude that the improvement in competitiveness may be related to the reduction of effort observed from the turn of the century (see Chapter 4) and the gradual increase in the status of European resources.

The question of competitiveness brings us inevitably to a related issue: the effects of a continued policy on subsidies. The generous and almost unconditional structural policy in fisheries for many years has contributed to create a 'culture of subsidy' that has done little to promote the economic self-sufficiency of the sector. As one distinguished representative of the European aquaculture sector once told this author, already in 1992, the high subsidies available for new aquaculture facilities have 'attracted low-quality entrepreneurs' to the sector, in reference to certain ill-conceived projects that received high subsidies only to fail shortly afterwards, due to poor technical design.

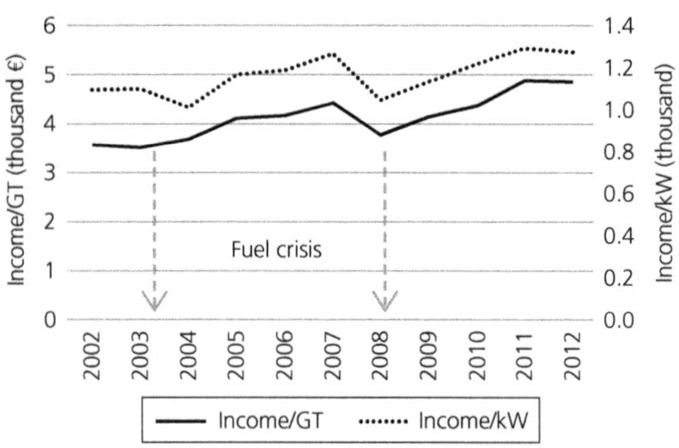

Figure 6.2 Evolution of competitiveness of European fleets since 2000. Source: © European Union. Joint Research Centre. The 2014 Annual Economic Report on the EU Fishing Fleet. JRC Scientific and Policy Reports. EUR 26901 EN – 2014. (See colour plate section for the colour version.)

It is very common to hear, in meetings with stakeholders and fisheries managers, that when the Commission suggests that public aid on certain kind of investments should be discontinued, this idea is immediately presented as the Commission wanting to 'ban' such investments. This shows a state of mind where certain investments are only conceivable if there is European structural money behind them. And it shows the commonplace that investment in the fisheries sector is not possible without subsidies. This is, arguably, the downside of the structural policy: that it creates a dependency from public aid that results in the need for that policy to perpetuate itself.

Scientists have also evaluated these effects in a number of cases. For example, the role of subsidies in aggravating the problem of fleet overcapacity has been studied. A concerted action project of EU fisheries economists concluded that: 'Any subsidy which increases revenues or lowers the cost of variable fixed inputs will tend to lead to increased fishing effort, which in the longer term means increased capacity' (Hatcher and Robinson, 1998).

A study focused on France underlined that, in connection with public aid:

> The evidence is, however, that subsidies allocated to solve short-term problems (generally blamed on easy but fallacious culprits) have done nothing to cure the root causes or to help the industry adapt to changing circumstances and have thus paved the way for a subsequent eruption of crisis. (Mesnil, 2008)

The author underlined that, beyond their sheer amount, the unconditional nature of these subsidies was at the origin of the phenomenon, where the subsidies were devised to continue business as usual, without any advice of their possible effects. This study only refers to France but it is clear that this Member State is not alone in this regard. Indeed, the debate on the need and opportunity for subsidies in the fishing industry is very old.

It is extremely difficult to analyse the effect of the policy on subsidies in the general positive trend shown on the competitiveness of European fleets since 2002. In any case, deepening this analysis, to have more insight into the factors affecting this improvement of competitiveness is a fascinating task for the future.

The 'resource rent' of the fishing sector in Europe

The question of subsidies is also one of interest for the whole of society. Over and beyond the effects of the subsidies on the economic performance of the fleets, there is another analysis of the effects of these subsidies that encompasses a wider consideration: what is the balance between what the sector provides to society as compared to what it receives from it. In the absence of economic data, this kind of analysis has been completely absent from the CFP (Frost and Andersen, 2006). Only in recent times the availability of economic data is allowing the first estimates of the resource rent generated by European fisheries.

Of course that analysis is very complex, particularly because it can involve many elements that are intangible, such as the preservation of a cultural heritage, the respect of people's dignity and so many considerations that clearly go beyond economic analysis. However, an economic part of that wider discussion can be provided through the notion of 'resource rent' of the sector.

As the resource rent concept is an indicator of what an economic sector (dependent upon the exploitation of a natural resource) contributes to society, its analysis in Europe encapsulates the very philosophy of fisheries management: should fisheries management be a system whereby the fishing industry, having the exclusive right to exploit a public resource, should produce a taxable economic surplus for the benefit of the whole society? Or, alternatively, should it be organised in a way that preserves jobs and traditions and fixes populations in their traditional activities?

There is no straightforward answer to this question and Member States have very different views about it that are ingrained in national traditions. Some Member States tend to consider the fishing industry as an economic sector that should produce a taxable surplus to society. Others consider that the preservation of the social and economic *status quo* is the objective even if the industry does not make a net contribution to society in economic terms. And in a number of cases, the same Member State has the first approach for their industrial sectors but the second one for their more artisanal ones.

The resource rent also encapsulates the vicious circle of fisheries management in Europe: if the European fishing industry does not generate taxable benefits for society, then this precludes the possibility of using any cost-recovery mechanism. But at the same time, the lower the resource rent, the higher the pressure to maintain a regime of high subsidies. The generally low resource rent from European fisheries in fact acts as a strong incentive for the system of subsidies to perpetuate itself.

The analysis of the resource rent of European fisheries is very recent and still tentative. However, it shows already that the fishing sector generates more wealth to society than it receives from it through subsidies. This would seem to demonstrate that the bad image of the European fishing industry as a highly subsidised sector is exaggerated. This is, however, very preliminary and is certainly important to deepen the analysis in the future (JRC, 2014).

Is there a social dimension in the CFP?

The 'invisible' part of the CFP

The social dimension of the CFP is perhaps the least visible of all its policy components. It has always been a commonplace that the CFP lacks a social pillar. It is also true that the 'social' angle of the CFP is very difficult to define. For some, it is about the lack of direct legislation on working conditions on board fishing vessels. For others, it is the consideration of biological and economic objectives over job preservation. For others, it is also about the – allegedly – insufficient consideration to the social ethos of fishing communities with a tradition and dependency from this activity.

Beyond the public image (which is very multifaceted) well-known social scientists have argued that the stance of policy-makers (not only in the EU but also at national level) is equivocal on social matters around the CFP: 'Though prepared to acknowledge the relevance of social concerns, they are strangely unwilling to incorporate explicit social objectives into the design of fisheries policy' (Symes and Phillipson, 2009).

This impression is true if we look at the core business of the CFP, the management of fishery resources, where the declared objectives are always related to the recovery of biological resources but they never target, for example, the preservation of the current level of jobs in the sectors concerned. However, if we look at the whole of the policy, including crucially the structural policy, it is easy to see the objectives and instruments of a purely social nature.

As we saw in Chapter 2, socio-economic considerations were an important part of the first attempt to establish the CFP. And, as we will try to explain, the policy has always included elements that seek to attain socio economic goals.

The objectives of the CFP, as expressed in the 2002 basic regulation and renewed in the 2013 basic regulation of the CFP, have been to achieve biological, economic *and social* sustainability. So the social dimension is part of the core objectives of the policy. Then what is the reason for the alleged lack of social objectives and instruments in the CFP? There can be several reasons for this.

The most obvious one is that the Union has limited competence on matters pertaining to social policy. Questions such as working conditions on board fishing vessels, safety on board and other similar questions of social policy are not EU competence. Consequently, the CFP has no legal basis to legislate on these areas. Yet, the very nature of the fishing activity, with a very high rate of occupational accidents, makes this lack of legal basis somehow frustrating for the social actors when looking at the CFP as a whole.

The best example is the question of the international conventions regarding working conditions and safety on board. In particular, the Fishing Convention of the International Labour Organisation (C188 of 2007) is of particular relevance in terms of working conditions in fishing vessels.[2] This convention lays down, *inter alia*, the maximum working time for fishermen on board vessels larger than 24 m. The Convention is not in force, since only five countries in the world have ratified it, none of them European. As a national competence, Member States, not the EU, are to ratify this convention.

Secondly, the most traditional instruments of the CFP do not 'look social' in nature, although their objectives do have a social component. The fact that the instruments of full European competence in the CFP are those of the conservation policy has traditionally given the impression that the CFP is a policy 'for the fish, not for the fishermen'. This image is a caricature which ignores the fact that there is hardly anything more social than ensuring a steady and abundant supply of fish to the fishermen. But a caricature repeated so many times ends up creating a powerful image of the CFP as a policy where Europe 'cares more about fish than about fishermen'.

This image is often misleading, since many decisions regarding the conservation policy have been taken ignoring the biological advice for the sake of minimising social impact. So even decisions that 'look' as if they were measures 'for the fish' have been taken with clear social objectives in mind, albeit on short-term considerations only. The image, in addition, ignores the fact that there has always been a structural policy accompanying the conservation measures and such policy, as we saw earlier, had many clear social objectives. In fact, arguably the structural policy has always tried to achieve socio-economic objectives rather than supporting the conservation policy.

But another element that has contributed to the image of a socially-blind CFP is the fact that the advice it has traditionally been based on has been almost exclusively biological advice. This was certainly the case in the first years, where ICES and STCF provided only biological advice to underpin policy decisions. While the biological advice was well-structured, based on evidence and presented publicly in a transparent manner, the consideration of social and economic considerations has lacked, for a long time, a similar scientific underpinning.

As a result, the social considerations in the CFP have been applied in a completely *ad hoc* manner, without a credible basis and have been also very often mixed with political considerations. Traditionally also, these socioeconomic considerations have been exposed by Ministers in Council, more than by the Commission, producing this apparent divide between a Commission who 'only listens to the scientists' and Ministers who 'only listen to the industry'. And also as a result there has been a certain belief that the 'social' arguments used by Member States at the time of decision making were political excuses (often based on upcoming elections) rather than real arguments. This has not helped the case of considering genuine social elements when taking decisions on the conservation policy.

Furthermore, the tendency of the CFP to concentrate on short-term decisions, at least until the gradual introduction of long-term plans, has not necessarily helped the serious consideration of social elements in decision making. In addition, the traditional lack of economic analysis and evaluation has turned these short-term consequences into an area of political, rather than scientific consideration.

Given the nature of fisheries management, decisions to recover biological resources always have negative short-term social consequences. As a result, all the short-term social considerations (but not serious economic analysis) around fisheries decisions have always been considered negative: any strategy to recover depleted resources implies less economic activity in the short term. However, if we analyse the long-term perspectives, indeed there can be increased economic activity when the resources recover to maximum sustainable yield levels after several years.

For many years this has been a vicious circle. The complete lack of economic analysis left the socioeconomic considerations in some kind of political, rather than scientific realm and this has contributed to discredit these legitimate considerations as short-term 'politics' rather than the legitimate considerations that they should normally be.

Of course, this has been changing gradually, notably through the development of economic advice through the STECF and, very specially, since the establishment of the data collection system and the introduction of the impact assessment as a previous step to all Commission initiatives. We will look at these questions in Chapter 12. Overall, the CFP has developed over the years a number of instruments for the consideration of social matters in its decision making. Still, the policy has never fully caught up with the negative image of a policy that is socially insensitive. We will try to demonstrate why this image is unfair.

Another factor that contributes, still today, to give the CFP a negative image in social terms is the fact that, over the last years, the profession of fisherman has become increasingly unattractive for young European people. In fact, there has been a tendency, at least in a number of Member States, to replace local labour with migrant labour, either from other Member States or also from third developing countries.

This is a complex phenomenon, which should deserve an in-depth analysis. A simplistic interpretation would be that this is the result of less attractive salaries than in the past. Considerations on the hardships of the job can also come into play, even though living and working conditions on board European vessels have improved dramatically thanks largely to the CFP. Cultural changes in European young people, who may be less willing to take hard jobs, or simply no longer consider fishing as the only alternative in their lives (as was the case for their grandfathers) may also be a factor.

At least one of the elements of these hypotheses seems to be demonstrated by evidence: the last report on the economic performance of European fleets shows that the incidence of foreign labour is lower in fleets with high profitability and high salary levels (see report on economic performance of EU fleets previously). But within the social issues at large, there is a particularly important aspect: the nature of the fishermen's activity and its implications for the CFP.

Working as a fisherman in Europe

Fishing activity is a unique professional occupation, in Europe and elsewhere, for a number of reasons:

- It is dangerous. The number of deaths is estimated to be around 24 000 *per annum* worldwide by the International Labour Organisation (ILO). In the EU, 160 deaths and over 2500 injuries were recorded in 1988 alone. The fatality rate is 15 times the national average in the UK, 8 times in Spain or 25 times in Denmark. A study for the UK shows that this fatality rate is not related to vessel size.[3]

- Employment on board is shrinking as a result of the general reduction of the fishing fleets but also as a consequence of further technical progress with the incorporation of automated systems on board. Crew sizes have generally been reduced to the minimum necessary for safety and technical reasons. The EU-15 workforce on board was reduced from 240 000 jobs in 1998 to 190 000 in 2003 that is, by 21% in 5 years!
- Professional qualifications are generally low, particularly for the coastal, artisanal fleets. And the sector has witnessed in recent years a net increase in the incorporation of foreign labour (varying from 3% in Portugal to around 10% in France, Spain and the Netherlands)[4] in a demonstration of the low attractiveness of these jobs to young Europeans from the most affluent Member States.
- The salaries are lower than the average in most Member States (SEC, 2011). And, in most small-scale fisheries, crew members do not have fixed salaries but a percentage of total net income of the vessel.
- The labour force is aging considerably, with 4700 fishermen still active beyond the age of 65 in 2006.[5] In addition, employment on board is largely male-dominated (over 95%) while fish processing has a 57% female workforce.[6]

These figures depict a sector characterised by low attractiveness for young Europeans, which in turn indicates generally poor salary and working conditions and, indirectly, insufficient profitability. When considering social objectives and notably that of 'preserving employment', one can certainly ask if it is that kind of employment that needs to be preserved.

The most recent reports on the economic activity of the fleets clearly establish a positive correlation between fleet profitability and the quality of the employment on board (JRC, 2014). If one wants the profession of fisherman to be attractive for young Europeans again, the need for high levels of profitability of the fleets, with everything that implies (abundant, sustainable fish stocks, fleet size in balance with resources) must be ensured.

The social elements of the CFP

Although the legal basis of the CFP does not cater for social policy legislation, there is indeed EU legislation outside the CFP on certain social aspects that are key for the fishing activity, such as safety on board. In this, mention should be made of two directives in this area:

- Directive 97/70 introduces in EU law the provisions of the 1993 Torremolinos Protocol, laying down safety standards for fishing vessels of more than 24 m (EC, 1998). These provisions are also extended to vessels landing in EU ports.
- Directive 93/103 introducers minimum safety and health requirements for workers on board fishing vessels of more than 15 m (EC, 1993b).

These directives, adopted in the 1990s, show a clear willingness of the Union to go as far as possible in legislating on social matters.

But beyond these examples of direct social legislation, many of the measures in the CFP have a clear social objective. In a non-exhaustive list we can include the following:

- The PESCA initiative, to provide financial support to local communities dependent on fishing, implemented in the period 1994–1999 and the subsequent policies on community-led local development (introduced in 2007 as axis 4 of the European fisheries Fund) and reinforced in 2014 in the context of the European maritime and Fisheries Fund. These initiatives have a clear social objective in mind, not only in terms of promoting the creation of jobs in fishery-dependent areas but, more importantly perhaps, by promoting a wide social participation (including spouses) in these projects, beyond professional fishermen.
- The inclusion of early retirement schemes for old fishermen, the support to young fishermen and so on, which have featured in different generations of the structural policy show the emphasis that the structural pillar of the CFP has placed on the social aspects of the policy.
- The policy of continuing to pay third countries (Morocco and Mauritania are perhaps the best cases) to let EU vessels fish in their waters has a clear objective: the support to the traditional activity of certain fleets that have no alternative

fishing grounds and constitute the social and economic lifeblood of their coastal communities. In addition, payment for temporary vessel laying-up for the periods where this activity is interrupted has a clear social objective.

- The traditional policy of financing the modernisation of vessels to increase safety. This objective has been so important and consistent as part of the CFP that even when the structural policy has limited the aid to the construction of new vessels or their modernisation, the improvement of the safety equipment and the improvement of working conditions in fishing vessels has always remained eligible for public aid from the different structural instruments.
- The importance given to community-led local development. This kind of instrument (the Axis 4 of the EFF, continued and reinforced in the EMFF) has a very clear social dimension: it focuses on local communities, it is reserved for them, ring-fenced from the traditional beneficiaries of the structural policy and it includes as potential beneficiaries not only fishermen themselves but also their families and other members of the local communities concerned.
- In the context of the new structural instrument adopted in 2014, the European Maritime and Fisheries Fund (EMFF: EU, 2014) a number of social measures were also introduced, such as the promotion of human capital (mainly through permanent training), job creation and social dialogue (Article 29), the aid to the establishment of young fishermen (Article 31), the support to the improvement of safety and working conditions on board (Article 32) and so on.
- The CFP is becoming less isolated from other policies related to the European seas through the development of the larger policy framework of the Integrated Maritime Policy (IMP), a recent policy initiative that focuses fundamentally on the development of economic growth and jobs based on maritime activities (the so-called 'Blue Growth') (COM, 2014b), with a clear economic and social objective in mind. This means that EU policy is no longer a policy just for the fishermen: both the Axis 4 and the IMP mean that the borders of the CFP are also active policy areas with a strong emphasis on growth and jobs for the local communities in maritime, largely fishery-dependent areas of Europe.

Overall, one can conclude that this 'invisible' part of the CFP has been and will continue to be, much more substantial than has ever been given credit for.

Notes

1 Article 9 of Regulation 2792/99.
2 www.ilo.org/dyn/normlex/en/f?p=NORMLEXPUB:12100:0::NO::P12100_ILO_CODE:C188.
3 MAIB Report on analyses of fishing vessel accident data (1999–2000) concerning the UK fleet.
4 European Parliament's report A5–0087/2001.
5 Study FISH/2004/4 –htpp://ec.europa.eu/fisheries/publications/studies_reports_en.htm.
6 European Parliament's report. 'The role of women in fisheries'. May 2008.

References

Capgemini Consulting (2014). Study on the implementation of Axis 4 of the European Fisheries Fund. Brussels, 25 July 2014. Capgemini consulting, Wageningen UR, Ramboll.

COM (2002). Communication from the Commission to the Council and the European Parliament. A strategy for the sustainable development of European aquaculture. Brussels, 19.9.2002. COM (2002) 511 final.

COM (2013). Report from the Commission. Sixth Annual Report on Implementation of the European Fisheries Fund (2012). Brussels, 18.12.2013. COM (2013) 921 final.

COM (2014a). Commission Decision of 3.10.2014 amending Decision C(2008)4993 approving the operational programme for Community assistance from the European Fisheries Fund in Ireland for the programming period 2007–2013 – C(2014) 7227 final.

COM (2014b). Communication from the Commission to the European Parliament, the Council, the European Economic and social Committee and the Committee of the Regions Innovation in the Blue Economy: realising the potential of our seas and oceans for jobs and growth. COM(2014) 254 final.

DECYZJA WYKONAWCZA KOMISJI (2014).z dnia 24.4.2014 r. zmieniająca decyzję C(2008) 6007 zatwierdzającą program operacyjny dotyczący pomocy wspólnotowej z Europejskiego Funduszu Rybackiego w Polsce na okres programowania 2007–2013 – C(2014) 2743 final.

Durchführungsbeschluss der kommission (2007). vom 5.12.2012 zur Änderung der Entscheidung K 6713 zur Annahme des operationellen Programms für die Interventionen der Gemeinschaft unter Beteiligung des Europäischen Fischereifonds in Deutschland für den Programmplanungszeitraum 2007–2013 – C(2012 8771 final.

EC (1993a). Council Regulation (EC) No 3699/93 of 21 December 1993 laying down the criteria and arrangements regarding Community structural assistance in the fisheries and aquaculture sector and the processing and marketing of its products. *OJEC* No L 346 of 31.12.93, pp. 1–13.

EC (1993b). Council Directive 93/103/EC of 23 November 1993 concerning the minimum safety and health requirements for work on board fishing vessels (thirteenth individual Directive within the meaning of Article 16 (1) of Directive 89/391/EEC) *OJ* L 307, 13.12.1993, p.1.

EC (1998). Council Directive 97/70/EC of 11 December 1997 setting up a harmonised safety regime for fishing vessels of 24 metres in length and over. *OJ* L 034, 9.2.1998, p.1.

EC (1999a). Council Regulation (EC) No 1263/99 of 21 June 1999 on the Financial Instrument for Fisheries Guidance. *OJEC* No L 161 of 26.6.99, pp. 54–56.

EC (1999b). Council Regulation (EC) No 2792/1999 of 17 December 1999 laying down the detailed rules and arrangements regarding the Community structural assistance in the fisheries sector. *OJ.* L 337 of 30.12.1999, p. 10–28.

EC (2006). Council Regulation (EC) No 1198/2006 of 27 July 2006 on the European Fisheries Fund. *OJEU* No L 223 of 15.8.2006, pp. 7–44.

EC (2007). Commission Regulation (EC) No 875/2007 of 24 July 2007 on the application of Articles 87 and 88 of the Treaty to de minimis aid in the fisheries sector and amending Regulation (EC) No 1860/2004. *OJ.* L 193. 25.07.2007, p. 6.

EC (2008). Council Regulation (EC) No 744/2008 of 24 July 2008 instituting a temporary specific action aiming to promote the restructuring of the European Community fishing fleets affected by the economic crisis. *OJ* L202, 31.7.2008, p. 1.

EEC (1976a). Council Regulation (EEC) No 101/76 of 19 January 1976 laying down a common structural policy for the fishing industry. *OJEC,* 19, L20, pp. 19–22 of 28.01.1976.

EEC (1976b). Council Regulation (EEC) No 100/76 of 19 January 1976 on the common organization of the market in fishery products. *OJEC,* 19, L20, pp. 1–18, 28.01.1976.

EEC (1977). Council Regulation (EEC) No 355/77 of 15 February 1977 on common measures to improve the conditions under which agricultural products are processed and marketed *OJ.* L 051 , 23/02/1977 p. 1–6.

EEC (1983). Council Regulation (EEC) No 2909/83 of 4 October 1983 on measures to encourage exploratory fishing and cooperation through joint ventures in the fishing sector. *OJ.* No L290 of 22.10.1983, p. 9–14.

EEC (1986). Council Regulation (EEC) No 4028/86 of 18 December 1986 on Community measures to improve and adapt structures in the fisheries and aquaculture sector. *OJEC,* 29, L376, pp. 7–24 of 31.12.1986.

EEC (1988). Council Regulation (EEC) No 2052/88 of 24 June 1988 on the tasks of the structural Funds and their effectiveness and on coordination of their activities between themselves and with the operations of the European Investment Bank and the other existing financial instruments. *OJEC* No L 185 of 15.7.88, p. 9.

EEC (1993). Council Regulation (EEC) No 2080/93 of 20 July 1993 laying down provisions for implementing Regulation 2052/88 as regards the financial instrument for fisheries guidance. *OJEC* No L 193 of 31.7.93, pp. 1–4.

EU (2013). Regulation (EU) No 1303/2013 of the European Parliament and of the Council of 17 December 2013 laying down common provisions on the European Regional Development Fund, the European Social Fund, the Cohesion Fund, the European Agricultural Fund for Rural Development and the European Maritime and Fisheries Fund and laying down general provisions on the European Regional Development Fund, the European Social Fund, the Cohesion Fund and the European Maritime and Fisheries Fund and repealing Council Regulation (EC) No 1083/2006. *OJ* No L 347, 20.12.2013, p. 320.

EU (2014). Regulation (EU) No 508/2014 of the European Parliament and of the Council of 15 May 2014 on the European Maritime and Fisheries Fund and repealing Council Regulations (EC) No 2328/2003, (EC) No 861/2006, (EC) No 1198/2006 and (EC) No 791/2007 and Regulation (EU) No 1255/2011 of the European Parliament and of the Council. *OJ* L 149, 20.5.2014, p. 1–66.

European Commission (2008). Guidelines for the examination of State aid to fisheries and aquaculture (2008/C84/06). O.J. No C84, 3.4.2008, p.10EEC (1970) Regulation (EEC) No 2141:70 OF THE COUNCIL of 20 October 1970 laying down a common structural policy for the fishing industry. *OJ.* No L 236 of 27.10.70, p.1–4.

European Court of Auditors (1994). Special report No 3/93 concerning the implementation of measures for the restructuring, modernisation and adaptation of the capacities of fishing fleets in the Community. *OJ.* C2, 4.1.1994.

Frost, H. and P. Andersen (2006). The Common Fisheries Policy of the European Union and fisheries economics. *Marine Policy* **30**(2006): 737–746.

Hatcher, A. (2000). Subsidies for European fleets: the European Community's structural policy for fisheries 1971–1999. *Marine Policy* **24**: 129–140.

Hatcher, A. and K. Robinson (eds) (1998). Overcapacity, overcapitalisation and subsidies in European fisheries. EU FAIR Concerted Action on Economics and the Common Fisheries Policy: perspectives for the future economic management of Europe's fisheries. Proceedings of the first workshop held in Portsmouth, UK, 28–30 October 1998.

Holden, M. (1994). *The Common Fisheries Policy. Origin, Evaluation and Future*. Fishing News Books. Blackwell Scientific Publications Ltd., 274 p.

JRC (2013). The Economic Performance of the EU Fish processing Industry (STECF – 13–31) JRC Scientific and Technical Policy Reports. EUR 26444 EN.

JRC (2014a). The Economic Performance of the EU Aquaculture Sector (STECF 14–18). JRC Scientific and Technical Policy Reports. EUR 27033 EN.

JRC (2014b). The 2014 Annual Economic report on the EU Fishing Fleet (STECF 14–16). JRC Scientific and Technical Policy Reports. EUR 26901 EN.

Markus, T. (2010). Towards sustainable fisheries subsidies: entering a new round of reform under the Common Fisheries Policy. *Marine Policy* **34** (2010): 1117–1124.

Mesnil, B. (2008); Public-aided crises in the French fishing sector. *Ocean & Coastal Management* **51**(2008): 689–700.

OECD (2003). *The costs of managing fisheries*. Published 17 March 2003. OECD Paris, 176 pp.

OJ (2008). Guidelines for the examination of State aid to fisheries and aquaculture (2008/C 84/06). *OJ* NO C84, 3.4.2008, p. 10.

Rashid Sumaila, U., A.S. Khan, A.D. Dyck, R. Watson, G. Munro, P. Tydemers and D. Pauly (2010). A bottom-up re-estimation of global fisheries subsidies. *J. Bioecon*. DOI 10.1007/s10818–010–9091–8. Published on-line: 18 August 2010.

SEC (2011). Commission Staff Working Paper. Impact Assessment accompanying Commission proposal for a regulation of the European Parliamant and the Council o the Common Fisheries Policy (repealing Regulation (EC) No 2371/2002). Brussels, 13 July 2011, SEC(2011) 891.

STECF (2013). Summary of the 2013 Economic Performance report on EU Aquaculture Sector (STECF 13-XX). Scientific, Technical and Economic Committee for Fisheries. JRC Scientific and Policy Reports, 2013.

STECF (2014). The 2014 Annual Economic report on the EU Fishing Fleet (STECF 14–16). JRC Scientific and Technical Policy Reports. EUR 26901 EN.

SWD (2014) Impact Assessment report accompanying the document: Commission regulation on the application of Articles 107 and 108 of the Treaty on the Functioning of the European Union to de minimis aid in the fishery and aquaculture sector. Brussels, 27.6.2014. SWD(2014) 203 final.

Symes, D. and J. Phillipson (2009). Whatever became of social objectives in fisheries policy? *Fisheries Research* **95**(2009): 1–5.

The World Bank and the FAO (2008). The Sunken Billions: the Economic Justification for Fisheries reform. The World Bank and the Food and Agriculture Organisation. 2008, 104 pp.

CHAPTER 7
The external dimension

The last frontier of the CFP: external resources

The external pillar of the CFP had a strong direct influence in the birth of the CFP. Although the first regulations on the management of fisheries under the CFP in 1983 (Regulations 170/83 and 171/83) did not mention at all the external dimension of the new policy, the considerations about the loss of fishing grounds in third countries were a fundamental factor in the political discussion leading up to the declaration of Exclusive Fisheries Zones at the end of 1976. Moreover, the establishment of the new policy certainly implied the need to establish bilateral and multilateral relations, notably with neighbouring countries, either to ensure mutual access to waters and resources or to jointly manage shared stocks. That means that the external pillar policy of the CFP existed from the inception of the policy, even before this was recognised in the first legal texts.

The best example is the Commission's communication that, in 1976, proposed the principles for the establishment of the CFP, based on the imminent perspective of declaring the 200-mile zones for the following year. This communication (COM, 1076) laid down the principles for an external policy in parallel to the proposal for a regulation laying down the basis for the CFP in European waters (EEC, 1976), clearly showing that the two initiatives were part of a package: there could be no CFP in European waters unless the external dimension was addressed at the same time.

The problems of the time were, to a considerable extent, the same as today in the context of the fishery agreement with Norway: the diversity of interests amongst Member States in terms of the balance between the fishing opportunities in third countries' waters and the concession to third countries in European waters.

In addition, other factors contributed to the development of this policy pillar as we know it today: the withdrawal of Greenland from the EEC in 1985 (and subsequently the need to agree on access to Greenlandic waters of historic fishing interests from the Community of 9) and, especially, the accession of Spain and Portugal, which brought to the Union a very important dimension of fisheries in third country and international waters. As the Union has enlarged the new Member States have brought in new issues and new challenges for the EU's external fisheries policy.

While the CFP developed, so did the world governance of fisheries and as an important actor in world fisheries, the Union had to play an important role in the way this governance was shaped. At the same time, the international developments had a strong influence in the way the CFP in European waters developed. Without the external dimension, the CFP would probably be an entirely different policy.

The Common Fisheries Policy: The Quest for Sustainability, First Edition. Ernesto Penas Lado.

The importance of the external sector

The external sector of EU fisheries is of fundamental importance to the CFP for a number of reasons:

- It represents an important source of economic activity and jobs in certain regions of Europe, where fisheries in international or third country waters are as old as many fisheries in European waters.
- It contributes to supply the European market in fishery products and partially compensates for the insufficient supply of domestic fisheries for Europe's consumption.
- It is inextricably linked with the domestic part of the CFP in those areas where the EU has shared stocks with neighbouring countries. This has considerably evolved following different enlargements.
- It turns the Union into an important and legitimate party in the multilateral governance of fisheries around the world, by virtue of the diversity and extension of European fishing interests and activity around the world.
- It is a policy that overlaps with other policies of the Union, in particular development (in certain areas of the world), foreign (especially in neighbouring countries) and trade policy.
- Finally, it has a considerable influence (for good or bad) on the development of the CFP itself. A significant number of initiatives under the CFP have not been generated internally but have been the result of agreements, or conflicts with third parties, either at bilateral or multilateral level. Conversely, in certain areas, the lack of initiative or commitment to effective fisheries management by third parties has contributed to delaying or discouraging action at the level of the Union.

Overall, the external pillar and the domestic management of fisheries in Europe have been closely interlinked and have influenced each other, even though such relationship has traditionally taken place more *de facto* than *de jure*.

A stand-alone policy pillar?

The policy applicable to the external fisheries sector of the Union has traditionally evolved largely outside the basic provisions of the CFP, as a separate policy. While in European waters the emphasis was put on the allocation of fishing rights among Member States, in the external pillar the main emphasis has been the preservation of the interests of European fishing in bilateral or multilateral contexts. This fact, together with the lack of a clear policy orientation in the first legal text of the CFP has allowed the development of this sector as a somewhat 'stand-alone' policy pillar of the CFP, until very recent times.

Although there was obviously a consistency in the approach regarding the domestic and the external sector, this was not incorporated into the legal texts of the CFP in the 1992 and 2002 reforms. It was not until the CFP reform of 2013 that the basic text of the CFP established provisions on how this policy should be conducted. Several factors may have contributed to this:

1. The external sector cannot be regulated in the same way as the rest of the policy. The Union can only negotiate and never impose its views on third countries.
2. Unlike the domestic policy, the external sector is related and very dependent upon other Union policies, such as the Trade, Development and Foreign policies. This makes this sector much less autonomous for policy makers.
3. There is a less clear relationship between the status of biological resources and the policy itself. In some occasions, the biological basis for this policy pillar is very weak (e.g. in the context of demersal resources off West Africa. Until relatively recently, little was known about the status of most demersal stocks in this area).
4. Very often, the issues to be managed were not related to the basic principles of fisheries management, such as the conditions allowing for sustainable exploitation. The struggle to defend the rights of European fishermen to benefit from the world's common resources was often far more important. The relationship between European activity and resources is not an exclusive one. If Europe stops fishing in a third country because of overexploitation of resources, there has always been a chance that other foreign fleets would take Europe's place, with no benefit for the resources and less opportunities for the local fishermen's communities.

5 Institutionally, this is a policy where the Commission has an enhanced role in decision making. As negotiator on behalf of the Union, the Commission has the responsibility to negotiate measures that would subsequently be backed by Council and/or Parliament. In the vast majority of cases, the deals negotiated by the Commission are at the end backed by the co-legislators, even when they may have preferred a slightly different outcome.

All these elements give the external pillar of the CFP a very specific nature as part of the CFP. However, this situation has changed and, as of 2013, this policy pillar is subject to common principles, consistent with those in European waters, in the context of the new basic regulation of the CFP (EU, 2013).

The internal and external CFP policy strands have strongly interacted and have obviously influenced each other. The internal policy choices of the CFP have considerably influenced the Union's position in many international *fora*. Conversely, many international developments have had a strong bearing on the development of the CFP in Europe. The relationship between the two policy strands is complex and multi-faceted.

The influence of the external sector in our policy has been both positive and negative. It has been positive when and where the third countries concerned have forced the Union to improve its own management. Negative when and where the lack of action by third parties has been an excuse for the Union not to take its responsibility. Conversely, internal problems have many times made the external action of the EU very difficult but when and where all Member States have agreed a common line in the internal policy, this has facilitated the role of the Union in international *fora*.

The global governance of fisheries

The European Union is a global player in the development of international fisheries law and in the development of multilateral fisheries governance. This important role is assured by several factors, *inter alia*: (1) the existence of direct fishing activity and/or economic interests from Europe in many fisheries around the world, (2) the size of the EU market for fish and its level of imports and (3) the fact that the Union represents many countries (thus playing a key role in the ratification of international instruments, where entry into force requires a minimum number of ratifications) and its tradition as a reliable and fair partner in international *fora*.

The development of international fisheries law

The international governance of fisheries concerns fundamentally the establishment of limitations to the right of fishing. And these limitations are of two kinds:

1 The territorialisation of the seas, that is, the geographical limitation to the right to fish, notably by the extension of coastal states exclusive rights from their coastlines and

2 The non-territorial limitations, that is the limitations to the total amount of fishing to be allowed for all those countries fishing in the high seas, beyond the jurisdiction of coastal States.

Put it another way: the limitations derived from the consideration of marine resources in the high seas, not as *res nullius* but as *res communis*. The history of international fisheries law is largely that of the confrontation between two conflicting notions: the freedom of the high seas and the complex maze of restriction imposed upon this freedom.

Today, this confrontation is settled in the context of the United Nations Convention on the Law of the Sea (UNCLOS). However, one can certainly distinguish two different degrees in which these disputes have been settled and on that basis the limitations to the freedom to access fishery resources can be divided in two categories: (1) the geographical limitations and (2) the quantitative limitations applicable to all.

The geographical limitations, that is, the rules concerning the rights of coastal States to exploit the resources in their area of jurisdiction (the Exclusive Economic Zone), have been well consolidated and by and large are widely respected. As we saw, these limitations are discussed at least from the seventeenth century (see Chapter 1). However, the non-territorial limitations, as such the need to limit the uptake of marine resources and to do so in a cooperative manner, is in fact less developed and, in particular, much less implemented in practice. These quantitative limitations originated notably through the Convention on the High Seas of Geneva in 1958, whose Article 2 outlines the general basis for such limitations: States must exercise the freedom of fishing with reasonable regard for the interest of other States (Koers, 1973).

However, this principle (enshrined in the current fisheries law, see next) is based on the notion of cooperation among States and where such cooperation is weak or inexistent, there is a big gap in fisheries governance.

As we saw, there have been two fundamental pieces of international legislation regarding the management of fishery resources: the United Nations Convention on the Law of the Sea (UNCLOS), agreed in 1982 and the subsequent United Nations Fish Stocks Agreement (UNFSA) of 1995 where part of the basic UNCLOS' fisheries legislation was developed. In both cases, the Union and the Member States played a key role in ensuring that international law reflected a balanced approach between the right of coastal States and those of long-distant fishing nations.

UNCLOS

Since 1982 and following years of disagreement and unilateral action, the United Nations Convention on the Law of the Sea (UNCLOS) has brought about some calm to the trouble situation with regard to the rights and obligations of States concerning the oceans and their resources. This convention represents a balance between the coastal States and those wishing to exploit the world's marine resources.

As we saw in Chapter 2, the negotiation of UNCLOS took a number of years in which the international governance of fisheries developed fundamentally through unilateral action. The text agreed in 1982 in fact had to catch up with those developments rather than preceding them. This is very important to bear in mind, because it is a precedent of how international law can evolve and many countries around the world have not forgotten this fact.

The process of preparation of the convention started in Geneva in 1958, when 86 States participated in a first conference that drafted four conventions, on subjects such as the territorial seas and contiguous zones, the high seas, fishing and the conservation of living aquatic resources of the high seas and the continental shelf (Farnell and Elles, 1984). But progress was very limited until 1967 (beyond the Convention of London of 1964 which enshrined the 12-mile territorial seas), when the UN General Assembly established an ad hoc committee to deal with the use of the seafloor beyond limits of national jurisdiction.

The work of that committee eventually led to the adoption of a declaration by the UN General Assembly convening a conference to adopt a convention on all matters relating to the law of the sea. That convention, held in 1973, was attended by 115 countries. From then on, 9 long years elapsed before the convention could be adopted on April 1982 and signed by 117 States on 10 December 1982. During these 9 years, many countries around the world declared unilaterally 200-mile Exclusive Economic Zones.

From the beginning of the process, the EU and its Member States have played a central role in developing this instrument. As some Member States of the Union are coastal States but also have a long-standing tradition of fishing in international waters, the Union represented a counter-balancing power (often together with Japan and other countries) against the intentions of many coastal States to gain further jurisdiction over international waters. In doing so, the Union ensured that the international governance of fisheries respect a balance between coastal and non-coastal States.

As indicated previously, the ratification of the instruments of international law by the Union represents a leap forward: the sheer force of 20-odd ratifications at once in the whole process of entry into force of these instruments is key: Member States ratified the United Nations Convention on the Law of the Sea of 1982 in 1998 (EC, 1998). Thereupon the convention entered into force in 1994 when the required minimum number of ratification was reached. Up to August 2013, a total of 166 countries had ratified UNCLOS.[1]

UNCLOS is still not ratified by certain coastal States such as the USA. Although a vast majority of countries around the world have done it, the fact that significant coastal States haven't done so, together with the recent history of unilateral action on questions of maritime jurisdiction, may cast some doubt as to whether UNCLOS is definitively consolidated international law, at least for certain aspects.

These non-ratifications may not necessarily be related to the fisheries part of UNCLOS. In the particular case of the United States, they have not ratified UNCLOS but have ratified UNFSA, which seems to indicate that the non-ratification of UNCLOS is not necessarily related to fishery considerations.

UNFSA

The United Nations Fish Stock Agreement (UNFSA) responded to the need to develop the fishery part of UNCLOS, in the understanding that the provisions of UNCLOS were not sufficient to address in particular the need for cooperative management of stocks straddling areas of national jurisdiction (i.e. stocks whose area of occurrence is partially within EEZs and international waters). The Agreement was signed in 1995 and entered into force in 2001.

The agreement clarified and consolidated the rights of coastal States in terms of the management of fishery resources in their Exclusive Economic Zones or straddling their EEZs and international waters, with a balanced combination of rights and obligations.

The negotiation of UNFSA took place in a context where UNCLOS was still not ratified by a number of coastal States. Although the 200-mile zones were enshrined in the 1982 UNCLOS text, a number of coastal States were still not satisfied that these zones would provide them with enough guarantees to fully benefit from the fishery resources within their waters, since a number of important fish stocks straddled the 200-mile zone and international waters and were thus amenable to exploitation by long-distance fishing countries.

Although UNFSA covered highly migratory fish stocks as well, it was the issue of the management of straddling stocks that mostly motivated the agreement. As we will see next, this agreement was, in its final stages, linked with the conflict over the management of an important straddling stock off Canada: the Greenland halibut. The tension between Canada and the European Union around the management of that fishery encapsulated the tension among coastal States and long-distance fishing countries that UNFSA tried to resolve.

For that reason, many of the features of that agreement were designed for the purpose of resolving this tension on straddling stocks. However, for the management of stocks outside EEZs (fundamentally highly migratory fish such as tuna or tuna-like species as well as a number of pelagic sharks) where the need for multilateral agreement among all involved is paramount for sound management, the Agreement remained more general, particularly in terms of rights and obligations by all States. Still today the multilateral management of these stocks remains a huge challenge for the global governance of fisheries, particularly in relation with the role of developing States and their access to the resources.

The European Union ratified this Agreement on 23 December 2003 and Member States did so gradually up to the most recent (Croatia) in 2013. Very much like UNCLOS, the Fish Stock Agreement has entered into force, although so far it had been ratified only by 81 countries by September 2013.[2] Again, certain important coastal States around the world such as China or Chile have not ratified UNFSA. Some of these countries, like Chile, are known to have practiced 'creeping jurisdiction' by taking national measures with certain effects on third country vessels outside their 200-mile EEZ.

The existence of international law in force is a necessary condition for good governance but certainly is not sufficient if and when implementation is poor. As we saw previously, the issues related to the territorial part of the law of the sea are well implemented: most countries in the world (with the exception of a number of Mediterranean countries, see Chapter 8) have declared Exclusive Economic Zones where they generally enforce the rights of access. On the contrary, on the issues related to the quantitative limitations to fish, much is still to be done.

Overall, the law of the sea represents a balance between rights and obligations. But, while all States are adamant to exert their rights, many are not so keen to fulfil their obligations under that framework.

This represents, in a nutshell, the biggest challenge for the world's fisheries governance.

UNFSA has the considerable merit of being a multilateral agreement that settles the traditional rift between coastal and long distance fishing countries. However, the fact that only 81 countries had ratified it by September 2013 (although some additional ratifications are under way), that is, half the number of ratifications of UNCLOS, casts some doubts as to whether this instrument is consolidated. After 20 years of its signature, the Agreement is now up for a possible revision, and this will require confronting its main challenges.

These challenges are mainly of two kinds:

1. Ensuring the effectiveness of the Regional Fishery management Organisations (RFMOs) that constitute the main instrument to turn UNFSA's principles into workable management solutions, and
2. Ensuring that developing countries can exert their rights and obligations under this Agreement.

We will discuss these challenges next.

Global governance

Another contribution to the global governance of fisheries has been the global initiatives at multilateral level such as the World Summits on Sustainable Development. Launched in Rio de Janeiro in 1992, these summits introduced certain objectives (although without specific target dates) for fisheries management. The notion of Maximum Sustainable Yield (MSY) as the best objective for the management of world's fisheries was reinforced there.

However, it was the second of these summits, held in Johannesburg in 2002, that contributed most crucially to global fisheries governance and to the CFP itself. The 2002 summit established, with the agreement of all EU Member States, an agenda for change that included a number of key elements, that one way or another have been included in the CFP and have been a source of inspiration and guidance in the external policy and, in some cases, also on the domestic policy of the EU. Crucially, in the fisheries Chapter, the summit incorporated as an objective a time target for the achievement of MSY in fisheries: 'maintain or restore stocks to levels that can produce the maximum sustainable yield with the aim of achieving these goals for depleted stocks on an urgent basis and where possible not later than 2015'.[3]

In addition, the summit also made important recommendations on issues such as, *inter alia*:

- The commitment to phase out subsidies that result in an increase in overcapacity.
- The Commitment to develop fisheries protected areas in international waters.
- To support the sustainable development of aquaculture.
- Encourage multilateral organisations to give due regard to the rights of coastal developing States.
- To take steps to implement the FAO plan of action on world fishing capacity.

These recommendations were complemented in 2012 in the Rio+20 summit (UN, 2012) attended by 192 States that, in a document titled 'The Future We Want', largely supported the objectives of previous summits but added objectives in terms of underlining the importance of the environment for development and, especially, insisted on the need for the UN objectives to be achieved, to improve governance, to provide for the necessary financial resources and to reinforce the UN instruments for action. However, for fisheries management, Rio+20 did not add substantially new targets as compared to those of Johannesburg in 2002.

Overall, these summits, while not producing legally binding results, have indeed set priorities that have had a clear impact in the CFP. The way to do so has been the initiatives of the Commission to translate these 'general political, non-binding objectives' into specific proposals for hard EU law. In other words, the Commission has ensured the influence of these international agreements on the CFP by taking them seriously and proposing to implement them in practice.

In addition, these international commitments provide also a fundamental guidance in establishing priorities in the external action of the EU around global fisheries governance.

Soft law: FAO

Other significant contribution to global fisheries governance is that provided by the Food and Agriculture Organisation (FAO) of the United Nations. FAO's contributions are soft law but in many occasions the recommendations of this organisation, which provides for a very good balance of interests between developed and developing States, have been instrumental in the development of global governance.

In particular, the International Plans of Action (IPOAs) have been very important in creating awareness about the need to tackle certain problems but also in providing workable technical solutions to handle them. FAO adopted, *inter alia,* four important international plans to tackle: (1) incidental catches of seabirds in long-line fisheries, (2) the conservation and management of sharks, (3) the management on fishing capacity (FAO, 1999) and (4) on illegal unreported and unregulated fisheries.

These plans have indeed had a bearing in the CFP notably by providing a basis upon which initiatives were taken at EU and multilateral level. These IPOAs have in fact been implemented by the Union: the IPOA sharks have given rise to a number of measures (see Chapter 11). The IPOA on seabirds was at the origin of the EU plan of action to protect these species (see Chapter 11) and the IPOA on fleet capacity has been a significant source of inspiration for the Union's fleet policy (see Chapter 5).

In addition, FAO has also adopted an IPOA on illegal, unregulated and unreported (IUU) fisheries, which has constituted a fundamental inspiration of the current EU policy against IUU fishing (see Chapter 9) (FAO, 2001).

FAO has also adopted other important documents such as the Code of conduct for responsible Fisheries, or the Port State Measures Agreement, all of which have represented important contributions that inspired the development of global ocean governance.

Future prospects

All these issues, one way or another, point in the same direction: the need to revise and improve global fisheries governance. This need is, incidentally, not restricted to international fisheries. On the contrary, a number of developments clearly indicate that the current order in the governance of the oceans may evolve substantially in the next decade. Issues such as (1) the melting of the Arctic ice cap, with consequent opening of areas for transport, oil exploitation and perhaps fisheries activity, (2) future action to fight global warming, (3) on-going initiatives to protect marine bio-diversity in the high seas, (4) the global fight against marine pollution and so on will require important changes in the world's ocean governance. And these changes will undoubtedly have a bearing on fisheries.

In addition, certain fishery-specific issues will have to be addressed in the future if the global fisheries governance is to meet its challenges.

Marine Biodiversity beyond areas of National Jurisdiction

This on-going initiative represents a new frontier in the development of the world's fisheries governance. The working group on Marine Biodiversity Beyond areas of National Jurisdiction (BBNJ) of the United Nations is examining this question with a view to a possible implementation agreement. The idea behind this is that the protection of marine biodiversity beyond areas under national jurisdiction lacks an appropriate legislative development, thus resulting in very few initiatives to protect large areas of the world's oceans.

It is premature at this stage to foresee the result of this initiative. However, from experience this author will advance a reflection. If this exercise fails, it may fuel claims that the only way to achieve the objective of protecting biodiversity in international waters is to extend the jurisdiction of coastal stats well beyond the current 200 miles. The success of this initiative may thus be important to consolidate the current *status quo* under UNCLOS.

Fleet overcapacity and global governance

One crucial point in the debate is the question of global overcapacity and the role it plays in terms of contributing to the multilateral management of fisheries.

In Chapter 5 we already looked into the relationship between fleet size and sustainability of resources, at EU level. The problem, obviously, is similar in the international arena. But with one important factor that seriously aggravates it: the general lack of agreement on the distribution of fishing rights.

We also saw in Chapter 5 how difficult it was at EU level to agree on a meaningful and effective fleet policy and was so despite the fact that the fishing rights are already allocated among Member States, not by one system but by two: the allocation of fishing possibilities under relative stability and the fleet capacity ceilings applicable to all Member States. At international level, however, this allocation is done only in a limited number of cases: in NAFO and NEAFC, in particular, while in other fisheries, such as those regulated by ICCAT, allocations by country exist for a limited number of stocks and even those allocations are frequently under pressure by new entrants that – legitimately – claim a right to fish as coastal States.

The problem of international overcapacity is particularly acute in cases where there are important fisheries in international waters (thus, highly migratory fish) but also in areas where the resources within the EEZs of coastal States are not managed effectively, in many cases in developing countries where there is insufficient fisheries management, thus resulting in significant levels of fishing activity including large-scale foreign vessels fishing there legally – or illegally.

This problem of overcapacity was addressed already by FAO, in the context of the International Plan of Action of fleet capacity.[4] However, this soft law approach has not allowed any specific, legally binding measures.

Other attempts to tackle this problem such as the Kobe process or the IATTC capacity ceiling mentioned previously have not solved the problem. For one fundamental reason: the lack of a definitive, stable agreement on the allocation of fishing rights.

We saw in Chapter 3 that the birth of the TAC and quota system in the CFP was inspired mainly by the need to resolve the distribution of fishing rights among Member States following the establishment of 'Community waters' in 1977. At global level, any attempt to limit the uncontrolled growth of global fleets in recent years (and very particularly by countries of the Far East, while Europe, Japan and the US self-restrained this development), is undermined by a complete lack of agreement on how to allocate the fishing rights. In particular, the tension between three blocks as follows:

1. The 'traditional players' (Europe, Japan, USA), with important markets for the fish
2. The emerging players, particularly far East countries (China, Taiwan, Korea) that have developed their fleets in recent years to unprecedented levels and
3. Developing coastal States, having a right to develop their fisheries but lacking technology, tradition and/or capital to do so.

Unless there is some kind of agreement on how to allocate fishing rights among these three blocks, it will be extremely difficult to arrive at any meaningful policy of limiting fleet capacity at international level. And this will, beyond any doubt, be a crucial question to tackle in the context of the world's fisheries governance.

The recognition of the Union as a world partner for fisheries governance

Today it is an undisputed principle that the management of international fishery resources falls under the exclusive competence of the Union and therefore it is the Union (represented by the Commission) and not Member States, who represent European fishing and negotiate in international *fora*. However, this was not at the beginning a question of principle but rather a pragmatic approach, that in certain cases took years to consolidate.

As we saw in Chapter 2, the joint declaration of EEZs was decided within the last quarter of 1976, to enter into force on the 1 January 1977. That meant that, within 2 months, the EEC had to solve several problems:

- Negotiating access to the newly declared EEZs of certain countries where European fleets had been traditionally fishing (Norway, Canada, USA, Russia etc.)
- Managing the – much reduced – presence of foreign fleets in the newly declared European EEZs (Russian, Polish, Spanish and other vessels were fishing in these waters prior to 1977).
- Handling existing bilateral agreements between certain Member States and third countries (e.g. between Denmark, Sweden and Norway).
- Establishing joint management measures with neighbouring countries for stocks shared between the EEZ on third countries and those of EEC Member States. The joint management of stocks in the North Sea was the best example.

Doing all that in 2 months was obviously impossible, leading the EEC to take a very pragmatic approach to negotiate all these arrangements: if those were to be done by individual Member States the difficulties would multiply. The only pragmatic solution was the let the Commission, on behalf of the EEC, do it for all.

But that implied another problem: that of the recognition of the EEC at the time as an international partner to negotiating agreements with third countries. While this was not a problem for certain States and the EEC could sign an agreement with Spain (then an accession candidate) and with Canada and the USA, other countries led by the refusal of the former Soviet Union to recognise the EEC, did not follow suit. The USSR, Poland and the former German Democratic Republic refused to sign mutual access fisheries agreements to avoid an international recognition of the EEC. This resulted in a loss of fishing opportunities for a number of fleets in the Europe of the time.

This recognition took place, *de facto*, sometime later, when the old ICNAF was replaced by NAFO and in that new context the USSR had to accept for the first time to negotiate with the EEC on an equal footing (Farnell and Elles, 1984). This was considered by the Commission as a breakthrough in the international recognition of the EEC of the time as an international partner.

Some authors have underlined the importance of this recognition for the European Commission, whose officials allegedly made a question of principle of defending the role of the institution as an acceptable partner in international relationships. In this vein, fisheries had a considerable importance, since the only areas where the Commission could claim that role at the time were the trade policy, the bilateral fisheries agreements and the participation of the CEE in regional fishery organisations (Lequesne, 2001).

The question of the international representation of the Union in multilateral fishery organisations remains a very interesting question, leading to varied situations, as we will see next. Fundamentally, the direct representation of the Union in these organisations is now clarified according to competence (Unions' exclusive competence or member States' competence), and this has interesting consequences.

For example, in areas of exclusive competence the Union counts only as one party. This is often resented by the fishing industry as a disadvantage. However, the long-standing practice shows that important countries such as the United States or Japan also have one vote only, and that does not prevent them from being very influential in the way multilateral decisions are taken.

It is often argued that exclusive competence limits the weight of the Union when it comes to ensuring the ratification of international agreements. When ratifying UNCLOS (mixed competence) the ratification by the Union means the ratification by its 28 Member States and thus 28 votes for the ratification of the instrument. However, when it comes to the FAO's Port State Agreement (see Chapter 9), that falls under exclusive EU competence on account of the trade measures contemplated, the Union's ratification represents only one vote and, as a result, the agreement is still today not in force (under mixed competence, the 28 votes of Member States would have implied immediate entry into force, since this requires only 25 votes).

Multilateral management of fisheries

The Union in Regional Fisheries Management Organisations (RFMOs): the question of membership

The Union is a contracting party to 14 RFMOs in the world (Table 7.2). In RFMOs dealing exclusively with fisheries management, the Union is the sole representative on the basis of its exclusive competence. There are, however, some cases where certain individual Member States are also members in their own right, in respect of overseas territories that are not part of the Union. Typical examples are Denmark on behalf of Greenland and the Faroe Islands, France on behalf of its *Territoires d'Outre-Mer*, or the UK with regard to its own overseas territories.

Table 7.1 shows that the EU is a major global player in fisheries management but the way to achieve that status has never been a smooth one. For many years, the legitimacy of the Union to defend the European rights to fish for common resources was questioned, either *de jure* or *de facto*.

Overall, the Union is the largest contributor to the budgets of these organisations. Without the EU

Table 7.1 List of RFMOs in which the Union participates as a contracting party or a cooperating non-contracting party.

Organisation	Status/membership
Convention for the Conservation of Antarctic Living Marine Resources (CCMLR)	Contracting party. Mixed competence with Member States
Convention for the Conservation of Southern Bluefin Tuna (CCSBT)	Cooperating non-contracting party
Inter-American Tropical Tuna Commission (IATTC)	Contracting party. Exclusive EU competence
International Commission for the Conservation of Atlantic Tuna (ICCAT)	Contracting party. Exclusive EU competence
Indian Ocean Tuna Commission (IOTC)	Contracting party. Exclusive EU competence
Northwest Atlantic Fisheries Organization (NAFO)	Contracting party. Exclusive EU competence
North Atlantic Salmon Conservation Organisation (NASCO)	Contracting party. Exclusive EU competence
South East Atlantic Fisheries organisation (SEAFO)	Contracting party. Exclusive EU competence
South Indian Ocean Fisheries Agreement (SIOFA)	Contracting party. Exclusive EU competence
South Pacific Regional Fisheries Management Organisation (SPRFMO)	Contracting party. Exclusive EU competence
Western and central Pacific Fisheries Commission (WCPFC)	Contracting party. Exclusive EU competence
General Fisheries Commission for the Mediterranean (GFCM)	Contracting Party. Mixed competence with Member States
North East Atlantic Fisheries Commission (NEAFC)	Contracting party. Exclusive EU competence
CECAF	Contracting Party. Mixed competence with Member States

contribution, some of them may collapse financially. While in most cases the participation of the Union in these RFMOs is directly related to the presence of EU vessels in the fisheries concerned, in some other cases the Union is a member or a cooperating non-contracting party even if there are at present no EU vessel participating directly in the fisheries concerned. However, the need and opportunity to ensure an active participation of the Union as a global player in fisheries governance justifies the presence of the Union in these fora.

The exclusive competence of the Union on the conservation and management of fish resources implies that international cooperation for the purpose of managing shared fishery resources is the responsibility of the Union and not of individual Member States. This situation is clear in Europe but it has been far from obvious for many of our international partners.[5] Obtaining full recognition of the role of the Union as the legitimate representative of the European fishery interest has been very often a difficult struggle.

There are several aspects to this question:

- The fact that many Regional Fisheries management Organisations (RFMOs) included in their founding text that membership was limited to 'States' and the Union did not qualify as such.
- In some cases too, the scope of certain multilateral organisations is not limited to fisheries management. Typically, environmental questions are also included and this means that the competence is mixed between the Union and its Member States; this leads to often complex responsibilitie of shared membership and voting arrangements.
- Over and beyond institutional questions of membership, a big challenge, particularly in RFMOs dealing with very distant fisheries, there has been a lack of enthusiasm by a number of partners about the rights and the legitimacy of European interests being represented.

Perhaps the best example of the difficulties associated with EU membership is the case of ICCAT.

This organisation, established in 1967, only admitted 'States' as members and, during many years, even after the CFP was established only some EC Member States were contracting parties while the EC was not. In fact, up until the mid-1990s, only France, Spain and Portugal were members of ICCAT and the EC was an observer. In addition, most measures decided in ICCAT were implemented through national law, not through CFP rules. In these circumstances, repeated attempts by the EC at the time to become a member, requiring a change of the ICCAT Convention to admit 'Regional Economic Integration Organisations' to adhere as full members, were blocked by certain ICCAT member countries.

Everything changed in 1996. That year, largely on the initiative of the US delegation, the idea of applying trade sanctions to countries not respecting the catch limitations on Bluefin tuna, agreed in 1994, was making headway. This posed a problem: France, Spain and Portugal could not decide trade sanctions that would subsequently have to be applied by the EC, on the basis of its exclusive competence on trade policy. Furthermore, some of the countries possibly subject to such measures might actually be EU Member States that were not yet members of ICCAT. France, Spain and Portugal could not possibly accept to apply trade sanctions against another EC Member State.

To avoid these problems, all delegations (and crucially the US delegation) understood that this new policy approach could only work if the EU was a full member of ICCAT. This change of attitude prompted a quick change of the ICCAT Convention, adopted in 1997. Upon this change, the Commission immediately deposited its instrument of accession to ICCAT and became an operational full member of ICCAT at the 1997 session itself.

Other case worth mentioning is the EU membership to the Inter-American Tropical Tuna Commission (IATTC). In this case, the interest of EU membership started with the participation of Spanish interests in the fishery, firstly through investments in central and South American countries in the mid-1970s.

In this case, the increased presence of the Spanish fleet in the 1990s, and their investments in countries of the region, even if perfectly legitimate under the Law of the Sea, was not welcome by some parties. In particular, the Mexican industry saw this presence as too a strong competition for their aging fleets. But the US delegation was also hardly enthusiastic. Just as in ICCAT, the Union could not become a full member because membership was reserved to 'States' in the original IATTC convention, which dated back to 1949. And the only Member State interested, Spain, could not accede because under exclusive competence Member States could not become members in their own right. Only France was a member on account of its status as coastal State, due to its sovereignty rights over the atoll of Clipperton, a non-EU French territory of the Eastern Pacific.

This situation led to a very curious situation: in the late 1990s, IATTC was negotiating a new convention, to adapt the old one to the evolution of international law (the UNFSA of 1995). In that context, the Commission managed to include 'regional economic integration organisations' as eligible for membership. However, until the entry into force of the new Convention, the question of membership required a provisional solution and this was agreed as the provisional membership of Spain on behalf of the Union (OJEC, 1999). Upon accession of the Union to the new IATTC in 2004, Spain's membership was withdrawn.

The membership of the Union, once achieved, is not without problems. In fact, as the Union is admitted as a member of an RFMO, it consequently has one vote. This has in the past been strongly resented by some Member States who believe that Union membership actually weakens the weight of Europe in these organisations. Attempts have been made in certain of these RFMOs to ensure that the Union would have as many votes as Member States interested but these have been rejected on account of exclusive EU competence.

In some cases, membership is made more complicated by the mixed competence of the RFMO in question. While most RFMOs deal exclusively with fisheries management, an area of exclusive competence by the Union, it is the Union and not its Member States who have the right for membership (with the exception of individual Member States on behalf of their non-Community overseas territories). However, certain RFMOs have also an environmental mandate beyond fisheries management. In these cases, since environmental policy is a mixed competence between the Union and Member States, both are members and have the right to intervene and decide on the basis of a declaration of competences that the Union presents prior to the meetings specifying whether each point of the agenda is for the Union or for Member States to decide upon. RFMOs with mixed competence are CCAMLR and GFCM.

Finally, the question of EU membership has in specific cases played an essential role in the constitution of certain RFMOs. The best example is the unsuccessful attempt to offset up a RFMO to manage fisheries in the South-Western Atlantic, off Argentina and the Falkland Islands.

In this case, the difficulty was that, as the Falklands were not EU territory, only the UK could represent them. But this was not acceptable to Argentina, since that would have amounted to recognition of UK sovereignty over these islands. The UK, in turn, could not agree with an RFMO where the Falklands would not be represented by the UK. To resolve the conundrum, the idea was floated of an RFMO with a mixed fisheries/environment mandate, thus requiring mixed competence. This would have implied that Member States, and not just the Union, would be members of the RFMO. Membership of other EU Member States together with the UK would politically 'dilute' the issue of UK membership. Each side could present the result as a win: the UK because they would be members in their own right and Argentina because they would recognise the UK membership only as one among other EU Member States on a mixed competence issue.

At the end, however, there was no deal and the area remains to this day unregulated in its international waters.

The relationship between the CFP and global fisheries management

The influence of third countries on the CFP goes in two directions: in some cases the third countries have an effect on the development of the CFP, for good or bad and in other cases the Union itself contributes to international management through its international leadership. The CFP might be quite different today without the influence of the external policy pillar.

The leadership of the Union

As we said previously, the external policy has often influenced the internal development of the CFP. After all, Union waters are not isolated and they straddle waters of other third States with which the Union has to agree common measures. It is important to underline that there has always been a tension between the need of the Union to take its responsibility in the management of fishery resources and the position of the third States with which the Union has common resources. The traditional question is this regard is: should the Union adopt fisheries measures (often resulting in qualitative and/or quantitative limitations of the activity) when third countries sharing the same resources are not doing the same?

The question is as old as the CFP and has no single answer. Of course, the Union has always tried to favour good fisheries governance both bilaterally and multilaterally but in a number of cases, it has not hesitated to take certain actions even before our partner countries were prepared to do so. A number of cases demonstrate this.

A good case in hand is the management of deep-sea stocks in the Atlantic. In the early 2000s, the scientific evidence on the need to protect deep-sea stocks became clear. The Union subsequently adopted measures in this regard: biannual catch limitations and an effort management plans.

In the early 2000s, scientific reports highlighted the extreme fragility of some of these species. Long-lived, with low fertility ratios, some of these species were considered extremely vulnerable to overexploitation. As they occur largely in international waters, their effective protection would require international agreement including other countries of the North East Atlantic.

Yet, cooperation from other countries, in the context of NEAFC was less than enthusiastic, notably as these fisheries were considered by certain parties as a possible new fisheries resource. The Union, however, did not hesitate to take the lead, by establishing unilaterally limitations of effort (EC, 2002a) and catches (EC, 2002b), even before NEAFC decided to follow suit.

In other cases, however, the leadership of the Union has been severely hampered by the lack of follow up by partner countries in the sea basin. The best case is, without doubt, the Mediterranean, where traditionally sluggish progress in GFCM has hampered progress on the management of European fisheries in that sea basin. We will refer to this specific problem in the next Chapter.

The contribution of RFMOs to the development of the CFP

The relationship between the CFP and the work of the RFMOs is bi-directional: the Union has been instrumental in bringing new initiatives to these organisations and, at the same time, the decisions taken in RFMOs have created very important precedents for the development of the CFP. In this section we will develop in particular this second question.

The participation of the Union has been instrumental in the way certain RFMOs have developed. In some cases, the Union's initiative has been crucial for the establishment (or the reform) of certain RFMOs. In other cases, in well-consolidated RFMOs, the Union has been in the driving seat for the adoption of new, forward-looking measures.

Many features of the current CFP have indeed originated in RFMOs. Just as examples, we can cite the following:

- Long-term plans. Before the CFP developed them internally, it crucially contributed to adopt them in certain RFMOs: the Atlantic swordfish recovery plan adopted by ICCAT in 1998, with a key contribution from the Union, was a clear precedent and a source of inspiration to future long-term plans in EU waters a few years later.
- Enhanced control. Some of the exceptional control measures adopted in NAFO and NEAFC have indeed had an influence in later developments of the Union's fishery control policy. Likewise, the need to ensure effective control of some of the Union's most important international fisheries have produced some of the biggest deployments of control effort in the Union. The Bluefin tuna in the Mediterranean is the best example.
- New species under TACs. Indeed, a number of species, caught partially in EU waters, have been subject to catch limitations and other conservation measures only as a result of the EU's participation in RFMOs. A good example is the albacore (*Thunnus alalunga*) fishery in the Bay of Biscay that used to be a very contentious, unregulated fishery until a TAC was adopted by ICCAT.
- Multilateral underpinning of the Union's efforts. This is the case in particular in the Mediterranean basin, where initiatives by the Union for its own fleets are often opposed if the same measures do not apply to foreign fleets in the same or neighbouring areas. In certain cases, advances in fisheries management is this area have been made possible only after GFCM succeeded to adopt these measures for all Mediterranean parties.
- Allocation exercises. In a number of cases, internal allocation of EU quotas among Member States has been agreed (or even re-discussed) only after certain decisions have been taken by RFMOs.
- RFMOs have also contributed to develop the fight against IUU fishing. As we will see in the next Chapter, the fight against IUU fishing through the development of trade sanctions to IUU countries started in ICCAT long before the Union established its own unilateral policy under the IUU regulation. These ICCAT actions started in practice in the late 1990s, when the Commission, following a recommendation by ICCAT, banned imports of Bluefin tuna from Belize, Honduras and Panama (EC, 1998). Other similar actions have followed since. These measures had a fundamental advantage: their multilateral nature, that made them politically and legally more resistant to possible challenges in the context of WTO. We will refer to that in Chapter 9.

These contributions have also being very varied in terms of their origin. Sometimes, the Union has used the RFMOs as the only way to make progress in the CFP. The best example is the Mediterranean, where the work of GFCM is essential to allow for progress on fisheries management in Europe, in a context where Member States concerned are reticent to make progress unless other countries of the Mediterranean basin follow suit.

In other occasions, decisions that the Union had to accept as part of a negotiation, under pressure from other contracting parties, have also established very good precedents for later developments in European policy. A good example is the very efficient system to control catches of Bluefin tuna, largely as a result of pressure within ICCAT to solve the problem.

What challenges for the RFMOs?

Important progress has been made in the work of RFMOs in recent years, as we saw previously. However, there are still very important challenges remaining. The fundamental ones remain related to developing countries.

Developing countries have little means to exert their responsibility as coastal and/or flag States. Their possibilities to collect scientific information and control the fishing activity are important limiting factors for them at national level but also for the successful work of the incumbent RFMOs. The Union has, in many cases, being prepared to use part of its development aid to palliate these weaknesses. For that to happen, the third countries concerned must also identify such questions as a priority, which has not always been the case.

The scientific advice

The basis for good scientific advice is an efficient data collection system. That is not always guaranteed in certain RFMOs, in particular because the coverage of data is often very incomplete: while certain fleets provide detailed data of all their activities, other fleets involved in the same fisheries do not provide data at all. This situation is related to compliance by certain industrial fleets, and to insufficient capacity building in some developing States.

While certain RFMOs have this question very well covered, the provision of scientific advice is quite a challenge for certain others. We have different situations in this regard:

- Certain organisations draw the scientific basis from the same source as the CFP: ICES, which advises NEAFC and NASCO.
- While certain RFMOs have scientific bodies made of scientists from the contracting parties, the IATTC have a permanent scientific staff that provide the advice to the contracting parties.
- Organisations such as NAFO, ICCAT or IATTC have very competent scientific bodies of their own. The challenge there is two-fold: how to ensure appropriate participation by EU scientists and how to build capacity in developing States for them to make substantial contributions to the scientific work.
- Organisations such as GFCM have a basic problem of very unequal coverage of scientific advice, with relative good coverage by the European side but very little from the southern rim of this sea basin.
- Organisations such as CECAF, where there is practically no scientific advice being provided at multilateral level, even though certain coastal states, such as Morocco, have competent marine research institutions with the capacity to provide such advice.

The enforcement dimension

One of the most common criticism of RFMOs is that they are 'toothless', that means, they don't have the capacity to effectively enforce their decisions. This presumed 'toothlessness' is used often by certain coastal States to claim that RFMOs are not the right answer to the issue of management of straddling or highly migratory fish stocks, so that the only answer to that question is to extend coastal States' jurisdiction beyond the 200 miles of their Exclusive Economic Zones.

In that vein, ensuring that the RFMO recommendations are effectively enforced constitutes an essential element for the success of these organisations but is also a guarantee of the stability of current international law.

In that regard, a number of initiatives have been taken within RFMOs in that direction:

- The control schemes agreed by NAFO or NEAFC constitute a model of how control can become an essential part of the work of a RFMO (see Chapter 9).
- The initiatives to identify IUU vessels by certain RFMOs have considerably helped in the fight against illegal fishing.

Overall, however, the enforcement part is a clear weakness of many RFMOs, particularly those with large membership of developing States that lack the financial means to exert their control obligations effectively. While international law foresees the financial aid to these countries as a fundamental element of the effective application of international law the implementation of this provision remains to this day largely unresolved.

The question of quota allocation

A fundamental challenge is the allocation of fishing rights. Particularly, for highly migratory resources, a precondition for good multilateral management is to find a balance between two legitimate rights and interests: those of developing States with little or no tradition of fishing, to develop this sector on the basis of the resources in their own waters and beyond and those of developed States with a tradition to fish in these waters.

This represents a major difficulty: developing States have a legitimate right to fish but they generally lack the tradition, the capital, the technology and know-how to do so. In addition, they want to fish to export to the markets of developed countries. In turn, these have the tradition, capital, technology and know-how and also have the markets for the fish. Asking them to transfer fishing rights for others to develop and encroach upon their own markets is not the easiest thing to do.

This challenge represents also a contradiction between the CFP in European waters and the international management of fisheries: while relative stability within European waters is well consolidated and stable over time, the external policy has to deal with a permanent tension to modify the international relative stability in favour of the new entrants to the fishery. Some fishing interests in European waters see their claims for higher quotas limited by relative stability but they have to face pressure for their own relative stability in international waters to be reopened.

The nature of the problem can be illustrated by the early attempts to adopt a tagging programme to study big-eye tuna in the Indian Ocean, as one of the first attempts in the IOTC to establish a sound scientific basis for the management of that key species. In 1999, these attempts by the EU and Japan were strongly resisted by a number of developing States even though the programme would be funded outside the IOTC budget and would not imply any financial burden to them. After difficult negotiations, the real reason for that apparently inexplicable opposition came out through the explanation of a representative from a developing country: if a research programme was funded, then scientists might discover that the resource was fully exploited or overexploited and this could be used by developed States as an excuse to limit the level of fishing to prevent the access to the fishery by developing States.

Resolving the issue of the fishing rights is, as this anecdote illustrates, a precondition for sound cooperative management of the resources. This question has been treated a number of times, notably in the context of tuna management RFMOs. The experiences there have been extremely illustrative of the difficulties to resolve the issue.

In ICCAT, the considerable difficulty of agreeing allocation keys among a very high (and ever increasing) number of contracting parties having a legitimate right to fish led to the establishment in 1998 of a working group to decide on allocation criteria. At EU level, the difficulty of that exercise led the Commission to prepare a communication in order to have a high-level discussion to obtain a mandate for the ICCAT working group. The question was so delicate that the document prepared by the Commission services in 1999 was considered confidential and not to be published. Only on the basis of such a document, Coreper could agree on a mandate for that difficult negotiation.

Eventually, the ICCAT working group adopted a list of criteria for future allocations that was so general and with so many criteria (in order to find a compromise) that in fact subsequent allocation exercises have been based on *ad hoc* negotiations and only very vaguely on the criteria defined by ICCAT. In any case the tension around existing allocation keys continues, and certain countries are still insisting on the need to re-open existing allocation keys to accommodate their demands.

In the IATTC, the question of fishing rights gave rise in the late 1990s to the establishment of a system of 'capacity quotas' per contracting party to the organisation. However, as new States claimed their legitimate interest to have a capacity quota, the only solution found was actually to increase the overall quota of available capacity, to levels that were not fully compatible with the balance fleet/available resources. This tendency continues today and current capacity ceilings are already in excess of what would ensure a sustainable fishery for yellow fin tuna in that region. This example illustrates how multilateral management is dependent upon the resolution of the question of the allocation of fishing rights.

Another experience is that of the so-called 'Kobe process': an informal process, promoted by Japan, to find elements for a global governance of highly migratory stocks (mainly tuna). This exercise was opportune and necessary because of the uncontrolled growth of tuna fleets in recent years, particularly by countries of the Far East. The Kobe process, that met three times, tried to find a solution by establishing caps to the development of fleets, so as to avoid a possible future collapse of these fisheries.

However, any attempt to cap fleet capacity encountered fierce resistance from a number of countries around the world, notably (but not exclusively) developing States. Their argument was that the capping would simply prevent them from increasing their participation in the fishery in the future. Developed States with a fishing tradition argued that they had the vessels, the technology, the tradition and the markets. The differences were such that the Kobe process failed to produce meaningful results.

This experience shows clearly that if a global capping of these fleets is to be established, it is necessary to find a new balance between the fishing rights of developing and developed States, something of extreme difficulty.

The role of developing countries

Many of these problems are one way or another linked to the difficulties by – at least some – developing States to fulfil their obligations as coastal and flag states according to UNCLOS, due to lack of economic means and, also often, lack of tradition and expertise. Finding a solution to this problem is key for the success of future of RFMOs.

Many developing States still see RFMOs as an instrument by rich countries to consolidate their economic leadership in the development of the fishery. That often translates to a low motivation for an active participation in the common efforts to manage stocks jointly. This motivation is linked to the question of fishing rights: without allocation of fishing rights there is little point in investing in fisheries management.

That means that the problem of 'empowering' developing States to play an active role in the multilateral management of fisheries within RFMOs is not just a matter of having the right human and financial means, it is also about having the guarantee of the economic rewards associated to significant fishing rights.

For a long time, the Union's development policy has considered capacity building in fisheries matters among its priorities. In some cases, these funds have been used by developing countries to promote their fishing industry as well as their administrative capacity to manage the fisheries concerned. This, however, remains largely voluntary. If and when developing States prefer to give other priorities to the EU's development funds, the Union cannot impose its priorities.

Fisheries outside RFMOs

There are also important fisheries that remain totally or partially outside the remit of current RFMOs. This is due to at least two reasons:

Certain species do not fall specifically into the scope of existing RFMOs and no single RFMO is formally responsible for their management. A clear case in hand is that of pelagic sharks, often caught in fisheries for tuna and swordfish but not included as such in all the tuna conventions.

Another problem is the distinction between the 'Convention Area' and the 'Regulatory Area' in some fisheries conventions. While the Convention Area includes international waters and the national EEZs where the straddling stocks occur, the Regulatory Area only covers international waters. Parties to the RFMOs have an obligation to cooperate in the Convention Area but the RFMO recommendations only concern the Regulatory Area.

This distinction is not a problem in certain RFMOs, such as NAFO, where the management of stocks straddling North American EEZs and international waters is regularly done in the NAFO meetings and according to NAFO rules. However, in NEAFC there is a completely different practice: here the management of certain stocks, particularly the widely distributed pelagic stock (mackerel, herring, blue whiting) is only dealt with by NEAFC after the cooperation among 'coastal states' has agreed on measures and allocation. In other words, the fundamental part of the work of NEAFC is done outside it, through the informal negotiations among 'coastal states' that completely lack rules and procedures. The practice for quite a long time has been that the lack of agreement on allocation, in some cases due to opposition by only one country, has delayed the adoption of much-needed management measure for several years.

Completing the network

Another fundamental challenge is to complete the network of RFMOs, so that all important resources occurring in international (or shared national) waters are covered by working multilateral arrangements to ensure sound joint management. In certain cases, there are, at least in theory, an RFMO with that mandate. Another thing, however, is whether it is fully operational. Until the late 1990s the Mediterranean was covered by a RFMO that was not operational. However, the Commission then led the way for an important revamping of the organisation.

Today, the most important case of the need to complete the network is that of CECAF,[6] that covers the fishery resources of Western Africa. These waters are subject to one of the highest rates of illegal and unregulated fishing in the world, with very negative consequences for the coastal populations of the countries concerned of Western Africa.

Bringing this organisation to a status of an effective body to ensure adequate management of fisheries in this area is of paramount importance in two respects: to ensure that the countries concerned can obtain the best sustainable benefits from their fisheries and, from an European perspective, to help resolve the problem of illegal fishing in that region and to allow the development of the Union's bilateral agreement policy on a sound, scientific basis, including the estimation of the 'surplus' resources.

Are resources improving in RFMOs?

As the CFP has clearly defined its objective (at least since 2002) as the achievement of sustainable fisheries, from a biological, economic and social perspective, it is legitimate to ask this question. The answer is not straightforward but we can try to summarise a few challenges in that regard.

Here, the general trend is positive, both in terms of the number of fish stocks being subject to meaningful management measures but also in terms of general trends in stock status. However, the contributions from different RFMOs are quite different:

- Certain RFMOs are successful in managing stocks and have been for quite a long time, generally with a key contribution from the Union. NAFO is probably the best example, with present recovery of some cod stocks and yellowtail flounder as well as stable management of Greenland halibut. Likewise, the success of ICCAT in recovering Bluefin tuna is a remarkable achievement.
- Certain RFMOs are gradually becoming more operational but still lag behind in terms of the number and importance of the stocks being managed. GFCM is probably the best example.
- Other RFMOs have most of the conditions to be successful but because of their structure and governance, are not always efficient in addressing sustainability issues: NEAFC, with its unique practice of taking decisions for international waters only after the coastal States have agreed through the informal arrangement of 'coastal States consultations', has demonstrated over the years its inefficiency to decide quickly, as demonstrated by the long and frustrating deadlock on the management of Atlantic mackerel.
- Decision making is another cause of concern. The requirement (or tradition) of consensus in certain RFMOs can considerably limit progress in management, since in some cases only one party opposes a decision and can effectively block progress.
- Finally, in some other cases, the absence of an operational RFMO is a crucial problem, both in terms of the overall multilateral management of stocks but also in the context of the estimation of the 'surplus' resources that the Union can negotiate to fish in third country's waters. The best example is CECAF[7] in Western Africa, which is constituted formally but is hardly operational at all.

Unlike the CFP in European waters, where the annual Policy Statement (see Chapter 4) summarises the situation of stocks in European waters, the resources under the management of RFMOs are not subject to the same kind of scrutiny. The differences between the two types of stock are obvious, in that progress in their management is not only the responsibility of the Union. But as the new policy agreed in 2013 is applied, the regular evaluation of the status of stocks under RFMOs could potentially be an important instrument to raise awareness about the need to reinforce the efforts of the international community to ensure a level of sustainability in the exploitation of these resources in accordance with the Union's and international management objectives.

Other neighbouring relationships. Coastal states of the Northeast Atlantic: the case of mackerel

Perhaps the most prominent issue in the European Union's neighbouring policy in fisheries concerns the management of widely-distributed pelagic stocks in the NEAFC area and, in particular, the case of mackerel.

As we saw above, because of the particular working practice of NEAFC, the decisions on the management of certain widely-distributed pelagic stocks (mackerel, horse mackerel, blue whiting, herring) are taken by NEAFC only after the so called 'coastal-states' (that is, the States that are coastal to the area of distribution of these stocks) agree on both catch levels and allocation keys. Depending on each stock and its geographic distribution, these can be the Union, Norway, Iceland, Greenland, the Faroe Islands and Russia.

The agreement among coastal states has proven extremely difficult to achieve over the last decades, with the recent case of mackerel being of particular relevance: the lack of agreement on the allocation of the stocks has lasted for several years and the parties set autonomous quotas that in conjunction were much higher than the catch levels recommended by science. This is due to a number of factors:

- The fact that these negotiations take place outside NEAFC (unlike similar decisions in NAFO or ICCAT) and thus outside its voting systems and dispute settlement procedures;
- The movements of the biomass of these species, which change from year to year, creating new 'coastal States' as the fish becomes abundant in the waters of countries where they did not occur before;
- The considerable value of these species in the context of the bilateral exchanges with Norway; and
- The position of certain coastal States (at different stages Norway and Iceland) that privileged strategic positions on quota sharing over sustainability.

There is no objective reason why the management of the mackerel, herring and blue whiting stocks in NEAFC should be any more difficult than, say, the management of Bluefin tuna by ICCAT or that of Greenland halibut in NAFO. And yet, the latter always arrive to an agreement on the policy at the annual meeting while NEAFC does not. This clearly shows that the current structure for decision making in the area of NEAFC, where decisions by this organisation are depending upon the agreement of this informal setting of 'coastal States' is not adequate to take timely decisions on fisheries management and needs to be changed.

Bilateral fisheries agreements

This part of the CFP is as old as the CFP itself. The shared nature of many resources with non-EU countries made it necessary, from the inception of the CFP, to ensure the management of these shared resources through bilateral cooperation with the Union's neighbours. This was the case of resources in the northern North Sea, shared with Norway that from the beginning of the CFP had to be managed through bilateral cooperation with that neighbouring country. Many years later, the accession of Baltic States to the Union turned the former multilateral management of fishery resources of the Baltic Sea into a bilateral cooperation between the Union and Russia.

These neighbouring agreements, in addition to the joint management of shared resources, can include exchanges of fishing possibilities in each other's waters, to benefit from the different commercial value of certain species in different markets.

On the other hand, the tradition and economic importance of fishing activities in the waters of third countries has led the need for an active policy to preserve such activity. If the third countries concerned do not have large fleets interested in mutual access arrangements with the Union, the continuity of European fishing is ensured by paying for access to the third countries waters.

These two cases are very different in their motivation, objectives and consequences. Let us examine them separately.

Mutual access agreements

Mutual access agreements were developed in the very earliest days of the CFP. Upon the declaration of EEZs by Atlantic Member States of the Europe of nine, it became immediately necessary to agree with neighbouring countries on the management of shared resources and on the mutual access to fisheries in each other's waters.

Other mutual access agreements have been negotiated traditionally with Iceland and the Faroe Islands, although these are today practically without effect.

Some of these agreements have had an important value in preparing the enlargement of the Union. From 1977 to 1985, the EEC of the time agreed the allocation with Spain of a number of licenses to allow access to the waters of the Community of nine, on the basis of its status as a candidate for accession. Likewise, four Baltic States (Poland, Estonia, Latvia and Lithuania) established mutual access agreements with the Union in the Baltic in the early 2000s, in preparation for their accession in 2004.

Norway

Among these agreements, by far the most important is that with Norway. The current agreement dates back to 1993 (OJEC) and, on that basis, annual consultations take place on two aspects: (1) the decisions on the management of joint stocks, particularly in the North Sea and (2) the exchange of fishing opportunities in each other waters. The agreed record of this consultation is normally implemented in the form of TACs, quota exchanges and, until the entry into force of the Lisbon treaty, through provisional technical measures implemented in the annual TAC and quota regulation.

Today, as many as the main stocks in the North Sea (excluding sole) and some in the west of Scotland are managed jointly and the Union can only implement its annual TACs for the concerned stocks after the total TAC levels and their allocation between Norway and the EU have been agreed in annual negotiations.

The agreement with Norway is a direct consequence of the declaration of EEZs by that country and Atlantic Member States of the Union on the 1 January 1977, which left the whole surface of the North Sea and the fish stocks therein, as shared stocks between the Union and Norway, requiring joint management.

As with the CFP itself, the most difficult question was initially the allocation of fishing opportunities between the two sides, which gave rise to important difficulties in 1977 and 1978. Initially, the EEC considered that the historical catch records should be the basis for the allocation but the evidence that many important stocks in the North Sea spent the early stages of their life in Norwegian waters reinforced Norway's claims and forced the EEC to eventually agree to a sharing out that, in essence, has been maintained up to this day.

Under this agreement, the EU and Norway agree on the joint management of a number of stocks (including some long-term approaches) and exchange fishing possibilities in each other's waters, according to the so-called 'cod-equivalents', a formula whereby, for example, a tonne of cod is equivalent to 10 tonnes of sprat.

The agreement with Norway is extremely influential in the CFP, for a number of reasons:

- It allows access to certain resources in Norwegian waters that are of paramount importance for some European fleets. Cod in Norwegian waters is the best example.
- The joint management of stocks, particularly in the North Sea has a high influence in the way the Union manages these resources.
- The Norwegian agreement also influences the bilateral negotiations between the Union and Greenland: certain fishing opportunities in that country are actually to be offered to Norway in the context of the exchange of quotas.
- This agreement also plays a role in the way the EU and Norway address the negotiations among North Eastern Atlantic coastal States.

The annual agreement with Norway is also very politically sensitive because it introduces elements that need to be politically balanced very carefully. The quotas the EU obtains in Norway are of interest for certain Member States in particular, while the resources Norway obtains in EU waters in return can be of interest to a different group of Member States. This produces a permanent tension to agree on an exchange that is seen as fair by all Member States. This has justified frequent annual discussions at the level of Council on the overall balance of fishing rights to be agreed with Norway.

But beyond these direct influences, the Norwegian case has been a permanent source of comparison with the CFP. The undoubtedly successful fisheries management in Norway has very often been used as an example of what the Union should do. These comparisons generally forget that what works in a scarcely populated and homogeneous country does not necessarily work in the much more complex situation of the Union. Despite these differences, Norway has been a source of inspiration and when the Union has addressed certain questions, such as the discard ban, the influence of the Norwegian experience has been obvious.

When referring to Norway, it is also important to refer to the case of Svalbard, a case of 'realpolitik' which like certain other non-legal arrangements, has demonstrated an extraordinary resilience over time.

The archipelago of Svalbard in the Arctic Ocean is surrounded by rich fishing grounds. This territory was recognised internationally as coming under Norwegian sovereignty by the Treaty of Paris in 1920, subject to free access by other signatory countries to its land and sea resources. Norway's initiative to declare a Fishery Protection Zone (FPZ) around Svalbard was a clear breach of the rights of the other signatory countries to the Treaty of Paris. However, geostrategic considerations came into play (as in the case of the cod wars with Iceland) and the important strategic position of Norway in the cold war, together with Norway's policy to use the FPZ to exclude Russian vessels from the area prevailed and the EEC did not repudiate the Norwegian declaration (Farnell and Elles, 1984).

This led to a situation where the Norwegian FPZ around Svalbard is neither recognised officially, nor rejected in practice and from the early stages of the CFP, the quotas available for the Union in these waters are always presented in a non-committal formula that implies *de facto* acceptance but not official recognition, in yet another demonstration of the importance and remarkable stability of certain non-legal arrangements in the CFP.

Russia

But beyond Norway, the most important case of a neighbouring bilateral relationship for the purpose of managing shared resources is that with the Russian Federation in the Baltic Sea. Following the accession to the Union in 2004 of four Baltic countries (Poland, Estonia, Latvia and Lithuania) the former multilateral body responsible for fisheries management in this area (the International Baltic Sea Fisheries Commission, IBSFC) had to be replaced by a bilateral agreement between the Union and the Russian Federation.

Indeed, an agreement was negotiated and initialled in 2009 and the Union subsequently ratified the agreed this agreement the same year. However, Russia delayed its ratification apparently on grounds of what the Baltic agreement would mean as a precedent for the Black Sea and in relation to political issues beyond fisheries. Finally, Russia ratified the agreement in 2012 and the bilateral talks to agree on common management measures started the same year.

The two parties have met repeatedly but have so far failed to agree on management measures. The reason is one that is very familiar for the CFP: the lack of agreement on the allocation of Baltic resources: while the Union considered that the old allocation keys agreed in IBSFC could be translated into a new key (by adding the quotas of the Baltic States to the Union quota, after accession), Russia considered that these historical allocation keys could not be directly translated into today's quotas.

To this day, lack of agreement on quota allocation prevents agreement on common management measures and the Union and Russia undertake fisheries management in the Baltic unilaterally in their respective waters. This is a clear demonstration of – yet again – the fundamental importance of agreeing on the allocation of fishing rights as a pre-condition for cooperative fisheries management to take place.

Agreements with trade concessions

These agreements do not exist today but some of them played an important role in the past. In particular, out of the agreements signed in the wake of the declaration of EEZs in 1977 the agreement with Canada was the only one where fishing rights were linked with trade concessions.

Theoretically, the place of the Union as the world's biggest market for fish (in particular after the accession of Spain) should give it an important leverage for that kind of agreement.

However, the EEC and its successors have not used such leverage to its full extent. There are certain reasons for this. One is the diversity of interests among Member States in terms of trade concessions: while some Member States prioritise the availability of tariff-free fish for the processing industry, others prefer a more protectionist policy to defend their catching sector. In addition, over the years the rules of the World Trade Organisation have made this kind of arrangement more difficult.

In fact, the Union only had another type of agreement of this kind, to promote the establishment of joint ventures in third countries. It was based on the on the previous experience of Spain before accession, where access to resources were negotiated against access to the Spanish market. At EU level, the agreement with Argentina, signed in 1993, established a number of joint ventures and was accompanied by tariff reductions for the main products of the agreements, in particular Argentinian hake (although such concessions had to be *erga omnes*; that is, applicable to all parties as required by GATT rules) (EEC, 1993). The agreement was eventually discontinued in the late 1990s and this type of agreement was never negotiated again.

Agreements with financial compensation

These agreements have always been a fundamental part of the CFP and one with many political ramifications. We can distinguish three main types of agreement:

1 The tuna agreements.
2 The traditional agreements with countries of western Africa.
3 The Greenland agreement.

The Union has always considered these agreements as a priority and the budget dedicated to this policy pillar clearly reflects the efforts of the Union, over the years, to do its best to guarantee the continuity of the traditional activity of EU fleets in third countries' waters.

Specific studies on these agreements have been referenced elsewhere and this book will not focus on this issue. However, their impact on the overall perception of the CFP, both in the Member States directly benefitting from this policy and from those that do not use them is very significant, depending upon who receives the fishing rights.

The continuation of these agreements over time has always been contentious. Many voices have criticised them, from different points of view:

- The use of public funds to ensure access of private operators. This has been considered a subsidy by some countries, for example, in the context of the World Trade Organisation.
- In certain cases the fishing opportunities have not been fully used and paying public money for fish that subsequently is not caught is understandably difficult to justify. Curiously, this may be in part the consequence of the EU policy itself: in the context of the negotiation of the second Lomé Convention in 1979, the Commission had to struggle to introduce in a joint declaration the idea that ACP countries[8] agreed not to treat the Community worse than other countries in terms of fisheries agreements (Farnell and Elles, 1984).
- The presumed overexploitation of third countries' resources by European fleets and thus the negative effects of these agreements on this livelihood of local fishermen. This argument has often been unsubstantiated but it has made headway in a large part of the public opinion.
- The alleged insufficient social benefits from the EU funds to the people of the concerned countries concerned (largely due to corruption in some cases).
- The fact that most agreements benefit only a few Member States, which makes them politically fragile.

- The allegation that countries will develop their own fleets and will eventually will not want these agreements; and the – again unsubstantiated – argument that the agreements prevent them from doing so.

These arguments are as old as the bilateral agreements themselves. Over the years, however, such arguments have been addressed by the Union, and criticism to these agreements is today much less than it used to be:

- Many of these questions have effectively been addressed over the years. In particular, the science basis for decisions on surplus stocks has been improved.
- On the utilisation of fishing opportunities, provisions have been included in certain cases to ensure that unused quotas can be re-allocated to Member States ready to use them.
- There is also the realisation that if European fleets leave these countries, the alternative will not necessarily be local fishing but rather large-scale fishing by other nations whose policies are not the same as the Union's. In fact, while certain NGOs used to be very critical of EU bilateral agreements, today they realise that the presence of European fleets is a much better alternative than that of certain other third countries.
- The agreement with Greenland represents an important balance in this policy. Most agreements benefit southern European Member States (in particular Spain) but the agreement with Greenland has always been one of the most expensive for the EU budget and yet it benefits only four Member States of the north. Largely considered development policy at the beginning, this agreement has been evolving in the sense of ensuring that the fishing quotas obtained in Greenland constitute more 'value for money' in terms of real fishing possibilities (either directly for European fleets or indirectly as fishing possibilities for Norway in the context of the bilateral agreement), thus in line with other agreements of this kind.
- In some cases, the vessel owners have provided an increased part of the total contribution of the Union. This is particularly the case of the tuna agreements, where vessel owners contribute to the overall financial package through the payment of licenses. This contribution is variable according to the agreement but in recent years it has been already in the vicinity of 10–20% of the total EU contribution. In some cases this percentage can be as high as 47%. This fact removes (at least partially) the criticism of publicly supported private activity.

With the new orientation of these agreements in the 2013 CFP reform (see Chapter 16) these problems were largely addressed and today the bilateral agreements with financial compensation have an unprecedented level of political support.

Private partnerships

The external sector of EU fisheries also has an invisible component: the private investment of European fishing interests in third countries, outside the framework of the bilateral agreements. This has always been significant but as the establishment of bilateral agreements becomes more and more difficult, the private arrangements have been growing in importance in recent years.

This activity has also been evolving over the years. For some time, the structural policy of the CFP actually considered an instrument to favour private investment: the aid to the establishment of joint ventures and joint enterprises.

In addition, there is another angle that is of relevance: the activity of EU-flagged vessels in third countries that do not have an agreement with the Union, through the payment of private licenses, for access to third country waters through private licenses, such as that of vessels fishing for squid in the waters of the Falklands or for tropical tunas.

The example of Namibia

The main example of this private activity outside an agreement is the case of Namibia and its national policy for fisheries development as of its independence in 1990.

Before independence, the extremely rich fishing grounds off Namibia were intensely exploited by a number of countries, including some EU Member States such as Spain and Poland. This activity was regulated by the International South East Atlantic

Fisheries Commission (ICSEAF), where 17 countries around the world established management measures for a number of stocks. The work of ICSEAF was instrumental to create in Namibia an awareness of the need to manage its rich fishing resources (Oelofsen, 1999).

Following independence, the Namibian government, conscious of the potential of its rich fishery grounds as a source of wealth and jobs for its people, declared an Exclusive Economic Zone of 200 nautical miles. Subsequently they expelled all foreign fleets from these waters (including fleets from countries which had provided substantial political and military support to their independence, such as Cuba) and set forth to develop their own national fishing industry. In line with this objective, they always refused to consider the possibility of a bilateral agreement with the EU, similar to those with other African countries.

Instead, from the early 1990s Namibia established its own policy, which was finally consolidated through the Marine Resources Act of 2000, establishing *inter alia* clear rules for the establishment of foreign investors, with requirements to ensure the *namibisation* of the industry. At this time, the Union still had a policy of support to joint ventures. This policy allowed the investment of a number of European companies in Namibia. These companies were established with local partners, employed primarily local labour and contributed to consolidate a Namibian fishing sector.

As a result, the Namibian fishing sector enjoys a very high percentage of Namibian labour and it processes in Namibian territory 85% of their landings, providing added value and jobs for the local people. 90% of the production is exported constituting 15% of the exports of the country, for a value of over US$500 million in 2005 (López Veiga and Sobrino Heredia, in press).

When we look at Namibia, this country represents a model of what our Fisheries Partnership Agreement policy aims to achieve: the promotion of local fisheries development that creates growth and jobs in the third country concerned, with the cooperation of European partners. Although the Namibian model also attracts criticism (because it is associated with foreign, large-scale fishing instead with local small-scale activity), it is clear that it stands out in comparison with most fishery development and management systems in developing countries in Africa, so many of them riddled with bad management and IUU fishing:

> Namibia's policy and legal framework for the marine fisheries sector has allowed the application of management strategies that are appropriate to Namibia's specific circumstances. The result has been the development of a business environment that has facilitated the growth of healthy fishing and processing industry that pays a fair price for the privilege of utilising Namibian's marine resources.[9]

The Namibian fishery is exceptionally well controlled with observer coverage and serious quota management.

The paradox is that this objective has been achieved without a bilateral agreement. True, the aid to joint ventures was instrumental in the establishment of some of the European industries in that country. But overall, the most important factor favouring this development was the legal guarantees and clear policy rules established by the Namibian government that proved essential in attracting healthy fisheries investment to the country. This example is worth keeping in mind for the future fisheries partnership of the Union with third countries.

The question of legal guarantees

There is a very important lesson to be learned from the Namibian experience: that the most important factor to favour healthy European fisheries investment in third countries is not necessarily the support of public money but the establishment of legal guarantees for investors.

Indeed, in a context where bilateral agreements may be increasingly difficult to negotiate, due to the increasing interest of coastal States to develop their own fishing industries, it is inevitable to assume that in the years to come much of the investment of EU fisheries interests abroad will have to take place as private investment, outside the framework of EU bilateral agreements.

Beyond Namibia, many other examples (notably by the tropical tuna industry) show that these private investments are taking place in some cases solely as access agreements against a license fee, or in other cases (e.g. tuna companies in Africa and Central and

South American countries) through investment in canning facilities on land. For many of these operators, the key factor taken into account for private investments is not the availability of public aid but rather the existence of legal guarantees in the third country concerned, as well as a favourable tariff regime for their products in the EU market.

EU private investment in fisheries is legitimate and positive both for Europe and for the third countries concerned. Establishing the conditions to ensure legal guarantees for investments, as well as adequate fisheries governance to clearly exclude these investments from being considered as IUU (see Chapter 9), may constitute a fundamental development in the external pillar of the CFP in the years to come.

Fisheries and international conflict: the Greenland halibut war

This experience has probably contributed more to European fisheries than many others. Some of its effects are still visible today.

In the early 1990s, there was growing evidence of the collapse of the cod stocks in North East Canada, an area regulated by the Northwest Atlantic Fishery Organisation (NAFO).

This evidence led NAFO to ban fisheries for cod in different NAFO areas between 1992 and 1994. As cod was a key species of interest for European fleets fishing in the NAFO regulatory area (outside the Canadian 200-mile EEZ), this closure meant that the fishery for Greenland halibut became the main resource that underpinned the presence of European fleets in the area.

Once the cod fisheries were closed, the political situation in Canada was such, that the necessary sacrifices for the population of Newfoundland could only be politically palatable if 'foreign overfishing' was also addressed. Accepting the stoppage of fisheries in Canada while foreign fleets continued to fish outside the Canadian 200-mile zone was difficult to present to the Canadian public opinion. For this reason, Canada targeted the main remaining foreign fishery in NAFO waters, once the cod fishery was closed: the fishery for Greenland halibut.

This fishery was in fact initiated in the 1980s by the fleet of the former German Democratic Republic. In the late 1980s and early 1990s other European fleets, primarily Spanish and Portuguese, developed this fishery – hitherto unregulated – very quickly following the loss of the Namibian fishing grounds for the Spanish fleet. In the case of the Portuguese fleet, this fishery became the alternative to the diminishing cod fishery.

The combination of a growing, unregulated fishery and the Canadian pressure to show to its public opinion that foreigners could not be fishing while their nationals had to stop doing so (Canada had no interest, adequate vessels or markets to fish for Greenland halibut), led almost inevitably to conflict.

The conflict escalated to unprecedented levels, with Canada sending warships to patrol the area. Eventually, the Vigo-based vessel *Estai* was seized by the Canadian authorities outside the Canadian 200 miles and brought to a Canadian port, in a clear breach of international law. The escalation was such that it prompted a political agreement to settle the matter, as the important EU-Canada relationships could not be 'damaged by a minor fishery issue'. This agreement consisted of a series of measures to bring the Greenland halibut fishery to a strict management regime and to incorporate a control system based on full observer coverage of all vessels which represented at the time the most strict fisheries control scheme in the world.

With certain variations, these unusually high standards of control have remained to this day, even if the risk of collapse of the stock is not an issue and the fishery has been non conflictual for the last decade.

One of the many lessons that can be drawn from this experience is the value of anticipating preventive solutions. That the conflict was brewing was obvious to everyone involved. In the face of it, the idea of establishing some kind of initial conservation measures (e.g. a possible freezing of fishing effort) was discussed. But the industry and Member States concerned were not forthcoming, as they saw the Canadian initiative as an attempt to 'kick them out of the place'. Consequently, absolute resistance was the line. This lack of action, in turn, fuelled the Canadian initiative by providing them with the alibi they needed: in the face of inaction by NAFO, unilateral action by the coastal State was the only solution.

In the end, probably nothing of what the EC had done could have modified Canada's position, as it was really driven by a domestic political agenda. But one can always speculate that, if some initial measures to regulate the fishery had been proposed by the EC side in NAFO, at least the internal position of the EU in the negotiation would have certainly be stronger and the result of such negotiation could have been better.

It is also important to underline that the conflict had a bearing on the on-going discussion regarding the UN Fish Stock Agreement of 1995. The political exploitation of the affair by the then Canadian minister, using illegal gear used by European fishermen in NAFO as part of a propaganda campaign certainly had a bearing on the final result, in terms of exposing foreign fleets fishing in international waters as the main scapegoats of the problem of overexploitation of fishery resources.

Eighteen years after that conflict, fishing in NAFO continues in a much less conflictual environment, including the continuation of the – strictly regulated – Greenland halibut fishery. But the effects of that conflict are still visible today. The strict control measures and the heavy investment of the Union in controlling that fishery are a clear reminder of how politically sensitive this fishery once was.

Notes

1 www.un.org/depts/los/reference_files/chronological_lists_of_ratifications.htm.
2 www.un.org/depts/los/reference_files/chronological_lists_of_ratifications.htm.
3 World Summit on Sustainable Development. Plan if implementation of the WSSD.A/CONF. 199/20.
4 www.fao.org/docrep/005/y3274e/y3274e0f.htm.
5 In this regard, it is relevant the recent ITLOS Advisory opinion on the request submitted by the sub-regional fisheries commission (SRFC), Case 21, 2 April 2015.
6 Committee for Eastern and Central Atlantic Fisheries.
7 Central and Eastern Atlantic Fishery Committee.
8 African, Caribbean and Pacific countries.
9 http://feru.sites.olt.ubc.ca/files/2013/03/Namibia_16_Nichols.pdf.

References

COM (1976). Commission communication to the Council. Future external fisheries policy and internal fisheries system. Brussels, 21.9.76. COM (76) 500.

EC (1998). Council Regulation (EC) No 1435/98 of 29 June 1998 prohibiting imports of Atlantic blue-fin tuna (*Thunnus thynnus*) originating in Belize, Honduras, and Panama. *OJ* L 191, 7.7.1998, p. 13–14.

EC (1999). Council Decision 98/414/EC of 8 June 1998 on the ratification by the European Community of the Agreement for the implementing of the provisions of the United Nations Convention on the Law of the Sea of 10 December 1982 relating to the conservation and management of straddling stocks and highly migratory fish stocks. *OJ* L 262, 8.10.1999, p. 23–23.

EC (2002a). Council Regulation (EC) No 2347/2002 of 16 December 2002 establishing access requirements and associated conditions applicable to fishing for deep-sea stocks. *OJ* L351, 28.12.2002, p.6.

EC (2002b). Council Regulation (EC) No 2340/2002 of 16 December 2002 fixing for 2003 and 2004 the fishing opportunities for deep-sea fish stocks. *OJ* L356, 31.12.2002, p. 1.

EEC (1976). Proposal for a Council Regulation (EEC) establishing a Community system for conservation and management of fisheries resources. OJ No. C255, 28.10.76.

EEC (1993). Council Regulation (EEC) No 3447/93 of 28 September 1993 on the conclusion of the Agreement between the European Economic Community and the Argentine Republic on relations in the sea fisheries sector. *OJ* L318, 20.12.1993, p.1.

EU (2013). Regulation (EU) No 1380/2013 of the European Parliament and of the Council of 11 December 2013 on the Common Fisheries Policy, amending Council Regulations (EC) No 1954/2003 and (EC) No 1224/2009 and repealing Council Regulations (EC) No 2371/2002 and (EC) No 639/2004 and Council decision 2004/585/EC. *OJ*. No L354 of 28.12.2013, p. 22–61.

FAO (1999). International plan of action for reducing incidental catch of seabirds in longline fisheries. International plan of action for the conservation and management of sharks. International plan of action for the management of fishing capacity. FAO. Rome, 1999, 26 pp.

FAO (2001). International plan of action to prevent, deter and eliminate illegal, unreported and unregulated fisheries. FAO. Rome, 2001, 24 pp.

Farnell, J. and J. Elles (1984). *In Search of a Common Fisheries Policy*. Gower, Aldershot, 213 pp.

Koers, A.W. (1973). *International Regulation of Marine Fisheries. A Study of Regional Fisheries Organizations*. Fishing News (Books) Ltd. 367 pp.

Lequesne, C. (2001). L'Europe Bleue: A quoi sert une politique communautaire de la pêche ? Presses de la Fondation Nationale des Sciences Politiques, 239 pp.

López Veiga, E. and J.M. Sobrino Heredia. *Inversiones privadas y acuerdos pesqueros, un ejemplo paradigmático desde el ángulo de la economía y el derecho: el caso de Namibia*. In press.

Oelofsen, B.W. (1999). Fisheries management: the Namibian approach. *ICES Journal of Marine Science* **56**: 999–1004.

OJEC (1993). Council Decision of 13 December 1993 on the conclusion of an Agreement in the form of an exchange of letters relating to the Agreement on fisheries between the European Economic Community and the Kingdom of Norway. OJEC No L 346, 31.12.1993, p. 25–29.

OJEC (1999). Council Decision of 10 June 1999 authorising the Kingdom of Spain to accede to the Convention establishing the Inter-American tropical Tuna commission on a temporary basis (IATTC). (1999/405/EC). *OJEC* No L 155 of 22.6.1999, pp. 37–38.

UN (2012). Report of the United Nations Conference on Sustainable development. Rio de Janeiro, Brazil, 20–22 June 2012. A/CONF.216/16.

CHAPTER 8
The mediterranean specificity

The CFP in the Mediterranean

A CFP *in* and *for* the North Sea and Atlantic

As we saw in Chapter 2, the CFP was born 'in' and 'for' the North Sea and the Atlantic. The way the Mediterranean was ignored in the two 1983 regulations is still striking today, although there were historical reasons to explain this. The Mediterranean has been considered as deserving a specific approach on fisheries management. This specificity is both the result of certain particular features of fisheries in this sea basin but also the result of the historical events leading up to the establishment of the CFP; events that were limited to the North Sea and the Atlantic. The specificity of the Mediterranean is both cause and effect of these historical events (Symes, 1999).

The CFP was initially a genuinely North Sea and Atlantic product. The reason is simple: the CFP was born to cater for the newly established 'Community waters' resulting from the declaration of Exclusive Economic Zones (EEZs) in the North Sea and Atlantic façade of the Community. These EEZs were not declared in the Mediterranean.

As a result, in the Mediterranean there was no such thing as new 'Community waters' whose resources would need to be allocated among incumbent Member States and thus no need for a TAC and quota system. In the absence of a common policy approach on TACs, the Mediterranean was completely ignored in the first regulation on technical measures (Regulation 171/83).

The Commission, however, realising that the complete absence of a conservation policy in this area could not be justified, proposed in 1990 the introduction of a common fisheries system for the Mediterranean tailored to the specific nature of the fishing activities in this region (SEC, 1990). However, the debate on that idea made very slow progress and the first reform of the CFP in 1992 still excluded the Mediterranean.

The proposal, however, produced as a result in 1994 the adoption of the first generation of technical measures for the Mediterranean (EC, 1994), although they did not substantially change the previously existing national measures at the time.

In addition to the original reasons why the CFP (or at least the conservation pillar of the CFP) did not apply to the Mediterranean basin, the difference in the approach has always been explained and justified as a result of the so-called 'Mediterranean specificity'.

This has produced an important historical gap: while the structural and market policies applied to Mediterranean countries from the beginning, the conservation policy only started in earnest 30 years later, when the first regulation establishing fisheries management plans was adopted (EC, 2006). Until then the CFP only provided structural support to the fisheries sector but no restriction on the fishing activity (the first and still only one was for Bluefin

The Common Fisheries Policy: The Quest for Sustainability, First Edition. Ernesto Penas Lado.

Tuna and only since 1988). This may have contributed substantially to the different attitude of all concerned when it comes to approaches to policy-making and cooperation that still today separate the Atlantic and the Mediterranean basins under the CFP.

The Importance of Mediterranean fisheries

From this, one may be tempted to think that the low priority traditionally accorded to Mediterranean fisheries may be somehow related to a lesser economic importance. However, this is not true: Mediterranean EU vessels represent about 24% of the total Community fleet in tonnage or 36% in engine power and about 51% in number. Small scale coastal fishing predominates: more than 34 000 vessels (~80% in numbers) are smaller than 12 m in length.

Total landings are 12% of EU landings in weight but the value of Mediterranean landings is important due to the predominance of coastal fisheries with a high percentage of daily landings that generally fetch high prices. In terms of jobs, it is estimated that around 40% of total EU jobs in the catching sector are in the Mediterranean, many of these are semi-professional and part-time fishermen. In terms of dependence from fisheries, some Mediterranean regions are among those with a higher dependence on the fishing industry.

These figures clearly underline the importance of Mediterranean fisheries in the EU as a whole. This, however, only makes it more difficult to understand how the Union has not given more priority to the sustainable development of these fisheries for so many years.

The Mediterranean specificity: myth or reality?

Apart from these original differences, there are other ones that, still today, determine a quite different policy approach in Mediterranean fisheries, as compared with Atlantic or Baltic ones. These differences have been considered obvious, to the extent that they have been considered by some authors as 'a sea of difference' (Symes, 1999). It is important to analyse the basis for these differences, also in terms of evaluating whether they are real or perceived and ultimately whether they can justify such a fundamental difference of approach on fisheries management.

One difference is the perception of a lack of scientific basis for policy making in this area. This perception is only partially true. As we will see in Chapter 10, a key factor in the Mediterranean has been the lack of two related factors: the combination of a clear 'client' (the TAC and quota system) asking for specific advice every year and the lack of an experienced scientific body (such as ICES) to provide it. In the Mediterranean there is and there has always been excellent fisheries research. But this research has not been organised (until relatively recently) in a centralised body that can establish standardised methods to deliver the advice. As a consequence, there is very abundant knowledge about fisheries in the Mediterranean but this does not result in a consistent, stable and standardised basis for the provision of the advice to managers.

There are also important differences in terms of the nature of fisheries and its protagonists. In the Mediterranean, there is comparatively more fishing in coastal areas, where management is local or national and less in the way of shared resources in international or Community waters. Therefore, there is less tradition of international management, less of a need for 'Brussels' to decide on fisheries matters. The fishing industry is also more artisanal and this results in less tradition of international negotiations, lobbying and so on.

Most traditional fisheries in the Mediterranean have targeted small fish in coastal areas, due to the narrowness of continental shelves, while the spawning stocks in deeper waters were not fished. This probably explains the stability of many Mediterranean fisheries even though they catch large amounts of juvenile fish. However, this has been changing in recent years, as deeper waters are also being exploited by larger vessels. This represents an important challenge for the future.

Contrary to extended belief, Mediterranean fisheries are not always less regulated than Atlantic ones but a number of them tend to be regulated in a very different manner: while Atlantic fisheries, given their nature of common resources among different countries, are managed multilaterally and thus under the CFP, the rather local, coastal nature of most Mediterranean fisheries has resulted in a deeply-rooted tradition of local management, very often in the form of self-regulation by the industry. For example, in many fishing ports there are self-regulatory measures such as the very strict timing for the fishing operations: operators all leave port at a given time and all come back at a given time too. These self-regulatory measures are often very strictly adhered to. This self-regulatory tradition means that the management of these fisheries should take that 'culture' into account to be successful (Valarié and Rey, 1999).

It is important to underline this fact: while the image (often the prejudice) of Mediterranean fisheries is that of unregulated fisheries, largely ignoring EU rules, the reality is that many of these fisheries are self-regulated by locally-decided measures; an altogether different approach. In planning new management measures, this history must be taken into account.

Another significant characteristic of these fisheries, particularly at the beginning of the CFP but much less so today, is the importance of the multilateral dimension. While the Atlantic façade of the Union was from the beginning all European and belonging to members or candidate countries, the Mediterranean basin is to a considerable extent non-European. This has obvious consequences in shaping up the policy, where many decisions could only be meaningful if they have a minimum of multilateral support in this basin. This is not an exclusive problem of the Mediterranean: the Baltic Sea was in a similar situation in 1983. Yet, multilateral cooperation in the Baltic Sea was good and effective through the work of the International Baltic Sea Fisheries Commission (IBSFC) while the multilateral cooperation in the Mediterranean, through its equivalent organisation, the General Fisheries Council for the Mediterranean (GFCM) has, until relatively recent times, been extremely ineffective.

Finally, it is also worth mentioning the effect of the structure of the Commission services on this policy. For a long time, the structure of the Commission department responsible for fisheries did not have a specific Directorate, not even a specific Unit, to deal specifically with Mediterranean issues. As a consequence, there was no continuity in the Commission's policy approach in this area. More often than not, the services responsible for the fishery management policy were always busy with other priorities and the Mediterranean came always as a second priority. It must also be said that the low demand for policy action by the Mediterranean Member States and stakeholders concerned did actually facilitate this phenomenon. It was not until 2008 that a Mediterranean directorate was created in DG Mare. This has changed that situation by ensuring continuity of the Mediterranean policy.

The status of Mediterranean fisheries

For many years, it was commonplace that the conservation status of Mediterranean fisheries was not well known. This was a very superficial image, as many reports already alerted about a worrying situation in many Mediterranean stocks. However, the lack of a consolidated structure to provide an overall image of stock status, on a systematic way, on the basis of standardised methodologies, has traditionally diluted the message and has left the question somehow in limbo.

This presumed lack of knowledge on the real situation of these fisheries has often been used to discourage the adoption of more ambitious management measures in this area. However, this situation has substantially changed in recent years and today the scientific community is already in a position to provide a systematic evaluation of stock status in this area (COM, 2014). And the results are not positive.

The current coverage of stocks in the scientific advice is improving in the western and central Mediterranean Sea but still insufficient in the eastern part of the sea basin, as shown in Figure 8.1. In any case, the level of uncertainty on the real status of conservation of these stocks remains substantially higher than in the Atlantic and Baltic Seas.

The current stock status was summarised by the Commission in the 2014 Policy statement (COM, 2014).

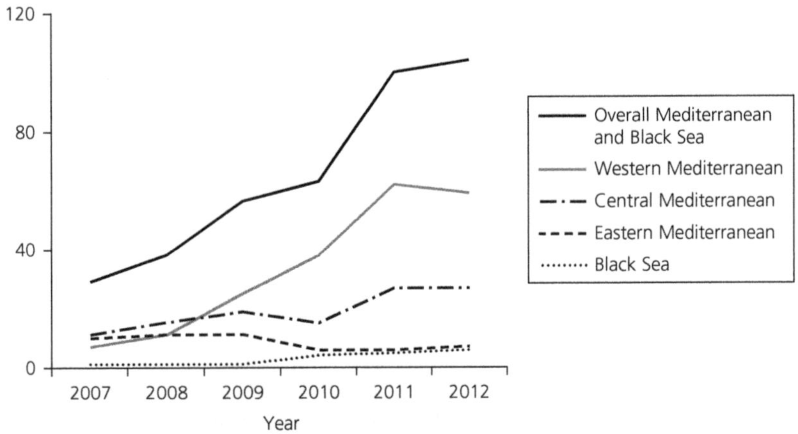

Figure 8.1 Evolution of the number of Mediterranean stocks with a scientific evaluation. Source: © European Union. Communication from the Commission to the European Parliament and the Council. Concerning a consultation on Fishing Opportunities for 2015 under the Common Fisheries Policy. Brussels, 26.6.2014 COM(2014) 388 final.

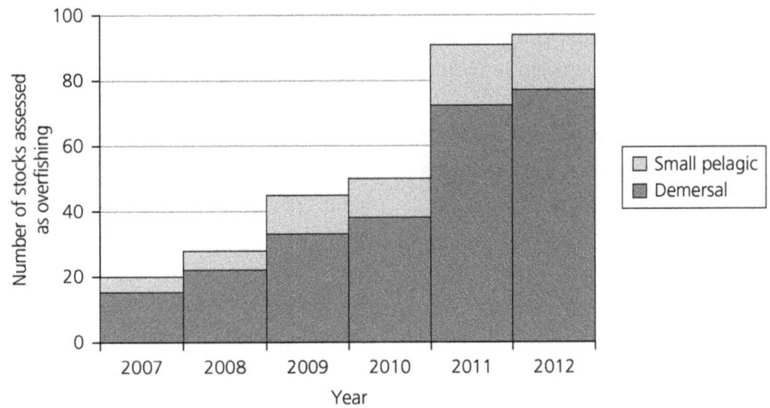

Figure 8.2 Number of stocks assessed as overfished from 2007 to 2012 by fishery in the Mediterranean and Black Sea. Source: © European Union. Communication from the Commission to the European Parliament and the Council. Concerning a consultation on Fishing Opportunities for 2015 under the Common Fisheries Policy. Brussels, 26.6.2014 COM(2014) 388 final.

Between 2007 and 2012 more than 300 assessments were performed, by the General Fisheries Commission for the Mediterranean (GFCM) and STECF. The number of assessed stocks rose from 29 in 2007 to 104 in 2012. Despite this, knowledge is still limited as the total number of stocks commercially exploited is considerably higher. The levels of exploitation are well beyond MSY targets. Out of 97 stocks, 91% were overfished. European hake, red mullet and deep-water rose shrimp are the most commonly overfished demersal stocks. Sardine and anchovy are the two common overfished small pelagic stocks. Figure 8.2 summarises the situation, showing a picture of increasing knowledge on stock status, and evidence of a very high level of overexploited stocks.

Other reports (European Commission, 2014a) also give important indications about the effects of this level of overfishing:

- Reducing fishing mortality will not be sufficient to reach MSY without changes in selectivity. This means that effort plans must be combined with effective measures to improve selectivity.

- The observed reduction in nominal effort in some EU plans did not result in a decline of fishing mortality. This is worrying, because it seems to indicate that the main instrument for the management of Mediterranean stocks, the effort management plans, are not delivering a reduction of fishing mortality.

This diagnostic shows the main elements of the problem and the challenge ahead for Mediterranean fisheries to achieve the objectives established in the new CFP. This raises the question as to whether the policy approach for the management of Mediterranean stocks (the effort management plans of Regulation 1967/2006) will suffice to reach MSY the targets between 2015 and 2020, as established by the CFP reform in 2013.

Some conclusions can be drawn from this advice:

- Firstly, the magnitude of the problem is now well known. There is no basis to claim that the lack of scientific knowledge is a barrier to effective management.
- Secondly, the reports points again to the need to improve size selectivity of Mediterranean fisheries. Although this was one of the declared objectives of the technical measures regulation adopted in 2006 (see next), one can wonder whether such regulation has produced the necessary effects so far.
- Finally, if the main management tool to manage Mediterranean fisheries will continue to be fishing effort plans, the report already warns that nominal reductions of effort, that are compensated by improvements in the ability of the vessels to catch fish, will not produce the desired effects.

In any case, this bleak picture must be seen under a particular light: although the general situation is bad, one must not necessarily talk about collapsing fisheries. The higher diversity of the Mediterranean ecosystem as compared to the Atlantic provides more stability in its fisheries (Bailly and Franquesa, 1999), and the collapse of individual stocks is probably less likely in the Mediterranean than in the Atlantic. In fact, the situation of Mediterranean stocks is bad but it has been perceived to be bad for a very long time, which indicates that the situation is that of stability within a very high level of exploitation, although this may also be affected by the expansion of fishing to deeper waters. In sum, the Mediterranean case is probably one of the so-called 'sustainable overexploitation', where the stocks are exploited at levels that do not provide high stable yields but that do no collapse either.

The economic data are difficult to interpret, due to important lack of data from a number of Member States but the overall picture is one of different trends for different Member States: in the period 2008–2011 the economic performance was stable in Italy, Slovenia and Cyprus, it improved in Bulgaria, Croatia, France and Malta and it deteriorated in Romania (European Commission, 2014b). But overall, there is no strong case to say that economic performance is systematically worse than in Atlantic fisheries.

This may be linked to the high price that many Mediterranean fish obtain in their local markets and may also be linked to relatively low fuel costs stemming from generally short fishing trips. In any case, the good economic performance of some Mediterranean fleets may also create a low priority for the improvement of stock status or for the adaptation of fleet capacities: if the economic performance is not bad and reasonably sustainable over time, there is little motivation to take seriously the threat of future stock collapse if the management does not improve.

Mediterranean fisheries may be a case of 'sustainable overexploitation', that is, fisheries that are exploited at levels too high to produce their highest sustainable yields but that are, however, stable because they are kept within safe biological limits, probably in some cases due to the relative protection of spawners in deeper waters.

Overall, the generally bad state of most Mediterranean resources, in the context of the CFP can also be interpreted as a proof of the relative success of the policy itself. If the state of the stocks is poor, one can certainly conclude that this may be because the conservation pillar of the CFP has not applied in the Mediterranean since 1983. The positive trend observed in the state of resources in the Atlantic and Baltic Sea in recent years can be associated with the – very slow and gradual – success of the policy in that area. Without a real conservation policy under the CFP until recently, the poor state of Mediterranean stocks is a demonstration of the 'cost of not having the CFP'.

A different policy approach

The Mediterranean specificity and notably the absence of TACs and quotas (with the only exception of those for Bluefin tuna, adopted in 1998 to implement an international agreement in the context of ICCAT) has often been explained as a result of the system of catch limitation by species not being adapted to the characteristic of Mediterranean fisheries, characterised generally by a higher diversity of species and more multi-species focus. In reality, this argument is only partially true, at least for three reasons:

1 Some Mediterranean fisheries, such as those for small pelagics (sardine, anchovy) are just as mono-specific as many fisheries in the Atlantic that are subject to TACs (e.g. anchovy in the Bay of Biscay). In fact, in the context of its evaluation of Mediterranean stocks (JRC, 2014), scientists have started discussing the value of possible catch limitations for certain stocks, such as small pelagics, which are caught in directed fisheries.
2 Certain demersal fisheries in south western Atlantic waters are not fundamentally simpler, or focused on fewer target species, than demersal Mediterranean ones. The best example is the fisheries in the waters to the west of the Strait of Gibraltar, which include a certain number of species subject to TACs, while fisheries of a similar complexity to the east of the Strait are not covered by TACs.
3 Certain Mediterranean areas (typically the Northern Adriatic) have a wide continental shelf, very much like the North Sea and with shared fisheries. In this area fisheries and their management can certainly be considered as similar to other Atlantic Areas. In fact it is not surprising that in the context of regionalisation under the reformed CFP, the Northern Adriatic is leading the way in developing cooperative approaches on fisheries management among the three Member States involved (Italy, Slovenia and Croatia).

Overall, the case against TACs in the Mediterranean, at least in certain fisheries, seems to be determined more by history and tradition than by objective differences with the Atlantic. Although there are real differences between Atlantic and Mediterranean fisheries, the so-called Mediterranean specificity remains somehow of a myth, that sometimes leads to requests for specific approaches with relatively limited justification in objective terms. A good example is that of the difficulties around the establishment of the Mediterranean Regional Advisory Councils (see Chapter 12). This RAC was the last one to be set up, as a result of a long-standing resistance by many stakeholders who wanted specific composition rules (and a lesser presence of NGOs) on account of this Mediterranean specificity.

In the context of the new CFP, where the long-term policy objectives has been clearly defined around maximum sustainable yield for all areas, with the scientific advice catching up on that of the Atlantic and with substantial recent progress on multilateral management in GFCM, there is much less of a case to talk about the Mediterranean specificity in the CFP.

There is, however, still an important difference in the formulation of the scientific advice which, together with the other differences mentioned, will probably justify a different approach to the conservation policy in this area as compared to the Atlantic: while scientists in the Atlantic still focus on the reduction of fishing mortality as the main condition for stock recovery, Mediterranean scientists focus more on the need to improve the size selectivity of the fisheries.

The jurisdictional regime of Mediterranean waters

As we referred to in the first section of this chapter, one essential characteristic that defines the Mediterranean approach to fisheries management is the shared nature of the Mediterranean Sea. Two geographical factors clearly justify this:

1 the fact that even today, after several enlargements, more than half the coastal States of the Mediterranean basin are non-members of the EU and

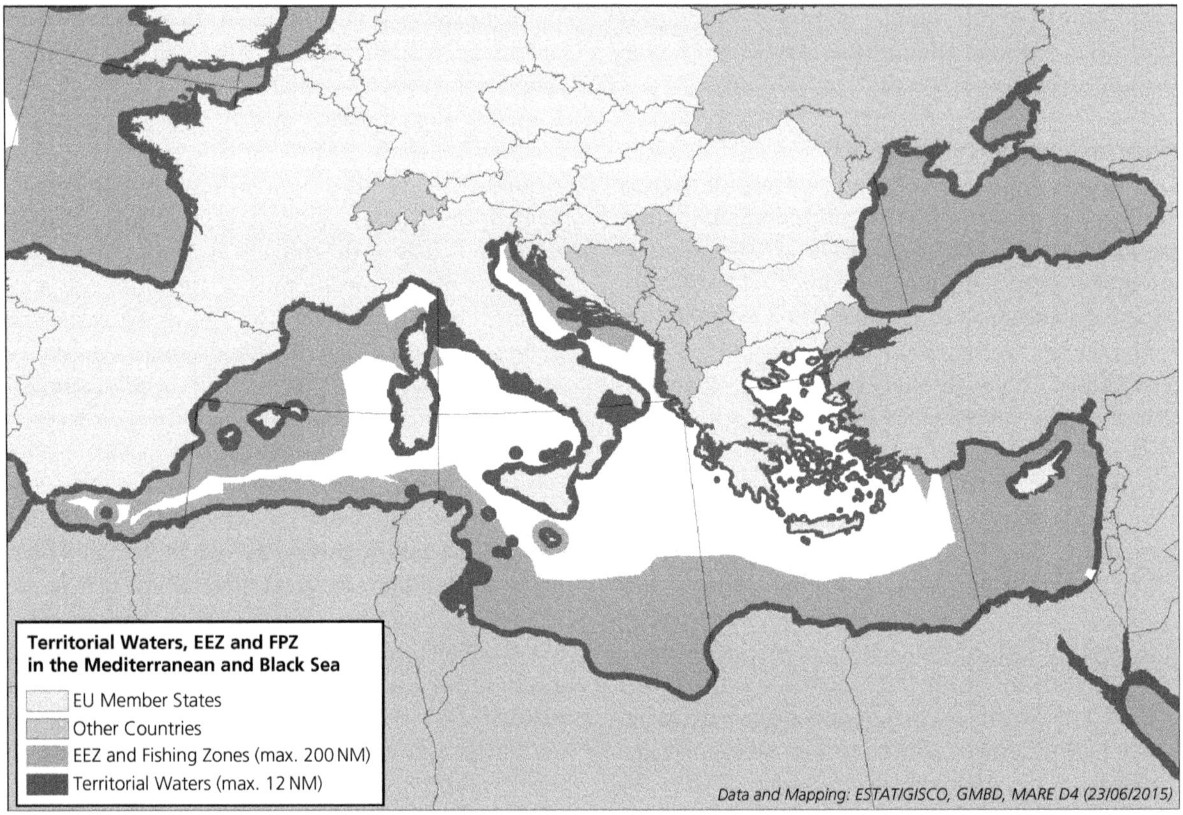

Figure 8.3 Jurisdictional regime of Mediterranean waters. Source: European Commission Services. Disclaimer: the areas represented are illustrative, not authoritative. (See colour plate section for the colour version.)

2 the existence of large parts of international waters in this basin, where rules and their enforcement are still weaker than in areas under national jurisdiction.

In these circumstances, progress on Mediterranean fisheries policy is still today strongly depending upon three developments:

1 the evolution of the status of sovereignty and jurisdiction of international waters in the sea basin,
2 the development of effective management by the two multilateral organisations with responsibility for fisheries management in this area: GFCM and ICCAT,
3 the willingness of EU Member States to bring about change in this context.

An open question is the changes that may be brought about in certain North African countries as a result of the regime changes in 2010–2012.

As we saw previously, the Mediterranean has not seen a concerted approach to the declaration of Exclusive Fisheries Zones as it happened in the Atlantic façade in 1977. Indeed, the Mediterranean is one of the few areas in the world where EEZs are not declared. Even in other semi-closed seas such as the Baltic, the whole surface of the sea is covered by national EEZs, through the agreement on median lines among the coastal States concerned.

Figure 8.3 shows the patchy situation of Mediterranean waters as regards their jurisdictional regime, with different, non-concerted approaches by the various coastal States. There is not even a common approach on territorial waters: while most coastal States have established them at 12 miles, as in the Atlantic, others have remained at 6 miles (Greece, Turkey) due to the proximity of neighbouring territories and the lack of agreement on such limits. As a result, there is an important part of the Mediterranean waters that are still international waters. This picture contrasts with that of the Atlantic, the North Sea and the Baltic (Figure 2.1 in Chapter 2).

A number of factors explain this in historical terms, these factors being largely relevant still today:

- the relative narrowness of the continental shelves (except in the northern Adriatic) and the relatively little amount of fishing beyond the 12 nautical miles;
- the difficult political situations, particularly in areas of the Eastern Mediterranean, making the necessary political cooperation needed for establishing median lines between the EEZs of neighbouring coastal States extremely difficult;
- The relatively small area of the Mediterranean covered by the – then – EEC Member States: only France and Italy, as opposed to the whole of the North Sea and western waters in the Atlantic.

In such scenario, concertation is the only way forward. This implies bilateral negotiations among neighbouring States to agree on the approach, timing and criteria to establish median lines. The need for concerted approaches on this point was already recognised by the Commission. In a Communication on a general strategy for the Mediterranean, published in 2002, the Commission stated that: 'the declaration of fisheries protection zones (FPZs) of up to 200 miles could be an important contribution to improving fisheries management'.

The EU set out to facilitate a process where Member States and then candidate countries, would debate on the need and opportunity to declare FPZs in a concerted manner (COM, 2002). It is important to underline that the Commission, conscious of the political difficulties that fully-fledged EEZs would entail, preferred a fisheries-only approach to remove part of the reticence by many countries to embark in such an exercise.

Further to this initiative, the Commission organised a conference on Mediterranean fisheries management (Venice, 25 and 26 November 2003), where the final declaration stated that:

> ...against the background of closer cooperation of all States benefiting from the biological wealth of the Mediterranean marine environment, we consider that the creation of fisheries protection zones permits the improvement of conservation and control of fisheries and thus contributes to better resource management and to our common commitment to combat IUU fishing.

The language may sound soft and non-committal but at the time it was a breakthrough, because it removed the impression that the mere mentioning of fishery protection zones would be taboo, an opinion very spread at the time, including the Commission services. It is also important to understand that the main motivation for the ministers to agree to such a declaration was the fact that only through the declaration of FPZs (or EEZs) could the coastal States tackle the problem of foreign vessels freely fishing for Bluefin Tuna and other valuable species in international waters of the Mediterranean basin.

The problem was, however, that there was no specific follow-up to this initiative. The possible political dynamic developed in Venice were never accompanied by subsequent action to give shape to it in practice. As a result, this initial political dynamics slowly died off, dominated by the difficulties of the approach. This has been an extreme example of how the lack of continuity in a policy approach ended up in the whole idea being forgotten.

As a result of a lack of a concerted action, different countries in the Mediterranean basin have a very different situation vis-à-vis the sovereignty and jurisdiction over their maritime waters.

Most Mediterranean States have established a 12-mile territorial sea. However, Greece and Turkey still have narrower territorial seas (6 miles) in the Aegean Sea. And even the very extension to 12 miles would be extremely contentious. As regards Exclusive Economic Zones and/or Fishery Protection Zones, the situation is extremely varied. In the absence of concerted approaches, some examples show this piecemeal approach:

- Malta established a 25-mile Fisheries Protection Zone in 1971.
- In 1997, Spain declared a Fishery Protection Zone of 37 miles under its jurisdiction beyond the territorial sea. This zone, however, did not extend into the more contentious waters in the Alborán

Sea, where the Spanish and Moroccan coasts are relatively close.
- Croatia and Slovenia agreed to submit their dispute on their EEZs to international arbitration.

The situation is only recently being discussed again. The Commission, conscious of the importance of resolving this issue, not only in fisheries but in the wider context of the integrated maritime policy, has reflected about the need and opportunity to declare concerted EEZs as an instrument to promote growth in the maritime economy. However, given the high political sensitivity of current developments in this sea basin, no initiative was launched.

In any case, by and large the lack of concertation on these approaches still makes the jurisdictional situation of Mediterranean extremely patchy and inconsistent, a real obstacle to take a more effective fisheries management approach at multilateral (and even at national) level.

The piecemeal approach to EEZs is explained by the difficulties summarised previously. However, these difficulties are not the same in all subareas of the Mediterranean basin. They concentrate in the Eastern Mediterranean. That allows for the possible consideration of sub-regional approaches to the question of national jurisdiction.

While it is easy to understand the difficulties of cooperating on possible EEZs in the Mediterranean coast of the Middle East, or in Cyprus, the situation in such areas as the Northern Adriatic or the Gulf of Lyons is completely different. In these areas there are no major conflicts that would impede agreements on the declaration of EEZs, including negotiated median lines.

By addressing this question at sub-regional level, in non-conflicting areas of the Mediterranean basin, progress could be made that is today not possible for the whole of the basin.

Fisheries management under the CFP

The CFP had to wait until 1994 to have the first regulation on technical measures in this area, 11 years later than in the North Sea and Atlantic.

The 1994 Regulation

Regulation 1626/94 (EC, 1994) was a side-effect of the first reform of the CFP in 1992. The idea came about then, when it was – again – obvious that the policy established in 1983 focused only on the North Sea and Atlantic fisheries. At the time, the impression was not only that the CFP (or at least the conservation pillar) did not apply to the Mediterranean but also that even at national level there was little transparency and no harmonisation of national rules.

The intention of this first regulation was mainly to gather information about national rules in order to consolidate them and, to the extent possible, harmonise them in a single piece of Community legislation. There was no attempt to base this exercise on the status of the Mediterranean fisheries, still considered largely unknown at the time.

The regulation did not establish a true regime for the management of resources, like the TAC and quota system of the Atlantic but only, as a first step, a series of basic rules to ensure selective fishing:
- The regulation imposed, for the first time, minimum landing sizes (MLS) at EU level, although such sizes were in a number of cases considerably lower than their equivalents in the Atlantic for the same species. For example, the MLS for hake in the Atlantic was 27 cm, while in the Mediterranean it was 20 cm. For horse mackerel, these MLS were 15 and 11 cm, respectively.
- The regulation also established the standard mesh size for demersal fisheries, which was fixed at 40mm, largely reflecting already existing national rules.

The regulation, however, was innovative in that it considered some new instruments that reflected a certain evolution from the distributive nature of the 1983 regulations in the Atlantic:
- It established rules on minimum distances from the coast for fishing with certain gear, in a forward-looking objective of protecting coastal areas. It also prohibited certain fishing techniques, such as St Andrews' crosses to gather red corals, due to their destructive nature.
- It established the obligation for Member States to declare fishery protected areas, although in a very generic way, without specific objectives.

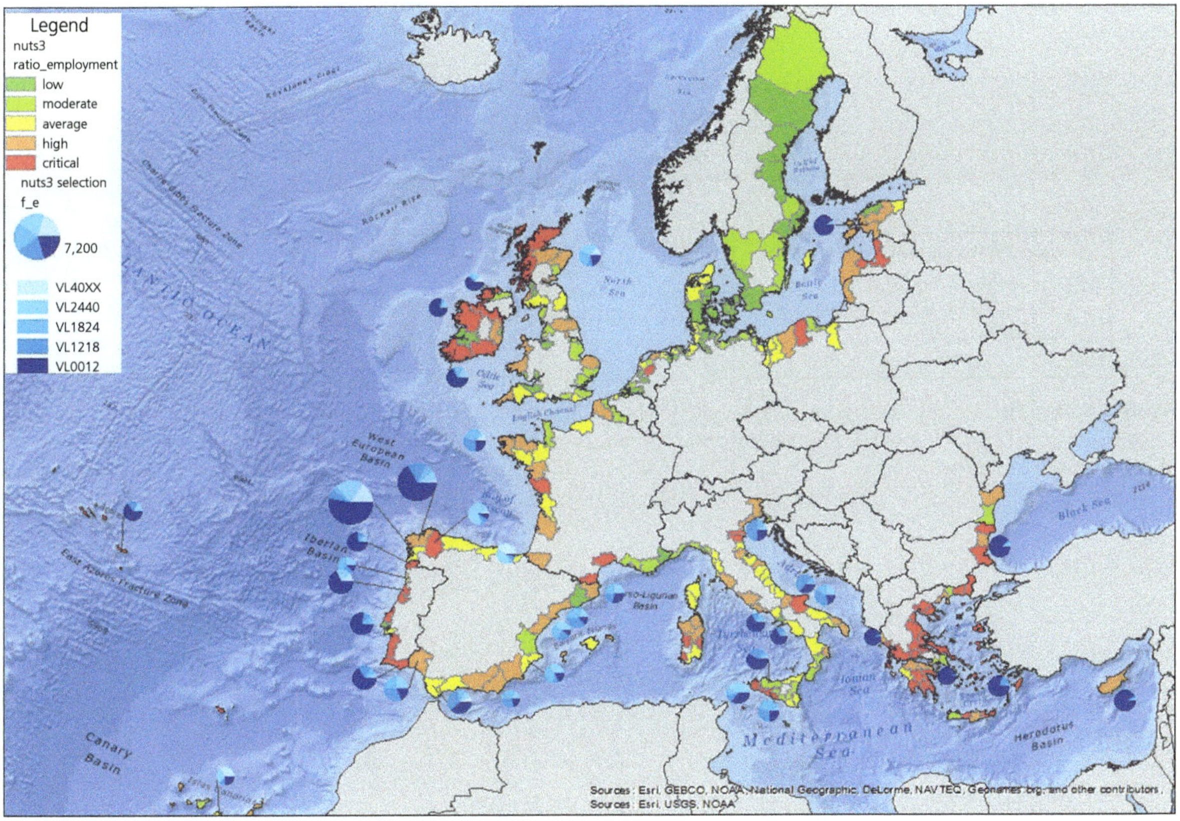

Figure 1.1 Fishery-dependent areas in Europe. © European Union. European Commission. Facts and figures on the Common Fisheries Policy – Basic statistical data – 2014 Edition. Luxembourg: Publications Office of the European Union. 2014 — 44 p.

The Common Fisheries Policy: The Quest for Sustainability, First Edition. Ernesto Penas Lado.

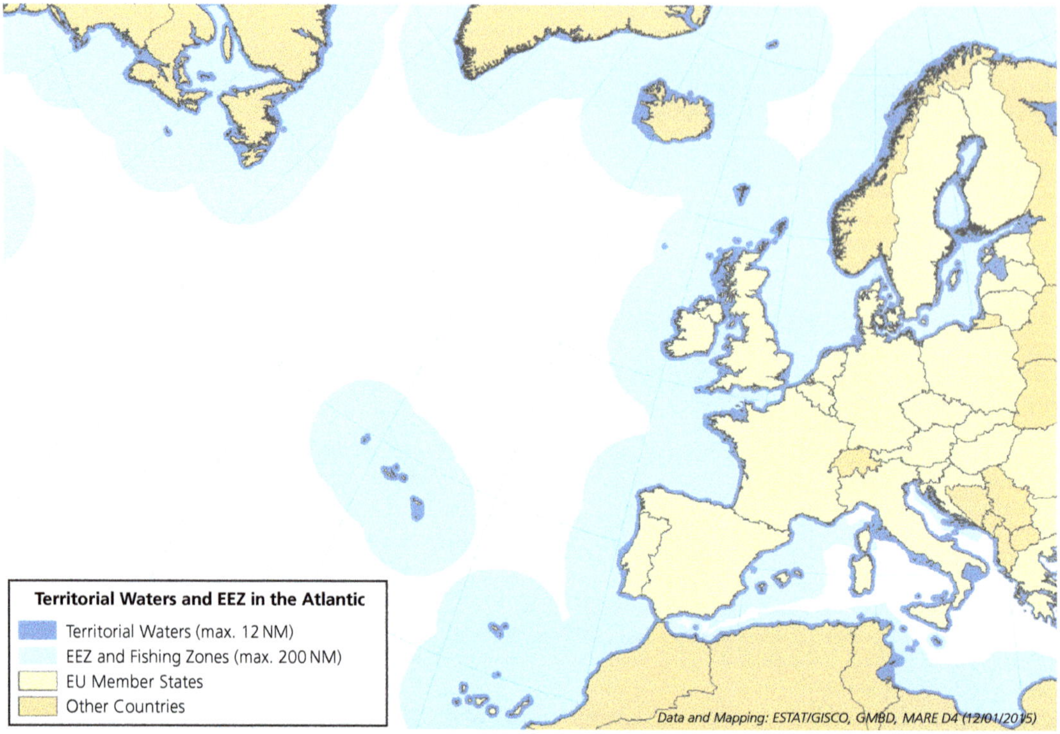

Figure 2.1 The effects of the declaration of 200-mile Exclusive Economic Zones (light blue) as compared with the previous 12-mile territorial seas (dark blue). Source: European Commission Services. (Disclaimer: the EEZs in this figure are illustrative, not authoritative.)

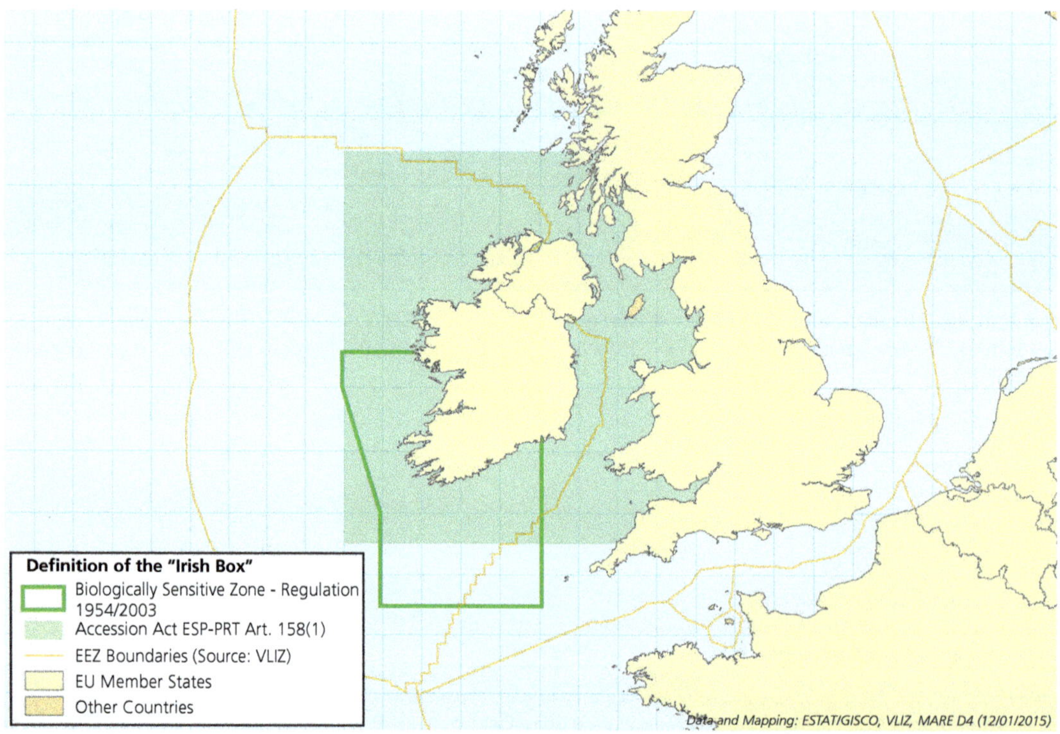

Figure 3.1 The biologically sensitive zone adopted in 2003 and the Irish Box of 1986. Source: European Commission Services. EEZ limits are indicative, not authoritative.

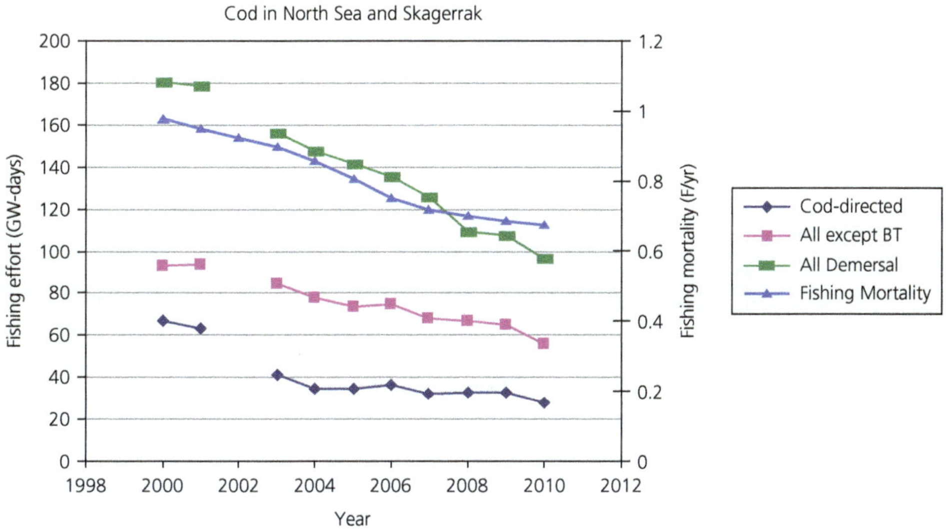

Figure 4.1 Time trends of fishing mortality and fishing effort (as total demersal effort, cod-directed effort or demersal effort excluding beam-trawls). Source: European Commission Services.

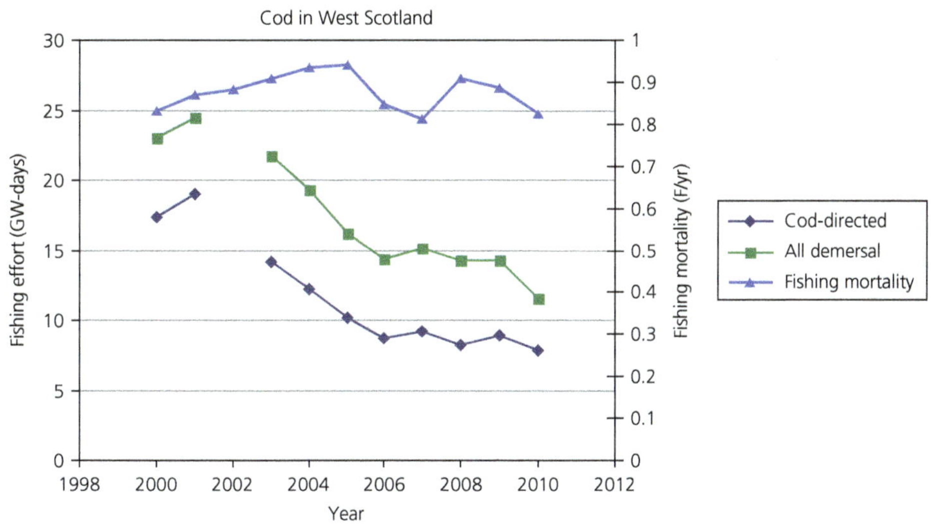

Figure 4.2 Fishing mortality and fishing effort of cod in the West of Scotland. Source: European Commission Services.

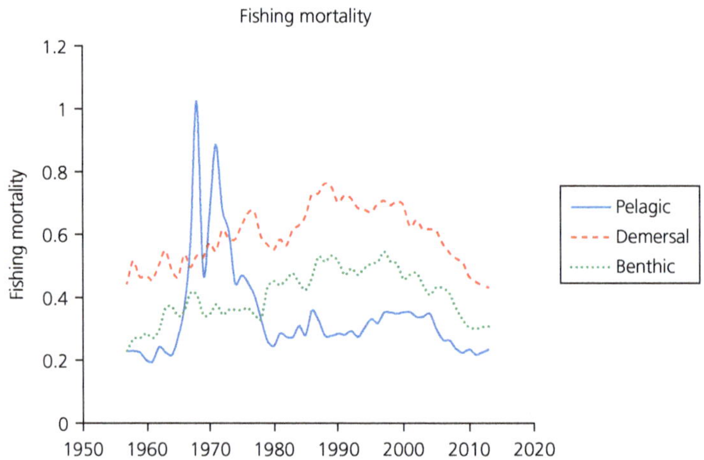

Figure 4.4 Evolution of the fishing mortality in European fisheries in recent decades. ICES (2013). Reproduced with permission.

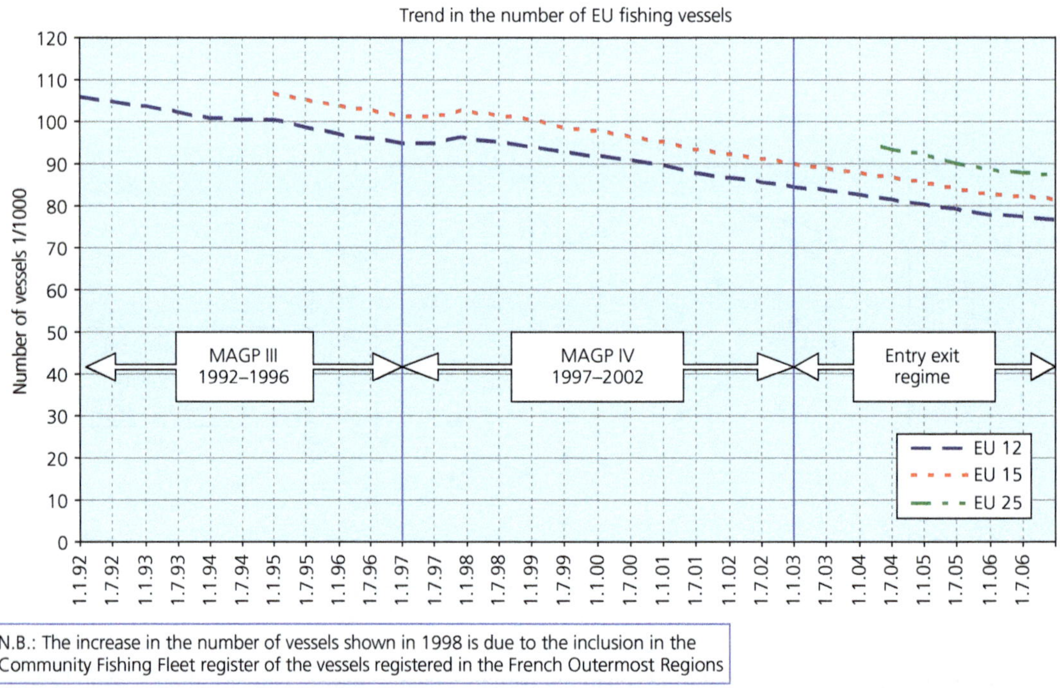

Figure 5.1 Evolution in the number of vessels in the European fleets in different periods and under different policy objectives. Source: European Commission Services.

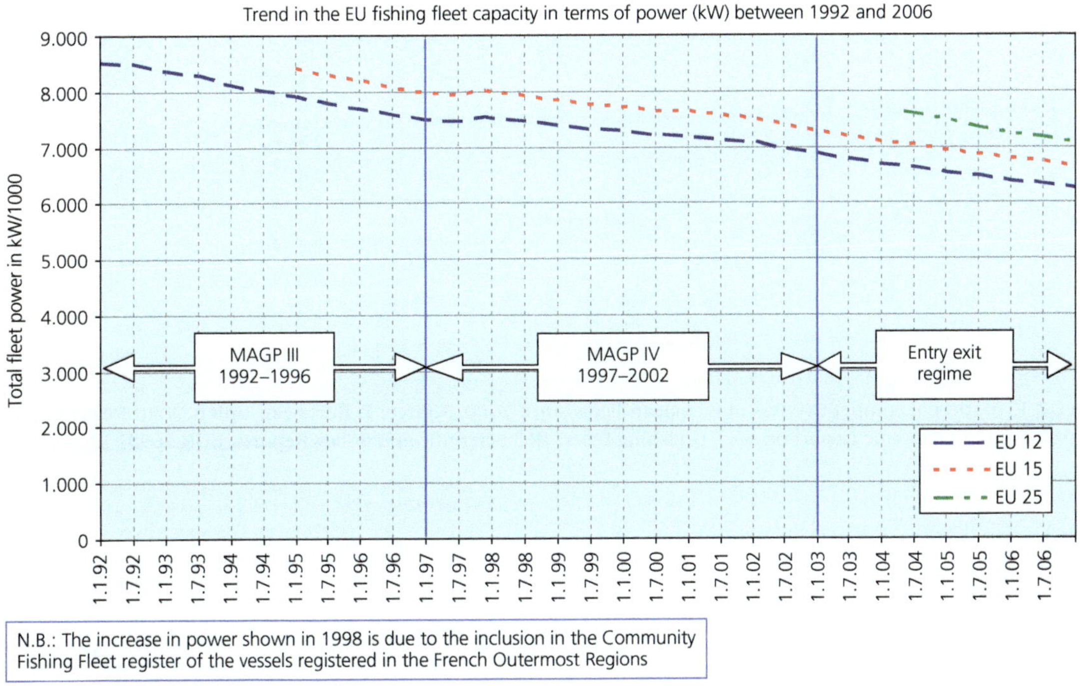

Figure 5.2 Evolution of fleet capacity, expressed in engine power (kW) in different periods and under different policy objectives. Source: European Commission Services.

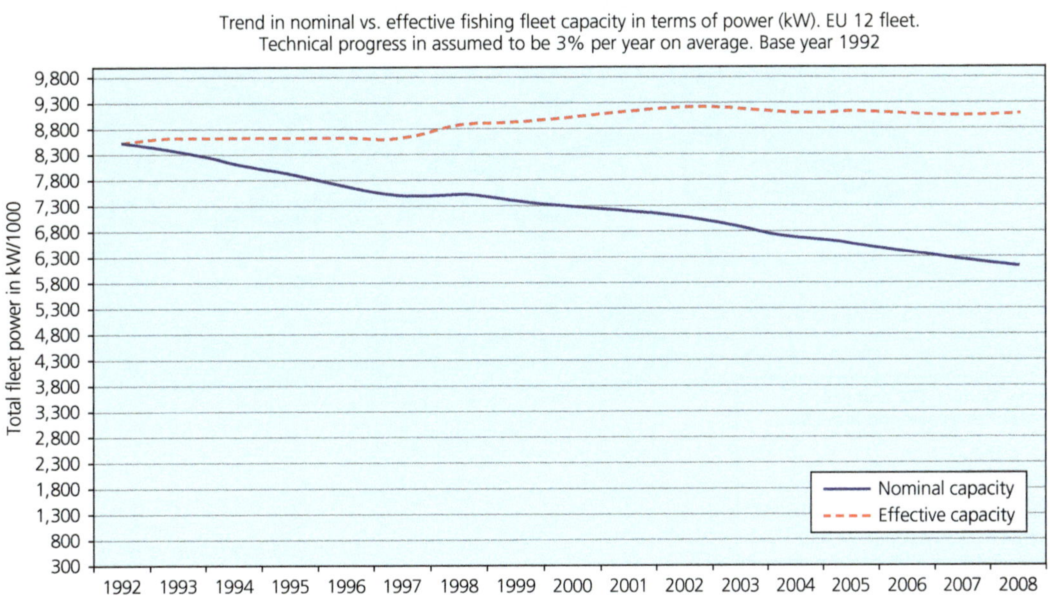

Figure 5.3 Comparison between the nominal reduction in capacity and the effective capacity taking into account technical progress. Source: European Commission Services.

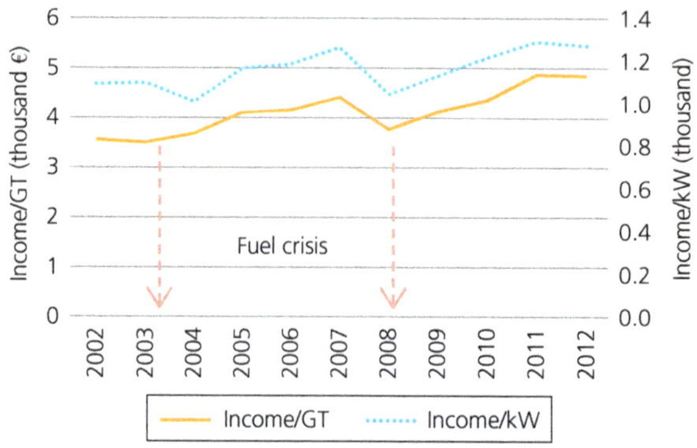

Figure 6.2 Evolution of competitiveness of European fleets since 2000. Source: © European Union. Joint Research Centre. The 2014 Annual Economic Report on the EU Fishing Fleet. JRC Scientific and Policy Reports. EUR 26901 EN – 2014.

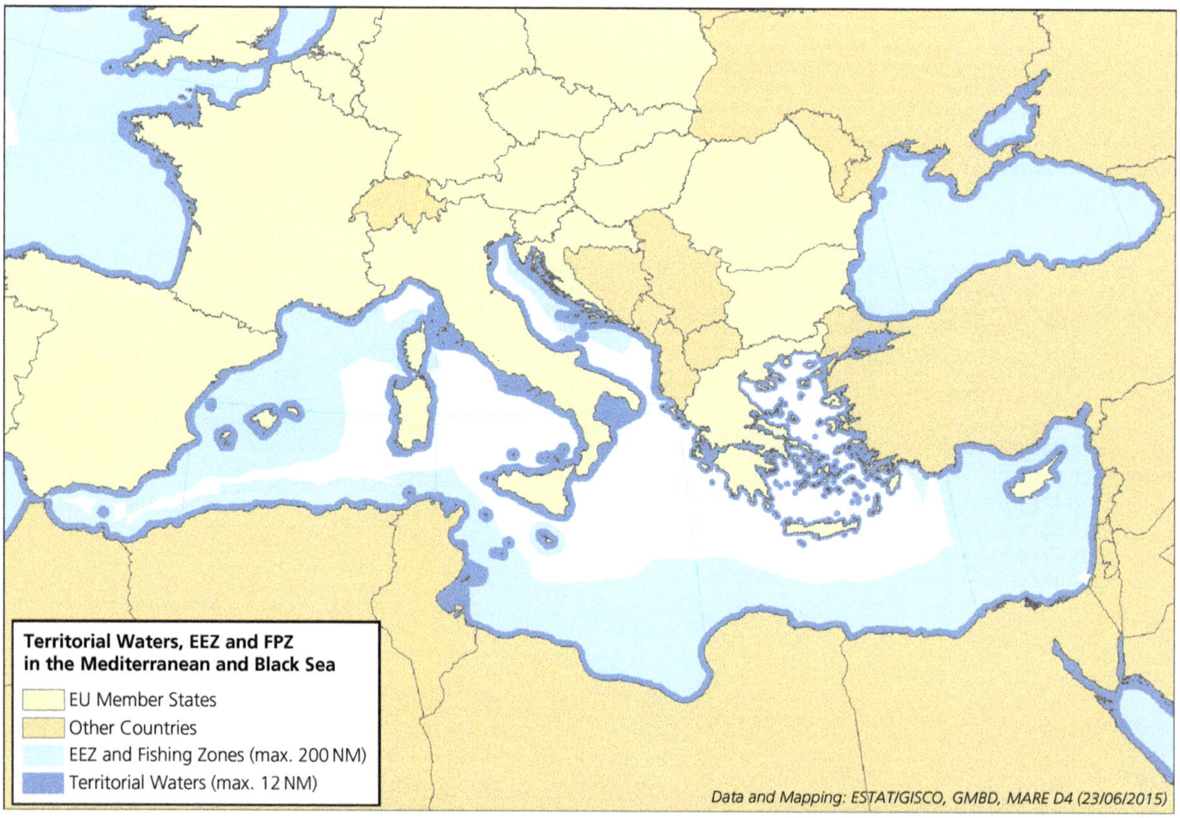

Figure 8.3 Jurisdictional regime of Mediterranean waters. Source: European Commission Services. Disclaimer: the areas represented are illustrative, not authoritative.

- It established the need to protect certain Mediterranean species (such as mammals) and habitats (such as coastal wetlands or *Posidonia* beds[1]). However, there were few mechanisms to ensure the delivery of these objectives.

Overall, this first Mediterranean regulation was a combination of technical measures, forward-looking environmental objectives and *ad hoc* adaptations to accession. However, what is most important is what the regulation did not cater for: the real management of Mediterranean fisheries. This regulation was thus a first step but one that did not put Mediterranean fisheries at the same level as Atlantic ones in terms of the objectives, the management methods and even the political priority to ensure their sustainable management.

Despite this character of 'regulating the status quo' that constituted the philosophy of this regulation, it was generally believed that enforcement of these rules was low. This, in turn, is a possible indication that the previous national rules were not well enforced either. Regulation 1626/94, in this regard, represented the *status quo*, not only in terms of existing rules but also in terms of their low level of implementation at the time.

Further developments also affected this regulation, in particular the ban on driftnets and the accession of Malta.

The implementation of EU fisheries law in the Mediterranean has also been difficult in connection with the ban on driftnets. As a result of the UN Resolution to ban large-scale driftnets and the conflict over the use of driftnets in the albacore tuna fishery in the Bay of Biscay (see Chapter 11), the use of large driftnets targeting tuna and tuna-like fish in the Mediterranean was banned in 1998. These affected in particular two kinds of Mediterranean driftnets: the *spadara* used to catch swordfish in Italy and the *thonnaille* used to catch Bluefin tuna in France.

Following the accession of Malta in 2004, the regulation was amended to include the conditions applicable to Malta's 25-mile management zone, at a time when no Member State of the Mediterranean had declared fishery protection zones beyond territorial waters. It also included rules for the management of Malta's main commercial species: *lampuki* (Dolphin fish, *Coryphaena* spp.). This was actually the 'communitarisation' of the existing management system of this stock in Malta prior to accession.

The 2006 Regulation

Today, the main instrument for fisheries management in the Mediterranean basin is the regulation adopted in 2006 (EC, 2006).

This started as a side-discussion in the context of the 2002 CFP reform. As happened 10 years earlier, the new basic regulation of 2002 did not substantially alter the *status quo* in the Mediterranean. However, at the time it was obvious that situation of Mediterranean fisheries was far from satisfactory. This prompted the Commission, in its Green Paper for the reform of the CFP in 2002 (COM, 2001), to clearly indicate that: 'Community policy in the Mediterranean has not met expectations: implementation of the technical measures Regulation has not been satisfactory, data are lacking and there has been slow progress on international cooperation'.

Consequently, the report recommended to: 'Promote the integration of the Mediterranean into the CFP through the improvement of scientific advice, the reviewing of the technical measures regulation No 1626/94 and the strengthening of the control and enforcement'.

Further to this document, a Mediterranean strategy (mentioned previously) was presented by the Commission in 2002, to identify the elements of a new approach for the management of Mediterranean fisheries (COM 2001). This strategy was very well received. It identified a number of actions necessary to improve fisheries management in this area, through a reinforcement of the multilateral management in GFCM and the adoption of specific fisheries management rules in close cooperation with the stakeholders, by building on their traditions and culture of management, particularly on the tradition of self-regulation of fishing time.

On that basis, the Commission services initiated a series of direct consultations with Mediterranean stakeholders to identify the main elements for such a regulation. The historical lack of a real fisheries policy in the Mediterranean made these consultations difficult and laborious at the beginning. The expectations from some of the stakeholders were somehow unrealistic: there was disappointment when the Commission clarified that the 'Mediterranean specificity' did not imply an altogether different CFP, including a specific structural policy in this area with more favourable co-financing rates.

However, the dialogue was slowly starting to bear fruit, when the decision was taken, in the summer of 2004, of bringing forward the presentation of the Commission proposal so it could be handled by the Italian presidency in the second half of that year.

The rushed proposal included a number of elements proposed by the stakeholders (such as the question of establishing a network of closed areas, or the use of local and regional fishing effort plans as the main management instrument instead of TACs). However, the presentation of the proposal well before the on-going dialogue with the stakeholders was completed had the effect of alienating the latter completely. As a result, the proposal was received with such hostility by the stakeholders that the institutional procedure of discussion was blocked and so were the further attempts by the Commission to restart the dialogue. This situation lasted more than a year (during which the initiatives by the Commission to restart the dialogue with the stakeholders were boycotted by many of them). The situation was only unlocked thanks to the good efforts of the European Parliament in the context of the preparation of their non-binding opinion on the proposal.

Eventually, after a long and difficult negotiation the regulation was adopted in December 2006. But this adoption, which included a number of concessions to accommodate specific requests by the industry, still suffered from the 'original sin' of the frustrated dialogue. Its implementation ever since has been slow and laborious. This regulation is still largely perceived by a substantial part of the industry as a top-down exercise against their will, even though, in objective terms, the regulation reflects a majority of the requests from the industry.

This case constitutes an excellent example of the importance of a confidence building in any consultation with stakeholders.

However, not all was negative in this process. This proposal actually contained a number of elements that were suggested by stakeholders, showing that such consultation was much more fruitful than it was given credit for. Also importantly, the proposal anticipated a number of the features that, at a later stage became more in line with some of the further developments of the CFP in general:

- The regulation established de-centralised management plans, based on effort, not catch limitations. This was done to build on the existing experience in Mediterranean fisheries, where catch limitations are not used but where the industry had a very well consolidated tradition of regulating fishing time, in many cases through self-regulation;
- It also established the obligation to create a network of protected areas, anticipating the decision taken in the CFP reform, to use these marine protected areas as a management instrument;
- The possibility for derogation from certain general rules based on well-justified arguments applicable to local conditions;
- The choice between two different kinds of mesh size considered equivalent in selectivity: the 50 mm diamond mesh or the 40 mm square mesh.

Overall, the regulation actually included most of the requests made by the stakeholders. But the sudden interruption of the dialogue with them in the preparation of the proposal was an 'original sin' that created an – unfair – image of this regulation as a top-down exercise that ignored the advice of those affected: an important lesson to learn.

Implementation and enforcement

Despite these positive elements and of all the efforts made to incorporate Member States and industry's concerns, the implementation of this regulation has been less than impressive. By 2013, only 21 management plans had been adopted (COM, 2014), and the implementation of the new rules on gear selectivity was still sluggish at best.

Since the main instrument for fisheries management in the 2006 Regulation is the implementation of the management plans based on fishing effort, the relative progress of fisheries management in the Mediterranean depends primarily, though not exclusively, on the implementation of these plans. If the objective of reaching MSY levels must be reached by 2020, it is clear that a special additional effort to have all important Mediterranean fisheries under management plans should be made. The recent scientific advice indicating that fishing effort in existing plans is not necessarily limiting fishing mortality (see previously) is a cause of concern that will have to be addressed.

Another important element in this regulation is the adoption of marine protected areas. Here, progress has been very limited.

In terms of enforcement, the Mediterranean regulation has been slow and difficult to apply. In particular, the question of the 40 mm mesh size in many fisheries has been questioned as inapplicable, when combined with the rules on twine thickness. Somehow, the enforcement of this regulation has suffered from the absolute priority given to the enforcement of the ICCAT rules on Bluefin tuna (see next) that has required the most important enforcement efforts from all involved, perhaps to the detriment of the remaining fisheries of the area.

Particular attention must be given here to the implementation of the ban on driftnets. Two cases are of particular relevance: the Italian *spadara* and the French *thonnaille* cases that have given rise to long legal struggles to ensure proper implementation.

Multilateral management in the Mediterranean

As we saw in Chapter 2, the non-declaration of EEZs was at the origin of the exclusion of the Mediterranean from the technical rules that constituted the CFP in 1983. But that does not explain all. Other measures could have been adopted in that area adapted to the geographical and jurisdictional characteristics of the sea basin. But something prevented it: the evidence that, without an effective multilateral framework to manage fisheries between European and non-European countries, any fisheries policy would be ineffective (Farnell and Elles, 1984).

Still today, some Mediterranean Member States argue that it is not possible to take certain management decisions unless they are shared with other countries of the Mediterranean basin. Recent history demonstrates that advances in the establishment of efficient multilateral fishery management mechanisms for the whole of the Mediterranean basin have been instrumental for the improvement of the Mediterranean part of the CFP.

The solution to this has been the gradual development of the corresponding regional fishery organisations: CGFM and ICCAT. It is interesting to see the very different influence of these two multilateral bodies. In one case (ICCAT) it is the multilateral dimension that has taken the lead (largely influenced by non-Mediterranean countries) to drive fisheries management by the Union. In the other case (GFCM) the slow progress in the multilateral dimension has often prevented the advancement of fisheries management in Europe. Although this division has changed considerably in recent years and is today quite different, however, it is important to look at how these situations occurred.

The case of GFCM

The multilateral dimension of Mediterranean fisheries has been a cause of (as well as an excuse) for the slow progress of fisheries management. While the multilateral management of fisheries has been effective in promoting sound management in certain areas (we have mentioned the positive example of NEAFC, IBSFC, ICCAT or NAFO) this has unfortunately not been the case, until recent times, in the Mediterranean.

The General Fisheries Committee of the Mediterranean (GFCM) was set up in 1949, under the auspices of the Food and Agriculture Organisation (FAO) as a dependent body (Breuil, 1999). The fact that the body, unlike most regional fisheries organisations, was depending on FAO's budget, together with the low priority given by many countries around the Mediterranean basin to fisheries management, led to an organisation that, up to the late 1990s, was almost inactive, certainly as far as fisheries management was concerned.

In the late 1990s the Commission, in the context of the Mediterranean strategy developed in the context of the 2002 CFP reform (COM, 2002), identified the lack of an effective multilateral management as one of the fundamental weaknesses of the CFP in this region. Consequently, the Commission took the initiative to change this state of affairs and took the lead to turn the GFCM into an organisation similar to those active at the time in other regions. This proved a difficult exercise, for several reasons.

- Firstly, most countries of the Mediterranean basin preferred to stay within FAO, to keep GFCM under the development-oriented umbrella of that organisation that has no powers to impose fisheries limitations for the purpose of conservation objectives.
- Secondly, the conflict concerning jurisdiction of waters between certain Member States and third countries (Greece and Turkey) made it very difficult to agree on any issue related to the definition of the regulatory area of the organisation.
- Thirdly, the unwillingness by many countries to increase their contribution to fund a stand-alone body, preferring instead a body entirely (though modestly) funded by the general FAO budget.

One key part of this modernisation was the establishment, for the first time, of a scientific advisory body: the Scientific Advisory Committee (SAC). Unlike in the Atlantic, where ICES has been for years the provider of scientific advice, the Mediterranean did not have a scientific body. The problem was also that since the CFP did not have a significant policy for managing resources (over and beyond the technical measures), the scientific community in the Mediterranean did not have a clear 'client' that would ask them on a regular and structured way the necessary advice for implementing of management measures. As a consequence, the Mediterranean scientific community, which included many excellent scientists, did not have the client-oriented structure to provide its advice and suffered from considerable dispersion in their approaches and methodologies.

Making GFCM operational, however, was another challenge. The slowness in developing a generally-accepted scientific advice has long been hampered by the image of the SAC as a Europe-dominated body and the lack of data from a number of countries of the sea basin has required approaches whereby European countries have invested in cooperation programmes with non-European coastal States to develop basic data series for scientific advice. COPEMED was the first of these programmes, funded by Spain to cover western and central Mediterranean. Today, it is also funded by the Commission and managed by FAO.[2] Others such as ADRIAMED followed to cover Adriatic fisheries, initially funded by Italy and, since 2007 also by the Commission and also incorporated as a FAO project.[3] At a later stage, the project MEDSUDMED was also set up to cover the fisheries of the Strait of Sicily.[4]

These projects have been instrumental to provide GFCM with basic data from non-EU Member States and their collaborative nature has contributed considerably not only to improve the knowledge basis for management but also to dissipate certain initial reticence by non-EU countries, who saw the revamped GFCM as too much of an 'European club'.

This has resulted in an impressive increase in the number of stocks assessed: from 2007 to 2012 the SAC has reported on over 220 stock assessments with respect to MSY and, from 2005 to 2014, GFCM has steadily increased the number and importance of the recommendations: 20 on fishery management, 7 on monitoring and control and 5 on data and information.[5]

Still, this progress is certainly not enough. A vast majority of fish stocks in the basin are still not subject to any multilateral management measure from GFCM. In addition, a specific area within its area of responsibility, the Black Sea, is almost absent from the activity of the organisation. The best evidence that more progress is needed is provided by the reaction by most Member States in connection with the MSY objective established in the 2013 CFP reform: Member States in this area underlined that to achieve the MSY objectives it is essential that other Mediterranean countries sharing these resources also commit themselves to delivering such an objective, something that only GFCM can deliver.

ICCAT: the case of Bluefin tuna

By far the most important issue in the management of Mediterranean fishery resources is that of Bluefin tuna. Bluefin tuna (*Thunnus thynnus*) is an iconic species extremely valuable economically, caught in the Mediterranean basin since ancient times. The history of its management in the multilateral context of ICCAT is perhaps the best example of the difficulties and challenges of fisheries in the Mediterranean basin.

As a highly migratory stock, the adult Bluefin tuna spends most of the time feeding in cold waters of the North Atlantic (as far north as Canada or Norway) while going to warm waters to breed. Scientists have recognised two stocks in the Atlantic: the western stock, that breeds in the Gulf of Mexico and the eastern stock (much more abundant) that breeds in the Mediterranean. Tagging has shown mixing of the two stocks. That means that, although the two stocks are primarily fished by the coastal States of each side of the Atlantic, the fishing activity in the Mediterranean (where most catches take place) is indeed an issue for all other countries of the North Atlantic, which can legitimately claim to have a stake in this fishery even if they do not catch this stock directly. This is for example the case of the US and Canada.

The management of Bluefin tuna: some historical background

Up until the accession of the EC to ICCAT in 1997, only Spain and France, out of the – then – four Mediterranean Member States were contracting parties of ICCAT: Greece never acceded as such and Italy only did so in 1997. This produced a very unusual situation where, as we saw in Chapter 7, the implementation of the conservation policy for tuna and associated species in the ICCAT area and notably the first catch limitations for Bluefin tuna agreed already in 1994 was done at Member State level, outside the CFP umbrella.

The two stocks of the species were subject to separate measures in ICCAT. In fact, the first catch limitations in ICCAT were adopted for the western Atlantic stock, of interest to the USA, Canada and Japan, while the – much larger – eastern stock remained unregulated. The evidence of migration between the two stocks (Bluefin tuna tagged in the west were captured in the Mediterranean and the eastern Atlantic), immediately mobilised the US and Canada, who then pushed for a management regime for the Eastern stock as well.

Largely as a consequence of the North American initiative, ICCAT adopted a catch limitation at its annual meeting of 1994. The evidence on the need for a reduction in catches was made clear by the Scientific Committee for Research and Statistics (SCRS), ICCAT's scientific committee. On this basis, ICCAT decided to apply a 25% reduction of fishing in a gradual manner, establishing the current catch level (as represented by the catches of 1993 or 1994 whichever the highest) as a limit, and to achieve this 25% reduction by the end of 1998.

The implementation of these measures at EU level was rejected in Council in December 1996 since it did not specify figures for the catch limits. However, in December 1997, after the accession of the Community to ICCAT the previous month, the Council agreed to incorporate this catch limitation into Community law, through a specific regulation concerning catch quotas for highly migratory fish (EC, 1998).

The application of this catch limitation, in practice, proved troublesome. Firstly, the 1994 recommendation did not precise specific catch limits but only referred to the catch levels as represented by catch data submitted by the Parties. The problem was that some countries actively fishing for Bluefin tuna were not ICCAT members and were not submitting catch data. Secondly, at the time of the adoption of the catch limitation, the catch level corresponding to 1994 was not known.

As a result, in 1998 a number of ICCAT Parties, including the European Community, proposed revisions of the historical catches and, consequently, of their reference limits.

In 1998 the second catch limitations were adopted. In this occasion, SCRS repeated its previous recommendation to reduce catches substantially. In order to provide an element of graduality, ICCAT decided to adopt a catch limit of 32 000 Tm in 1999 and of 29 500 Tm in 2000. An essential element in this agreement was the fact that the catch limit of 1999 should be, in practice, lower than the agreed 32 000 Tm, on the basis of the application of the penalties for quota overshoot corresponding to the 1997 catches. The penalty to be paid by the Community for its excess fishing in 1997 was 4,029 Tm. Therefore, the agreed Community quota of 20 165 MT effectively corresponded to a quota of 16 135 MT. This is the figure assigned to the Community in the implementation regulation for 1999 (EC, 1999).

Initially, the catch limitations for Bluefin tuna in ICCAT were certainly not a success story in fisheries management. The difficulties in relation with catch statistics were particularly acute in the case of this stock. The reported catches were already well in excess of the agreed TACs. But the main problem was that of unreported catches. It was difficult to ascertain the real magnitude of these catches but by all accounts the real catches could well be in excess of 50 000 Tm.

To add to the problem, the establishment of a system of penalties to those parties having exceeded their quotas and, especially, the real application of these penalties to some ICCAT parties (including the Community) may have had some detrimental effects in the quality of the data. If data collection is already difficult and costly in a fishery as complex as that of Bluefin tuna, the existence of penalties for those that overshoot their quotas may not encourage good collection of data, since better statistics would have contributed to larger penalties for the countries concerned.

If the application of this measure at ICCAT level was difficult, its application in Community law only added to these problems. When ICCAT adopted this recommendation, two Member States with Bluefin tuna fisheries (Italy and Greece) were not ICCAT members did not participate at ICCAT meetings and did not cooperate with the organisation. As a consequence, their reported level of catches used as a reference for catch limits, were allegedly not correct.

When the Council of Ministers of December 1997 adopted the regulation 65/98, it included provisions to the effect that the Community should bring about a revision of historical catches and that the Community quotas should be modified accordingly.

In 1998, as we have seen, ICCAT established a new catch limitation, applicable to 1999 and 2000. This time, however, this catch limit was already negotiated with the Community as a contracting party to ICCAT and therefore the ICCAT agreed for the first time on a Community quota. In order to implement this quota in 1999 and beyond, the Community had to establish an allocation key. This was done in Regulation 49/1999 referred to earlier.

The allocation established by Council regulation 49/1999 was contested by Italy, mainly on grounds that the basic allocation key would not be fair and that the penalties accepted by the Community in ICCAT should not be applicable to Italy, on the basis of its date of accession to ICCAT. Italy, subsequently, took the case to the European Court of Justice in 1999. The Court ruled in favour of the Council so that Italy's allocation was consolidated as adopted.[6] Despite thus Court ruling, other Member States (i.e. Greece) continued questioning the allocation for years.

Recent developments: from international embarrassment to a success story

For a number of years afterwards, the difficulties related to the control of this fishery became a major source of embarrassment for the Union. As the main stakeholders, the failure to properly control catches of this stock created a very bad image of the Union in the international *fora* and seriously hampered its negotiating position. As the Bluefin tuna attracted more attention by NGOs and the public at large as an iconic species in deep trouble due to overexploitation, redressing this situation became a high priority for the Union.

This prompted the adoption of a recovery plan for Bluefin tuna in ICCAT, where the Union had a decisive contribution. This plan, adopted in ICCAT in 2006, was initially implemented provisionally in the TAC and quota regulation of the following year and was consolidated in EU law in 2007 (EC, 2007). The plan has subsequently been updated following further ICCAT recommendations, requiring a new plan being adopted by the EU in 2009 (EU, 2009). Further ICCAT recommendations in 2012 and 2013 require further updates.

The plan was an example of how the political importance of an iconic fishery and the influence of foreign countries (in particular the United States) in raising awareness about the need to ensure that Europe would fully take its responsibility on this fishery led to a series of measures that would otherwise have been unacceptable. Provisions on by-catch, on recreational fishing, and an extremely detailed control system were part of this plan.

But in the initial stages the control of the agreed catch limitations was not perfect, and this put the Union under pressure as the main stakeholder of the fishery. The situation was made even more complicated through the development of the practice of fattening, whereby the tuna caught by purse-seiners is towed to floating cages where they remain (being fed and thus fattening) until they are marketed. This practice introduced a considerable additional complexity in the management and control of the fishery, including the difficulty to cross check the catches with the sales several months after, when the wild fish caught had considerably gained weight.

The Commission deployed an unprecedented effort in human resources to establish an effective control system of catches, in particular by purse-seiners, the holders of the highest percentages of national quotas in the Mediterranean Member States and the suppliers of the fattening cages. This system included also very strict rules on the operation of the caging practices that have become the rule for the catching and fattening of this species over the last decade, with traceability systems to ensure that the fish fattened coming out of the cages really corresponds to the wild fish caught and towed to the cage, in an unprecedented level of monitoring and control of the fishing and processing activity. In this system, the European Fisheries Control Agency (EFCA) plays an important role (see Chapter 9).

The control system established, requiring the reporting of catches day by day, has become a good precedent for what international management can do to control a fishery properly. The good control of the catches has resulted, for the first time in many years, in a rather positive trend in the biological status of the stock. In addition, ICCAT adopted a recovery plan for the stock that, in combination with much more effective control, clearly seems to be working, as the stock has started to show a significant recovery.[7]

Today, the management of Bluefin tuna, at least in the Union, is one of the best-enforced fishery management systems in the world. What was once an important source of embarrassment for the Union and for the CFP has become one of its biggest success stories.

The Black Sea: the new challenge of the CFP

Following the accession of Bulgaria and Romania in 2007, the CFP opened a new front: the management of fishery resources in the Black Sea.

Although generally considered as part of the Mediterranean, the Black Sea is quite different from the main Mediterranean basin in terms of oceanographic conditions and the fisheries resources occurring in its waters. Specially, the Black Sea represents a challenge that the CFP has never had to deal with: being a minority player in the overall management of fishery resources of a sea basin.

The Black Sea is a semi-closed sea, with an inflow of fresh water that far outweighs evaporation, thus provoking a net outflow of water to the Aegean Sea. Also as a consequence, surface water is less saline and thus it has a lower density and does not mix easily with deeper, denser and more saline water. As a result, it is estimated that 90% of the water mass of the Black Sea is anoxic due to lack of contact with the atmosphere. Marine life and thus fishery resources are concentrated in the 10% of oxygenated surface water. This has obvious consequences for the type of fisheries and their distribution.

The state of fish stocks in the Black Sea is highly deteriorated. In the absence of an agreed multilateral conservation policy, a number of important stocks collapsed in the 1950s and 1960s. In addition to overfishing, the collapse of the stocks is considered by some authors to have been caused by negative environmental consequences, such as the massive growth of populations of invasive jellyfish, probably introduced by accident in the ballast waters of shipping in the 1990s and facilitated by a 'trophic cascade' following the collapse of the apex predators (Daskalov, 2002).

EU waters cover a small proportion of the area of the Black Sea and EU fisheries represent a relatively small part of the total fishing activity, with Turkey,

Russia and Ukraine representing the majority. In addition, this is an area where there is no tradition of cooperation among the different coastal States. As a result, multilateral management of the concerned stock is missing. This puts the measures under the CFP under a particularly difficult strain: whatever the Union does to manage resources will both have a limited impact on the resources and can be easily opposed on grounds that self-limitations will benefit the Union's neighbours in the sea basin.

Although the Black Sea is covered by the GFCM convention, this body cannot play a significant role in managing Black Sea stocks because some important coastal States of this sea basin are not members of GFCM: while Bulgaria, Rumania and Turkey are GFCM members, Russia, Ukraine and Georgia are not. The absence of three key coastal states in GFCM prevents this organisation from being the adequate multilateral forum that is necessary to manage Black sea stocks. Unfortunately, EU attempts to set up an effective multilateral system to manage joint stocks have not made headway. Recent political tension in the region will certainly not contribute to any process of multilateral cooperation any time soon.

In these circumstances, it is not surprising that the management instruments adopted for this sea basin were very limited: only TACs for two species, turbot and sprat were set. As we saw in Chapter 4, TACs were conceived fundamentally as a distributive instrument. In the Black Sea and short of multilateral management of the shared resources, the agreed TACs have at least the value of allocating fishing rights only between Romania and Bulgaria. Whether these TACs are effective management instruments, it is not easy to tell but one can easily conclude that until effective multilateral management is established in the Black Sea, the TACs in that area are at least a pre-condition to a future effective management, as was the case in the Atlantic.

Notes

1 *Posidonia oceanica* is a marine phanerogam that constitutes large seagrass beds in coastal Mediterranean areas. They are areas of high diversity and spawning grounds for many fish species.
2 www.faocopemed.org.
3 www.faoadriamed.org.
4 www.faomedsudmed.org.
5 www.Gfcmonline.org.
6 Case 120/99 Italy/Council. Judgement of the Court (5th Chamber) of 25 October 2001 in case C-120/99.
7 www.iccat.int/Documents/Meetings/Docs/2014_BFT_ASSESS-ENG.pdf.

References

Bailly, D. and R. Franquesa (1999). Social and Economic Indicators for Fisheries Management in the Mediterranean. In *Europe's Southern Waters: Management Issues and Practice*, D. Symes (ed.). Blackwell Science, Oxford. pp. 141–153.

Breuil, C. (1999). The GFCM and the management of Mediterranean fisheries. In *Europe's Southern Waters: Management Issues and Practice*, D. Symes (ed.). Blackwell Science, Oxford. pp. 154–169.

COM (2001). Green Paper on the future of the Common Fisheries Policy. Brussels, 20.3.2001; COM (2001) 135 final.

COM (2002). Communication from the Commission to the Council and the European Parliament. A Community action plan for the conservation and sustainable exploitation of fisheries resources in the Mediterranean Sea under the Common Fisheries Policy. Brussels, 9 October 2002. COM (2002) 535 final.

COM (2014). Communication from the Commission to the European Parliament and the Council concerning a consultation on Fishing Opportunities for 2015 under the Common Fisheries Policy. Brussels, 26.6.2014. COM (2014) 388 final.

Daskalov, G.M. (2002). Overfishing drives trophic cascade in the Black Sea. *Marine Ecology Progress Series* **225**: 53–63.

EC (1994). Council Regulation (EC) No 1626/94 of 27 June 1994 laying down certain technical measures for the conservation of fishery resources in the Mediterranean. *OJ* L 171, 6.7.1994, p. 1–6.

EC (1998). Council Regulation (EC) No 65/98 of 19 December 1997 fixing, for certain stocks of highly migratory fish, the total allowable catches for 1998, their distribution in quotas to Member States and certain conditions under which they may be fished. *Official Journal* L 012, 19/01/1998 p. 145–148.

EC (1999). Council Regulation (EC) No 49/1999 of 18 December 1998 fixing, for certain stocks of highly migratory fish, the total allowable catches for 1999, their distribution in quotas to Member States and certain conditions under which they may be fished. *Official Journal* L 013, 18/01/1999 p. 54–58.

EC (2006). Council Regulation (EC) No 1967/2006 of 21 December 2006 concerning management measures for the sustainable exploitation of fishery resources in the Mediterranean Sea, amending Regulation (EEC) No 2847/93 and repealing Regulation (EC) No 1626/94. *OJ* L 409, 30.12.2006, p. 11. *OJ*. No 36 of 8.2.2007.

EC (2007). Council Regulation (EC) No 1559/2007 of 17 December 2007 establishing a multiannual recovery plan for Bluefin tuna in the Eastern Atlantic and Mediterranean. *OJ* L340, 22,12,2007, p.8.

EC (2009). Council Regulation (EC) 302/2009 of 6 April 2009 concerning a multiannual recovery plan for Bluefin tuna in the eastern Atlantic and Mediterranean, amending Regulation (EC) No 43/2009 and repealing Regulation (EC) 1559/2007. *OJ* L96, 125.4.2009, p.1.

European Commission (2014a). Assessment of Mediterranean stocks – Part 1 (STECF-14–17). JRC Scientific and Policy Reports. Scientific, Technical and Economic Committee for Fisheries. EUR 26955 EN. Joint Research Centre.

European Commission (2014b). The 2014 Annual Economic report on the EU Fishing Fleet (STECF 14–16). JRC Scientific and Technical Policy Reports. Scientific, Technical and Economic committee for Fisheries. EUR 26901 EN. Joint Research Centre.

Farnell, J. and J. Elles (1984). *In Search of a Common Fisheries Policy*. Gower, Aldershot. 213 pp.

JRC (2014). Scientific, Technical and Economic Committee for Fisheries (STECF). Assessment of Mediterranean Stocks – part 1 (STECF-14–17). EUR 26955 EN.

SEC (1990) *Outline of a Common Fisheries System in the Mediterranean*. Brussels, 25 July 1990. SEC(90) 1136.

Symes, D. (ed.) (1999). *Europe's Southern Waters: Management Issues and Practice*. Fishing News Books. Blackwell Science, Oxford, 198 pp.

Valarié, P. and H. Rey (1999). Can common fisheries policy in the Mediterranean ignore its past? In *Europe's Southern Waters: Management Issues and Practice*, D. Symes (ed.). Blackwell Science, Oxford. pp. 129–140.

CHAPTER 9
Enforcing the CFP

The control pillar of the CFP

Why enforcement matters

Like any other policy, the CFP is as good and effective as it is implemented and enforced. From its inception, the question of the control of the CFP has been an important issue, albeit one considered always with care and prudence.

Effective enforcement is an essential question when it comes to evaluating the CFP, for a number of reasons:

1 Firstly, bad enforcement clearly undermines the core business of the policy and prevents the adopted measures from providing results. This is particularly crucial for the conservation policy. Even a good policy, if not properly enforced, can fail, thus giving the impression that the policy in itself is inadequate. This has long been a problem for the CFP. Of course, this can have an alternative reading: that if the policy is not well enforced is because the policy itself is not good. In other words, it is the lack of 'buy-in' of the policy by Member States and stakeholders that makes it difficult to enforce. This is an interesting idea but the experience shows that many measures, even when fully accepted by administrations and stakeholders, have been poorly enforced anyway.

2 Secondly, poor enforcement forces policy makers to introduce additional restrictive measures, thus resulting in an increasingly unpopular policy. This is very characteristic of the CFP. Failure to enforce a reasonable decision gives rise to a deterioration of the resource, thus resulting in the need for more restrictions than would have been necessary otherwise. A tradition in the CFP has been that to compensate for poor enforcement, stricter rules (i.e. lower TACs) have been necessary, or additional rules have been adopted with the objective of filling the gaps. At times, this has produced a 'spiral' of bad enforcement and an increasingly complex policy that has proven very difficult to break.

3 Thirdly, poor enforcement creates an impression of lack of a level playing field, which does not contribute to encourage enforcement. Fishermen always have the impression that their own authorities are controlling them much more than their counterparts in neighbouring Member States. If these differences are genuine, they can constitute a serious negative incentive for Member States to be strict on control with their own nationals.

4 In the international field, poor enforcement has sometimes undermined the image of the Union and has weakened its negotiating position in multilateral and bilateral negotiations.

Overall, one can certainly contend that the bad image of the CFP that we have referred to in this book is to a certain extent attributable to poor enforcement of the existing rules and not necessarily always to the rules themselves.

The Common Fisheries Policy: The Quest for Sustainability, First Edition. Ernesto Penas Lado.

The question of competence

As we saw, the CFP and particularly its core business, the management of biological fishery resources, is one of the few exclusive competences of the Union. However, the control of this policy is a competence of Member States. This paradox is at the very origin of many of the weaknesses observed over the years in the implementation of the policy.

In 1983 when the conservation part of the CFP was set up, the Commission considered that an integrated European policy would require an integrated control system, where the Commission, better than any national control system, would ensure the necessary level of adherence to the new rules in a way that would ensure a level playing field. However, this proposal was not accepted by the member States of the time, who did not want the Commission to become an enforcing agency for the new CFP (Lequesne, 2001). Since then the role of the Commission is not to control directly the behaviour of fishermen but to monitor the way Member States fulfil their obligations to enforce the policy under their own responsibility. That is, the role of the Commission is not to control the fishing activity but to 'control the controllers' in Member States.

This division of responsibilities has few exceptions but they are worth mentioning: in the case of distant fisheries, where the control of Member States over their national vessels is very expensive and where control has become a very important issue, requiring a very intensive level of control activity, Member States have exceptionally accepted a higher role for the Union (and particularly Commission inspectors). The case of NAFO in 1988 where Council authorised the Commission to assign Community inspectors to the control scheme in the NAFO Regulatory Area is an example of how Member States, in certain circumstances, can accept that Commission officials may have a more direct role as inspectors (EEC, 1988).

Over the years, the Commission has had at its disposal a number of legal instruments to act against Member States that do not fulfil their obligations. Some of these instruments are of a cooperative nature. Others involve the possibility of taking the Member States concerned to the Court of Justice for a possible sanction. The traditional *modus operandi* is that the Commission has favoured the cooperative approaches first and only applied the more punitive approaches as a last resort (Long and Curran, 2000). Such state of affairs continues today, albeit with new legal and financial instruments.

Any attempt to question this basic divide in competence, for example, to give the Commission more direct inspection and sanctioning powers, has always been rejected by Member States. The analysis of strengths and weaknesses of the control system of the CFP cannot be understood without taking this into account.

The historical development of the control system of the CFP

A very good account of the development of the control pillar of the CFP was provided by Ronan Long and Tony Curran in 2000. In the CFP, indeed like in any other policy, the relative merits of the policy in achieving its own objectives depend largely on the way it is controlled and enforced. One way or another, the question of the control of the CFP has always been an issue and one can certainly argue that if the CFP, with all the shortcomings of its design, was really well controlled and enforced, it would certainly be a much better policy.

1982: The first regulation

In this case, the need to adequately control the instruments agreed to run the fisheries policy were part of the general considerations leading up to the birth of the policy. The first regulation on fisheries control was adopted in 1982, prior to the official birth of the CFP (EEC, 1982). At that time, catch limitations were in fact already applied but there were allegations of significant misreporting of catches, particularly of catches landed in other countries. This created an additional political difficulty to the already difficult discussions on quota allocation that had been going on since 1977. Building trust among the seven Member States involved to the effect that, whatever quota allocation was agreed, it would be adequately enforced, became a pre-condition to reaching the final deal on allocation in 1983.

This regulation, therefore, established the basic obligations to skippers and Member States particularly as regards the recording and declaration of catches (notably the logbook) to ensure a good control of national quotas. The Commission was a central point where the information was received and processed. At the time, with seven Member States and a relatively low number of species, this was feasible. However, this created from the very beginning a tradition of the Commission handling all this information in a centralised manner. Later on, with many more Member States and species involved, such handling has become extremely complex and difficult manage for the Commission services.

The 1982 regulation had to overcome a considerable reticence by national inspection services as regards cooperation. In fact, cooperation worked well with respect to the control of foreign vessels in Community waters but cooperation among Member States for the control of their respective fisheries was limited, particularly on the crucial question of sharing information, question where Member States were extremely zealous of maintaining national sovereignty (Farnell and Elles, 1984).

Although the initial control provisions gave the Commission a central role in monitoring catches it did not provide it with the powers to close the fisheries and, given the delay in processing data on catches, the system was not effective in closing fisheries when the total TAC was exhausted. In addition, the issue of the follow up of infringements of the new rules of the CFP through the national sanction systems was not even discussed. This system was such that certain authors already had the foresight to advance that: 'progress towards a control system which matches the Community's apparent commitment to conservation is unlikely to occur before substantial overfishing of a sensitive stock discredits the present system' (Farnell and Elles, 1984).

This statement was to prove right a number of years later.

In any case, this first control regulation was a fundamental building block of the basic deal on the CFP in 1983. In being so, it established a precedent that has been repeated many other times in the CFP: that unless good control is assured by everybody, it is difficult to agree on certain measures due to the lack of trust among the different Member States.

1987: The second regulation

In 1987, following the accession of Spain and Portugal, a second control regulation was adopted to tighten up the provisions of 1982 (EEC, 1987). Again, the digestion of the Spanish fleet into the CFP had a strong bearing in this development. This regulation addressed the weaknesses of the initial one, in particular by:

- Increasing the responsibilities of Member States vis-à-vis the verification of the declarations and registration of catches;
- Sending more frequent catch reports to the Commission;
- Enlarging the list of species subject to reporting (extending it to non-TAC species) and, crucially;
- Establishing the basic empowerment for the Commission to inspect Member States and to close fisheries once the quotas have been exhausted.

This regulation opened a new era in the management of TACs: the Commission actually issued countless closures of the fisheries, when quotas were consumed. This approach was adopted at a time when the level of complexity of the policy still allowed for a number of functions to be centralised in Brussels. As happened with so many other issues in the CFP, this created a tradition of central handling of control issues that later on, as the Union and the policy itself expanded, became increasingly difficult to manage.

1993: The third regulation

On the occasion of the 1992 reform, the Commission had an opportunity to evaluate the control of the CFP and propose significant improvements, in the framework of its 1991 communication on that reform (Commission des Communautés Européennes, 1991). At the time, in a context of policy continuity, the Commission tried to be non-controversial in its public evaluation of the control of the policy. However, it did already clearly see the weaknesses of the system. As a result, the evaluation of the Commission was apparently contradictory: it did not openly criticise the control system but it did propose a number of improvements in it, including quota penalties, licence withdrawals, financial measures, fines and so on. In addition, in specific, less political reports on control questions, the critical views were voiced much more strongly (see next).

So while the Commission's rhetoric of the time was not openly critical, at least in the main political documents, the ideas suggested to improve the system, and the more technical, specialised reports already implied the full awareness of the need to address the considerable weaknesses of the control system.

These weaknesses were addressed in the new control regulation adopted the following year: Regulation 2847/93 (EEC, 1993). This regulation was still very much focused on the control of the catch limitations, thus reflecting the still strong emphasis on the distributive nature of the policy: the control of TACs and quotas still represented the most important part of this new regulation. The main reason was that poor enforcement of quotas was considered a *de facto* breach of relative stability.

In that vein, the regulation introduced a number of new elements, *inter alia*:

- The extremely complex nature of the capture, landing and trade of fish made it necessary to improve the whole system of traceability of catches. New provisions were introduced on the need to trace all landings and their transport. This also included imported fish. New technologies, notably the extended use of computerised methods to speed up the transmission of data and allow for cross-checking of different data were introduced.
- The need for the Commission to be able to inspect Member States without prior notice was introduced. This was necessary to address the problem of Member States 'preparing' the visit of the Commission inspectors in a way that they could cover up their weaknesses.
- The enlargement of control instruments to other areas such as the market policy, the new effort management stemming from the new MAGPs (see Chapter 5), aquaculture or technical measures.
- Crucially, the obligation for Member States to have a system of sanctions that would deter misbehaviour by fishermen. This was an essential element of the new regulation. Although the Commission tried to harmonise these sanctions, strong opposition from Member States did not allow this to happen. However, the principle was adopted that Member States should have a penalty system for fisheries infractions that would have a significant deterrent effect on offenders.

This new control framework was a breakthrough at the time. However, a number of problems somehow undermined its effectiveness:

- Firstly, the regulation required a number of implementing provisions to be fully applicable. In practice, however, only 4 out of 20 of these implementing regulations were actually adopted.
- Secondly, the regulation itself was amended many times, thus creating a legal basis that was both incomplete and permanently re-opened for discussion.
- Thirdly, as we will see next, the practical implementation of this new regulation was less than perfect in a number of cases.

Overall, the 1993 represented an important step forward in terms of the control instruments for the CFP but insufficient implementation substantially undermined its potential positive effects.

Further developments

In the 2002 reform, the issue of control did not feature as a very high priority. It is true that in the Green Paper of 2001 (COM, 2001), the shortcomings in the control of the CFP were cited as one of the motivations for reform: 'Poor enforcement of decisions actually taken has also contributed to overfishing'. However, the Green Paper only dedicated one of its 40 pages to this question, to suggest the need to advance in the coordination of national control efforts among Member States and to examine the possibility of setting-up a common inspection structure to make that possible. This vague recommendation, however, led to the establishment of the European Fisheries Control Agency (see next) a few years later.

Regulation 2847/93 was modified many times and its basic rules were not fully developed into the necessary implementation regulations by the Commission. The result was a legal framework for the control policy that was incomplete, dispersed in many texts, inconsistent and difficult to understand, let alone to apply. This eventually led to the preparation and adoption of a unified, reinforced text that was adopted in 2009: Regulation 1224/2009. We will refer to this regulation later in this chapter.

The evaluation of the control policy

The control policy of the CFP has for a long time been considered complex, expensive and not conducive to producing the desired results. This fact is addressed every time the policy is evaluated. The language of this analysis may have changed over the years but the analysis has been remarkably stable.

The first warning about the problem of insufficient control of the CFP came shortly after the policy was established. Already in 1986, on the occasion of its first report on the enforcement of the CFP the Commission indicated that: 'Without proper enforcement, conservation would be threatened and international fishery relations would suffer' (COM, 1986).

This statement has proven to be right very many times over the last three decades. Time and time again, the poor enforcement of the CFP has undermined the sustainability of fisheries and has often embarrassed the Union in the international arena.

The report indicated that not all Member States had established the necessary measures to ensure adequate control of the catch reporting system to ensure that national quotas were adhered to. Significant weaknesses were also observed in the way Member States undertook control of technical measures on board fishing vessels. As a consequence, the Commission proposed that the powers of the Commission to ensure adequate enforcement be reinforced.

It is surprising to observe that, 27 years after such report, the opinion of the Commission remains surprisingly similar. That after 30 years of the establishment of the TAC and quota system and, despite many important improvements, its control is still considered insufficient.

A second report on control and enforcement was issued by the Commission in 1992, coinciding with the first reform of the policy the same year. In this report (SEC, 1992) the Commission was again critical of the way Member States were enforcing the CFP. However, at the time the Commission tried to avoid being very openly critical with individual Member States and preferred to concentrate on the systemic deficiencies of the system, as a more conciliatory approach. Indeed, the Commission identified at the time three weaknesses in the system:

- Lack of understanding of the policy. This referred in particular to fishermen's lack of 'ownership' of the policy but also, importantly, to a general lack of understanding by the legal system and the public opinion of the importance of ensuring that the policy be well implemented.
- Lack of adequate instruments for the Commission. In particular, the lack of possibilities for the Commission inspectors to conduct autonomous inspections in Member States. As a consequence, it was very difficult for the Commission to generate the evidence necessary to use its legal powers against offending Member States.
- Inadequate national control structures and lack of coordination among Member States. In addition to the well-known weaknesses of national control systems, the Commission pointed its finger, for the first time, to the fact that, given the transnational nature of many fishery operations (such as vessel from a Member State landing their fish in another Member State) the lack of real coordination among national control systems did actually prevented an efficient control of these transnational activities.

This report made a fundamental contribution to what would be, soon afterwards, the next generation of control rules of the CFP: Regulation 2847/93, which tried to address all three issues mentioned previously, although it did so only in a limited manner.

For the 2002 CFP reform control and enforcement were also mentioned in clearly critical terms in the Commission's Green Paper (COM, 2001). Three weaknesses were specifically identified at that stage:

- The current arrangements are insufficient and cannot ensure a level-playing field across the Union;
- The organisation of monitoring and control is fragmented. Better coordination and optimal use of monitoring and inspection are needed;
- A satisfactory follow-up of infringements has not been achieved.

However, at the time the reform efforts concentrated on other issues, particularly the suppression of financial aid to the construction of vessels and the new emphasis on the recovery of the stocks, in the wake of the cod crisis. Despite the very critical view of the Commission, the improvement of the control system was not a priority in that context.

A more fundamental step in the awareness of the control problems of the CFP came through the report of the Court of Auditors in 2007[1] that unveiled the low level of enforcement of the policy, notably by Member States and concluded that the CFP has serious problems of enforcement by Member States. The Court considered the current control system of the CFP 'inefficient, expensive, complex and it does not produce the desired results'.

It is very important to underline that the evidence that the CFP was a policy seriously hampered by poor enforcement was one of the main factors leading the Commission to promote an ambitious reform of the policy. In fact, the Commission largely agreed with the conclusions of the Court and this fact prompted a proposal for a major reshuffling of the control regulation of the CFP, which was subsequently adopted in 2009.

The Court of Auditors report was one fundamental driver of the following and to date most far-reaching evaluation of the control policy, published by the Commission in 2008 (COM, 2008). In this communication, the Commission identified the 'drivers of for non-compliance' as follows:

- overcapacity of the fleet;
- lack of consistency in intensity and criteria of the control efforts, resulting in inefficiency and lack of trust;
- complex legal framework and
- limited and ineffective control mechanisms on the part of the Commission.

On the basis of this analysis, the Commission set forth to introduce a number of key improvements in the control system of the CFP, including *inter alia*: systematic risk analysis as the basis for control, strategic programming, extended use of modern technologies, standardised and coordinated control methodologies or the introduction of traceability systems for the fish caught. In addition, the Commission identified the need to give further roles to the European control agency, to enlarge the instruments at the disposal of the Commission (including cooperative instruments such as action plans), developing economic sanctions for non-compliance and other ideas.

Also in the wake of the Court of Auditor's report, the Commission was more open than ever before in recognising publicly the deficiencies of the control of the CFP. In the preparation of the 2013 reform, the Green Paper of 2009 put considerable emphasis on compliance as one of the five 'structural deficiencies' of the CFP (COM, 2009a). This analysis went much further than ever before in identifying poor compliance (as a complex, multi-faceted problem, including lack of industry buy-in) as a fundamental shortcoming of the CFP.

The reform, however, did not make specific proposals on control, given that a new package of control measures was being adopted in parallel. However, it indirectly contributed to this goal notably by introducing more elements of ex-ante conditionality in the context of the structural policy, thus introducing an altogether new instrument to tighten up the enforcement side of the CFP reform (see Chapter 16).

The difficulty of enforcing the CFP

There are many reasons why the control of the CFP is a challenging question and these reasons concern Member States as well as the Union itself. Let us examine these reasons.

The legislation

One fundamental reason why control is generally difficult is legislation itself. And this has two main components: the complexity and the administrative burden it implies.

The complexity

The complexity of many of the rules of the CFP (in particular from 1993 to 2009, in a context of an increasingly complex and dispersed regulatory framework) has very often rendered them difficult to understand, let alone enforce. This has also given rise to abundant problems of interpretation. In Chapter 13 we will refer to the issue of the complexity of many CFP regulations.

In addition to being complex, very often CFP rules have been difficult to control in practice. Traditionally, when agreeing on management measures, the issue of controllability was not taken into account by the legislator, often resulting in rules that are extremely difficult, or extremely burdensome to enforce.

Another element has been the dispersion of the control regulations themselves. For many years, these were extremely dispersed in a plethora of regulations and amendments, making it extremely difficult for fishermen and even inspectors, to be fully aware of all the intricacies of the control rules applicable. This is, however, no longer the case after the new regulation of 2009 pooled together in one single text all the pre-existing rules.

Administrative burden

The legislation has also been very quickly evolving, including the incorporation of new technologies and new instruments. Although the Commission has provided financial aid to face these challenges and while the introduction of new technologies has always been gradual, it is true that the control legislation has not been stable over the years and national systems have been in a permanent quest to catch up with new developments, rather than consolidating their basic systems. This heavy (and ever growing) administrative burden imposed by legislation has long been resented by Member State as one of the difficulties of the CFP.

Of course these arguments can and often are used as excuses. Despite these difficulties, control in some cases has been much worse than would be justified, even taking these problems into account, as we will refer to next. After all, the control of fisheries is burdensome by definition: if the reduction of administrative burden is taken too far as an argument, the result would be fisheries without control.

Member States' control

There is traditionally no doubt that the control activity by Member States has been less than perfect in a number of cases. The Court of Auditors has made this fact crystal clear in its report (referred to previously) published in 2007 that represented a watershed in terms of the EU's awareness about its own weaknesses in controlling one of its most integrated policies. This report contributed considerably to creating the basis for an ambitious reform of the CFP in 2013. But why was Member States control of the policy weak?

One first reason is the multinational natures of the policy, in which national authorities often consider that they would be unfair to their own fishermen by controlling them very strictly if other Member States don't do the same with theirs. This is somehow a practical effect of the 'tragedy of the commons', in which Member States take the position of individuals: it is not worth being very tough on our nationals, only for other fishermen from other countries to benefit from our effort and sacrifice.

Control of the fishing activity is a national competence. If there is one complaint universally heard from the industry, is that they always feel more strictly controlled by their own authorities than their counterparts from other Member States. Likewise, national control authorities always tend to believe that they control their own nationals strictly, while other Member States don't necessarily do the same. The – real or alleged – lack of a level playing field of control comes usually as the ideal excuse to avoid controlling strictly the activity of their vessels. This clearly calls for further coordination and common approach on control matters. The coordination of efforts among different services and the development of synergies among them was achieved only relatively recently, thanks notably to the establishment of the European Fisheries Control Agency (EFCA).

The cost of control, both for the industry and the national administrations, is considerable and is not always easy to be identified as a national priority by budgetary authorities. As a consequence, the control means to implement the CFP have always been scarce. This has been aggravated by the lack, for a long time, of effective mechanisms to pool resources and increase the cost-effectiveness of the activity. The zealous independence of national control systems has never facilitated cooperation among control services, not only among services of different

Member States but even within Member States among different maritime control services.

Acting on control is not politically palatable, particularly in fisheries shared with third countries that can be blamed as scapegoats. When elections approach, politicians tend to avoid giving bad news to their fisheries constituencies.

Finally, there is indeed a genuinely national dimension to the problem of enforcing TACs. Member States have very different ways of allocating their national quotas to their national fishermen. In some cases, Member States distribute these quotas through individual transferable quotas. In other cases, quotas are distributed among fleet segments in a rigid manner. In other cases, they are distributed to Producer Organisations, who in turn allocate them among their members. In other cases, finally, there is no distribution and the national quotas are consumed under 'Olympic fishery' conditions, that is, individual fishermen fish freely until the whole quota is exhausted. These systems of national allocation often make it extremely difficult to control catches and, in addition, in certain cases seriously contribute to increasing discarding.

The Commission's role

As for the European Commission, although it always insists on the need for adequate compliance, the practical action to discipline Member States in that regard has often in the past been less intensive than necessary, largely due to the inadequacy of the traditional instruments. Taking a Member State to the Court of Justice is an extremely heavy, time-consuming procedure that must be based on solid evidence, not always fully available and that produces results only after a number of years.

The Commission has had at its disposal certain instruments to discipline Member States not fulfilling their obligations on control and enforcement, but these instruments have traditionally been weak and burdensome and, therefore, their usefulness has been limited. In practice, in the area of fisheries, there has only been one case in which a Member State was sentenced to pay a financial penalty. The judgement in question was delivered by the Court of Justice in 2005 for non-compliance with a 1991 judgement concerning a case in which the first infringements had been recorded in 1984, that is, 21 years earlier (Commission des Communautés Européennes contre République Française, 2005)!

These limitations have been addressed in recent years through two important new instruments: the new control regulation of 2009, which established a number of new instruments for the Commission to use and more recently the conditionality on administrative capacity to control the CFP included in the new structural instrument, the EMFF (see Chapter 16).

The new control regulation adopted in 2009 has provided the Commission with a much wider palette of instruments both cooperative and punitive. The Commission's limitations are, therefore, no longer what they used to be and now the Commission has an unprecedented number and variety of measures at its disposal to exert its powers to ensure good enforcement of the CFP.

Whatever the instruments available to the Commission, a fundamental question is the political will to use them. This requires a strong conviction of the need to use these powers as an instrument to improve the CFP, beyond short-term political difficulties and in the face of opposition to do so.

The industry's attitude

It has long been a common presumption that the European fishing industry does not 'buy-in' to the rules of the CFP. It follows that when industry's acceptance is not there, enforcement is low. Certain authors have underlined this phenomenon, considering that fisheries are an example of Max Weber's analysis between the sense of equity and abidance with the law, at least in certain cases where fishery rules are perceived as deeply unfair to specific groups of fishermen (Lequesne, 2001). This view however, is less and less true as the CFP develops increasingly effective stakeholder consultation bodies such as the RACs (see Chapter 12). In any case, low enforcement on account of lack of agreement on the measures is probably more an excuse than a real justification: if fiscal policy depended on the taxpayers' buy-in, it is likely that few people would pay taxes.

Typically, the peer pressure (exerted within the industry to ensure all vessels respect the rules applicable, to ensure a level playing field) in the CFP only exists among fishermen from different Member States, who often complain about poor enforcement by their neighbours. Occasionally, this pressure may exist within a Member States between fleet segments with different and conflicting interests. But it is extremely rare that the peer pressure is exerted within a Member State and within a given fleet segment.

Curiously, there are examples of good enforcement of rules through the industry's own enforcement but those tend to be associated with schemes of self-management by the industry (particularly in the Mediterranean) as well as in some cases of fisheries strictly operated under rights-based management. In both cases, this industry-led enforcement corresponds to cases where the rights to fish are clearly ring-fenced to a limited (and rather homogeneous) number of operators that have a strong sense of ownership of the resources they exploit.

This problem is not easy to resolve but one can expect that the increasing role of stakeholders in shaping up the policy, notably through the Advisory Councils, should work in favour of increasing the buy-in to the policy by the industry.

The issue of sanctions

As we have seen, Member States are under a general obligation to control the CFP. This obligation, according to the current control regulation, entails the obligation of establishing a system of sanctions to fisheries offenders that will ensure a deterrent effect (Article 89 of Regulation 1224/2009).

As Long and Curran already indicated, the national sanctioning systems have traditionally been different among Member States. Historical attempts to harmonise them have always failed. According to a study by the Court of Auditors[2] the maximum level of sanction imposed by Member States for fisheries offenses varies between €170 and 120 000. Of course, some of these differences are justified: a Member State with large-scale fishing should contemplate higher sanctions than another that only has small-scale fishing.

This situation is currently being addressed by the Commission as a matter of priority. The new obligations under the 2009 control regulation are clear. Compliance with them is essential, not only to ensure a good implementation of the CFP but also to reinforce the reputation of the Union in international *fora*, where poor enforcement of European and international obligations would seriously undermine its status, both legally and politically in the context of the world's fisheries governance.

Control as a key factor in the development of the CFP

A very typical development in the CFP has been that the weaknesses of the control system have had a bearing on the way the legislation has evolved: in a number of cases, legislation has been adopted to address control weaknesses. That means that failure to control properly the existing regulations has led to the adoption of additional layers of regulation. This, in turn, has increased the administrative burden of national administrations, thus making control even more difficult. And, as control is more difficult, this prompts new layers of obligations, in an apparently endless vicious circle.

The relationship between a policy and its implementation is always two-sided. It can certainly be argued that the CFP, if properly enforced, would be a much better policy that is given credit for. However, the opposite is also true: if the policy was better designed, it may actually be easier to enforce.

Finally, the policy itself can be responsible for its difficult implementation. The complexity of certain regulations can also be considered a factor that has made control difficult. This complexity is largely the result of the decision-making process that we will see in Chapter 13.

It is also important to underline the importance of control as a driving factor for the whole of the CFP. In particular, many developments in EU fisheries legislation are the direct result of control failures of the existing rules.

There is perhaps no better example than that of the development of effort management in long-term plans.

A case story: control and effort management

The best example of how poor control influences EU legislation is the use of fishing effort as a management instrument that we saw in Chapter 4. It must be borne in mind that the main reason why effort limitation was introduced was the repeated advice from scientists that TACs were not really lessening the fishing mortality of the cod stocks. This recommendation deserves a careful analysis.

There are at least two reasons why this recommendation was made:

1 A first one was related to the multi-species nature of most fisheries for cod. The effect of this is that, even if the TACs for cod were reduced considerably, much higher TACs for other associated species (other gadoids, Norway lobster etc.) would in fact nullify the effects of the reduction of these TAC reductions.
2 But there was another reason, more difficult to recognise openly: the control of the actual level of catches being rather less than perfect, the TACs themselves were considered an inefficient way of controlling the total outtake of the cod. So another, reinforcing instrument was necessary: the effort. This debate was also influenced by the regime applicable since 1986 to the Spanish fleet. Upon accession, even though the Spanish fleet was subject to a system of catch quotas like any other fleet, additional effort management measures (the maximum of 150 standard vessels operating simultaneously) were imposed. This historical example had an important influence in the decision to go ahead with a double system of catch limitations and effort limitations to address the challenge of the recovery of the cod stocks.

Since then, the effort system has been implemented in a very patchy way, as seen in Chapter 4 and many voices claim this management instrument should be abandoned. In this context, it is worth bearing in mind that the effort management systems in long-term recovery plans may have never existed if the control of the previously existing instruments had been adequate.

Financing the control of the CFP

Fisheries control is very expensive. The combination of an increasing number of regulations to control under the CFP with the increasing number of obligations for Member States has made the issue of how to finance the national fisheries control systems an important issue. Although control is a national competence, pressure has always been high from Member States to have financial support from the Union in order to facilitate the necessary investments in control infrastructure.

This infrastructure has two different aspects, worth distinguishing: (1) the infrastructure of national control administrations and (2) that which exists on board the vessels of private operators. The EU financing of that infrastructure has covered both. The Union has co-financed such equipment as patrol vessels and aerial control platforms operated by national administrations. It has also covered entirely the purchase of control equipment to be operated on fishing vessels such as satellite-tracking systems (vessel monitoring systems, or VMS) and the more recent electronic logbook (electronic reporting system, or ERS). This indicates that the CFP has considered the control infrastructure as a burden that would have to be co-financed by the Union budget, thus excluding any form of cost-recovery.

The financing of control means has been implemented by the Commission until 2013, when it was largely de-centralised in the context of the new financial instrument of the CFP.

The traditional support

The high cost of control has given rise, from the inception of the CFP, to claims for the Commission to provide financial aid to Member States. Since the beginning of the policy, the Commission recognised the need to support Member States' efforts to control the CFP, with a particular emphasis on the case of Ireland, a small Member State with a large EEZ to control vessels from different Member States (Farnell and Elles, 1984). Since then financial aid has been provided to Member States under different formats.

In recent years, the Commission has provided this financial aid under the so-called 'second financial instrument' (EC, 2006), including for a number of priority actions, such as the installation of electronic logbooks, with an extremely high co-financing rate of up to 90%

The use of these funds has been very variable and, particularly in recent years, many of the actions for which funding were obtained have not been executed. Part of the problem is the financial crisis affecting a number of Member States since 2008: when it came to reducing public spending, fisheries control has been a rather easy choice in some cases. This shows one clear difficulty of the control of the CFP: the low political priority being given to this question in national budgets, reflecting a general lack of understanding of the economic effects that poor enforcement may have on the overall economic performance of the sector.

The support under EMFF

Under the new EMFF (EU, 2014), the traditional aid to purchasing control means has been transferred into the shared-management system. In any case, however, the Commission keeps a part of this budget to finance control means (e.g. patrol vessels) to be used jointly by several Member States in the context of regional cooperation on control.

It is important to underline that the EMFF established a minimum amount of the EMFF funds that Member States should spend on control: at least a total of €580 million should be spent on control and enforcement purposes by Member States, that is, 10.1% of the total expenditure under shared management.

In addition, the structural policy under the new EMFF, for the first time, has considered as eligible the expenditure related to the reinforcement of the administrative capacity (including running costs) of Member States' control systems, in order to address the question of the low absorption and also to allow Member States to pass the test of *ex-ante* conditionality on control (see Chapter 16 on EMFF). This clearly shows that improving control has not necessarily been in all cases a high political priority by Member States.

It is meaningful to remember that this amount was actually agreed on insistence of the European Parliament, since Member States in Council preferred a lower figure, on account of the difficulties to spend the money under the traditional system, even though such system granted them a 90% co-financing rate from the Commission on the most important expenditure.

Is cost-recovery possible?

As cost-recovery we understand the (partial) funding of the control expenditure by the industry itself. This idea is already in application in a number of countries around the world. For example, already in 1999 the cost of fishery management (largely control) recovered from industry was 50% in New Zealand, 37% in Iceland and 23% in Australia, as part of cost-recovery schemes designed for efficiency improvement and the application of the 'user pays' principle applied in connection with rights-based management (OECD, 2006).

One can certainly question why the control equipment that is a pre-requisite for vessels to have the right to fish (for example, the vessel monitoring system) has to be paid for by the European taxpayer. And indeed the Commission considered this possibility as an option in the context notably of its proposal on rights-based management related to the fleet policy in the 2013 reform: if operators should pay to acquire fishing rights from other operators, part of this income could eventually be dedicated to improving control (or scientific data, for that matter).

The fact that the idea of rights-based management was turned down in that policy revision by both co-legislators did not provide the right context for this idea to make headway. However, as the levels of profitability and competitiveness of the fleets increase (see Chapter 6) the idea becomes more feasible and could certainly be part of future debate.

The new control regulation

Why a new instrument?

This regulation was adopted in 2009, prior to the last reform of the CFP. However, in a number of ways this regulation can be considered as an anticipation of the CFP reform, in that it introduced new mechanisms and approaches on control in line with the latest policy developments. It is important to rebuild the story of the need and opportunity of the new control regulation, as this new legal basis provides an opportunity to address the traditional weaknesses of the control system of the CFP.

Prior to the presentation of this regulation, the Commission undertook a diagnostic of the weaknesses of the system (COM, 2008). The result of this diagnostic can be summarised as follows:

- An inadequate legal basis. The legal framework was extremely fragmented, with different elements scattered in several regulations. Moreover, the legislative development was incomplete, with only 4 out of 20 possible implementing regulations having been actually adopted;
- Very insufficient control of TACs and effort quotas, particularly due to the lack of instruments and mechanisms to cross-check data on catches and landings with those of transport, marketing and so on;
- Absence of general control standards, thus giving rise to different standards in different areas and fisheries;
- Lack of a level playing field, in particular the lack of harmonisation of sanctions among Member States;
- Inadequate strategic programming of the control activity, also giving rise to control priorities often disconnected from the risk areas;
- Absence of effective powers for the Commission to act quickly in case of clear non-compliance by Member States.

These questions were addressed in the new control regulation, Reg. 1224/2009 (EC, 2009), soon complemented by its implementing regulation, Reg. 404/2011 (EU, 2011).

Regulation 1224/2009

Regulation 1224/2009 made a series of fundamental contributions to the EU fishery control system:

- It reunited all provisions on control into one legal instrument and it completed all the necessary implementing rules just 2 years later, through regulation 404/2011. It provided a clear, complete and unified legal basis for the control policy; something much needed for many years;
- It enlarged the field of application of the control activity, including such questions as transport, markets, traceability of the catches; thus providing for instruments to control the fishing activity 'from net to plate';
- It filled important gaps in the previous legislation that were used to undermine the efficiency of the control activity;
- It introduced the notion of programming of control, based on the notion of risk assessment;
- It created the conditions for a level playing field, notably by establishing harmonised inspection procedures;
- It created new instruments to enhance the control capability of the system, for example by introducing new obligations concerning the control of catches, such as the introduction of the electronic logbook or to improve the measurement of engine power;
- It contributed to alleviate the administrative burden by expanding the use of electronic means, replacing paper-based systems;
- Last but not least, it provided the Commission with a much wider choice of mechanisms to act against Member States that do not fulfil their control obligations. These instruments are both punitive (for example, financial sanctions) and cooperative (for example, action plans).

This new approach also has consequences for the Commission: as control by Member States become more complete and systematic, the role of the Commission evolves towards a more systematic evaluation of Member States' performance on control, thus changing a more traditional piecemeal approach to a more systematic audit of Member States' systems and practices.

The implementing regulation 404/2011 completed, for the first time, all the provisions of the basic control rules with their corresponding implementing rules, thus providing for the first time a complete set of primary and secondary rules. At present, however, the implementing rules are being adapted to the Treaty of Lisbon.

The new control regulation, together with the consolidated work of the control agency (see next), gives the EU an ideal legal framework to address all the control weaknesses identified by the Court of

Auditors in 2007. In addition, since 2014 it is accompanied by additional measures to ensure adequate enforcement of the CFP: the *ex-ante* conditionality and the system of interruptions/suspensions of structural funds laid down in EMFF (see Chapter 6).

For the time being, it would be premature in these pages to draw clear conclusions on implementation, as many provisions of the new regulations actually entered into force on a step-by-step basis in 2010, 2011 or 2012.

Future perspectives

In 2015 the Commission should elaborate a report (as required by the regulation) on its implementation. Despite all the important contributions of the new control regulation, the timing of its adoption (2009), coinciding with the launching of the exercise on the last reform of the CFP (see Chapter 16) inevitably leads to an important question: how does this regulation compare with the reformed CFP?

Regulation 1224/2009 is the last important regulation of the CFP adopted just prior to the new emphasis of the reform on de-centralised, simplified, results-based legislation. The control regulation, in this regard, is the last of the 'old fashioned' regulations, in that it is still highly detailed, prescriptive and not adapted to regionalisation.

In the context of the new CFP adopted in 2013, where many rules will be de-centralised, adapted to specific sea basins, it is inevitable to ask the question of whether these regionalised approaches would require regionalised, sea basin-specific control schemes.

Already today there is a clear tendency to favour certain control instruments over others on a geographical basis. The best example by far is the increased use of CCTV cameras on board, favoured by Member States in certain fisheries in certain regions but fiercely opposed in others. This control and monitoring methodology is very promising to deliver fully-documented fisheries but it is not adapted to all fisheries (in very multi-specific fisheries it is very difficult to distinguish what species come in the catch and are retained or discarded). But every attempt to use this technology meets resistance in Council as many Member States think it can be used as a precedent for their own fisheries. This fear derives clearly from the tradition of legislating horizontally the control of all fisheries, where all control instruments are compulsory in all areas.

The question is therefore whether it would be possible to consider control obligations that could differ for different regions and fisheries, so as to also adapt the control itself to the regional characteristics of fisheries. This possibility, of course, is delicate, because Member States are (logically) extremely adamant to ensure a 'level playing field' among EU fishermen of different regions.

If such level playing field is also a fundamental principle in other areas (such as the conservation policy) and this has not prevented the evolution of the policy towards regionalised approaches, one can certainly wonder whether the future control policy could not also be based on a more regionalised, results-based approach where control instruments would also be adapted to local conditions, in a context of harmonised objectives and standards.

It is to be noted that Regulation 1224/2009 was adopted just before the entry into force of the Treaty of Lisbon. It was, from that point of view, the last of the significant 'old regime' regulations to be adopted. This is important because this regulation, although somehow considered part of the overall package of reforms of 2013, does not take on board the orientations adopted during that reform process, notably the ideas of simplification, regionalisation, results-based management and so on. This may produce a gap between the new instruments of the CFP after 2013, notably the regional approaches to management and the more classic, horizontal structure of the control regulation.

All these considerations led to the Commission to include this regulation under the 'Refit' project starting in 2015, whereby a number of regulations will be evaluated to see if they are 'fit for purpose'.

The international dimension of the control of fisheries

The control of fisheries is not only a domestic issue of the CFP. On the contrary, the role of the Union and its ability to control its fleets in international fisheries has often been problematic and always an important 'reputational issue' for the EU. Alleged or real shortcomings in controlling EU fleets in international fisheries have given rise to important conflicts. Any weakness of the Union in ensuring strict adherence of EU vessels to international rules regularly undermines the negotiating position of the EU in international *fora* on fisheries management.

In Chapter 7 we referred to the conflict between the Union and Canada over the fishery for Greenland halibut in the NAFO Regulatory Area. It is important to understand that any weakness in our control of the European vessels in this fishery was exploited skilfully by Canada in their media campaigns against 'foreign overfishing'. In Chapter 8 we also touched upon the control of the Bluefin tuna fishery in the Mediterranean as a fundamental question of credibility of the Union in international *fora*. These are just examples of how the ability of the Union to control its fisheries has an important bearing on the Union's credibility and on its ability to negotiate internationally.

For that reason, it is not surprising that the international control schemes associated to the activity of EU vessels in international waters have enjoyed a high priority from the beginning of the CFP and have contributed substantially to establish precedents on the use of new control instruments.

The best example of this took place even before the CFP as such was established: in the original Joint Enforcement Scheme established by the old ICNAF in the 1970s (see Chapter 2) for the fishery off Newfoundland, contracting parties, including European countries that later on became members of the EU, accepted a system of mutual inspection of their fishing vessels at sea as the basis for taking judicial proceedings against any of their vessels which had broken the agreed rules (Farnell and Elles, 1984).

This precedent, however, was not followed in the control system in European waters, showing that Member States are ready to go much farther in pooling national sovereignty when engaged in long-distant fisheries in international waters where it would be too expensive to have full inspection coverage with their own means alone.

Beyond the development of the control part of the CFP within Europe, the control policy has been considerably shaped by multilateral developments, as we saw in Chapter 7. The participation of the Union (and the activity of European fishing vessels) in certain Regional Fishery Organisations has led to the development of specific control schemes that in some cases go beyond the basic requirements of control in EU waters. These schemes have been sometimes the result of very contentious conflict over the fishing activity in certain areas, and the reinforcement of control measures has played a key role in resolving the conflict. This is clearly the case of the control scheme in NAFO, developed in its current form largely as a result of the Greenland halibut conflict in 1995 (see Chapter 7). In such cases, reinforced control is the best way to re-establish the lost trust and cooperation among the parties concerned.

In other cases, however, reinforced control has been a key factor in the cooperation for the management of certain shared stocks, where the difficulties in agreeing to the allocation of fishing rights of key stocks has led to the adoption of very tight control mechanisms. These have been instrumental in building trust among the parties concerned that, after difficult negotiations on the allocation of catches, all parties will respect their allocations. This has led, for example, to a very sophisticated scheme of control in the fisheries regulated by NEAFC (EU, 2010).

NAFO

The control scheme in NAFO played an important role in defining the control policy of the Union as of 1983. We saw in Chapter 2 that ensuring the respect of quotas allocated was part of the political deal leading up to the agreement on the CFP in 1983. In fact, this agreement also comprised the need to ensure adequate control of the catch quotas agreed in NAFO, as part of the tendency of the EEC of the time of taking responsibility for the international obligations of the emerging CFP. This resulted in the NAFO control scheme being incorporated into EU law (EEC, 1983).

This initial scheme was considerably reinforced following the Greenland halibut conflict of 1995 (see Chapter 7). Every year, the effectiveness of the scheme is evaluated by a dedicated body (the standing Committee on International Compliance, STACTIC) and new measures are introduced regularly, such as the right to de-route vessels case of serious infringements (introduced in 2006) or the additional reporting requirements for the Greenland halibut fishery introduced in 2007. In addition, the participation of the Union in the control of this fishery is coordinated by the European fisheries Control Agency (EFCA, see next), including in the past the deployment of a dedicated patrol vessel.

NEAFC

This scheme, developed gradually over the years, and still subject to adjustments today, is a complete set of rules and obligations, including reporting of catches, communication, inspections at sea, obligations for the vessel's master, notifications and so on. Meaningfully, it also includes provisions on the practical implementation of FAO's Port State Agreement (still not in force, see Chapter 7) in a demonstration of the forward-looking and ambitious approach agreed in the context of NEAFC on control matters.

Adherence to these schemes enjoys a high priority for the Union, since the continuity of the activity of European fleets in distant waters, and in general terms the credibility of the Union in international negotiations are largely dependent upon the Union's fulfilment of its obligations in these schemes. As a good example, the annual work programme of the European Fishery Control Agency (EFCA, see the following) the coverage of the control schemes of NAFO, NEAFC and ICCAT (for Bluefin tuna) enjoys permanent priority, while certain fisheries in European waters are still today not covered in practice by the work of the agency.

Other cases

In Chapter 8 we referred to the control system applicable to the Bluefin tuna fishery. All these examples show that the international obligations on control matters are taken by the Union as a matter of priority. The three plans, NAFO, NEAFC and Bluefin tuna are permanent priorities for the Union, and this is clearly reflected in the annual programme of the EFCA.

Other examples of very strict control schemes is that of the fishery for tuna in the eastern Pacific, regulated by IATTC (see Chapter 7), where the implementation of the parallel agreement to protect dolphins (the AIDCP)[3] led to a system where all vessels of more than 400 GT must have an inspector on board.

Control and the international 'level playing field': the IUU regulation

Control in a global context

When we talk about control, we must bear in mind that the European Union imports around two thirds of the fish we consume. So the question is what about the fish we import? Is it sustainable? Does it come from sources that respect fishery management rules? Thus the question that many industry representatives and Member States refer to as the 'level playing field': that the fish we import in the European market be subject to the same standards on sustainability that are applied to European fishermen.

One of the main traditional difficulties of the CFP is that any effort to raise the standards of European fishing in conservation, environmental protection, enforcement and so on automatically encounters the argument of the 'level playing field': the CFP demands from European fishermen very high standards that fishermen from abroad do not necessarily respect and, yet, they have easy access to European fish markets.

This must be put into perspective, since a high proportion of fish imports into the EU market come from developed countries (Norway, Canada etc.) whose standards of conservation and control are certainly not lower than those imposed by the CFP to European fishermen. The problem is that of imports from countries that have significantly lower standards in the way they control their vessels and/or their waters.

The problem of IUU fishing is very serious around the world. It has been estimated that the amount of fish illegally caught would be between 11 and 26 million tonnes, worth between US$ 10 and 23 million.[4] This means that beyond the level playing field the problem of IUU fishing can seriously undermine the chances of developing States to develop their own legitimate activity. In addition, some IUU activities involve European operators.

For all these reasons, the Union has a clear responsibility to act against IUU fishing on a global scale. However, the awareness about the need to take an active role against IUU fishing is relatively recent. In the Communication on a new fisheries policy in 2001, the IUU was still absent from the analysis and the objectives.

Such awareness increased in the mid-2000s, largely as a result of the growing evidence about the imports to the EU market of fish obtained through IUU activities, representing a clear unfair competition to legitimate EU operators. It is also true that international campaigns, some of the carried out by certain NGOs, contributed also substantially to raise awareness about this problem.

The precedents

The idea of using trade measures to enforce fisheries measures is not new. Before the adoption of the IUU regulation, measures had indeed been taken in this regard, both at unilateral and multilateral level.

At unilateral level, the best example is the development of legislation by the United States: the 'dolphin safe' legislation that we will refer to in Chapter 11.

At multilateral level, significant precedents are those related to the implementation of the United Nations resolution on the banning of Large Scale Driftnets. This resulted in certain initiatives, largely conducted by the United States, for its practical implementation, such as the Driftnet Impact, Monitoring Assessment and Control Act of 1989 (Long and Curran, 2000).

At EU level the first precedents applied were the multilateral measures agreed by ICCAT as of the 1990s: the import ban of Bluefin tuna from Panama, Belize and Honduras adopted in 1998, following an ICCAT resolution of 1997 (EC, 1998). Other similar measures have been applied for other stocks and countries since then. These measures had the fundamental merit of being multilateral and, by their own nature, much more difficult to challenge in the framework of the World Trade Organisation.

The problem of IUU fishing has been growing in recent decades. This worrying development, which hurts developed and developing States alike, led FAO to elaborate and adopt already in 2001 an Action Plan against IUU fishing,[5] which identified this question as an essential tool in the quest for sustainable fisheries in the world. This plan established clear responsibilities by all States, flag State responsibilities and port State measures. It also recognised as an instrument the adoption of 'internationally-agreed market-related measures' in accordance with the international trade rules under the World Trade Organisation.

It has been extremely difficult to turn the FAO plan of action into legally binding, internationally agreed rules, the only tangible result being the lists of IUU vessels elaborated by a number of RFMOs. One of the reasons for this is the reticence expressed by a number of developing countries that often see the trade measures as a potential instrument by developed nations with large fish markets to apply protectionist measures on account of presumed IUU activity. Furthermore, developing States realised that to avoid such measures they should invest in effective fisheries management including effective control of their vessels and their waters, something that is not always easy to deliver.

The 2009 Agreement on port State measures to prevent, deter and eliminate illegal fishing is the first attempt to adopt a legally-binding global instrument focusing specifically on combatting IUU fishing. This Agreement requires port States to prevent entry into, or use of, ports by vessels that have engaged in IUU fishing. The Agreement, signed in 2009, has not yet entered into force, due to the low level of ratifications so far (only 11 out the 25 necessary).[6] The European Union ratified this instrument on 7 July 2011. It is noteworthy that since such agreement concerns fishery measures and possible trade restrictions, both questions exclusive competence of the Union, the ratification by the European Union of this agreement represented only one vote.

Although not in force, port State measures are implemented in some cases. For example, the NEAFC control and enforcement system already applies port State measures as laid down in the text of the Agreement.[7]

In these circumstances, the fight against IUU fishing around the world is still primarily the result of unilateral action by a few trading blocs in the world. In this context, the EU is leading the way with an IUU system that is, beyond doubt, the most advanced in the world.

The IUU Regulation of 2010

The EU is the largest import market of fish in the world. This gives the EU a special responsibility in the global fight against IUU fishing: if the EU does not take measures, whatever other countries would do would be severely undermined. This led the Commission to take the initiative and present a communication establishing a strategy to prevent, deter and eliminate IUU fishing. The strategy proposed a number of multilateral actions as well as cooperative instruments but, specially, it proposed to: 'Entitle the Community to act unilaterally in order to identify and establish black lists of states hosting flags of non-compliance and vessels responsible for IUU fishing, and enact trade measures towards them' (COM, 2007).

This led to the adoption of Regulation 1005/2008 (EC, 2008). This regulation establishes the basis for the Union to identify a third country as 'non-cooperating' and to eventually apply trade restrictions to their fishery products. Likewise, it allows the identification of a 'black list' of IUU vessels that cannot land their fish in EU ports.

The regulation has three main motivations:

1 Illegal fishing is an unfair competition to European fleets that have to be subject to a very demanding control regime.
2 Illegal fishing seriously undermines the chances of developing States to develop their own local industries. IUU activity does not benefit developing countries but rather those who abuse their weaknesses.
3 In a global fisheries governance where, as we saw in Chapter 7, both flag and coastal States adamantly exert their rights, while some are less enthusiastic about fulfilling their obligations, the fight against illegal fishing is an important contribution to re-balance the situation.

The objective of this regulation is not to apply trade sanctions. In fact, the ideal outcome of the implementation of this regulation would be that the third countries concerned would improve their behaviour in terms of ensuring the respect on international law and fisheries management rules by the vessels flying their flag, so as to make trade sanctions unnecessary. The legal instrument should act as a deterrent.

Likewise, the philosophy of this regulation is not protectionist. The emphasis is not in establishing difficulties to importing fish from third countries, or to obtain a competitive advantage to EU operators. For this reason, the IUU regulation considers the possibility of identifying as IUU vessels flying the flag of EU Member States.

The regulation imposed that all fish entering the EU market should be accompanied by a 'catch certificate' where the flag State should certify that the fish imported comes from legal sources. The purpose of the instrument is to make the exporting country responsible of the certification. The catch certificate is a condition *sine qua non* to import fish into the EU market but it is not a guarantee that the products originate from legal sources.

The definition of IUU activity needs a clarification. Contrary to certain views according to which a vessel committing a serious infringement is automatically an IUU vessel, the definition in the EU regulation indicates that the IUU activity is that where the infringements are serious, repeated and, specially, are not punished by either the flag or the coastal State. This is important, because the IUU regulation does not try to replace the primary responsibility of flag States to discipline their vessels, or the coastal State responsibility to control its waters. What it does is to intervene when these two mechanisms do not work. Then and only then the IUU instrument comes into play.

The adoption of the regulation by Council was not easy. Two difficulties had to be overcome: the opposition of certain EU interests (part of the processing industry saw this regulation as an obstacle for their supply of raw materials) and the allegation that it would be burdensome for administration and the industry alike. In both cases the allegations have proved to be wrong: the implementation of the IUU has not undermined the supply of (legal) raw materials for the European industry and the presumed administrative burden is not generally seen as a problem in most cases.

Implementation

The Commission services have worked intensively since the entry into force of this regulation and have investigated abundant evidence on IUU activities coming from different sources: governments, NGOs, and so on. The results of these investigations can be classified as follows:

- Many vessels (both from the EU and third countries) were investigated. However, the need to ensure full legal guarantees (including a contradictory procedure with the flag State and the vessel owner) was so complex that the Commission decided not to pursue the implementation of this instrument, as having an excessively high cost/efficiency ratio. A lesson learned after very intensive work.
- Among the countries investigated, under a bilateral and confidential dialogue, a number of them reacted positively to the Commission's indications on the weaknesses to address as coastal and/or flag States. For these countries, the Commission was satisfied with the progress achieved and did not pursue the case further. These are the real success stories of the IUU regulation: not the countries pre-listed or listed as non-cooperative but on the contrary the countries that were not listed because they adequately addressed their responsibilities.
- Other countries, on the contrary, did not adequately address their weaknesses and were subject to the two-tier system: a pre-identification by the Commission (the so-called 'yellow card' in football terms) that is, the warning that a full identification can occur if the weaknesses are not addressed and, for those that did not react properly to the warning, the identification as non-cooperating countries, with subsequent trade measures imposed by Council (the so-called 'red card').
- Other important elements of the implementation have been the training organised by the Commission in third countries in terms of the use of the catch certificate. This is in more recent times being complemented with 'capacity building' actions with African developing countries, to help them obtain the know-how necessary to properly exert their responsibilities as coastal and flag States. This educational part of the IUU regulation is very important to ensure that the instrument is not just punitive: it is also an instrument of cooperation to help developing countries discharge their obligations as coastal and flag States under UNCLOS and other relevant instruments of international law.
- The international cooperation is also an important angle to the IUU policy. Today, there is no multilateral IUU policy, beyond the identification of IUU vessels by RFMOs. The IUU regulation remains a bilateral instrument between the EU and the third country concerned. And so is the US IUU policy. However, the more the IUU policy is multilateral, the more effective it will be. This has led the Union to promote cooperation with the other large fish markets in the world (the USA and Japan) to promote cooperation on IUU matters, through the signature of a Memorandum of Understanding.

The first deliverables of this regulation were adopted in 2012, when the Commission pre-identified eight countries as possible IUU, giving them 6 months to react before they would be proposed for identification as IUU countries (Commission Decision, 2012). Of these eight countries, three were actually identified as non-cooperating and the Commission proposed for them the adoption, by Council, of the trade sanctions foreseen in the IUU regulation (Council Implementing Decision, 2013). These sanctions were subsequently adopted by Council for Belize, Cambodia and Guinea Conakry (Council Implementing Decision, 2014a).

More countries have been pre-identified, such as the Philippines and Papua New Guinea (Commission Decision (2014), and others are in the pipeline.

However, much more important than the sanctions themselves, is the change of behaviour in the countries concerned as a result of the Commission's investigation in the IUU regulation framework. This clearly shows that the ultimate objective of this regulation is not to curtail trade but to promote change in the behaviour of flag States in the direction of exerting their responsibilities as flag states. There are abundant proofs of this:

- The list of three identified countries has been quickly reduced to two. Belize reacted positively to the black-listing and quickly adopted management measures that justified its de-listing relatively soon afterwards (Council Implementing Decision, 2014b).
- Of the eight initially pre-identified countries, five were de-listed also from the pre-identification, following the adoption of national measures to better control their vessels
- Last but not least, many countries investigated by the Commission services have never been even pre-identified, as they reacted to the investigation phase by adopting measures in the right direction. This was probably the best success of the IUU regulation, its effect of promoting the improvement of behaviour by a number of countries in terms of their management of the fishing fleets flying their flag.

More recent experiences are also noteworthy. Two important fishing countries, Korea and Philippines, were pre-identified in 2014. However, the awareness of the threat of a possible subsequent identification caused an extremely positive reaction in the two countries, which developed new legislative instruments and invested in establishing an effective mechanism to control their vessels. Following these developments, the two countries were de-listed in early 2015 (EC, 2008, 2015).

These two cases are the best demonstration of the positive effects of the IUU regulation: even without trade sanctions, the pre-identification acted as a very strong political incentive for the countries concerned to substantially improve the control and enforcement of fishery rules by their vessels around the world. The actual trade restrictions are just the exceptions, not the rule, and these cases set the tone of what the fight against IUU fishing should be all about: a system to deter this kind of activity through cooperation, with trade sanctions only as a last resort possibility where everything else has failed.

Another angle to the implementation is the action by Member States to effectively control their borders to detect the possible import of fish without adequate guarantees. In this regard, implementation is still patchy and can be improved. Curiously, among the Member States that take such implementation more seriously there are some land-locked ones.

Overall, the implementation of the IUU regulation so far has been a spectacular success in terms of the improvement in fisheries management and control by a number of countries around the world, and this success in some cases goes even beyond fisheries.

Although the IUU is and will likely remain an instrument related to fishery management, the experience of its implementation shows that it can also provide indirect benefits on other areas. The best example is the case of Indonesia, a country that has not been pre-listed by the Commission on account of their excellent cooperation to resolve their IUU problems and the practical steps they have taken in that direction. Those steps have also involved the dismantling of IUU vessels where slave labour was being used, so the IUU regulation has indirectly contributed to successfully address a serious problem of human rights.

Vessel lists

All the previously discussed implementation refers to the identification of countries. However, the regulation also contemplates the possible listing of vessels. This possibility has been implemented only partially:

- In the case of IUU vessels identified by Regional Fisheries Management Organisations (RFMOs), the Commission publishes regularly an updated list of such vessels that cannot land their fish in European ports. Since the vessels concerned are identified by multilateral bodies, the establishment of trade restrictions to their catches is very straightforward.
- On the contrary, for the vessels identified unilaterally by the Commission, a system of legal guarantees is necessary before any listing can take place, including the opportunity for the vessel owner and the flag State to present allegations to the case.

Overall, the procedure is very time consuming and this led the Commission to consider that it would be much more cost-efficient to concentrate only on countries, in addition to the RFMO-identified vessels.

Possible future developments

The future fight against illegal fishing will have to explore a number of synergies and beyond the positive results of the implementation of the 2010

IUU regulation, there is little doubt that the fight against illegal fishing will remain a fundamental pillar in the future global governance of fisheries. And one where the European Union, as the World's Number 1 import market of fish, will undoubtedly have a key role to play. Some of these developments may be:

- Developing an electronic catch certificate to replace the current paper document. This should allow for example, cross-checking catch certificates between countries, thus preventing current practices of using copies of the same certificate to sell fish in different markets.
- Exploiting synergies with other EU policies, such as those for the protection of consumers, notably in terms of the IT instruments.
- Developing further the 'multilateralisation' of the fight against IUU. Beyond the bilateral cooperation established or to be developed with like-minded States, the preparation of the global catch certificate to be prepared in FAO will be a very good opportunity. Furthermore, the ratification (requiring only 25 countries) of the Port State Agreement would be an important step towards recognition of trade restrictions as an instrument to fight against illegal fishing.
- Enhance capacity building in developing States. It is important to underline that the main victim of IUU fishing is the development of a healthy, local fishing sector in these countries. Capacity building means helping these States to fulfil their obligations under UNCLOS as coastal and flag States so as to avoid being listed. But it also means helping them develop a genuine fishing industry with real economic links with the country. Something that is often undermined by the IUU activity.
- Continuing to ensure that EU Member States fulfil their obligations as coastal and flag States as according to EU legislation. The legitimacy of the IUU action against third countries is based on the EU fulfilment of its control obligations.

Overall, with the 2010 regulation, the fight against illegal fishing has only started. As we saw in Chapter 7, the world's fisheries governance is characterised by a balance between rights and obligations but while the rights are exerted by all, the obligations are not necessarily fulfilled by everybody. In that context, a reinforcement of the IUU policy could be an essential contribution to redress that situation.

Towards integration of the control of fisheries

One of the key factors to improve the control of fisheries in Europe and to improve its cost-effectiveness is to develop all possible synergies among policy pillars, among Member States and with other initiatives that touch upon the control of marine activities. There are three major areas under development in that regard: the coordination of Member States' control efforts under the European Fisheries Control Agency (EFCA), the on-going project to develop synergies among Europe's different maritime surveillance systems: the Common Information Sharing Environment (CISE) for maritime surveillance in Europe and the introduction in the new structural policy of conditionality conditions: public money will not be given to offenders.

European integration: EFCA

The need to improve coordination in control was clearly identified in the context of the CFP Reform of 2002. This materialised sometime later in the decision to establish the European Fisheries Control Agency (EFCA), which set up its headquarters in Vigo, Spain, as decided in a European Summit in 2004.

The mandate of EFCA is: 'to organise operational coordination of fisheries control and inspection activities by Member States and to assist them to cooperate so as to comply with the rules of the Common Fisheries Policy...' (EC, 2005).

The Agency does not have power to fine individual operators, or to sanction Member States. Its role is to coordinate the control action by Member States, by providing a mechanism where significant gains of cost-efficiency can be achieved.

At the moment of establishing the Agency, a fundamental debate took place concerning the powers to be given to the new agency. Ideas such as delegating all control functions from the Commission to the Agency were floated. And so was the possibility that it would have inspectors with an entitlement to inspect directly the fishing activities of the Member States. This would have provided a boost to the level playing field in control and the sense of fairness. However, most Member States opposed this idea, theoretically on legal grounds: a number of Member States alleged that in their national legal systems only authorised national agents from their own country could impose sanctions on their own fishermen.

However, the idea that inspectors from a third country (even neutral EU agents) would have the right to initiate an infringement procedure that their own judiciary would have the obligation to process, was also a politically unpalatable idea for many Member States, who mistrusted other Member States' inspectors and considered that this competence is a *regalian* one that cannot possibly be surrendered.

Of course, inspectors from any Member States have the right to impose sanctions against all fishermen operating in their EEZ, as part of the jurisdiction regime of these areas. But any possibility that neutral, Union agents would have the right to initiate sanctioning procedures was completely ruled out. This was somehow a lost opportunity, because EU agents could have provided reassurance to Member States of a real level-playing field in control: those who believe they control their nationals much more than certain neighbours would be reassured that their own neighbours would also have to accept infractions triggered by neutral agents, not just their national agents. And those who believe they are discriminated against by other Member States' inspection systems should be reassured if the control was exerted by neutral not national agents from the coastal states where they fish.

Finally, the role of the Agency is not to control directly but to coordinate the inspection activity of national inspectors, the only ones having the right to initiate sanction proceedings to fishermen. In this context, the main contribution of EFCA is to substantially increase the cost-efficiency of the control, mainly by pooling together the inspection platforms and coordinating methodologies and priorities.

Beyond cost-efficiency and despite these considerations on the lack of EU agents as inspectors, the coordination of inspections and pooling of means does contribute to improve the level playing field. This element that was crucially missing for years in the CFP is a clear success of EFCA: by implementing the Joint Deployment Plans (JDPs), where inspectors from different Member States share inspection platforms and agree on risk analysis and objectives, nobody can argue that control in a given area is discriminatory or is carried out under different standards. However, this important contribution to a level playing field must then be complemented by Member States in terms of ensuring adequate follow up of infringements, an area outside the scope of EFCA.

The best example of the gains in cost/efficiency are shown by the implementation of certain JDPs, whereby concerned Member States pool the inspection platform so that they can cover the same number of fishing operations under a much lower cost. Today, EFCA runs a number of JDPs and, over time, it should cover all main EU fisheries.

It is also worth mentioning the educational work of EFCA, notably though the development of the *core curriculum*, a tutorial for fisheries managers around Europe, as a fundamental text book for training of inspectors. In addition, EFCA also plays a key role in training of inspectors in Europe, as well as abroad, including capacity-building initiatives in certain African countries as of 2013.

Another important contribution of EFCA is the provision of intelligence on what is happening in the fishing grounds. EFCA's periodic reports are gradually including more information (not broken down by Member States, to avoid any 'naming and shaming' that is clearly outside the Agency's remit) on the number of infractions of various kinds, thus providing crucial information on how the legislation is being implemented, what works and what does not work, what is respected and what is not. This intelligence is crucial for the Commission to know what the weaknesses in the control of the CFP are, and even what the problems of the legislation itself are, as a precondition to propose corrective action.

Last but not least, EFCA has significantly contributed to develop a culture of cooperation among Member States. While the control authorities have a long tradition of mistrust between different Member States, the operation of the JDPs has clearly contributed to build trust and cooperation among them and that, alone, is in itself a considerable contribution to the improvement of control but also to the establishment of a cooperative culture that is such an important part of the regionalised approached favoured by the last reform of the CFP in 2013.

The role of EFCA is complementary to that of the Commission. While EFCA ensures coordination of the inspection activity, it does not participate in the evaluation of compliance by individual Member States, which is an exclusive responsibility of the Commission. While EFCA evaluates compliance on a generic basis, allowing the EU institutions to identify problems, it cannot get involved in 'naming and shaming' of Member States. That would not be compatible with its core task of promoting cooperation among them.

Overall, EFCA represents an important step forward in the establishment of a long-established objective of the CFP: the creation of a 'culture of compliance' in the fisheries of the Union. While much is still to be done, it seems clear that the fisheries where control is coordinated under the JDPs have much higher standards of quality and efficiency.

Sectorial integration: the opportunity of CISE

The establishment of EFCA is by no means the end of the process of integration of control means. In the context of the newly developed Integrated Maritime Policy (IMP) the Commission identified clearly the need to streamline and increase the cost/efficiency of the different systems of maritime surveillance existing in the Union. Not only different Member States have different maritime surveillance mechanisms, often with a very low level of coordination. Even within Member States, different maritime surveillance systems exist, having a total or partial responsibility on fisheries inspection, or having the potential for that and yet little has been done historically to coordinate them, develop synergies among them and, in short, increase their efficiency/cost ratio. This is the objective of the project Common Integrated Surveillance Environment (CISE), an idea that the Commission launched in 2009 (COM, 2009b).

The CISE idea underlines that Europe has a multitude of maritime surveillance systems. Over and beyond fisheries, the fight against crime, illegal immigration, smuggling, piracy and other objectives is carried out by many surveillance bodies, some military, other civil, with their own means, their own legal base, their own protocols for the use of their means and their information. These different surveillance systems are not necessarily coordinated. Indeed, coordination among them is in a number of cases very weak even among different services of the same Member State. Let alone among bodies of different Member States.

Creating an EU body (an EU Coast Guard) is an attractive idea that has been floated by many people, it is certainly a far-reaching one that may materialise one day. In that context, what can be done to improve cost-efficiency is to create a system where the information generated by all these surveillance systems can be shared and used by all, thus producing an important gain in productivity as well as important potential savings for public administrations in all Member States.

To put this idea in practice, the Commission proposed a roadmap which provides guidance on the different steps to achieve that goal (COM, 2010). Progress in this roadmap is slow and difficult, given the complexity and variety of the systems and the legal difficulties along the way. However, there seems to be no doubt about the potential to increase cost/efficiency among European maritime surveillance systems in this initiative.

Policy integration – control and structural policy: conditionality

The idea of using financial instruments to improve control by Member States is not new. Withholding EU funds to those Member States that clearly do not fulfil their obligations under the CFP was in fact an instrument included in the control regulation of 2009 (Article 103 of Council Regulation 1224/2009, see previously). But the concerned provision concerned was never applied. Part of the problem was that this provision was introduced after the European Fisheries Fund was adopted, for the period 2007–2013. By the time these provisions entered into force, the national operational programmes under the EFF were already being implemented and any suspension of funds would have been legally difficult to justify and politically delicate to put into force.

In this context, the adoption of the new structural fund for fisheries, for the period 2014–2020, (the EMFF regulation 508/2014, see Chapter 16), provided an opportunity to reinforce such mechanisms. In this case, the relationship between compliance and access to public funds will be constructed according to three conditionality mechanisms.

Ex-ante conditionality (Article 9 and Annex IV of EMFF)

The idea of the ex-ante conditionality is that, in the framework of the negotiation of the Operational Programme under the EMFF, Member States should demonstrate that they possess the necessary administrative capacity to ensure adequate control of the CFP. If they fail to demonstrate this, then the availability of the EMFF funds will require the commitment to develop a programme to address and resolve the identified weaknesses by 2016. If the Member State concerned fails to implement this plan, then financial suspensions can take place.

This mechanism has the advantage that it can be used in a cooperative way, not just a punitive one: by requiring good administrative capacity to control the CFP, the Commission can impose the obligation to address the weaknesses without necessarily suspending the access to the funds. This is also aided by the fact that, as we will see in Chapter 16, the eligible actions to improve control do actually include even running costs, so Member States have no excuse whatsoever if they cannot deliver on their control obligations.

Conditionality on compliance with policy objectives (Article 102 of EMFF)

This mechanism allows the Commission to interrupt or even to suspend payments in case that a Member States has failed to enforce 'the CFP rules that are essential to the conservation of marine biological resources'. The mechanism is to be used in two phases: suspension, followed by time for Member States to make refute the allegations and eventual suspension if the Member State has not justified its failure to comply. This mechanism is very important in terms of compliance within the EU. If properly applied, it could become a very strong incentive to promote better compliance with the rules of the CFP. A delegated act (in preparation) will lay down the specific rules to implement this mechanism.

Admissibility of applications for structural aid (Article 10 of EMFF)

This is a mechanism that applies to individual beneficiaries. In effect, beneficiaries having committed serious infringements to the rules of the CFP are not admissible for aid. This is a very powerful instrument to promote good compliance. The specific rules to implement this provision will be laid down in a delegated act.

The EMFF is in fact the only one of the five structural funds that will have such specific mechanisms to promote compliance with the policy, over and beyond the general ex-ante conditionality rules applicable horizontally to all five funds: the macroeconomic conditionality as well as those on the respect of public procurement rules and so on. These instruments, together with all those referred previously in the new control regulation provide an extremely powerful set of instruments to allow for a high level of compliance with the CFP.

Notes

1 Special report no. 7/2007 on the control, inspection and sanction systems relating to the rules on conservation of Community fisheries resources together with Commission replies.
2 The European Court of Auditors' Special Report No 7/2007 on the control, inspection and sanction systems relating to the rules on conservation of Community fisheries resources. ECA/07/35. Brussels, 4 December 2007.
3 Agreement on the International Dolphin Conservation Program, entered into force in 1999 and ratified by the EU.
4 www.mrag.co.uk/services/illegal-unreported-and-unregulated-fishing.
5 www.fao.org/fishery/ipoa-iuu/en.
6 www.fao.org/fishery/psm/en.
7 www.neafc.org/scheme.

References

COM (1986). Report by the Commission to the Council on the enforcement of the Common Fisheries Policy. COM (86)301.

COM (2001). Green Paper on the future of the Common Fisheries Policy. Brussels, 20.3.2001 COM (2001) 135 final.

COM (2007). Communication from the Commission to the European Parliament, the Council, the European Economic and Social Committee and the Committee of the Regions. On a new strategy for the Community to prevent, deter and eliminate illegal, unreported and unregulated fishing. Brussels, 17.10.2007. COM(2007) 601 final.

COM (2008). Communication from the Commission to the European Parliament and the Council on the proposal for a Council Regulation establishing a Community control system for ensuring compliance with the rules of the Common Fisheries Policy. Brussels, 14,11,2008. COM(2008) 718 final.

COM (2009a). Green Paper. Reform of the Common Fisheries Policy. Brussels, 22.4.2009. COM (2009) 163 final.

COM (2009b). Communication from the Commission to the Council and the European Parliament, the European Economic and Social Committee and the Committee of the Regions. Towards integration of maritime surveillance: a common information sharing environment for the EU maritime domain. Brussels, 15.10.2009, COM (2009) 538 final.

COM (2010). Communication from the Commission to the Council and the European Parliament on a draft roadmap towards establishing the common information sharing environment for the surveillance of the EU maritime domain. Brussels, 20.10.2010. COM(2010) 584 final.

Commission Decision (2012). Commission Decision of 15 November 2012 on notifying the third countries that the Commission considers as possible of being identified as non-cooperating third countries pursuant to Council Regulation (EC) No 1005/2008 establishing a Community system to prevent, deter and eliminate illegal, unreported and unregulated fishing (2012/C 354/01).

Commission Decision (2014). Commission Decision of 10 June 2014 on notifying the Third Countries that the Commission considers as possible of being identified as non-cooperating Third Countries pursuant to Council Regulation (EC) No 1005/2008 establishing a Community system to prevent, deter and eliminate illegal, unreported and unregulated fishing 2014/C 185/02 and 2014/C 185/03.

Commission des Communautés Européennes (1991). Rapport 1991 de la Commission au Conseil et au Parlement sur la Politique commune de la Pêche. SEC (591) 2288. Bruxelles, le 4 Décembre 1991.

Commission des Communautés Européennes contre République Française (2005). Arrêt de la Cour (grande chambre) du 12 juillet 2005. Commission des Communautés Européennes contre République Française. Manquement d'État - Pêche - Obligations de contrôle mises à la charge des États membres – Arrêt de la Cour constatant un manquement – Inexécution – Article 228 CE – Paiement d'une somme forfaitaire - Imposition d'une astreinte. Affaire C-304/02.

Council Implementing Decision (2013). Commission Implementing Decision of 26 November 2013 identifying the third countries that the Commission considers as non-cooperating third countries pursuant to Council Regulation (EC) No 1005/2008 establishing a Community system to prevent, deter and eliminate illegal, unreported and unregulated fishing (2013/C 346/02). *OJ*. C 346 of 27.11.2013, p. 2.

Council Implementing Decision (2014a). Council Implementing Decision 2014/170/EU of 24 March 2014 establishing list of non-cooperating third countries in fighting IUU fishing pursuant to Regulation (EC) No 1005/2008 establishing a Community system to, prevent, deter and eliminate illegal, unreported and unregulated fishing. *OJ* L91, 27,3,2014, p. 43.

Council Implementing Decision (2014b). Council Implementing Decision of 15 December 2014 amending Implementing Decision 2014/170/EU establishing a list of non-cooperating third countries in fighting IUU fishing pursuant to Regulation (EC) No 1005/2008 establishing a Community system to, prevent, deter and eliminate illegal, unreported and unregulated fishing as regards Belize (2014/914/EU). *OJ* L360,17,12,2014, p. 53.

EC (1998) Council Regulation(EC) No 1435/98 of 29 June 1998 prohibiting imports of Atlantic Bluefin tuna (*Thunnus thynnus*) originating in Belize, Honduras, and Panama. *OJ* L 191, 07.07.1998, p. 13–14.

EC (2005). Council Regulation (EC) No 768/2005 of 26 April 2005 establishing a Community Fisheries Control Agency and amending Regulation 2847/93 establishing a control system applicable to the common fisheries policy. *OJ* L 128, 21.5.2005, p. 1.

EC (2006). Council regulation (EC) No 861/2006 of 22 May 2006 establishing Union financial measures for the implementation of the common fisheries policy and in the area of the Law of the Sea. *OJ* L 160, 14.6.2006, p. 1.

EC (2008). Council Regulation (EC) No 1005/2008 of 29 September 2008 establishing a Community system to prevent, deter and eliminate illegal, unreported and unregulated fishing, amending Regulations (EEC) No 2847/93, (EC) No 1936/2001 and (EC) No 601/2004 and repealing Regulations (EC) No 1093/94 and (EC) No 1447/1999. *OJEC* L286 of 29.10.2008, pp. 1–32.

EC (2008). Notice of information of the termination of the demarches which a third country notified on 10 June 2014 of the possibility of being identified as a non-cooperating third country pursuant to Council Regulation (EC) No. 1005/2008.

EC (2009). Council Regulation (EC) No 1224/2009 of 20 November 2009 establishing a Community control system for ensuring compliance with the rules of the common fisheries policy, amending Regulations (EC) No 847/96, (EC) No 2371/2002, (EC) No 811/2004, (EC) No 768/2005, (EC) No 2115/2005, (EC) No 2166/2005, (EC) No 388/2006, (EC) No 509/2007, (EC) No 676/2007, (EC) No 1098/2007, (EC) No 1300/2008, (EC) No 1342/2008 and repealing Regulations (EEC) No 2847/93, (EC) No 1627/94 and (EC) No 1966/2006. *OJEU* No L343 of 22.12.2009, pp. 1–50.

EC (2015). Notice of information of the termination of the demarches which a third country notified on 10 June 2014 of the possibility of being identified as a non-cooperating third country pursuant to Council Regulation (EC) No. 1005/2008 establishing a Community system to prevent, deter and eliminate illegal, unreported and unregulated fishing (2015/C 142/05). *OJ* C 142, 29.4.2015, p. 6.

EEC (1982). Council Regulation (EEC) No 2057/82 of 29 June 1982 establishing certain control measures for fishing activities by vessels of the Member States. OJ No L 220 of 29.7.82, p. 1–5.

EEC (1983). Council regulation (EEC) No 1728/83 of 20 June 1983 concerning the management and control of certain catch quotas for vessels flying the flag of a Member State and fishing in the Regulatory Area defined in the NAFO Convention. OJ L169, 28.6.1983, p. 9.

EEC (1987). Council Regulation (EEC) No 2241/87 of 23 July 1987 establishing certain control measures for fishing activities. *OJ* No L 207 of 29.7.87, p. 1–7.

EEC (1988). Council Regulation (EEC) No 1956/88 of 9 June 1988 adopting provisions for the scheme of joint international inspection adopted by the Northwest Atlantic Fisheries Organisation. *OJ* L175, 06.07.1988.

EEC (1993). Council Regulation (EEC) No 2847/93 of 12 October 1993 establishing a control system applicable to the common fisheries policy. *OJ* No L 261 of 20.10.93, p. 1–16.

EU (2010). Regulation (EU) No 1236/2010 of the European Parliament and of the Council of 15 December 2010 laying down a scheme of control and enforcement applicable to the area covered by the Convention on the future multilateral cooperation in the North-East Atlantic fisheries and repealing Council Regulation (EC) No 2791/1999. OJ L348, 31.12.2010, p. 17.

EU (2011). Commission Implementing Regulation (EU) No 404/2011 of 8 April 2011 laying down detailed rules for the implementation of Council Regulation (EC) No 1224/2009 establishing a Community control system for ensuring compliance with the rules of the Common Fisheries Policy. *OJ* L 112, 30.4.2011, p. 1–153.

EU (2014). Regulation (EU) No 508/2014 of the European parliament and of the Council of 15 May 2014 on the European Maritime and Fisheries Fund and repealing Council Regulations (EC) No 2328/2003, (EC) No 861/2006, (EC) No 1198/2006 and (EC) No 791/2007 and Regulation (EU) No 1255/2011 of the European Parliament and of the Council. *OJ* L 149, 20.5.2014, p. 1.

Farnell, J. and J. Elles (1984). *In Search of a Common Fisheries Policy*. Gower, Aldershot. 213 pp.

Lequesne, C. (2001). *L'Europe Bleue: A quoi sert une politique communautaire de la pêche?* Presses de la Fondation Nationale des Sciences Politiques, 239 pp.

Long, R.J. and P.A. Curran (2000). *Enforcing the Common Fisheries Policy*. Blackwell Science, Oxford. 379 pp.

OECD (2006). *Using market mechanisms to manage fisheries: Smoothing the Path*. Executive Summary: 11–16. OECD, Paris.

SEC (1992). Report on monitoring implementation of the Common Fisheries Policy. SEC(92)394 final.

CHAPTER 10
The scientific advice for the CFP

The CFP, a science-based policy

From its inception, the conservation policy of the CFP has been a science-based policy. This never meant that decisions were made solely on the basis of scientific advice but that scientific advice on the state of the stocks was a basis upon which other considerations could be introduced by fishery managers. In fact, already in 1979 the Commission established the Scientific and Technical Committee of Fisheries (STCF) as the scientific advisory body for the CFP, although a fundamental source of advice for this policy was and continues to be, the International Council for the Exploration of the Sea (ICES), at least for the Atlantic, North Sea and Baltic fisheries.

The use of scientific evidence as a basis for fisheries management is relatively recent in history. Fisheries science as such started to develop in Northern Europe and North America in the early 1950s and the influence of scientific advice in management still took a number of years to fully develop. Today's methods of evaluating fishery resources are still fundamentally based on the seminal work carried out by Ray Beverton and Sydney Holt in the 1950s (Beverton and Holt, 1957). Other key contributions were the work of ICES (see next) and perhaps especially the early work of scientists in ICNAF[1] since the establishment of this organisation in 1954. ICNAF scientists, particularly during the first half of the 1970s, were at the cutting edge of innovative fishery analysis, providing for assessments and supporting studies which served as the basis for catch limitations, fishing effort and the world's first attempt at multispecies management (Anderson, 1998).

The fact that this initial science was fundamentally biological was from the beginning a problem for decision makers. For years, any decision to manage a fishery was considered a political decision, taken on the basis of common sense at best. Even in the previous example of ICNAF, the world's best scientific advice did not prevent failure in persuading member countries to agree on effective management measures. Turning management decisions into a process led by scientific evidence represents a very slow process that is not yet entirely achieved, particularly in areas where the scientific evidence is weak.

The way in which the scientific advice has evolved and the way in which such advice has been integrated into the decision making process is one of the most interesting stories of the development of the CFP.

The sources of fisheries research to underpin the CFP

In Europe there is a well-established network of national and regional fisheries science laboratories (often covering also oceanographic and marine ecological research), with permanent staff funded by Member States who do surveys at sea, collect data from the fishing industry and develop mathematical models. These laboratories are the producers of the basic fisheries science that underpins the CFP.

The Common Fisheries Policy: The Quest for Sustainability, First Edition. Ernesto Penas Lado.

It is important to underline that the providers of scientific advice (ICES, STECF, see next) only produce advice, not science as such. ICES and STECF can only provide their advice on the basis of the data collected and research carried out by these institutions.

The Union has long provided for support to fisheries and aquaculture science, through its Research programmes. The Fisheries and Aquaculture Research programme (FAR), established by Council in 1987, was implemented by the Directorate General for Fisheries of the time as part of the management of the CFP and provided funding for research in these areas for the period 1988–1992 (Council Decision, 1987), with a budget of 30 million ECU.[2] This programme allowed the Commission to fund research necessary for the implementation of the CFP.

These funds were later integrated in the general research programmes of the Commission. The Third Research Framework Programme included the Agriculture and Agro industrial research programme (1991–1994) that dedicated 20% of its 333 million ECU budget to fisheries and aquaculture research.

Since then the fisheries and aquaculture research has been part of the overall research framework programmes of the Commission and always with a relatively stable level of funding. For example, in the Sixth Framework Programme (research programme covering the period 2000–2006) fisheries and aquaculture-related projects have been funded for a total budget of €277.2 million of which 47 projects (€61.6 mn) for fisheries and 74 (€92.7 mn) for aquaculture. This support will continue in the future: the programme *Horizon* 2020 (for the period 2014–2020) foresees the funding of research on fisheries and aquaculture for that period.

However, this important support does not suffice. Fisheries science is not (or not necessarily) about scientific breakthroughs but rather it is about good biological knowledge of the species and the fisheries and, crucially, about historical series of data to study trends. This kind of research is often considered as 'routine' or not innovative. This fact must be combined with the clear trend, all around Europe in the last 10–20 years, for national fisheries research institutes to move from permanent budgets to ensure routine research, to a system where funding must be obtained from different sources for specific programmes. This type of system is less amenable to the kind of long-term routine research for the provision of the annual advice on management.

Ensuring the continuity of time series is thus essential for a good scientific basis to underpin the CFP. This is thus a fundamental challenge and the answer to it is to be found in the data collection regulation that we will refer to next.

The sources of advice for the CFP

On the basis of the data collected in Member States, by Member States' scientists and all the other sources of fisheries and marine ecosystem research, the advice for managing the CFP is traditionally provided for by certain scientific bodies, notably ICES and the STECF.

ICES

The International Council for the Exploration of the Sea (ICES) was established in 1902 and ever since then it has been the main provider of scientific advice for fisheries in Northern and Western Europe (Rozwadowski, 2002). On the invitation of the Norwegian Government an International Conference met at Christiania in May 1901 to revise and complete the proposals formulated at the Stockholm Conference in 1899 for a combined hydrographical and biological exploration of the North Sea and adjoining waters in the interests of the sea-fisheries.

This initiative represented for the first time in Europe the idea that fisheries management should be based on scientific evidence. Constituted by Atlantic European and North American States, ICES remit is to coordinate the marine science efforts of its members (including oceanographic research). It also provides annual advice on the state of fish stocks and their exploitation under an agreement with the EU.

ICES covers marine resources in the North Atlantic (except in the area of NAFO) including the North Sea and Baltic Sea. Likewise, its scope extends from fisheries science to marine ecology. However, it does not cover the Mediterranean Sea.

Beyond the CFP, ICES is also the provider of scientific advice to a number of bilateral arrangements and multilateral bodies where the Union is concerned: the science basis for the bilateral management of joint stocks with Norway is provided by ICES. The same applies for joint management of Baltic resources with Russia. Multilateral bodies such as NEAFC, or NASCO, also base themselves on ICES advice.

The importance of ICES goes beyond the provision of scientific advice. By having clear and regular 'clients' (and among them, the CFP) ICES has developed a methodology to deal with their requests, notably through standardised analytical methods. This standard methodology has two very important side effects:

- It contributes to train young scientists by familiarising them with these standard methodologies;
- It coordinates research among Member States' scientists: in order to make their contribution to the advisory process, European scientists have used ICES as a forum to determine the kind of data and information necessary for the advice. The effects of the lack of such mechanism, applied for decades, is evident in the shortcomings of Mediterranean fisheries science, where no such clear 'client' has existed until recent times.

Over the years ICES has been gradually increasing the scope of its work. Scientific questions related to environmental issues, or technical measures, have been gradually incorporated into ICES' remit, reflecting the evolution of the requests from its clients. At the time of writing, an expansion to economic and social analysis is being considered.

STECF

The STECF (Scientific, Technical and Economic Committee for Fisheries) was formally set up (as STCF) by the Commission in 1979 (EEC, 1979) and consolidated in 1983, in the first basic regulation establishing the conservation policy of the CFP (Regulation 170/83, see Chapter 2). ICES had many clients beyond Europe and the CFP. When establishing the CFP, even though the fish stocks concerned by this policy in 1983 were basically evaluated by ICES, the EEC of the time felt the need to have a specific advisory body that would be immediately at the disposal of the CFP to provide advice that ICES, on the basis of its wider remit, or its priorities agreed with non-EU members, might not be able to provide as needed.

Initially, the mandate of STECF was limited to biological and technical advice. Over time, however, STECF has proven ever more useful as the CFP has developed instruments requiring scientific advice that is not covered by ICES. For example, STECF provides much needed advice on such grounds as:

- *Economic advice.* As the need for economic advice has grown in the CFP (see later in this Chapter) STECF has been the only body able to produce it. The original STCF of 1983 did not include economic advice. It was in the first CFP reform in 1992 that an amendment to the STCF was agreed to include the 'E' of Economic in the name and the remit of this advisory body. This inclusion was not smooth at all, since the idea of the economic advice was seen by some as introducing 'politics' into science. In addition, the economic advice has taken time to take off, due to the shortage of time series of data and to the fact that biologists and economists within STECF spoke a very different language for a long time.
- *Aquaculture and data collection* as of 2005. Of particular important is STECF work in the data collection framework, notably by analysing national plans to provide the basic data needed to prepare the advice by all concerned scientific bodies.
- *Long-term management plans and effort management.* As a number of these plans and certainly their effort management component are exclusively EU policies, not shared by other clients of ICES, the necessary advice on these issues has been provided by STECF.
- *Specific requests.* Many specific requests for advice, for example following statements by Council during TAC negotiations, are often handled by STECF, since its timing and priorities, unlike those of ICES, are determined by the Commission.

To be able to deliver this kind of advice, the Commission has structured this body to ensure a wide coverage in terms of subject matters (including biologists, economists, gear technologists etc.) and geographical areas: the over 30-strong composition of its plenary reflects clearly the needs of its one and only client: the CFP.

But STECF goes beyond the work of the plenary. In fact, the working groups to prepare the advice are open to participants beyond the members of the plenary. In this way, the STECF has the capacity to assemble a very wide variety of experts from all across Europe.

Over the years, STECF has become a fundamental instrument of the CFP. Its contributions on areas not covered by ICES have allowed the CFP to break new grounds in fisheries management. Its reputation as a neutral, reliable provider of advice has also grown over many years. In some key cases, where the provision of scientific advice was both technically difficult and politically sensitive (e.g. on anchovy in the Bay of Biscay in the mid-2000s), STECF has provided neutral and credible advice that has indeed contributed to find effective management solutions.

The credibility is such that Member States often want to use STECF to revise advice from ICES that they consider wrong or insufficient. This credibility is at the origin of the decision by Council, in the context of the CFP reform of 2002, on the requirement that the Commission should base its proposals on advice from STECF, even when such advice was already available from ICES, thus producing an unnecessary duplication of efforts. As we will see, this was corrected in the CFP in 2013 in order to avoid that this requirement will duplicate the advisory work of the two scientific bodies.

ICES and STECF are fundamentally complementary. Although there has been in the past a certain degree of overlap particularly, during many years, STECF revised ICES advice, sometimes providing for different recommendations and possibly creating confusion. This overlap has been removed in the CFP reform of 2013 (see Chapter 16) and this will allow the scientific advice to be streamlined, as ICES will provide the regular advice on Northeast Atlantic stocks and STECF will provide the more CFP-specific requests.

Other scientific bodies

The external policy of the EU has different and varied sources of scientific advice. In particular, while certain RFMOs rely on ICES for advice (NEAFC, NASCO, and the former IBSFC), others have their own scientific bodies: NAFO and ICCAT[3] are good examples. And while these specific bodies are constituted by scientists from the contracting parties, there is one body, the IATTC,[4] where the scientific advice is prepared by the permanent scientific staff of the organisation itself.

A special and particularly difficult case is that of the scientific advice for the bilateral fisheries agreements. Here, the advice can be obtained from the relevant scientific bodies of the RFMOs (where these exist, are operational and concern straddling or highly migratory fish stocks). But the problem arises when handling resources in areas where multilateral cooperation is weak (the case of West Africa and CECAF is the most clear example, as we saw in Chapter 7) or where the stocks concerned are discrete stocks that are not subject to multilateral management.

The basic data for science and advice

Fisheries science and advice are extremely dependent upon the availability of good basic data on catches of commercial vessels and on surveys. Unlike other scientific areas, dominated by bold theoretical developments, or by spectacular breakthroughs, the scientific basis for fisheries advice is relatively stable, dependent upon methodologies that have been developed some time ago and which have evolved relatively slowly. They can only provide results on the basis of good time series of core data according to certain fundamental parameters, notably on the biology of the fish, the amount of fish being killed by the fishing activity and the amount of fishing activity being exerted on the stocks.

This is no rocket science; it is a science whose results still depend on the availability of these basic data. No amount of new thinking, new models and new hypothesis can replace the basic information. This knowledge, to take into account natural fluctuations (inevitable when dealing with natural resources) can only be reliable when it extends over a certain number of years, thus constituting historical data series where trends can be identified over and above annual fluctuations.

All this means that the scientific basis for the CFP has to ensure that the key time series of basic data are collected on a permanent and systematic basis. This has required specific instruments as the CFP developed.

In the past, the basic historic series of data tended to be ensured (though not in all cases) by national fisheries research institutions, whose funding was relatively stable and not dependent upon temporary grants for research projects. However, this situation has changed in some institutions in the last two decades.

The tendency to replace the funding of research laboratories from permanent budgets to specific research lines has in some cases affected the continuity of time series. An additional problem has been that of the relatively low attractiveness of fisheries science for talented young European scientists. This may be largely due to the low curricular value of many of the publications related to the production of the fisheries advice. These factors have produced a limited and often shrinking human base for fisheries science in Europe; in some key areas of fisheries research, there is almost no-one working.

To try to respond to this problem, the Commission, in the 1990s, used the funding of 'biological studies' to allow for the continuation of key time series of data whose funding was no longer assured in national fisheries research institutions. This solution was only a partial one: it could not ensure the continuity of historical data series on a systematic basis. This, in turn, produced in many cases a net loss in the level of available time series for fisheries. Not only were the time series not completed; some were even discontinued.

The CFP needed a more systematic way to address this question. The solution was the establishment of a specific instrument to co-finance with Member States the collection of the basic data necessary for the provision of advice for the CFP. This led to the adoption in 2000, of the first data collection regulation (EC, 2000).

This regulation has applied ever since and has been incorporated into the main body of the basic texts of the CFP in the reform of 2013: the new basic regulation (1380/2103) established the obligation for Member States to collect the data necessary for the scientific advice of the CFP and the new structural fund, the EMFF, provided financial support for Member States to collect the biological and economic data to implement the CFP under more favourable terms and conditions than ever before (see Chapter 16).

The importance of the data collection must be stressed. The continuation of time series of fisheries parameters is hardly a high priority in any of the current systems to fund science in Europe. Hence the need to have a specific instrument that would guarantee that the necessary routine data continue to be collected and that time series are continued.

There is, however, a caveat. The provision of a specific budget for the data collection has ensured that time series are continued but parallel attempts to broaden the resource base of fisheries research in Europe have failed, since the additional funds available for fisheries research institutes has often been used to replace other sources of funding, rather than to increase the resources for research. This remains a challenge for the future.

The processing of the data

The success of the CFP in recent years in ensuring a steady supply of basic data cannot be overstated. However, with this success comes a challenge: the increase in available data, when handled by a scientific community with limited human resources and overstretched agendas, is giving rise to a new phenomenon: the inability of European science to process all the information and fully incorporate it into the scientific advice.

This may be aggravated by a management system that requires as a priority the provision of advice on a specific basis: the quantitative assessment of catch levels. Certain authors have warned that the focus on these deliverables may lead to a certain disregard of basic fish biology in the evaluations (Pawson, 2008). If this is the case, then the question is how the new biological data provided by the data collection is going to be used in the assessment workflow.

It has been argued that the incorporation of new knowledge in the preparation of management plans has been slow compared with the rate at which the new knowledge is accumulated (Apostolaki et al., 2008). In the context of the increasing accumulation of new knowledge from the data collection, the way in which such knowledge will be processed and incorporated into the scientific assessment process constitutes a new and important challenge for the scientific community and for the CFP.

The evolution of the scientific advice

The scientific advice for the CFP has evolved substantially since the inception of the CFP. This evolution is the result of several factors:

- The evolution of fisheries science, particularly in relation to the development of new methodologies and an evolving notion of the role of scientist's vis-à-vis fisheries managers.
- The changing demands of the main 'client' of the scientific community: the fishery managers under the CFP, following the gradual introduction of new instruments and policy approaches.
- The search for solutions to fill gaps in the knowledge of the state of stocks.

This has led to an evolution of the scientific advice along several lines:

- The change in the long-term objective: from sustainability to maximum sustainable yield. While the traditional advice catered for the rather loose objective of fisheries management in the past, the introduction of the MSY concept as a long-term objective has forced scientists to produce advice on MSY levels for an increasing number of stocks.
- The gradual introduction of long-term advice in addition to the annual advice, in order to provide advice to long-term management plans.
- The introduction of multi-species considerations in the advice. This evolution is, however, still in its infancy.

In addition to this, the scientific community has provided advice on many other questions related to the CFP: environmental questions (including the very recent on the Marine Strategy Framework Directive), technical measures, the management of data-poor stocks and so on.

From safe biological limits to maximum sustainable yield

The advice for the initial CFP in the 1980s was based on the annual nature of the TACs and therefore focused on the attainment of management objectives in 1 year. Initially, ICES actually provided MSY advice, at least for certain stocks. However, this kind of advice was generally not welcomed by Ministers in Council, who asked for more flexibility, notably to be able to achieve the objectives in more than 1 year.

At the time, the policy had not yet specified a clear long-term objective for the management of the stocks, so the advice was not directed to the achievement of any such long-term objective but to ensure that the fishery was sustainable – in 1 year.

This biological sustainability was, for years, represented by the notion of minimum biologically acceptable level (MBAL) and objective that established catch levels as those necessary to achieve sustainable exploitation from a biological point of view, without prejudging whether fishery managers would wish to achieve other, more far reaching goals, such as maximum sustainable yield.

The notion of MBAL was replaced, years later, by the notion of 'safe biological limits'. These constitute a minimum standard for management: whatever the fishery managers may set as their management objective, the advice on safe biological limits are the levels of exploitation with an acceptable level of risk (typically around 5%) of the stock being depleted. These limits are expressed through the so-called 'reference points', than in turn can be based on spawning stock biomass or in fishing mortality, as follows:

Limit reference points are maximum values of fishing mortality (F_{lim}) or minimum values of biomass (B_{lim}) which must not be exceeded. Otherwise, it is considered that it might endanger the capacity of self-renewal of the stock. However, the uncertainties associated to the estimation of limit reference points led to the consideration of precautionary reference points. These points, expressed also in terms of fishing mortality (F_{pa}) or biomass (B_{pa}) represent a more restrictive approach: they should be used to trigger management decisions even before the limit reference points are reached.

The reference points were the result of the adoption of the Precautionary Principle, proposed by FAO in the Code of Conduct for Responsible Fisheries of 1995.[5]

With the notion of reference points, scientists also drew a clear line between science and management: by providing advice on safe biological limits, scientists established the minimum standard to guarantee sustainability. But they did not prejudge whether managers should aim at other management objectives, such as maximum sustainable yield, maximum economic yield and so on. That is not a decision for scientists.

This kind of advice was logical when the CFP did not have a clear long-term objective. In that case, scientists would only provide advice on how to ensure biological sustainability. Anything beyond that was outside their remit. This served the CFP well, until the policy itself decided that a more concrete, long-term goal was necessary: the maximum sustainable yield.

In any case, these biological reference points remain useful for management. Although they are no longer the long-term objective of the conservation policy, they still represent important references of levels that cannot be trespassed for their high risk. Even if the new management under the CFP targets MSY levels, the references are still used in management plans, for example to trigger the adoption of special measures if the stock biomass of the stock concerned falls below B_{pa}.

From short to long term advice

The advice provided by ICES has been evolving over the years, along with the evolution of the management policy of its main client: the CFP. It is interesting to note how the advice has been formulated as a result of what the CFP required, or what scientists in ICES believed was needed by the CFP.

For many years, as the CFP concentrated on decisions on annual catch possibilities, the advice by ICES was geared to provide advice for such annual decisions. But, as we have seen, the CFP has traditionally lacked a clear-cut long-term objective for the management of the stocks. This means, in particular, that the CFP had never been clear enough in terms of how the biological and the economic factors leading up to TAC decisions should be combined.

The adoption of the UN Fish Stocks Agreement in 1995 (see Chapter 7) enshrined the precautionary approach to fisheries. As a response to this, scientists had to combine the annual nature of the advice with the implementation of the precautionary approach. This combination led to a situation where scientists provided advice on how to ensure biological sustainability – in 1 year. Beyond that, whether fisheries managers in the CFP wanted to go beyond mere biological sustainability and go towards objectives based on economic parameters, was not ICES' affair.

In this scheme, it was also important to underline that scientists did not see themselves as managers; it was clearly not their job to compound biology with economics, let alone with politics.

The problem of this kind of advice was, fundamentally, that by implying that the objective of management was to achieve safe biological limits in one year (in the absence of long-term plans, this was the only possible assumption) then the resulting advice was, very often, for a very dramatic reduction of TACs from one year to the next and even, in a number of cases, for a complete closure of the fishery.

Perhaps the best example is the advice for North Sea cod, in 2002, when the collapse of the stock was already obvious but the long-term plan was not yet adopted. In effect, that year ICES' advice was:

> Given the very low stock size,..., ICES recommends a closure of all fisheries for cod as a targeted species or by-catch. In fisheries where cod comprises solely an incidental catch there should be stringent restrictions on the catch and discard rates of cod, with effective monitoring of compliance with those restrictions (ICES, 2002).

After the adoption of long-term plans, ICES produced advice on the annual effects of the implementation of such plan. But this kind of advice continued for years for stocks not subject to multiannual plans.

Such closures or dramatic reductions of annual TACs were politically unacceptable, leading Fisheries

Ministers to ignore the advice and establish TACs at much higher levels than those recommended by scientists. This state of affairs has characterised the CFP for many years and is at the origin of the lack of trust between scientists, fisheries managers and industry. This, in turn, produced a very negative image of the policy: one where the decision makers ignore scientific advice.

Indeed, this situation was not the scientists' fault. Rather, it was the inevitable effect of a management system based exclusively on annual decisions, without a framework allowing for gradual recovery of overfished stocks over a number of years. In these circumstances, scientists had no option but to recommend that measures be taken to bring stocks within safe biological limits within the time frame for which the advice was requested: one year.

This was one of the reasons that justified the need for multiannual management plans. This state of affairs had to be changed. And it could only be changed by a new kind of approach by policy-makers: by shifting the emphasis away from annual decision making to long-term multiannual plans, it would be possible to have advice recommending a step-wise multi-annual strategy to achieve the management objective.

As the Union started proposing and adopting long-term management plans, the scientific community also started changing the way they provided the advice, largely by replacing a simple, one-year advice for recommendations on long-term approaches, in cases where the short term advice was clearly difficult to assume for decision makers, or where measures in one year would not suffice to achieve sustainable levels. A good example is the advice for Irish Sea sole in 2007:

> It is not possible for the stock to reach B_{pa} in one year – a complete closure of the fishery in 2008 would bring SSB above B_{lim} in 2009. A management plan for effort reduction that can be phased in over a number of years and implemented in conjunction with technical conservation measures should be considered. (ICES, 2007).

This clearly shows the evolution of the scientific advice, from a simple recommendation for a closure to that on a management plan based on a gradual recovery. This evolution is fundamentally the result of the new emphasis of the Union in considering long-term approaches rather than just annual – dramatic – measures.

Moving from single species advice

Commercial fish species do not live in isolation. Even species with a strong schooling behaviour interact with other species in the sea. Especially, fish species are not caught in isolation. Many fisheries catch a combination of species and this combination is practically impossible to avoid. However, TACs are traditionally applied as if the stocks concerned were isolated.

The introduction of multi-species considerations in the CFP required the evolution of the advice from single-species to multi-species. It is important to clarify what we understand by 'multi-species' advice, as opposed to the traditional single-species advice provided for the TAC and quota system since 1983. In fact, this can imply two different things:

1 Mixed-fisheries advice takes account of technological relationships, that is, the way different species are caught together in a given fishery. One very well-known relationship of this kind is that between sole and plaice in the beam-trawl fishery in the southern North Sea: it is practically impossible to fish one of these two fish without the other.
2 Multi-species advice takes account of the evaluation of trophic ecological relationships between two or more species. A classic example is the relationship in the Baltic Sea between a predator (cod) and its prey (herring and sprat) and the effect of a higher or lowers fishing mortality of the cod on the abundance of the prey stocks.

One of the first and most important approaches in this regard came as a result of the implementation of the cod recovery plan in the North Sea. As we saw in Chapter 4, this plan had a major flaw: it was based on the need to protect one stock (cod) but it had obvious consequences on other stocks. This clearly identified the need to have a better idea of how to manage the whole assembly of species caught together with cod.

The first attempt to produce multi-species advice took place in the North Sea in the late 1980s, where the knowledge of fisheries is probably the best in

Europe (Symposium on Multispecies Models Relevant to Management of Living Resources, 1989).

Later on, ICES started to produce a matrix where the technological relationship among stocks was evaluated through a semi-quantitative scale. This type of multi-species advice is easier, since it requires the examination of catch compositions in the gear concerned, a kind of information that is easier to obtain. More recently, ICES has been able to advice on the relative effects of different levels of fishing mortality on the mortality of the seven main stocks associated in the fisheries (ICES Advice, 2013).

The second type of relationship is more difficult. It requires data on predator-prey relationships (which in turn require data on stomach content of prey species) which are not always available. But most crucially, it requires a better understanding of the relationship between predator and prey at the adult age. In some cases, the predator/prey relationship can be inversed: herring and sprat can also eat cod eggs and larvae. Assessing which of these relationships will prevail and how can be extremely difficult. In addition, the ecological relationship can be very variable according to different areas and it is sometimes difficult to establish clear predator/prey relationships that will be valid for the whole of the management area of the species in question.

This kind of research, however, has stagnated in recent years. The reason is that the multispecies models are multivariate in nature. Using them would require managers to make – difficult – decisions on fleet structure, gear type, fishing effort and catch limits in a very detailed way. The management challenges were such that have not been met, largely due to concerns about these models affecting relative stability.

The challenge is to see if the elements of the reformed CFP in 2013 (regionalisation, discard ban and further emphasis on multispecies approaches, see Chapter 16) will provide a new impetus for this kind of research and advice. The question is, then, whether the scientific community will be able to provide such advice. Indeed, the scientific community is fully aware of the need for this evolution (Reeves et al., 2008), and has been working on different kinds of multispecies models for 80 years (Pinnegar et al., 2008). However, the current level of the advice is very variable for different areas and, while the North Sea is relatively well modelled, the Baltic (theoretically simpler) is proving difficult to model, and the western waters lag behind in the level of knowledge on the technical and trophic relationships among the different species and their environment. This constitutes another major challenge: as the deadline to deliver the objective of the new CFP as reformed in 2013 (MSY by 2020) gets closer, the need to provide such multispecies advice becomes urgent.

The case of 'data poor' stocks

The quality and in some cases the very availability of scientific advice has always been an important issue in the CFP. Indeed, the CFP has always suffered from a problem of shortage of scientific advice for certain stocks, in particular for the annual TAC and quota decision.

The problem of 'data poor' stocks, for which ICES did not provide specific management recommendations, has traditionally been very high: on average, around 40% of the TACs in EU waters have consistently been set without scientific advice. What is worse, this figure has been static for a number of years.

There are reasons for this: in the last decade, the human and financial resource base for fisheries research in Europe has not increased. Although the EU added funds to support data collection, Member States reduced the national budgets in proportion, so there has been no net increase in the total investment on fisheries research. As a result, there has been little or no possibility to invest on research for those 'data poor' stocks.

In recent times, however, a new approach has been taken on this issue. As we see in Table 10.1 the total number of stocks for which there is no advice has been dramatically reduced to 14, from around 40 in previous years.

This reduction actually does not respond to new data being available but to a new approach by the scientific community. On request from the Commission, ICES started to provide the so-called 'data-poor advice', whereby, in the absence of analytical advice, ICES studies trends in catches and at least provides quantified advice on sustainable catches using the partial information that is available.

Table 10.1 Evolution of the completeness of the scientific advice for TAC setting, according to Commission's policy statement of 2014. Source: © European Union. Communication from the Commission to the European Parliament and the Council. Concerning a consultation on Fishing Opportunities for 2015 under the Common Fisheries Policy. Brussels, 26.6.2014 COM(2014) 388 final.

	Number of fish stocks											
Advice for year:	**2003**	**2004**	**2005**	**2006**	**2007**	**2008**	**2009**	**2010**	**2011**	**2012**	**2013**	**2014**
Stock where stock size and fishing mortality can be forecast	40	34	40	31	29	30	34	36	36	40	46	49
Stocks where a quantified scientific advice concerning fishing opportunities is available	59	52	54	65	61	62	63	55	55	47	77	71
Stocks where no Scientific advice is available	31	40	39	29	35	34	33	42	40	44	9	14

These figures refer to the Atlantic. In the Mediterranean, where only in recent years the advice is being presented in a more systematic way, the percentage of stocks with a high degree of uncertainty is higher than 90% (COM, 2014a).

The challenge here is, however, that for certain stocks of secondary economic importance, there will probably never be resources to study them, so chances are that certain stocks will remain fundamentally unknown for the foreseeable future. The CFP will have to accept this and apply proxies when available, or simply base the management of the fishery on stocks for which a minimum amount of knowledge will be available.

The thinking here is that if a stock is not important or valuable enough to attract a directed fishery, it is unlikely to be overfished as a by-catch so long as the target stocks are fished sustainably.

The precautionary principle

Whatever the improvement in the scientific advice, there will always be a level of uncertainty in the advice and, in the case of secondary species, the low level of scientific information may not be entirely resolved. In these circumstances, the CFP has enshrined the implementation of the precautionary principle.

As enunciated by FAO in the 1995 Code of Conduct for Responsible Fisheries, the Precautionary Principle means that the limitations, uncertainties or lack of data for the assessment or for the estimation of parameters, cannot be justification for not applying regulation measures, especially when there is information that the stocks are overexploited.

The implementation of this principle, however, has always been difficult. The CFP has long suffered from what one could call an 'inverse precautionary principle'. It has been common practice that when scientific evidence was not clear but there were signs of the need for management action on a stock, this lack of clarity has been used to avoid the adoption of measures for 'lack of scientific evidence'. This situation clearly has a negative effect: not only ignores warnings about the deterioration of stocks but also provides a negative incentive to produce that scientific evidence in the first place. And then the situation perpetuates itself as a vicious circle.

This is largely a heritage of the first years of the CFP, where some authors have summarised the problem as follows: 'within the political process, the advice is often simply ignored or its alleged ambiguity is used as a pretext for compromise' (Symes, 2010).

Yet, it is important that the data poor advice is applied. Not least because if it is not, there is a risk that fishermen or national administrations will see the insufficiency of scientific data as an advantage in terms of the management measures negotiated. In other words, if the CFP is to be a policy based on science, it is necessary that the system creates a real incentive for the data and scientific information to be there in the first place.

The advice in the Mediterranean

From the beginning of the CFP, the case of the Mediterranean was different. There, the absence in the first 10 years of a real conservation policy under the CFP in that area resulted in the absence of a clear 'client' for standardised and consistent provision of scientific advice as we saw in Chapter 8. In addition, there was no equivalent for ICES in the Mediterranean. This has been, for many years, a vicious circle: since there was almost no conservation policy in the Mediterranean, there was no need for an efficient provider of scientific advice. But in turn the absence of an efficient provider of advice has hampered the development of a conservation policy in the area. This vicious circle is only recently starting to be broken.

In Chapter 8 we saw the effects of the lack of a systematic approach to the provision of scientific advice. This matter has in recent years evolved very positively, notably thanks to the work of STECF and the parallel work of the Scientific Advisory Committee of GFCM.

The real breakthrough happened in 2012, when for the first time STECF produced a general picture of the status of around 100 Mediterranean stocks using a standardised methodology. This was consolidated in subsequent years and in 2014 the Policy Statement (referred to previously) described, for the first time ever, the status of Mediterranean stocks with the same presentation as that for Atlantic stocks, although the evaluation methodology was different.

Science and policy making: an often difficult relationship

The relationship between scientists and then other actors of the CFP has often been difficult and controversial but has also evolved quite substantially over the years. Let us look at some cases.

Scientists and policy-makers: a different language

Traditionally, the CFP has witnessed an almost permanent rift between the scientists and policy makers (let alone the industry). This is as old as the CFP itself. Indeed, during the period of discussion of the CFP (1977 to 1983) the Commission tried repeatedly to convince Council to adopt catch limitations based on the scientific advice. However, in the absence of a clear framework for the consideration of the advice and since decision makers were concentrating on defending national interest in terms of their share of fishing rights, the scientific advice was often considered as an unrealistic and inopportune basis for decisions. In fact, ICES' recommendations at the time were considered by some as an 'inflexible conservationist approach' (Farnell and Elles, 1984), being based on a MSY criteria which, at the time, were very far from current conditions and whose attainment was seen as impossibly idealistic.

The difficult relationship has also been often the subject of scientific literature, which sometimes depict decision makers as pursuing only short-term victories and scientists as a result not feeling responsible on the outcome of fisheries management (Daw and Gray, 2005).

However, over the last decade there has been quite a substantive evolution in the way fisheries science is perceived by decision makers.

Still, in the early 2000s, it was common to listen to Ministers claiming higher TACs on the basis of short-term political arguments, for example that a given stock was essential in their constituency and that they had elections coming up. This was of course never done in public.

Some years later, after the annual publication of the Policy Statement, the argumentation started to change substantially and, in 2008 a Minister claimed that instead of a politically-based increase in the TAC during the Council, he undertook to provide the new scientific evidence proving that the stock would deserve to be classified in a different category (as according to the Policy statement of that year) and therefore subject to a different approach in establishing the annual TAC. In other words, the more traditional short-term political argument was changed for a science-based approach based on the systematisation proposed by the Commission. This was a historical break-through.

This breakthrough came about after a slow learning curve and in this learning period, two challenges had to be met: the two reasons that one way or another call into question scientists' independence: (1) the use of scientific advice as a the only political option, that is, the use of scientific advice to adopt automatically political decisions without further considerations and (2) the use of scientific evidence for enforcement purposes.

These two challenges deserve some further explanation.

Science and political decisions

Although political decisions take science into account, they are not limited to these scientific considerations. Otherwise, if and when the scientific advice would automatically translate into political decisions, then scientists would be in fact making political decisions, something they don't want to do and they should not do.

However, there is a recent trend that in some cases tends to blur the boundary between scientists and decision makers: as the later are often criticised for 'ignoring science' there can be a trend by decision makers to follow the scientific advice to the letter, without further (legitimate) considerations. This can give rise to an automatic translation of scientific advice (only biological) into political decisions. This is something that the scientific community does not appreciate and that can indeed give rise to worse science.

This has always been an issue in the CFP and an issue that is dependent upon the request for advice by the Union. For example, if scientists are requested to provide a single figure for an annual TAC then the recommendation by scientists amounts to a political decision. If, on the contrary, scientists are asked for different options with different levels of risk, then fisheries managers can decide on the basis of the risk they are prepared to accept.

A case in hand is that of the advice for the TAC of salmon in the west of Greenland in the framework of NASCO. This TAC is of a very particular nature: the salmon of the West of Greenland hatch in European and North American rivers and spends 2 or 3 years as adults in the feeding grounds of west Greenland before going back to their rivers to spawn. The difficulty to handle this TAC (whose fish is that?) made it necessary to find an 'objective' method to fix the annual TAC.

In 1994, NASCO adopted a model that, when fed by estimates of certain basic biological parameters provided a resulting TAC. The idea was that when ICES provided the annual advice, it would estimate the parameters for the model and NASCO would easily and automatically agree on a TAC level. It looked like the perfect, objective solution. But things did not turn out that way.

The following year, when ICES delivered its advice, it did estimate the parameters of the model but then went on to make a recommendation for a TAC level that did not correspond to the implementation of the model. This was obviously based on the consideration that the model was not necessarily perfect. But one can certainly speculate if scientists, when confronted with the estimation of the parameters to feed the model, did have the impression that, by providing precise estimates for these parameters, they were *de facto* being asked to take a political decision on the resulting TAC level. And they refused to do so.

But not all this is negative. In certain cases, scientists have been asked to provide advice knowing that their advice may automatically result in a political decision. Yet, they provided such advice in a remarkably objective and responsible way. The case of the advice on the possible re-opening of the anchovy fishery in the Bay of Biscay, after the closure of the fishery following its collapse in 2005, was an extraordinary achievement by the STECF, in that it provided an advice that was accepted by the different and conflicting interests affected.

In any case, the need to base decisions on 'science' has gradually become over the years increasingly palatable in political terms. The frequent allegation by fisheries managers that their decisions were 'based on science' has made considerable headway and has become increasingly acceptable politically. This is a positive development but it has to be considered with care, given that fisheries science is subject to considerable uncertainty, and that allows for very different political positions to be allegedly 'based on science'.

Fisheries science and enforcement

There is another question that may undermine the independence of the advice: the temptation of using scientists' hypothesis, projections and estimations for control and enforcement purposes.

Scientists often refer to the evidence or suspicion about, for example, TACs being overshot, or other rules being ignored. They do so because these considerations are important to understand and judge the advice. In connection with this, in a number of cases fisheries managers have had the temptation of using this kind of advice to substantiate cases of non-compliance against Member States. This has always been a very dangerous course of action, for two main reasons:

1 Firstly, scientists' hypothesis and assumptions do not necessarily constitute reliable proof in front of a judge. This does not mean that scientists are wrong; just that the burden of proof when evaluating aggregate data is different to the burden of proof in a legal case. Normally, to build a case against a Member State on the control of the CFP it is necessary to provide proof to a high standard.
2 If scientists' analyses are used in a disciplinary procedure against their Member State, there is a strong chance that scientists will be put under pressure by their authorities not to provide such analyses. This can produce a level of self-censorship which can result in a considerable loss of information being provided by the advice.

A case in hand is that of Bluefin tuna in ICCAT, in the mid-1990s. At the time, the statistics of catches of this species were done by the scientific body of ICCAT: the Scientific Committee for Research and Statistics (SCRS) and involved a high degree of guess work by the scientists, given the weakness of many data collection systems and the extreme complexity of the gears and landing sites involved in the catch of this stock.

In 1994, ICCAT adopted a system of penalties to contracting parties catching more than their allocated quotas, in the previous years. To verify the respect of quotas, ICCAT had no choice but to use the catch statistics produced by its scientific committee, SCRS (largely dependent upon guess work by scientists). The result was immediate: certain scientists, under pressure not to reveal statistics that would result in quota penalties for their countries, simply stopped making assessments of over-quota catches and reported to SCRS only those catch levels that were fixed as official statistics. This resulted in a considerable loss of quality of the estimated catches, which no longer represented the best available information.

The lack of credibility of catch statistics of Bluefin tuna has been a problem ever since, up until recently when a special effort was made by the Commission to control catches to an unprecedented level of quality.

Science and industry: from mistrust to cooperation

The traditional mistrust between scientists and the fishing industry has also been a problem for the CFP. The lack of trust by the industry has frequently resulted in denial of the conservation problems raised by scientists.

This has gradually changed through a gradual process of confidence building in which the Commission has played a key role. Since the mid-2000s, most of the expert groups of STECF have been open to the participation from observers of the fishing industry. Likewise, ICES has practised an open policy towards industry and NGO observers since. This has certainly improved the level of trust and has allowed the industry to have a platform where its opinions on technical issues could be taken into account when assessing the data.

Before the industry was admitted to scientific meetings, they had a picture of a process that was driven by national or other interests, likely to be biased in favour of one fleet or another. Once admitted, the industry discovered a process that was driven by data and common standards, where great efforts were made to ensure objectivity and fairness. The credibility of the science improved greatly. While some attempts by the sector to alter conclusions did take place, these were very few and largely unsuccessful.

However, this initiative is not enough. Industry observers at scientific meetings more often than not provide anecdotal information that scientists cannot easily integrate into their fishery evaluation models. A new kind of approach was necessary and that was the 'industry-science partnerships' whereby industry could establish a more systematic way to provide useful information to scientists.

Such partnerships were developed from the 1990s (Armstrong et al., 2008), in the case of the preparation of advice for severely depleted stocks, where the need for industry buy-in of the scientific evaluations was paramount. Some of these partnerships have been very successful in increasing the trust of the industry in the advice, and the new financial instrument of the CFP (the EMFF, see Chapter 16) has included these partnerships as actions eligible for financial aid.

This is beyond an area where considerable progress has been made in the last 10 years. While scientists were traditionally strongly criticised by the industry (in particular when they recommended strong reductions in fishing) the progress in recent years can be summarised by the statement of an important industry representative during the seminar on the presentation of the scientific advice to stakeholders organised by the Commission in September 2014: 'we have no issue with the scientific advice'.

When comparing such a statement with the industry's arguments only a few years back, it is clear that we have come a long way.

The future scientific advice: new methodologies and new demands of the reformed CFP after 2013

In Chapter 16 we will look into the reform of the CFP in 2013. This reform brought about new objectives and mechanisms that will imply an important extra demand on the scientific community in Europe, both in terms of the amount of advice needed and on its diversity. If anything, the new CFP after 2013 is clearly more dependent than ever before from adequate scientific advice:

- The policy now has very specific objectives for management: maximum sustainable yield for all stocks. This implies that the scientific community will be asked to provide MSY estimates or proxies for all regulated stocks in European Commission.
- The implementation of the discard ban implies that the real levels of discarding will have to be better evaluated, to ensure that the TAC policy associated to the discard ban does not result in an increase in fishing mortality.
- The multispecies approaches will require a new type of advice, beyond the traditional single species considerations.
- The development of regionalisation requires that scientific advice be made available also to this process. This creates a demand for advice on regional approaches.
- The scientific bodies (mainly ICES and STECF) are also required to provide advice on other areas (environmental policy, for example), thus resulting on extra pressure on their limited resources.
- The increasing importance of stakeholder consultation and the increasing public interest in the policy require that the advice be explained. This will require a more intensive dissemination of the advice to non-experts.

Responding to these new demands will not be easy. The data collection with increased financial means should be a key factor but for that the priorities will have to be clearly established and duplications avoided. Moreover, the system of provision of the advice will need to be streamlined, to make optimal use of the human and financial resources. In particular, the advice from ICES and STECF will have to avoid unnecessary duplications.

Another important challenge for the future scientific advice will be the effective incorporation of new methodologies. Some of these may not affect substantially the traditional quantitative advice on annual TAC levels but they will be increasingly important in the assessment of the new policies, in particular the incorporation of environmental considerations and the combination of the CFP with the associated objectives of environmental legislation, particularly the Marine Strategy Framework Directive (see Chapter 11).

The use of such methodologies as new developments in fisheries acoustics, telemetry, molecular genetics, isotope microchemistry and computationally intensive modelling can provide new light into traditional fishery assessment (Metcalfe et al., 2008).

Economic science in the CFP

Traditionally in the CFP, the expression 'scientific advice' is equated to biological advice. Even if economic advice is part of the remit of the STECF since 1992, it is true that fisheries economics has lagged behind biological fishery science for a long time and is only now slowly starting to catch up.

The lack of economic analysis traditionally prevented the discussion of the economic consequences of fisheries decisions. It has been argued that this lack of economic information has prevented the identification of dissipation of resource rent in European fisheries. However, the same authors recognised that the fleet policy was a demonstration that the CFP was indeed sensitive to the bio-economic theory (Frost and Andersen, 2006).

That means that the CFP has been run for many years without discussing the place of the fisheries sector in the context of its contribution to the whole of the European society. Only in recent years this analysis is starting to be done (see next).

Economic versus biological science in the CFP: closing the gap

As we have seen, the initial scientific advice for managing the CFP was exclusively biological advice on the conservation status of fish stocks. This situation was relatively unchanged during quite a long time. Even if economic analysis was needed, it was not available, to the level of completeness, methodological harmonisation and availability of historical data series as those available for biological advice.

For many years, this has been a vicious circle: there was no economic data to inform the political decisions under the CFP but as these decisions were taken without reference to such advice, the need to provide economic advice on a regular structured basis was not created either. This vicious circle was broken through several actions:

- Firstly, the CFP reform of 1992 changed the remit of the existing STCF by turning it into STECF. The addition of the 'E' (which stands for 'Economic') to the committee and the inclusion of fisheries economists in its composition took place in 1992.
- In the early stages after 1992, the use of some research programmes to promote cooperation among fisheries economists. At the beginning, fisheries economists in Europe did not even know each other. A number of research programmes allowed for the community of fisheries economists to meet and develop common methodologies.
- The establishment of time series of data on economic parameters, in the context of the Community support framework for the data collection for fisheries management (see previously). This started in 2007.

Economic science is thus catching up with biological science but the historical gap is such, that it will still take time to have an equally developed scientific basis for decision making on biology and economics.

In addition, the economic advice is also difficult to incorporate into decision making for one reason: it is still very difficult to model the long-term economic effects of management measures. While biological science, in the best cases, can make projections of catch levels in the long term as a result of management strategies, economic science still has trouble to provide comparable scenarios with the desired level of reliability. This is so because certain economic factors, having a crucial effect on the structure of costs for the industry, such as oil prices, are practically impossible to predict in the mid and long term. If economic analysis is used only to measure short-term losses, then there is a genuine risk that this would be used to avoid taking difficult measures in the short term, on account of the immediate effects on profits or levels of employment.

An additional difficulty is that while biological information can be collected quickly using surveys at sea, the current methods for economic analysis require the closure of accounting periods and lengthy data transmission. In practice, biological advice can be given 3–4 months after data are collected but for economic data the lag is of the order of 2 years. This greatly limits the usefulness of such analysis for short-term management decisions.

Finally, one must bear in mind that economic science is a social science and thus intrinsically linked to politics. As we saw in Chapter 6, different governments can have different political objectives regarding the economics of their fishing sectors, from an objective of economic self-sufficiency and high competitiveness, to another objective of social engineering to distribute the wealth of fisheries as evenly as possible even at the cost of low competitiveness. All these objectives are legitimate and respectable but they can be so different that the same economic analysis becomes positive or negative depending upon them. In other words, what some can consider 'economic rationality', in terms of profitability and economic self-sufficiency of the industry, others consider politically undesirable. For this reason, it is much more difficult to obtain an economic advice that would lead to the same policy decisions in all Member States at once.

Overall, the economic advice available is still largely descriptive and is used to inform decision makers of the economic situation of the sector (two years in the past) and to project the economic effects of certain policy decisions. Against this background it is difficult to see how economic advice will one day have the same widely accepted value as the biological advice.

The evolution of the economic advice for the CFP can be summarised as corresponding to four phases:

1 A first phase, from 1983 to 1992, in which economic information was not only missing completely, but was not even mentioned in the CFP.

2 A second phase, from 1992 to roughly 2007 in which the word 'Economic' was introduced in the name of the scientific advisory body of the Commission (from then on the STECF), meaning that the CFP aspired to have economic advice. However, this remained largely a theoretical objective, since the lack of systematic data prevented the preparation of significant and reliable economic advice for the CFP.

3 A third phase from 2007 to recent times with the inclusion systematic collection of economic data allowing for the establishment of the first descriptive reports of the economic situation of the fishing and related sectors in Europe, but largely without analysis of cause/effect relationships. In this phase, the Commission also introduced the preparation of an Impact Assessment as a previous requirement for new legislation. The Impact Assessments have since then increasingly included economic analysis, as economic data started to emerge.
4 A fourth phase, barely started, in which the economic analysis should deepen its capacity to relate the description of the economic situation with the policy decisions, so as to become a useful instrument for decision making, alongside with the biological advice.

Let us look briefly at the two last phases.

The Annual Economic Reports

These reports are produced by the Commission, based on the work provided by the STECF, since the mid-2000s. Today, the Commission services produce regularly a report on the profitability of the European fleets, another one on the aquaculture sector and a third one on the fish processing sector, as we saw in Chapter 6.

STECF is providing such advice, based on data from the data collection regulation. In recent years, the Commission's Joint Research Centre also provides an important contribution. While ICES has considerably diversified its areas of activity (marine environment, etc.) economic science is still not in its remit. At this moment the provider of economic advice to fisheries management is the STECF.

This annual report was started only when there was enough data from time series to evaluate trends. In fisheries economic science, like in its biological counterpart, annual snapshots are of little value; only trends (a 'trend', typically, requires at least a 5-year time series) can provide reliable information on what happens in the real world and why.

Ever since its preparation by the Commission, this annual economic report has revealed a series of extremely interesting findings, as we saw in Chapter 6 (European Commission, 2014).

This economic report must evolve, in future years, into a much more powerful tool for policy making. Given the recent availability of data, the time series are still short and it is therefore difficult to draw meaningful conclusions from the – short – observed trends. However, as the data collection regulation ensures the continuity of the collection of economic data, in the future the annual economic report will be able to study trends in a much more reliable way.

Finally, another important future challenge is to provide these analyses without the current 2-year delay, to be more useful for management. For example, the 2014 report referred to previously included as the most recent data the 2012 data. For this kind of advice to be more useful for decision making, it would be necessary to include more recent information, as it is the case for the biological advice that often includes results from surveys carried out only a few months before the delivery of the advice itself. The methodology to do this exists already. The next challenge is to implement it in practice.

Economic analysis as a tool for fisheries management

At present, the economic picture described cannot clearly be explained as the result of policy making, in most cases. The analysis of the causes behind the trends observed is in its infancy. For example, it has not been assessed so far if the positive trend observed in the profitability of the European fleets observed in the period 2008–2011 is the result of the general improvement of the biological status of the stocks, a fall in the price of fuel, the removal of fleet capacity or other factors.

It would be extremely important to use economic information as a real driver of policy making within the CFP. In this regard, there are two promising precedents that show the way to possible future developments:

1 The use of some data on economic performance of fleets exploiting stocks that have reached MSY levels, as included in the last versions of the annual Policy Statement. Although merely descriptive, this is an illustration, for policy makers and industry alike, that fisheries managed at MSY level tend to be more profitable. For the time being

these are simple examples but the economic analysis should be able to deepen this kind of analysis in the years to come.

2 The use of economic data in the Impact assessment that must precede all significant proposals for management measures. This contribution is important to gauge the economic consequences of adopting different policy decisions and should be a substantial contribution to eliminate the false image of the CFP being a policy 'for the fish and not the fishermen'.

Ideally, the economic data could, one day, build bio-economic models for fishery management. This will be of particular relevance and interest in the context of future multispecies models, where the different options for managing complex, mixed fisheries could be analysed not only in terms of the achievement of MSY for the leading stocks but also in terms of the economic benefits of different exploitation and management patterns. The contribution of economic science to the impact assessment work of these plans should be fundamental.

Likewise, the use of economic analysis should allow in the future for cause/effect relationships to be drawn between, for example, the structural policy and the economic performance of the sector and to evaluate the economic effects of the policy and suggest possible future changes. The structural policy for fisheries is supposed to address market failures, but in practice the decisions on the structural policy have been taken without an in-depth analysis of the real market failures in EU fishing, and without serious economic analysis of the effect of existing structural instruments in addressing the structural problems of the sector.

This would eventually change the decision making on the structural policy from a purely political discussion into a more objective, science-based approach, so as to be able to measure, for the first time, its real impact of the structural policy on the achievement of the CFP objectives as laid down in the Treaty. A better economic science basis should provide a basis for a more evidence-based approach for the next generation of the structural policy after 2020.

Another emerging area where economic advice will become necessary in the future is the convergence of the CFP with other instruments or policies such as the Marine Strategy Framework Directive (see Chapter 11) or the Commission policy on 'blue growth' (COM, 2014b). These three elements together will lead to the future need to make difficult decisions on the best use of the maritime space and resources. This will make it necessary the study of the trade-offs between different options for the use of this environment, including also the externalities of the different uses, which implies *inter alia* the development of estimations of the so-called 'ecosystem services' in the European seas.

Finally, there is also an important ground to develop in terms of the study of the economic factors underlying the competitiveness of the European fishing and aquaculture sector. In a world where we can only expect further globalisation and market openness, it is essential to better understand the key factors (in EU or national policy) that promote or undermine competitiveness.

All these elements point to a future where, thanks from the continuity of time series of socio-economic data in the context of the data collection regulation, and through in-depth analysis of the drivers of economic change in the European fishing sectors, the economic advice will eventually become an essential part of the advice for the CFP, on an equal footing with the biological advice.

Notes

1 International Commission for North Atlantic Fisheries, later replaced by the Northwest Atlantic Fishery Organisation (NAFO) in 1979.
2 European Currency Unit.
3 ICCAT's scientific body is the Scientific Committee for Research and Statistics, SCRS.
4 Inter-American Tropical Tuna Commission.
5 www.fao.org/docrep/005/v9878e/v9878e00.htm

References

Anderson, E.D. (1998). The history of fisheries management and scientific advice – the ICNAF/NAFO History from the end of World War II to the Present. *Northw. Atl. Fish. Sci.* **23**: 75–94.

Apostolaki, P., G.M. Pilling, M.J., Armstrong, J.D. Metcalfe and R. Forster (2008). Accumulation of new knowledge and advances in fishery management: two complementary processes? In *Advances in Fisheries Science. 50 Years from Beverton and Holt.* A. Payne, J. Cotter and T. Potter (eds). Blackwell Publishing. Cefas, pp. 229–254, 546 p.

Armstrong, M.J., A.I.L. Payne and J.R. Cotter (2008). Contribution of the fishing industry to research through partnerships. In *Advances in Fisheries Science. 50 Years from Beverton and Holt.* A. Payne, J. Cotter and T. Potter (eds). Blackwell Publishing. Cefas, pp. 63–84, 546 p.

Beverton, R.J.H. and S.J. Holt (1957). *On the Dynamics of Exploited Fish Populations. Fishery Investigations.* London. Series II, XIX, 533 pp.

COM (2014a). Communication from the Commission to the European Parliament and the Council. Concerning a consultation on Fishing Opportunities for 2015 under the Common Fisheries Policy. COM (2014) 388 final.

COM (2014b). Communication from the Commission: Innovation in the Blue Economy: realising the potential of our seas and oceans for jobs and growth. Brussels, 13.05.2014. COM(2014) 254/2.

Council Decision (1987). Council Decision of 19 October 1987 adopting Community research and coordination programmes in the fisheries sector for the period 1988 to 1992. *OJ* L 314, 4.11.1987, p. 20–23.

Daw, T. and T. Gray (2005). Fisheries science and sustainability in international policy: a study of failure in the European Union's Common Fisheries Policy. *Marine Policy* **29**(2005): 189–197.

EC (2000). Council Regulation (EC) No 1543/2000 of 29 June 2000 establishing a Community framework for the collection and management of the data needed to conduct the common fisheries policy. *OJ* L 176, 15.7.2000, p. 1.

EEC (1979). Commission Decision 79/5721/EEC of 8 June 1979 relating to the institution of a Scientific and Technical Committee for Fisheries. *OJ* L156, 23,6,1979.

European Commission (2014). The 2014 Annual Economic report on the EU Fishing Fleet (STECF 14–16). JRC Scientific and Technical Policy Reports. Scientific, Technical and economic Committee for fisheries. EUR 26901 EN. Joint Research Centre.

Farnell, J. and J. Elles (1984). *In Search of a Common Fisheries Policy.* Gower Publications, Aldershot.

Frost, H. and P. Andersen (2006). The Common Fisheries Policy of the European Union and fisheries economics. *Marine Policy* **30** (2006): 737–746.

ICES (2002). Report of ICES Advisory Committee on Fishery management, 2002. Copenhagen, 21 May – 30 May 2002. Copenhague, 9 October – 17 October 2002. Part I.

ICES (2007). *Report of the ICES Advisory Committee on Fishery Management 2007.* Book 5: Celtic Sea and West of Scotland. ICES.

ICES Advice (2013). *Mixed-fisheries advice for Subarea IV (North Sea) and Divisions IIIa North (Skagerrak) and VIId (Eastern Channel).* ICES Advice 2013, Book 6.

Metcalfe, J.D., Righton, D.A., Hunter, E. Neville, S. and D.K. Mills (2008). New technologies for the advancement of fisheries science. In *Advances in Fisheries Science. 50 Years from Beverton and Holt.* A. Payne, J. Cotter and T. Potter (eds). Blackwell Publishing. Cefas, pp. 255–279, 546 pp.

Pawson, P. (2008). The contribution of science to management of the North Sea cod (Gadus morhua) and UK seabass (*Dicentrarchus labrax*) fisheries: can we do better? In *Advances in Fisheries Science. 50 Years from Beverton and Holt.* A. Payne, J. Cotter and T. Potter (eds). Blackwell Publishing. Cefas, pp. 155–183 546 pp.

Pinnegar, J.K., Trenkel, V.M. and J.L. Blanchard (2008). 80 years of multispecies fisheries modelling: significant advances and continuing challenges. In *Advances in Fisheries Science. 50 Years from Beverton and Holt.* A. Payne, J. Cotter and T. Potter (eds). Blackwell Publishing. Cefas, pp, 325–357, 546 pp.

Reeves, S., Marchal, P., Mardle, S., Pascoe, S., Prellezo, R., Thébaud, O. and M. Travers (2008). From fish to fisheries: the changing focus of management advice. In *Advances in Fisheries Science. 50 Years from Beverton and Holt.* A. Payne, J. Cotter and T. Potter (eds). Blackwell Publishing. Cefas, pp. 135–154, 546 pp.

Rozwadowski, H.M. (2002). *The Sea Knows No Boundaries. A Century of Marine Science Under ICES.* International Council for the Exploration of the Sea, Copenhagen, in association with University of Washington Press, Seattle and London. 410 pp.

Symes, D. (2010) cited in Schweiger, L. (2010). The evolution of the Common Fisheries Policy: Governance of a Common-Pool Resource in the Context of the European Integration. Institute for the European Integration. *Austrian Academy of Sciences. Working Paper* No 07/2010. November 2010, 88 p.

Symposium on Multispecies Models Relevant to Management of Living Resources (1989), 2–4 October 1989. *GeoJournal.* April 1989, **18**(3): 334.

CHAPTER 11

Fisheries and the environment

Environmental performance of the CFP

The CFP, like all other EU policies, must integrate environmental concerns, as established by Article 11 TFEU,[1] which reads: 'Environmental protection requirements must be integrated into the definition and implementation of the Union's policies and activities, in particular with a view to promoting sustainable development'.

This does not mean that the CFP is in any way hierarchically dependent on environmental policy. It means that when managing fishery resources, the CFP must also take into account environmental concerns. However, the relationship between fisheries and environmental policy goes beyond this basic requirement. In fact, fisheries and environmental issues very often overlap in the marine environment and in recent years fisheries issues have drawn increasing attention from the environmental action groups and the public at large. For this reason, there is a close relationship between these two policies and this relationship is increasing over time.

The integration of environmental concerns in the CFP

The question of the integration of the environmental concerns in the CFP has always been elusive and often controversial, leading the Commission to produce a number of guidance documents on this and related topics, such as the 1999 Communication on nature conservation and fisheries management (COM, 1999), or the one on elements for an integration strategy (COM, 2001). These papers set the tone of what would be a growing awareness to make the CFP an environmentally friendly policy. But it is interesting to trace the evolution of the integration of environmental concerns in the basic text of the CFP over the last decades.

Environmental objectives did not feature in the first instruments of the CFP. The 1983 basic regulation was completely silent on the need to ensure the achievement of environmental objectives.

In 1992, the new basic regulation[2] included a reference to the objectives of the policy as:

> ...to provide for the rational and responsible exploitation on a sustainable basis, in appropriate economic and social conditions for the sector, taking account of its implications for the marine eco-system and in particular taking account of the needs of both producers and consumers.

As we can see, there was a reference to the effects of fishing on the environment but clearly as a second priority in comparison to economic and social factors and to the interests of producers and consumers.

On the occasion of the second reform of the CFP in 2002, the integration of environmental objectives into the CFP was made more explicit in its legal basis (EC, 2002). In particular, this reform clearly spelled out as its objective the achievement of: 'environmental, economic and social sustainability' in European fisheries, clearly for the first time putting economic, social

The Common Fisheries Policy: The Quest for Sustainability, First Edition. Ernesto Penas Lado.

and environmental objectives exactly at the same level of priority. This formula, in fact, was repeated for the 2013 reform, as we will see in Chapter 16.

Beyond this very general objective, the Commission tried to translate the generic objective of integrating environmental concerns in the CFP into operational conclusions, through a new Communication (COM, 2002). And, meaningfully, the services of the Commission created by that time a new Unit in the Directorate General for Maritime Affairs and Fisheries to deal specifically with this question.

This communication set out priority and complementary objectives. Among the first, the Commission proposed a general reduction of fishing pressure on European stocks as a means to also achieve environmental objectives. The reduction of discards and the elimination of subsidies that enhance capacity completed the priorities. In addition, that communication also proposed a number of measures that anticipated future policy developments.

It is to be noted that in two out of these three main objectives, the fisheries and the environmental objectives were perfectly consistent. This should be no surprise. Fisheries and environmental policy should reinforce each other: reducing fishing pressure to ensure sustainability of fisheries also has positive effects on marine ecosystems and their biodiversity. Conversely, the preservation of healthy marine ecosystems from man-made disturbance (pollution, drilling etc.) also contributes to ensure the sustainability of fishery resources.

This communication, however, was not easy to translate into specific policy, in particular because it did not contain indicators that could be used to measure progress in achieving the objective of integrating environmental concerns. For this reason, a new communication was presented in 2008, where the principles and objectives of this communication were translated into specific indicators of environmental performance of the CFP (COM, 2008).

These indicators require data for their meaningful utilisation. And, since 2009, the Data Collection regulation includes the collection of data to estimate these indicators.

At the same time, ICES has started to provide preliminary results on the first three indicators: (1) the conservation status of species, (2) the proportion of small fish in the catch and (3) the average maximum length of the fish. Other indicators require data series that have not been traditionally available, so the evaluation of results will still take some time. When these historical data series become available, the question of the integration of environmental concerns into the CFP will be done on a regular, science basis and that would be the right moment to evaluate the degree of success of such integration.

In any case, the relationship between the two policies has traditionally been controversial. But this is changing and changing fast. As we will see next, the integration of the two policies and the synergies between them have been clearly increasing in recent years.

What has been achieved

The CFP has already incorporated a very substantial number of measures whose objective is clearly environmental, rather than related to fisheries management. A number of these instruments can be mentioned:

- The CFP has made considerable progress in the key areas mentioned in the 2002 communication: fishing pressure has been reduced (see Chapter 4), aid to new vessel construction has been discontinued and discards have finally been subject to an active no-discard policy, adopted in 2013;
- The CFP is already the carrier of measures to protect certain marine habitats of high biodiversity from the effects of fishing, in the framework of the implementation of the Natura 2000 network under the Habitats Directive (Directive 43/92: EEC, 1992a). Cases such as the Darwin mounds, the seamounts around the Azores, the Porcupine bank and so on are very good examples. These measures were introduced gradually through the TAC and quota regulations and, after the Treaty of Lisbon (when technical measures can no longer be included in that annual regulation), they have been consolidated as an integral part of the technical measures regulation (EU, 2013a);
- The CFP has also adopted measures to protect certain species that are killed accidentally in some fisheries: the regulation to protect cetaceans

through the compulsory use of pingers in gillnet fisheries is also a good example (EC, 2004a);

- The CFP has introduced the prohibition to catch certain endangered species, in the context of the annual decision on fishing possibilities. Following the example of the ban on catching basking shark (*Cetorhynus maximus*) in the early 2000s, the annual TAC and quota regulation has included, over the years, an increasing number of species entirely protected from fishing. In the regulation for 2015, as many as 28 species are banned from fishing, in different areas of the world (EU, 2015). In certain cases, the CFP has even taken measures to protect the food source of certain wild species. The closed area to forbid fishing for sandeel in the North Sea is a measure to protect the food for the kittiwake seagull, an endangered species (Article 29a of Regulation 850/98);
- Other regulations have imposed obligations on Member States that imply a combination of fishery restriction measures and habitat restoration measures. This is the case of the regulation on the management of European eel (EC, 2007a);
- In terms of governance, the CFP has turned environmental NGOs into mainstream stakeholders, having their presence assured in the new consultative bodies of this policy: the Regional Advisory Councils and, with a reinforced presence, the Advisory Councils of the 2013 reform (see Chapter 16);
- Finally, the CFP has also adopted action plans to enhance the adoption of mitigation measures to reduce the accidental catches of seabirds (COM, 2012).

Overall, most of the elements of the 2002 strategy are totally or partially incorporated into the CFP.

Fisheries and environment: a difficult relationship

Despite these achievements, fisheries and environmental policy have not always been hand in hand and the synergies between the two could have been much better exploited and developed. Very often, environmental policy has appeared as hostile to the fishing industry and environmentalists have also often considered fishermen as environmentally unfriendly. The fact that the two policies overlap in a number of areas has only aggravated the difficulties of conciliating the two.

These difficulties have manifested themselves in different contexts and with different topics. Differences of approach appear between fishermen and environmentalists. This is very particularly the case in relation with large scale fishing, the usual target of environmentalists' campaigning, while small scale fishing enjoys a reputation of low impact on the environment that makes it much less contentious in the environmental debate, even if the scientific evidence for that preference is not necessarily clear.

These discrepancies have often touched upon other areas under the purview of the CFP, such as aquaculture development, which has traditionally been seen by environmentalists as an environmentally unfriendly activity.

Beyond these differences, it is also the case that different departments of the Commission and also different departments of Member States have often conflicting views about the relationship between fisheries and aquaculture, on the one side and environmental conservation on the other.

Meaningfully, in the context of the stakeholder consultation bodies of the CFP (see Chapter 12 on Regional Advisory Councils) one of the main problems is that of reconciling the views of the fishing industry with those of representatives from environmental NGOs.

Despite these traditional difficulties, a lot has been achieved in terms of making fishing activity more environmentally friendly. One can certainly argue that the CFP has made more efforts to be environmentally friendly that it is often given credit for. Such credit, however, has been increasingly recognised in recent years and the difficulties outlined here are certainly much less significant than they used to be in the past.

Fisheries and the environment as complementary policies

Although the CFP is related to other EU policies, the relationship between fisheries and environmental policy is of particular relevance and has traditionally been difficult and often contentious. This is not surprising: ultimately, fisheries management is about balancing the need and urgency to ensure sustainability of biological resources and the natural environment around them with the socioeconomic consequences.

In addition, there are many issues where the two policies overlap. The way the two policies interact and handle these overlapping responsibilities has a considerable influence on the outcome of the CFP itself.

There are many examples of issues where fisheries and conservation policy overlap. The best example is that of the protection of certain species that are a by-catch of the fishing activity and, in some specific cases, even target species. Pelagic sharks are the best example.

The interaction between fisheries and environmental policy has been difficult in the past. There is, however, and evidence of that emerges all the time: when fisheries managers fail to address a conservation problem for a species or a group of species, the issue ends up becoming an environmental one and environmental bodies tend to take over what fisheries managers have not been able to deliver. Let us look at some examples.

Pelagic sharks

The case of pelagic sharks is the most clear. Pelagic sharks have been secondary species in many European fisheries for a long time. In recent years, the improvement in the treatment of shark meat on board has allowed a higher economic return from shark catches and this in turn has changed the perception of sharks by parts of the industry, from secondary species to be avoided to real target species.

However, the preservation of sharks (whose low fecundity makes them very vulnerable to overexploitation) has in the past been more a question of preservation of endangered species than one of management of commercial species. As a result, they were not a priority for the organisations dealing with fisheries management.

Since the late 1990s ICCAT has been collecting statistics about pelagic sharks, caught as by-catch of certain fisheries for tuna but also being targeted directly in certain pelagic long-line fisheries for swordfish. However, other priorities in ICCAT always prevented progress on this file. As a result, as fisheries managers did not take serious steps for the management of these stocks, the environmental dimension took over and the first initiatives to protect sharks did not come from fishery organisations but from environmental ones.

An additional complication has been the issue of shark finning. This practice, cutting and landing the shark fins (for the very lucrative far-East markets for shark-fin soup, a delicacy traditionally served in wedding parties) while discarding the shark carcass, was very common among long-liners targeting tuna for the sashimi market. The issue became, already in the late 1990s an important issue in terms of the possible effects of this practice on the conservation of the stocks concerned. What turned the question into an issue of massive public interest was the animal welfare dimension: pictures showing the return to the sea of live sharks with their fins cut hopelessly sinking to the bottom of the sea incapable of swimming had an immediate impact on public opinions.

Under these circumstances, what was in principle a question of management of exhaustive fishery resources, gradually turned into a question of preservation of biodiversity and subsequently a matter of animal welfare. The Union, in this regard, took measures to ban shark finning already in 2003 (EC, 2003a), even if this practice was not common among European vessels, which had learned to use the whole shark carcass for different purposes and generally landed everything.

However, allegations that this regulation was not being controlled properly led to a tightening up these rules, by introducing the obligation to slice and fold the fins to leave them attached to the carcass until landing (EU, 2013b).

Whaling

The ultimate example is, however, the management of whales in the framework of the International Whaling Commission (IWC). Although in principle this organisation was created to manage the whale populations, in fact the policy agreed at IWC since the establishment of a moratorium on whale hunting in 1986 has been to preserve whales irrespective of the possibilities to exploit sustainably at least some of the species.

The status of whale populations is evaluated regularly by the IWC[3] and the picture is mixed. Some species, such as blue whale, are clearly recovering, although their overall abundance is still low. Humpback whales in the North Atlantic seem to have recovered to perhaps pre-exploitation levels, with 12 000 individuals. Fin whales in this area number around 35 000 individuals and, in the central North Atlantic and the west of Greenland they seem to be in a healthy state. Minke whales are also abundant, numbering around 180 000 individuals in the North Atlantic. The North Atlantic right whales on the contrary, shows little sign of recovery its populations amounting to between 300 and 350 individuals.

This evaluation can compare with that of most important fish stocks. However, the moratorium on commercial whale hunting continues with a strong support from the EU countries, in a demonstration that whales in Europe are no longer considered an exploitable resource but a natural asset to be preserved irrespective of their status of conservation. Under this perspective, the IWC has turned into a whale conservation body, rather than a whale management one, even though the scientific basis for a management strategy has long been ready. Consequently, this is considered today, *de facto*, an environmental policy that is handled by the Environmental Department of the Commission services.

Eels

Another example of the interaction between fisheries and environment is the management of European eels. This extraordinary species was on the brink of extinction in the late 1990s. According to ICES, the arrival of glass eels to European rivers in the late 1990s was just 1% of the abundance 20 years earlier.[4] This led the Union to adopt a regulation to recover this species (EC, 2007a).

Since the species is catadromous (it reproduces in the sea but spends most of its lifetime in fresh water) part of the life cycle of the species was outside the scope of the CFP and this required a very specific approach to management. In addition, it was clear that the collapse of the populations was a mixture of different factors: (1) excessive fishing, (2) the degradation of the river systems through pollution or the construction of barriers (dams and locks) and (3) possibly certain diseases. Consequently, the solution to the problem was and still is today, a mix of measures on fisheries and on environmental policy.

The 2007 regulation was a rather unique legislation, in that it established a directive-like obligation for Member States to deliver a result, giving each of the them the freedom to find the best combination of different measures such as limiting the fishing of eels (either the juveniles, glass eels or the adults, silver eels), recovering the escapement levels of eels in rivers through the removal of physical barriers to the migration of eels, or the use of re-stocking for the purpose of recovering the wild population.

As the conservation of this stock had a major trade component (most catches of glass eels in Europe were exported alive to the Chinese market) the issue soon became an environmental one when the Union proposed and obtained the listing of European eel in list II of CITES.

The management of eels constitutes therefore a unique combination of restriction of fishing, environmental restoration and trade measures. The last scientific reports point to a recovery in the number of glass eels coming back to European rivers although the status of the stock continues to be critical.[5] The Commission has analysed the implementation of the current management plans (COM, 2014), concluding that the positive trend in recruitment cannot yet be attributed necessarily to the effects of the management plans. It was also critical with the practical use of re-stocking.

It is very difficult to foresee the future of this file, since the long life span of eels (around 20 years) makes it very difficult to check if the current management regime is really conducive to a full recovery of the European eel stock. In any case, the management of European eels has opened up a new style in managing natural resources, through a rather unique mixture of de-centralised fishery management and environmental and market measures. The evolution of the eels must be followed very closely, not only to ensure that the apparent recovery in recent years is confirmed and consolidated but also to see if this alternative approach to the management of an important natural resource does indeed provide satisfactory results.

Marine mammals

We will refer next to the ban on driftnets, a measure taken in the 1990s to address the problem of the mortality of dolphins caused by that gear but the adoption of the ban on driftnets had another consequence: it was obvious that driftnets were not the only gear producing mortality on cetaceans. Other gear, notably gillnets, also produced such mortality. From this perspective, even a rather heavy-handed measure such as the ban on driftnets would not in itself address the problem of the mortality of endangered cetaceans. This led the Commission to propose and Council to adopt a regulation laying down the compulsory used of mitigation measures in certain fisheries (EC, 2004a). The solution was to require in gillnet fisheries (where dolphins are widely reported to get entangled and choke to death) the use of acoustic deterrent devices, the so-called 'pingers', devices that produce a sound that scares dolphins and prevent them from getting closer to the gillnets.

The implementation of the pingers regulation was subject to an analysis by the Commission in 2009 (COM, 2009). This analysis showed that the regulation was implemented in a very variable way, with some Member States having made genuine efforts to reduce cetacean by-catches, and others clearly lagging behind. Despite the patchy implementation, no action was taken against Member States not fully implementing this regulation. The reasons are that such a stand-alone regulation has not been a priority for national inspection systems.

A further report was prepared in 2011 by the Commission (COM, 2011). The conclusions were in fact quite similar to those of the earlier report. It is important to draw lessons from this. True, a number of Member States have not been very forthcoming in implementing the regulation, let alone reporting on such implementation. However, the existence of many other priorities and problems to solve has led to the de facto consideration of this regulation as a secondary priority.

In addition, the horizontal one-size-fits-all nature of this instrument is also resented as out of tune with the recent trend towards more region-specific approaches in the CFP. These considerations were in fact leading reasons that contributed to develop a new approach for the future, in the context of the 2013 CFP reform: the future integration of measures to protect marine mammals within the framework of future long-term management plans, thus transforming the current stand-alone approach into a full integration of this objective into the mainstream instruments for fisheries management (see Chapter 16).

Seabirds

This is a similar case to that of marine mammals. Many seabirds, which feed on fish, are caught and killed in fishing gear when trying to catch their prey. In some cases, this mortality can contribute to seriously endangering the species, as in the case of albatross.

This led to the adoption by FAO of an international plan of action to reduce the by-catches of seabirds.[6] This was subsequently turned into an EU action plan (Communication COM, 2012, 665, see previously) where a number of mitigation measures are identified. The real value of this strategy is not the communication itself but the way in which it should be implemented: in the context of future management plans, issues related to seabird by-catch should be considered and mitigation measures implemented as appropriate, in a de-centralised way, adapted to the specific problems identified and to the characteristics of the fisheries concerned. For a number of years the Union did not take specific measures in this regard. The reason being that the problems of seabird mortality were associated mainly with the use of surface long-lines in fisheries for tuna, notably carried out by fleets from the Far East. When setting and hauling the long-line at daylight, seabirds were attracted by the bait and got caught by the hooks. The main European fleets using surface long-liners in international waters (the swordfish long-liners) set their gear at night, so it was considered that they produced much lower mortality.

It is important to underline that, unlike previous cases, the Commission chose here a soft-law instrument: the action plan contains a number of principles and recommendations but it is not a legally binding instrument in itself. However, it includes a commitment to consider that, in future long-term management plans for fish stocks, mitigation measures to reduce seabird by-catches will have to be considered if and when relevant. In such cases, the mitigation measures would become legally binding in the relevant cases. Ensuring that the future multispecies management plans include such measures (in all likelihood, under regionalised approaches) would be the challenge to turn this soft law instrument into an effective one.

Environmental questions as drivers for fisheries decisions

This is an area of overlap between the two policies that has always existed but that has been growing in recent years. Today it is somehow difficult to know the origin of certain fisheries regulations that 'look' inspired by fisheries considerations alone. But in fact some of them may have originated, or being fundamentally affected, by environmental considerations. This trend is inseparable from that of the growing interest of environmental NGOs in fisheries policy in recent years.

The interest of environmental NGOs in marine conservation issues is relative old. In the first years, however, it concentrated on the defence of certain endangered, iconic marine species such as whales or dolphins. Not surprisingly, campaigns against whale hunting (leading to the moratorium adopted by the IWC in 1986) and the initiatives on 'dolphin safe' tuna fishing in the eastern Pacific or the campaigns against large scale driftnet fishing (known to kill marine mammals) were the main focus in the 1970s and 1980s.

Gradually environmental NGOs have extended their field of interest into fisheries activities proper. Interest in the commercial fisheries for iconic fish has been growing steadily over the last decade, with cod, Bluefin tuna and various shark species attracting much attention.

The best example to illustrate the influence of environmental considerations in the CFP is the conflict over the use of driftnets in the Bay of Biscay, in the 1990s.

The seasonal fishery for albacore tuna (*Thunnus alalunga*) was conducted, primarily by Spanish and French fishermen, from June to September in a large area from the Azores to the Bay of Biscay, coinciding with the annual migration of this species. Up to the 1990s, the fishery was carried out with such traditional gear as trolling lines and pole and line with live bait. These gears were very labour intensive and their high running costs led France to develop more industrial types of fishing for this stock, notably driftnets and pelagic trawling.

The driftnet fishery soon proved very efficient and made headway in France, with other Member States (UK, Ireland) also developing this fishery which, at the time, was completely unregulated, either at Community level or in the framework of ICCAT. The development of the driftnet fishery, in an unregulated context (and thus, without relative stability) altered the *status quo* and led to intense competition among the gears used and the Member States using them. This competition was aggravated by the fact that the main end market for the fish was the same: the canneries of northern Spain.

This gear conflict unfolded in an international context where the UN had taken initiatives against the use of large-scale driftnets which, in certain areas of the world, were known to be responsible for the incidental mortality of dolphins: the UN Resolution 44/225 of 15 March 1990, followed by two further Resolutions (45/147 of March 1991 and 46/125 of February 1992) seeking a moratorium and recommending a substantial reduction in drift netting in the short term. The question was also discussed in the agenda of the UN Conference on the environment and development (UNCED) held in Rio de Janeiro in 1992.

This led to a first measure to ban driftnets of more than 2.5 km in length (EEC, 1992b). However, this legislation had a number of derogations (some driftnets were authorised up to 5 km) and proved soon very difficult to enforce, leading the Commission to consider that the rules were inadequate. Considering that the *status quo* was not an option the Commission took the view that a ban on all oceanic driftnets for catching tuna, swordfish and other highly migratory species was necessary (COM, 1994) and in April 2004 introduced a proposal to that effect.

The new proposal met considerable opposition by Member States having developed the driftnet fishery in recent years, even if the proposal was carefully crafted to affect only the kind of driftnets that produce dolphin mortality and not the small-scale, coastal ones that have no significant effect on them. At the time, Member States still considered this issue a gear conflict rather than an environmental issue. As such, some Member States saw no reason to ban a gear that operated in a – then – unregulated fishery.

However, as the evidence of dolphin mortality in tuna driftnets became known to the general public, the picture finally changed dramatically. The UK government, up to then contrary to the driftnet ban on account of the defence of the legitimate interests of its fishermen, changed position when British public opinion, traditionally very sensitive on animal welfare issues, was made aware of the dolphin mortality. What had been up to then a fishery issue suddenly turned into an environmental one. The UK shifted its position and favoured the driftnet ban. This altered the political deadlock in Council in favour of the ban and in 1998 this was finally adopted, four years after the Commission proposal (EC, 1998).

Importantly, the evidence that certain driftnets in the Mediterranean were also responsible for dolphin mortality (the *thonaille* for Bluefin tuna and the *spadara* for swordfish) led to this ban to apply to the Mediterranean as well. The fact that the issue had originated around a gear conflict in the Bay of Biscay created a certain sense of unfairness among the Mediterranean fishermen concerned: they saw themselves paying for somebody else's problem and the implementation of this ban would become a problem in the following years. However, the environmental dimension of the whole issue made it inevitable to address the problem in the Mediterranean as well. This led to the adoption of buy-out schemes, whereby the Union provided for funds to help the fishermen concerned find alternatives to the need to abandon driftnet fishing.

This was the first time in the CFP that a given fishing gear was not just regulated or limited but simply banned and the reasons why it was banned were clearly environmental. This also set a precedent for the banning several years later of salmon driftnets in the Baltic, where they were responsible for incidental mortality of the Baltic harbour porpoise, a cetacean on the brink of extinction. This ban was adopted in 2005 (EC, 2005a). Here too, like in the Mediterranean, the users of the driftnets have always resented the decision, particularly since the abundance of harbour porpoise is so low in the Baltic, that many fishermen claim they have never seen one.

Implementing environmental law through the CFP

The cross-cutting nature of environmental policy makes it necessary to implement such policy through the CFP, when dealing with environmental issues affected by the fishing activity. The interaction between fisheries and environmental policy at EU level is made complicated because of the different competence level: while environmental policy is a mixed EU/Member State competence, fisheries policy under the CFP is an exclusive competence. This means that, when implementing environmental policy in the marine environment, Member States cannot take fisheries decisions at national level affecting other Member States: the CFP has to come into play.

There are a number of examples of this but the most clear being the implementation of the Natura 2000 network in the marine environment.

Natura 2000 and the CFP

The implementation of the Natura 2000 network in the marine environment is an essential part of the implementation of the Habitats directive (EEC, 1992a) and the Wild Birds directives (originally adopted in 1979, recently codified in 2009) (EC, 2010).

In principle, the implementation of environmental law requires that the different elements of this policy be handled according to sectorial competences between the Union and Member States. This means that, when any issue of implementation of environmental law interacts with fishing, which as an exclusive competence, the implementation must take place through the mechanisms of the CFP. In other words, the CFP intervenes only in a given and specific moment within the overall workflow of the implementation of Natura 2000, notably downstream of the main decision to declare a zone as part of the network. However, the CFP had never created a specific mechanism to do so.

This created a contradiction: while Member States were under an obligation to implement Natura 2000, including measures to protect the zones from other human activities, they could not adopt fisheries measures to protect these zones as such measures are an exclusive competence of the Union.

The Natura 2000 areas in the marine environment require action by Member States, potentially with respect to a number of human activities. The preservation of a sea bottom area with high biodiversity may require protection from such activities as oil and gas drilling, gravel extraction, passage of pipelines and so on. In addition, it may require protection from certain types of fishing activity. But while the former activities are to be regulated by Member States themselves, protection from fishing is an exclusive competence of the Union and therefore has to be carried out under the CFP.

The transition between environmental protection under Member State responsibility and the protection from fishing under the CFP has not been smooth. And, until the reform in 2013, the CFP did not have any particular methodology to do it.

The best example is the adoption of the protection regime for the so-called 'Porcupine Bank', to the west of Ireland, in December 2007. In this case, Ireland notified to the Commission its intention of declaring this area as a protected area under Natura 2000 because of the existence of deep-water coral beds (notably of the species *Lophelia pertusa*). The Commission, subsequently, proposed the prohibition to fish in the areas concerned, plus a buffer zone. To facilitate control, Ireland proposed that all trawling be prohibited in this area.

The proposal was examined by the Fisheries Council in December 2007 and was initially rejected by Member States, mainly on grounds of its lack of proportionality: all trawlers were excluded, including pelagic trawlers, which by definition would never touch the bottom of the Porcupine Bank. An agreement was possible only once pelagic trawling was excluded and specific and proportionate control measures were established.

This case illustrates the mismatch between the obligations for Member States under the habitats directive and the decision-making process under the CFP: there was a genuine risk that, after a Member State had decided to protect an area under Natura 2000, the CFP would have not adopted the necessary measures to protect the area from fishing.

Despite these difficulties, the CFP has been quite efficient in incorporating marine Natura 2000 sites in the *acquis* of the policy: in 2003 and 2004 the Commission already brought protection to the Darwin Mounds through emergency measures (EC, 2003b) and this protection was extended for another 6 months (EC, 2004b) until it was consolidated by Council in 2004 (EC, 2004c). In 2005, the seamounts around the Azores islands were protected (EC, 2005b). The case of the Porcupine Bank in the West of Ireland was adopted in 2007 (EC, 2007b) and an area to the North of Spain ('el Cachucho') was adopted in 2009 (EC, 2009).

The case of measures within the 12 miles of Member States represented a simpler approach, as in the case of the so-called 'Voor Delta' (EC, 2008a). Here, the implementation was allowed under the provisions of Articles 8(3) and 9(2) of Regulation 2371/2002, that is, under national rules. The reason was simple: within 12 miles only the vessels from the coastal State can fish, plus those vessels from neighbouring Member States authorised to fish between the 6 and 12 miles under the derogation to the 12-mile regime (see Chapter 2). The Netherlands consulted the national fleets and those with a right to fish between the 6 and 12 miles and could adopt a protection regime without the need for a Council decision.

For a number of years, these measures were repeated every year in an annex to the annual TAC and quota regulation. However, as of entry into force of the Treaty of Lisbon, which clarified – and clearly limited – the kind of measures that could be adopted by Council alone around annual TACs (Article 43(3) TFEU) this possibility disappeared. This has produced a period where the implementation of Natura 2000 sites under the CFP did require co-decision.

This has been resolved only under the 2013 CFP reform, where a specific, fast-track procedure was adopted to implement fisheries measures in Natura 2000 sites, through a regionalised procedure ending up in a Commission delegated act (Article 11 of Regulation 1380/2013). However, it must be clear that this provision only caters for the implementation of fisheries measures under the CFP, not any other measure to protect these areas from other human activity.

The ecosystem approach

The implementation of the ecosystem approach is an obligation under the CFP. The principle in itself has been agreed by the Union in the framework of international conventions, such as the Convention on Biological diversity or the United Nations' World summit on Sustainable Development of 1992. These translate into the inclusion of the obligation for the CFP on the 'progressive implementation of an ecosystem approach to fisheries management' which is included among the objectives of the CFP ever since 2002, in its basic Regulation 2371/2002 and reiterated in the 2013 reform.

The ecosystem approach is a notion that everyone agrees with but that everyone understands in a different way. Turning that notion into something workable is, however, a major challenge. Many authors have proposed specific ways to do this, with an extremely wide array of possible approaches.

For example, certain authors have applied this notion through the use of fleets as the centre of the study, as a way to put together the fleets and the marine ecosystem interact to address the ecological, economic and social pillars of the sustainable development of fisheries, using Celtic and North Sea case studies (Gascuel et al., 2012).

Other authors have made very interesting studies of the practical consequences of the application of this notion in legal and management terms (Sobrino et al., 2010).

Overall, the most crucial question around the definition of the ecosystem approach in the CFP is the problem of defining the boundaries of the exercise, and the balance between the scope and the feasibility of the approach:

- A relatively narrow interpretation (commercial fisheries and effects on non-target species) is in principle easily workable, but if leaves out many other elements (physical oceanography, climatic questions, effects on species other than iconic by-catch species etc.) that can be considered as part of the marine ecosystem.
- A very wide scope (considering all biological, physical and socioeconomic elements around fisheries) would be more complete, but the required level of information involved would be such that the feasibility of the approach would be compromised.

Overall, the tendency around the CF has been to consider an incremental approach, that is, to start with the known elements of the CFP (the management of commercial species) and gradually add new elements into the picture (notably effects on by-catch species, effects on the marine communities etc.) as the information on these additional elements becomes available.

Implementation in the CFP

The ecosystem approach has been defined in many ways. For example, the Convention on Biological Diversity defines it as:

> The ecosystem approach is a strategy for the integrated management of land, water and living resources that promotes conservation and sustainable use in an equitable way. Application of the ecosystem approach will help to reach a balance of the three objectives of the Convention. It is based on the application of appropriate scientific methodologies focused on levels of biological organization which encompass the essential processes, functions and interactions among organisms and their environment. It recognizes that humans, with their cultural diversity, are an integral component of ecosystems.

In turn, FAO describes the ecosystem approach to fisheries management as:

> An approach to fisheries management and development that strives to balance diverse societal objectives, by taking into account the knowledge and uncertainties about biotic, abiotic and human components of ecosystems and their interactions and applying an integrated approach to fisheries within ecologically meaningful boundaries.[7]

These two examples illustrate the considerable differences in the scope of the approach. The ecosystem approach is the best example of a notion that everyone agrees on but where few would agree on what exactly it means in practice. There are many different attempts to turn this general notion into workable proposals. Some authors have summarised the different guidelines proposed (Garcia and Kochrane, 2005).

The Commission has also tried to turn this general notion into a workable set of more concrete proposals in the specific context of the CFP. In 2002, it presented a Communication (SEC, 2001), based largely on a FAO meeting held in Reykjavik on this topic four months earlier.

In 2008 the Commission prepared a new communication (COM, 2008) on the implementation of the ecosystem approach on a wider perspective, largely as a result of the development of two new policy areas: the establishment of an Integrated Maritime Policy (IMP) and the adoption of the Marine Strategy Framework Directive (MSFD).

In this communication, the Commission introduced the notion that the ecosystem approach was not something that could be implemented in a sector alone but that a cross-sectorial approach was necessary. The development of the Integrated Maritime Policy and the adoption of the Marine Strategy Framework Directive provided the appropriate opportunity. The communication developed three ideas:

1 That quite a substantial initiatives within the CFP already corresponded to an implementation of the ecosystem approach (see the first section of this chapter);
2 Explore how the CFP links with the implementation of the cross-sector instruments such as IMP and MSFD and
3 Show the measures that will have to be taken under the CFP in the future to implement the objective in practice.

In that context, the core actions envisaged were:

- The overall reduction in fishing pressure;
- The protection of sensitive marine habitats;
- The protection of sensitive species and;
- To prevent important distortions in the marine food webs.

These proposals were largely endorsed by the European Parliament in a resolution on this topic adopted in 2009 (European Parliament Resolution, 2009).

By and large, the objectives established in the 2008 communication are being addressed. Although much remains to be done, the reduction of fishing pressure is a reality since the 2000s (see Chapter 4). The protection of sensitive habitats has considerably progressed considerably through the establishment of marine Natura 2000 areas, the protection of marine mammals and seabirds is also subject to different initiatives (see previously) and the non-disturbance of food webs has a few examples. The question of the distortion of the marine food webs has not been specifically addressed in the CFP. However, this is a clear example of the cross-sectorial nature of the ecosystem approach: the Marine Strategy Framework Directive (MSFD, see next) has included a specific objective in this regard, to be delivered by 2020.

Overall, the ecosystem approach is slowly, gradually being implemented. But further action is required and, in this regard, the implementation of the MSFD will represent a new frontier in the implementation of the notion of ecosystem approach, not just in relation to fisheries policy but in connection to all EU maritime policies. In this regard it is interesting to look at the different approaches and levels of action that can be envisaged from these three perspectives: fisheries, environment and integrated maritime policy.

From a fisheries perspective, the ecosystem approach can be conceived as an incremental process where the initial approach of managing individual species should evolve overtime to different levels of integration: (1) multi-species approaches based on technological relationships among species, (2) consideration of effects of fishing on by-catch species and habitats (3) integration of economic and social factors in the management and (4) integration of food web considerations in the management. These steps can be taken as the scientific basis to underpin them develops over time.

From an environmental point of view the approach is the opposite but complementary: rather than an incremental process, the MSFD tries to achieve a sustainable management of the ocean resources taking into account all factors that may affect the ecosystem, and trying to move forward in all aspects in parallel.

Finally, from the perspective of the integrated maritime policy, the recent initiative on maritime spatial planning (see next) can be an important contributor to the ecosystem approach, by providing for a framework where all considerations on the uses of the maritime space can be discussed and planned in an integrated manner.

Managing the ecosystem?

The implementation of the ecosystem approach will ultimately bring a new challenge to fisheries managers: the temptation to 'manage the ecosystem'. The consideration of species interactions, such as predator-prey relationships can have obvious consequences on the way we manage our fisheries. But these decisions can also have political implications, since the ecosystems are never equally distributed among all the parties concerned. The best example is the agreement between the Union and the Russian Federation on shared management of resources in the Baltic Sea.

The negotiations revealed a very different view of the consequences of the ecosystem approach: the Russian waters of the Baltic do not have much cod but are more abundant in the prey species: herring and sprat. Russia insisted to have a say in the management of the western cod stock (entirely in EU waters and therefore a discrete, not a shared stock). The bottom line of the discussion was that by having a say on the management of cod, larger TACs would be favoured, with the result that the cod will be more heavily fished and its preying on the herring and sprat will be reduced, thus allowing for a higher biomass of the latter that are the only species under quota to be caught in Russian waters.

Beyond this rather anecdotal case, the issue will appear, at some stage in the future when the Union engages in multi-species approaches. As we indicate in Chapter 4, the consideration of predator/prey relationships among important commercial species of fish may lead to the temptation to manage fisheries in a way to maximise the exploitation of the most valuable species, through the management of the marine ecosystem, that is, by overfishing the predator (if the prey is the most important species) or the other way round by underexploiting the prey so as to allow more food for the predator as the target species. This could be politically difficult, as fishing rights for predator and prey fish are not equally distributed, so different scenarios would not be politically neutral. Beyond that, this poses also a scientific problem: will the scientific advice ever be reliable enough to take the risk of 'managing the ecosystem' in that way?

In addition, a further question can be posed. To a considerable extent, we are already managing the ecosystem but in a way that does not necessarily correspond to scientific criteria. For example, marine mammals are highly protected but sea bottoms are not (except in specific areas of high diversity). But one can ask the following question: what is more important for the preservation of the marine ecosystem: the preservation of dolphins or that of the worms of the ocean's muddy bottoms?

The question is not farfetched. Scientists have underlined the importance of marine benthic communities and their links with commercial fish species (Rees et al., 2008). These benthic communities can be essential as providers of food and spawning grounds, and their importance in a scientifically based ecosystem approach cannot be ignored. In this vein, the implementation of the ecosystem approach in fisheries will require further knowledge on ecosystem interactions and the influence and role of different ecosystem components that could, eventually, suggest possible initiatives that today may sound counter-intuitive.

The Marine Strategy Framework Directive

The most important instrument at EU level to implement the ecosystem approach in the European seas is that represented by the Framework Directive on Marine strategy (MSFD) (EC, 2008b).

The directive intends to provide an integrated framework to protect the European seas, by fixing conservation standards on a number of areas. These standards correspond to the notion of 'good environmental status', defined as:

> the environmental status of marine waters where these provide ecologically diverse and dynamic oceans and seas which are clean, healthy and productive within their intrinsic conditions and the use of the marine environment is at the level that is sustainable; thus safeguarding the potential for uses and activities by current and future generations.

The MSFD represents the opposite approach to the one implemented around the CFP: instead of an incremental process of gradual addition of new elements into the picture, the MSFD starts with the whole picture, and sets out to reach objectives in all areas that are relevant for the management of the marine ecosystem. Of course these differences are originated in the different starting point of the fisheries and environmental policies: for the CFP this is the fishing activity, for the MSFD it is the whole sustainability of the marine environment.

Despite these differences of approach, the MSFD and the CFP should ideally fit seamlessly. The objective of ensuring marine ecosystem that are kept healthy and productive, to ensure sustainable exploitation is fully compatible with and conducive to the achievement of the objectives of the CFP. It also enshrines the principle of regionalisation, based on the need to act jointly at the level of sea basins, in a way that anticipates the regionalisation of the CFP.

The directive translates the notion of good environmental status in 11 'descriptors' on how to reach this general goal (see Table 11.1). Some of these descriptors are unrelated to fisheries. Others, however, touch upon questions related to the CFP:

- Descriptor 1 requires in fact that the CFP will contribute to the preservation of these areas by preventing fishing activities that may damage marine areas of high biodiversity. The preservation of deep-water coral beds (notably *Lophelia pertusa*) is the best example. The protection of sea grass beds (*Posidonia oceanica*) in the Mediterranean is another one. But there can be many others. The implementation of this descriptor may lead to the

Table 11.1 Qualitative descriptors for determining good environmental status. © European Union. Directive 2008/56/EC of the European Parliament and of the Council of 17 June 2008 establishing a framework for community action in the field of marine environmental policy (Marine Strategy Framework Directive) OJ L 164, 25.6.2008, p. 19–40.

1	Biological diversity is maintained. The quality and occurrence of habitats and the distribution and abundance of species are in line with prevailing physiographic, geographic and climatic conditions.
2	Non-indigenous species introduced by human activities are at levels that do not adversely alter the ecosystems.
3	Populations of all commercially exploited fish and shellfish are within safe biological limits, exhibiting a population age and size distribution that is indicative of a healthy stock.
4	All elements of the marine food webs, to the extent that they are known, occur at normal abundance and diversity and levels capable of ensuring the long-term abundance of the species and the retention of their full reproductive capacity.
5	Human-induced eutrophication is minimised, especially adverse effects thereof, such as losses in biodiversity, ecosystem degradation, harmful algae blooms and oxygen deficiency in bottom waters.
6	Sea-floor integrity is at a level that ensures that the structure and functions of the ecosystems are safeguarded and benthic ecosystems, in particular, are not adversely affected.
7	Permanent alteration of hydrographical conditions does not adversely affect marine ecosystems.
8	Concentrations of contaminants are at levels not giving rise to pollution effects.
9	Contaminants in fish and other seafood for human consumption do not exceed levels established by Community legislation or other relevant standards.
10	Properties and quantities of marine litter do not cause harm to the coastal and marine environment.
11	Introduction of energy, including underwater noise, is at levels that do not adversely affect the marine environment.

protection from fishing of certain marine areas of high diversity;

- Descriptor 2 has a bearing on the question of the possible introduction and impact of new species for aquaculture. This will have implications for the possible introduction of alien species for aquaculture and questions about the implementation of this descriptor in closed-circuit aquaculture plants must be discussed;
- Descriptor 3 coincides with the objective of the common fisheries policy. In this regard, the objectives for fisheries management as adopted in the 2013 CFP reform ensure consistency with this objective;
- Descriptor 4 is also relevant for the CFP, as the fishing activity exerted on certain species may alter the balance of marine food webs. This may raise extremely interesting and difficult questions. For example, whether the protection of certain species while others are exploited may pose the question of the effects on the ecosystem of differentiated exploitation patterns and their effect on the overall balance of the marine ecosystem;
- Descriptor 6 may also be relevant, when certain fishing activities may alter the integrity of certain sea bottoms. Here, a key question is to determine what are the sea bottoms that require protection from disturbance. In particular, whether the sea bottoms that have been trawled for decades will actually qualify for this objective.

By and large, the fisheries-related descriptors do not collide with the objectives of the CFP. Rather, they reinforce each other in a number of cases. Crucially, they coincide in descriptor 3 dealing with the exploitation of fishery resources. In fact, fully fledged implementation of the MSFD should be beneficial for the CFP, in that fisheries yields and sustainability do require healthy ecosystems, well protected from other man-made disturbances that may affect its productive potential.

There will be, however, certain areas where the conciliation among different descriptors will be difficult. A good example can be descriptor 6 on the preservation of floor integrity. Depending upon the practical implementation of this objective (will it apply to sea bottoms that have been trawled for a long time? Or only to pristine areas largely already protected under Natura 2000?), this descriptor may have a bearing on others, particularly number 3 on the exploitation of commercial stocks.

The implementation of the MSFD represents an excellent opportunity to inscribe fisheries in a wider framework where the preservation of a healthy, productive ocean is the key objective. If these issues on conciliation of certain descriptors are adequately resolved, its implementation will represent a very important contribution for a better CFP in the future. Likewise, the combination of the MSFD and the new directive on maritime spatial planning (see next) provides an ideal opportunity to establish a clear framework where all activities in the European seas can be managed sustainably, exploiting resources to their full potential and reducing conflict among users.

So far, the implementation of this Directive seems to be rather slow, notably because the practical definition of good environmental status is very elusive in some cases and also because the cooperation among Member States within sea basins is still in a preliminary stage. In addition, the collection of the basic data to implement it is costly and in a period of financial difficulties this is a major drawback.

But even more importantly, the very objective of the directive probably requires further clarification. While its declared aim is clearly environmental but with a clear indication that the European seas should remain open for economic activity, the ultimate balance between nature conservation and economic development is not necessarily clearly delineated. Just as an example: how ambitious should be the descriptor 6 on sea-floor integrity in a Europe where energy dependency requires a huge investment in the search for alternative energy sources? Further clarification of the balance between these two objectives will probably help a practical implementation of this fundamental legislation.

The notion of 'balanced harvest' and the ecosystem approach

In recent years, a group of scientists are developing a new notion, the 'balanced harvest' that implies that if the human intervention on the marine environment was proportional to the productivity of the different components of the ecosystem, this would contribute to improve ecosystem structure and function (Garcia et al., 2012). The idea is scientifically attractive but its practical implementation would call into question certain well-consolidated considerations, notably on the wisdom of protecting by-catch species or small fish.

Many, particularly in the environmental NGO community, consider that one of the fundamental elements of the ecosystem approach in fisheries consists on the protection of by-catch species, that is, on the very selective nature of the fishing activity that should remove only a very limited and selected part of the marine ecosystem. The notion of balanced harvest would be the opposite: that the fishing should be limited but unselective, so as to affect proportionally the different components of the marine ecosystem as the best way to preserve its natural balance.

The implications of this notion are very far-reaching. Today, the removal of marine life through fishing is very selective, both in terms of the species and in terms of the sizes of these species. Long-standing policies such as the protection of marine mammals, or the protection of juvenile fish would actually be called into question if the notion of balanced harvest made headway. Likewise, balanced harvest could actually call into question the way relative stability is formulated. And on these three issues Europe is not likely to change policies any time soon.

However, the ideas behind that notion are interesting and at the very least raise a fascinating scientific debate. It would be of interest to develop further the scientific evidence of what the promoters of the balanced harvest notion are claiming, before evaluating its possible consideration in possible future policy changes.

The ecosystem approach and maritime spatial planning

The implementation of the ecosystem approach is also strongly related to the notion of spatial planning in the marine environment. The linkage between these two questions has been already studied in Europe, showing the potential of marine spatial planning to make the implementation of the ecosystem-based management of the seas a much more tangible process (Douvere, 2008).

Actually, it can be argued that the ecosystem approach in the marine space will inevitably require, as it is the case on land, a certain specialisation of the space. On land, it is obvious that different activities, with different degrees of compatibility, cannot all coincide in the same or even neighbouring physical space. Consequently, the development of human activity in a context of preservation of key natural functions requires specialising the space through such instruments as land planning.

In the oceans, this need may be less obvious, due to the dynamic nature of the oceans and their low level of human utilisation as compared to land. However, for a number of reasons what will happen in our oceans will have to increasingly resemble the physical organisation of the uses of the space in a way similar to the process that has already taken place on land. In the seas, too, certain activities are incompatible with each other (for example, oil drilling and the avoidance of sea-bottom disturbance), many of such activities taking place in fixed locations. Likewise, the amount and significance of human activity in the seas has sky-rocketed in recent years and very notably in connection with energy development. And, finally, the potential overlap of different activities with different degrees of compatibility will increase accordingly. This situation can only be resolved through an increasing use of spatial planning instruments.

In that regard, the recently adopted Directive on Maritime Spatial Planning (EU, 2014) constitutes an opportunity to ensure that the maritime space is organised in a way that can favour investment in creating economic growth and jobs while, crucially, ensuring that the basic natural balance of the sea be respected. Spatial planning is something that a number of Member States were active on, at national level. However, a number of activities to be developed in the European seas (energy production and marine electric networks, navigation routes, the preservation of natural habitats and many other important questions) straddle the limits of natural jurisdiction and require concerted approaches with neighbouring Member States in the same sea basin: Hence the need and opportunity for European legislation.

The practical implementation of this directive poses a number of interesting challenges. The establishment of the appropriate stakeholder consultation mechanisms is an important one, given the diversity of interests involved. But the real challenge is not to find the mechanism but rather to find the necessary motivation to participate in the process. Two extreme cases can illustrate this:

Powerful sectors like the energy industry may not see a need to engage in this process, preferring to defend their interests against other users of the maritime space based on their sheer financial muscle and political influence. Weaker sectors like fisheries may not want to engage for fear of this process simply leading up to losses of fishing grounds.

An interesting question in this context is what role will the fishing interest play in the context of a 'specialised' marine environment? Will fisheries interests be better defended in comparison with other economic ones (energy, transport, tourism) in an integrated framework? This author will argue that a well-planned marine environment should favour the interests of the European fishing industry, particularly as opposed to other economic interests, such as energy, transport or tourism, for a number of reasons:

- The absence of this kind of framework would be conducive to the 'law of the jungle' where the most powerful economic interests (certainly not the fishing industry) will tend to prevail;
- The consultation and concertation process inherent to all spatial planning can provide a level playing field that may otherwise not be available;
- The ultimate goal of providing for healthy ecosystems will always be for the benefit of fishermen, as healthy ecosystems will be a pre-condition for a sustainable fisheries activity.

Overall, the instrument of marine spatial planning should be considered by the fishing sector as an opportunity, rather than a threat, in the context of the future European seas that, for sure, will be under ever-increasing pressure from different potential users.

Notes

1 Consolidated version of the Treaty on the Functioning of the European Union; *OJ.* C 326, 26.10.2012, p. 47.

2 Article 2.1 of Council Regulation 3760/92.

3 http://iwc.int/status.

4 Report of the ICES Secretariat – March 2010 – Review Service: Evaluation of eel management plans

5 www.ices.dk/sites/pub/Publication%20Reports/Advice/Popular%20advice/eel-eur_popular.pdf

6 www.fao.org/docrep/006/x3170e/x3170e02.htm.

7 www.fao.org/fi/glossary/.

References

COM (1994). Communication from the Commission to the Council. The use of large scale driftnets under the Common Fisheries Policy. Brussels, 8,4,1994. COM (94) 50 final.

COM (1999). Communication from the Commission to the Council and the European Parliament – Fisheries management and nature conservation in the marine environment. COM(1999) 363 final.

COM (2001) Communication from the Commission to the Council and the European Parliament – Biodiversity Action plan for Economic and Development Co-operation. COM (2001) 162 final.

COM (2002). Communication from the Commission setting out a Community action plan to integrate environmental protection requirements into the Common Fisheries Policy. Brussels, 28.5.2002 COM (2002) 186 final.

COM (2008). Communication from the Commission to the Council and the European Parliament. The role of the CFP in implementing an ecosystem approach to marine management. Brussels, 11.4.2008. COM (2008) 449 final.

COM (2009). Communication from the Commission to the European Parliament and the Council. Cetacean incidental catches in fisheries: report on the implementation of certain provisions of Council Regulation (EC) No 812/2004 and on a scientific assessment of the effects of using in particular gillnets, trammel nets and entangling nets on cetaceans in the Baltic Sea as requested through Council regulation (EC) No 2187/2005. Brussels, 16.7.2009. COM (2009) 368 final.

COM (2011). Communication from the Commission to the European Parliament and to the Council on the implementation of certain provisions of Council Regulation (EC) No 812/2004 laying down measures concerning incidental catches of cetaceans in fisheries. Brussels, 21.9.2011. COM (2011) 578 final.

COM (2012). Communication from the Commission to the European Parliament and the Council. Action Plan for reducing incidental catches of seabirds in fishing gears. Brussels, 16.11.2012. COM(2012) 665 final.

COM (2014). Report from the Commission to the Council and the European Parliament on the outcome of the implementation of the Eel Management Plans, including an evaluation of the measures concerning restocking and the evolution of market prices for eels of less than 12 cm in length. COM (2014) 640 final.

Douvere, F. (2008). The importance of marine spatial planning in advancing ecosystem-based sea use management. *Marine Policy* **32**(2008): 762–771.

EC (1998). Council Regulation (EC) No 1239/98 of 8 June 1998 amending Regulation (EC) No 847/97 laying down certain technical measures for the conservation of fishery resources. *OJ* L171, 17.6.1998, p. 1.

EC (2002). Council Regulation (EC) No 2371/2002 of 20 December 2002 on the conservation and sustainable exploitation of fisheries resources under the Common Fisheries Policy. *OJ* L 358, 31.12.2002, p. 59.

EC (2003a). Council Regulation (EC) No 1185/2003 of 26 June 2003 on the removal of fins of sharks on board vessels. *OJ* L 167, 4.7.2003, p. 1.

EC (2003b). Commission Regulation (EC) No 1475/2003 of 20 August 2003 on the protection of deep-water coral reefs from the effects of trawling in an area north west of Scotland. *OJ* L 211, 21.08.2003, p. 14–15.

EC (2004a). Council Regulation (EC) No 812/2004 of 26 April 2004 laying down measures concerning incidental catches of cetaceans in fisheries and amending Regulation (EC) No 88/98. *OJ* L 150, 30.4.2004, p.12.

EC (2004b). Commission Regulation (EC) No 263/2004 of 16 February 2004 extending for six months the application of Regulation (EC) No 1475/2003 on the protection of deep-water coral reefs from the effects of trawling in an area north-west of Scotland. *OJ* L 46, 17.02.2004, p. 11.

EC (2004c). Council Regulation (EC) No 602/2004 of 22 March 2004 amending Regulation (EC) No 850/98 as regards the protection of deep water coral reefs from the effects of trawling in an area north west of Scotland. *OJ* L 97, 01.04.2004.

EC (2005a). Council Regulation (EC) No 2187/2005 of 21 December 2005 for the conservation of fishery resources through technical measures in the Baltic Sea, the Belts and the Sound, amending Regulation (EC) No 1434/98 and repealing Regulation (EC) No 88/98. *OJ* L 349, 31.12.2005, p. 1.

EC (2005b). Council Regulation (EC) No 1568/2005 of 20 September 2005 amending Regulation (EC) No 850/98 as regards the protection of deep-water coral reefs from the effects of fishing in certain areas of the Atlantic Ocean. *OJ* L 252, 28.09.2005, p. 2–3.

EC (2007a). Council Regulation (EC) No 1100/2007 of 18 September 2007 establishing measures for the recovery of the stock of European eel. *OJ* L 248, 22.09.2007, p. 17–23.

EC (2007b). Council Regulation (EC) No 41/2006 of 21 December 2006 fixing for 2007 the fishing opportunities and associated conditions for certain fish stocks and groups of fish stocks, applicable in Community waters and, for Community vessels, in waters where catch limitations are required. *OJ* L 15 of 20.01.2007, pp. 1–213.

EC (2008a). Commission decision of 11 June 2008 on the confirmation of measures proposed by the Netherlands for the conservation of marine ecosystems in the VoorDelta area. *OJ* L332, 1.12.2008, p. 1.

EC (2008b). Directive 2008/56/EC of the European Parliament and of the Council of 17 June 2008 establishing a framework for community action in the field of marine environmental policy (Marine Strategy Framework Directive). *OJ* L 164, 25.6.2008, p. 19–40.

EC (2009). Council Regulation (EC) No 43/2009 of 16 January 2009 fixing for 2009 the fishing opportunities and associated conditions for certain fish stocks and groups of fish stocks, applicable in Community waters and, for Community vessels, in waters where catch limitations are required. *OJ* L 22, 26.1.2009, p. 1.

EC (2010). Directive 2009/147/EC of the European Parliament and of the Council of 30 November 2009 on the conservation of wild birds. *OJ* L 20, 26.01.2010, p. 7–25.

EEC (1992a). Council Directive 92/43/EEC of 21 May 1992 on the conservation of natural habitats and of wild fauna and flora. *OJ* L 206, 22.7.1992, p. 7.

EEC (1992b). Council Regulation (EEC) No 345/92 of 27 January 1992 amending for the eleventh time Regulation (EEC) 3094/86 laying down certain technical measures for the conservation of fishery resources. *OJ* L 42, 18.2.92, p. 15.

EU (2013a). Council Regulation (EU) No 227/2013 of the European Parliament and the Council of 13 March 2013 amending Council Regulation (EC) No 850/98 for the conservation of fishery resources through technical measures for the protection of juveniles of marine organisms and Council regulation (EC) No 1434/98 specifying conditions under which herring may be landed for industrial purposes other than direct human consumption. *OJ*. L 78, 20.3.2013, p. 1.

EU (2013b). Council Regulation (EU) No 605/2013 of the European Parliament and of the Council of 12 June 2013 amending Council Regulation (EC) No 1185/2003 on the removal of fins of sharks on board vessels. *OJ* L 181, 29.6.2013, p. 1.

EU (2014). Directive 2014/89/EU of the European Parliament and of the Council of 23 July 2014 establishing a framework for maritime spatial planning. *OJ* No L 257, 28.8.2014, p. 135.

EU (2015). Council Regulation (EU) 2015/104 fixing for 2015 the fishing opportunities for certain fish stocks and groups of fish stocks, applicable in Union waters and, for Union vessels, in certain non-Union waters, amending Regulation No 43/2014 and repealing Regulation (EU) No 779/2014. *OJ* L22, 28.1.2015, p. 1.

European Parliament Resolution (2009). European Parliament Resolution of 13 January 2009 on the CFP and the ecosystem approach to fisheries management (2008/2178(INI)). *OJ* C 46 of 24.2.2010, p. 31–37.

Garcia, S. and K.L. Kochrane, (2005). Ecosystem approach to fisheries: a review of implementation guidelines. *ICES Journal of Marine Science* **62**: 311–318.

Garcia, S.M., J. Kolding, J. Rice, M.-J. Rochet, S. Zhou, T. Arimoto, J.E. Beyer, L. Borges, A. Bundy, D. Dunn, E.A. Fulton, M. Hall, M. Heino, R. Law, M. Makino, A.D. Rijnsdorp, F. Simard, A.D. Smith (2012). Reconsidering the consequences of selective fisheries. Policy forum. *Science* 335: 3 March.

Gascuel, D., G. Merino, R. Döring, J.N. Druon, L. Goti, S. Guenette, C. Macher, K. Soma, M. Travers-Trolet and S. Mackinson (2012). Towards the implementation of an integral ecosystem-based management of European fisheries. *Marine Policy* **36**(2012): 1022–1032.

Rees, H.L., J.R. Ellis, K. Hiscock and S.E. Boyd (2008). Benthic communities, ecosystems and fisheries. In *Advances in Fisheries Science. 50 Years from Beverton and Holt.* A. Payne, J. Cotter and T. Potter (eds) pp. 358–398. Blackwell Publishing, Oxford. Cefas, 546 pp.

SEC (2001). Commission Staff Working Paper – The Ecosystem-based Approach to Fisheries Management (EAFM): possibilities and priorities for international co-operation/SEC (2001) 1696.

Sobrino, J.M., E. López Veiga and A. Rey (2010). *La integración del enfoque ecosistémico en la política pesquera común. Tirant Monografías.* Tirant lo Blanch, Valencia. 286 pp.

CHAPTER 12

The stakeholders

The CFP and the 'ivory tower' effect

The CFP has traditionally suffered from a reputation of being far removed from the real world and that the policy is run by unaccountable 'Brussels bureaucrats' that have no idea about the reality in the fishing grounds. Whether or not this image is a distorted, biased one, or whether it corresponds to reality to some extent, can certainly be argued at length. But even if this is more an image than a reality, the perception is there, so it has to be addressed.

Initially, there may have been a basis for that negative image. For years, the services of the Commission did not speak directly with the stakeholders. In many instances, the Commission was seen as an inaccessible 'ivory tower' where bureaucrats with little or no knowledge of fisheries would take fundamental decisions that would affect the livelihood of fishermen. The fact that, initially, the Commission had very few people who had experience in the fisheries field (see Mike Holden's views in Chapter 2) also contributed to the image.

But the problem is deeper. Even if the Commission has expertise and knowledge of the real world (which, since a certain time, has always been the case) the policy itself is perceived as being designed and negotiated without a real dialogue with the stakeholders. And indeed this has been the case for many years. The result of this is a very low level of 'ownership' of the policy by the fishing industry.

This diagnostic, however, must be seen against the background of the existence of the Advisory Council for Fisheries and Aquaculture that, under different names and structures, has existed since 1971. So it is not true that the CFP has been run without consultation with stakeholders. It is important to analyse why, if the CFP, from an early stage, had a consultative body with stakeholders, the image of a policy far removed from reality has never disappeared.

One reason is that, beyond this consultative body, which traditionally concentrated on general, horizontal questions, consultation with stakeholders on specific issues was generally understood by Member States as their prerogative. Up to 1990s attempts by the Commission to speak directly to stakeholders were seen by some Member States with mistrust. The Commission, in this regard, was seen as an ivory tower and this was clearly seen by stakeholders as a problem but not necessarily so by national administrations. The ivory tower effect has in fact for many years been somehow aided by Member States' attitudes.

Stakeholder consultation, on the contrary, is relative old in Europe at the level of its individual countries, where a variety of systems have been set up along the years. These different systems reflect national traditions and cultures and there is certainly no one-size-fits-all solution to this question. In any case, the experience of countries (both EU and non-EU) in stakeholder consultation shows certain trends.

The Common Fisheries Policy: The Quest for Sustainability, First Edition. Ernesto Penas Lado.

The participation in stakeholders' bodies must strike a balance: they should be open to all but not rigid in terms of composition if the demand is not there. A study claimed that a stakeholder body set up in Norway did not work initially because of a participation requirement that did not correspond to demand: 40% of female representatives (Mikalse and Jentoft, 2008).

The expectations must be set right: if stakeholders believe that all their recommendations will be automatically followed, they will be quickly discouraged and will not invest in producing the advice. The experience of most stakeholders systems around the world shows that there is a period of several years before the stakeholders build enough confidence to make the system work. That implies that expecting quick results is not realistic.

The Advisory Committee for Fisheries and Aquaculture

A historical background

The idea of providing the CFP with a stakeholder consultation body is, in fact, as old as the CFP itself. As we saw in Chapter 2, the first elements of the CFP were the first structural and market regulations for the European fishing sectors adopted already in 1970, prior to the first enlargement. Shortly afterwards, the first Advisory Committee on Fisheries was established in 1971 through a Commission decision (EEC, 1971).

This first committee was obviously focused to advising the Commission on matters pertaining to structural and market aid, the only two policy pillars existing at the time. Its composition was very varied: 45 members form the fishing and processing industry, cooperatives, traders and workers. But no NGOs were included. However, consumers were part of the participants.

This first committee was soon modified in 1973, fundamentally to take into account the enlargement of 1972, thus changing the composition but basically respecting the initial orientation and remit (EEC, 1973).

It is noteworthy that the structure and composition of the committee was not fundamentally altered, after the birth of the CFP in 1983 but only adjusted to take into account the enlargement of 1986. The main purpose of the committee, to be a consultation body to consult notably with the professionals, was remarkably unchanged following the adoption of the two basic regulations of the CFP in 1983.

A more substantial change took place in 1997, when the committee was adapted to include aquaculture in its remit (EEC, 1997). But it was not until 1999 that the Committee took its definitive shape as ACFA (Advisory Committee for Fisheries and Aquaculture) (EEC, 1999). This new ACFA constituted four working groups (fisheries, aquaculture, markets and structures) and, for the first time, its composition actually included environmental NGOs as full members, as perhaps the clearest sign of the opening of the Commission to other societal interests in Europe as legitimate stakeholders of the CFP.

Since then the consultative body has evolved in the sense of favouring the establishment of European umbrella organisations, that is, the structure has favoured the representation of those sectors that can unite into pan-European organisations. This was done to ensure that, in an ever-expanding EU, the work of ACFA would not be affected by an excessive number and dispersion of viewpoints. The idea was, the, that by favouring the establishment of European-wide stakeholders organisations, ACFA would more easily be able to conciliate different industry positions, so as to offer the Commission more widely agreed and clear advice on policy orientation.

However, this orientation may have contributed to turn ACFA into a lobby more than an advisory body. Indeed, the difference between the two functions has never been fully clarified, and this has worked against the image of AFCA in the long run. In addition, the lack of clarity on the role may have also affected the level of accountability of ACFA representative's vis-à-vis the grassroots they are meant to represent. In this regard, it has been argued that industry representatives have often used their status to 'demonstrate that they have been in Brussels' in front of their own associates, rather than to bring views about the real world to the Commission (Lequesne, 2001).

The contribution of ACFA

ACFA has provided useful advice on the CFP over the years, including in areas such as markets, aquaculture and so on. However, its existence did not prevent the need for the establishment of the RACs and, as these were set-up, ACFA lost to a considerable extent its utility and its discontinuation in 2013 has not provoked major complaints from stakeholders.

In fact, when the CFP reform was discussed in 2002, both the Commission and Member States agreed that an effective stakeholder consultation was a missing pillar of the policy. The fact that everyone arrived at that conclusion despite the existence and the work of ACFA is perhaps the most important indicator of the shortcomings of this committee. It is important to understand why.

- Firstly, the structure of ACFA was clearly not adapted to the kind of detailed discussion on technical details that has been a trademark of the CFP from its inception;
- Secondly, the functioning of ACFA being accompanied by an active policy to promote the integration of the industry into European 'umbrella' organisations, such as COPA-COGECA or Europêche. These organisations, not surprisingly, tended to dominate the debate. But this came at a cost: the need to conciliate positions among very different and often conflicting interests made their recommendations often too general and therefore of limited value;
- In addition, the kind of recommendation issued sometimes crossed the line between sound advice and 'lobbying' for industry's interest, perhaps because the very objective of ACFA was never sufficiently clarified in this respect;
- Last but not least, because ACFA was never in fact an exclusive source of advice and influence. Very often, the industry concerned by a certain question managed to present its views to the Commission directly, without having to compromise with other colleagues in ACFA. Very often, this input was judged more specific and perhaps more useful than the more generic recommendations from ACFA. By presenting their views outside ACFA, parts of the industry perhaps also contributed to undermine this committee. And by readily receiving and responding to this kind of input, the Commission did so too.

In 2008, the Commission carried out a study of the effectiveness of ACFA through an external consultant (COWI, 2008). The study concluded notably that although certain Working Groups (Aquaculture) have been very effective in providing useful advice, the general impression was that ACFA's advice was rather limited. In particular, the study concluded that rather than contributing on technical grounds to the preparation of Commission proposals, ACFA acted mainly to convey political positions.

This impression has indeed become commonplace, whether it is fair or not and has contributed ultimately to its discontinuation in 2013. In any case, the main factor that determined the end of ACFA was the setting up of the Regional Advisory Councils (RACs). Unlike ACFA, the RACs were created at a time when the need to ensure effective input from stakeholders was very high in the political agenda and this was reflected in their structure and mandate, much better adapted to the real needs for stakeholder input in the CFP. It is not surprising that, when duplications or overlaps between the two bodies appeared (for example when ACFA and RACs gave opinions on annual TACs) there was a tendency to privilege the input from RACs as more meaningful.

It is true, however, that the RACs initially did not cover issues such as aquaculture or fish markets, arguably two of the areas where the advisory work of ACFA has been of particular relevance and value. This was recognised in the 2013 CFP reform, not by keeping ACFA to deal with these issues but by creating new Advisory Councils to deal with them (see Chapter 15).

For the future, the question of the stakeholder consultation on issues not covered by the RACs will have to be addressed. Although there will be, arguably, fewer such issues (due to the establishment in the 2013 reform of Advisory Councils on two of the issues covered by ACFA Working Groups: aquaculture and markets) there still may be a need for a horizontal stakeholder consultation body beyond the future ACs. A possible example of such a need is a possible fuel crisis, like in 2008, where the Commission needed to discuss possible instrument on a horizontal basis and chose to do so with ACFA, as that discussions in parallel with seven RACs would have been simply unworkable.

Any future replacement of ACFA would have to have certain characteristics, to learn from the experience of the extinct body:

- Have a clear scope of issues to deal with, to avoid any kind of duplication. In particular, it should avoid duplicating discussions of the ACs on the conservation policy.
- It should have a very clear mandate, to ensure that no duplication takes place with Expert Groups established by the Commission in connection with the preparation of implementing regulations.
- It should concentrate in providing 'reality check' on the Commission's ideas, rather than on expressing political positions.
- It should establish a wide representation of stakeholders, ensuring in particular that those usually less represented (small-scale fishing) have an opportunity to have their voice heard.

For the time being, however, and pending the definitive shape of the new Advisory Councils (ACs) established by the CFP reform of 2013, the need for this new body is still not entirely clear and if the new ACs ensure sufficient coverage of stakeholder consultation in the new CFP, this new body may not even be established at all.

Regional Advisory Councils

The establishment of the RACs did not happen overnight. In fact, already before the CFP reform of 2002 there were different initiatives to start filling the gap of stakeholder consultations, through informal settings: the so-called 'regional workshops'.

The precedents: regional workshops

In the 1990s, the evidence of the need to address specific problems of fisheries management affecting specific stakeholders made it clear that the consultations in the framework of ACFA were not specific enough and that a more focused kind of stakeholder consultation was necessary. This gave rise to the organisation of informal 'regional workshops'. Once more, the workshops organised to consult stakeholders on the preparation of the cod recovery plan established the precedent that was then followed by many other cases and topics.

The increasing complexity and difficulty of many of the questions at stake, from international negotiations to discussions on the first management plans, clearly underlined the need for early consultations with stakeholders by the Commission, in order to ensure that, before Member States would bring their own points of view, the initial proposal from the Commission would already take stakeholders' input into account.

These first experiences were not smooth. The lack of tradition of dialogue and perhaps the different expectations from the various participants made these first meetings very difficult. At the time, there was a considerable gap between the stakeholders (mainly the fishing industry) and the Commission. Both spoke very different languages and communication was difficult. The industry had typically a defensive, critical attitude with the Commission. Given the lack of specific channels for the industry's views to be fed into the EC system, the industry did not make positive proposals but rather criticised whatever the Commission proposed.

Filling this gap required fundamentally two elements: (1) providing for a channel where the professionals could organise themselves to make positive contributions to the policy and (2) providing for time to build confidence between the two sides. The first element would be the formal constitution of the Regional Advisory Councils (RACs) the second being the patience, continuity and determination to make them work.

The Commission reported in 1999 on the experience gathered in these regional workshops (COM, 1999). Already at that stage, the issue of the expectations was underlined as a major difficulty: 'a feeling of frustration was created because the participants understood that the meetings could not intrude in the decision-making process'.

Without that expectation, stakeholders wondered if it was worth investing in these meetings. In its evaluation, the Commission was very reticent about taking these workshops a step forward and simply undertook to address the question of the stakeholder consultation in the context of the future reform of the policy in 2002.

The main factor that led to a new, more institutionalised approach to stakeholder consultation was the cod crisis (see Chapter 4) where the need to prepare a plan to save the cod stocks led to a series of regional workshops where the political need to obtain stakeholder buy-in in what would be a very difficult political exercise (given the important sacrifices required) became obvious to all.

The establishment of the RACs

The Regional Advisory Councils (RACs) were established soon after the 2002 reform, through Council Decision 2004/585 (EC, 2004). Much discussion took place as to the role to be given to these bodies. In particular, while the mainstream opinion was that they should have a purely advisory role, for both legal and political reasons, there were also strong voices claiming that a real overhaul of the CFP should imply a truly bottom-up governance system whereby the RACs would have real executive powers. The final decision to give RACs an advisory role was resented by some as a lost opportunity (Gray and Hatchard, 2003).

- In fact, there were several good reasons why the RACs were not given executive powers. Firstly, legal reasons: the Treaty at the time (and still today under the Lisbon Treaty) did not foresee any mechanism for devolution of decision-making powers on exclusive competence policies to any kind of regional body.
- Secondly, in political terms, the idea of regional decision-making bodies was resented by many as an open door to a re-nationalisation of the CFP and was received with considerable mistrust, if not open hostility.
- Thirdly, it was also argued that no policy in the Union was actually run by the stakeholders themselves. Why should fisheries be an exception? Nobody could provide an argument in that regard.

The procedure for the establishment of the RACs was bottom-up by design. It was understood that if the establishment was closely directed by the EU institutions they would lose part of their legitimacy in terms of representativeness. This of course had a cost: that the establishment of the different RACs in different areas was very different in timing. Consequently, the role they have played in the CFP has also been variable, particularly in the first years.

The first to be established was the North Sea RAC (set up in November 2004). The North Western waters RAC followed in September 2005, the Baltic Sea RAC in March 2006, the Long Distance RAC in March 2007, the South Western waters RAC in April 2007 and finally the Mediterranean RAC in September 2008.

The growth phase

During the expansion phase, from 2004 to 2008, the activity grew steadily as new RACs came into existence and they started their work. The number of RAC recommendations grew very quickly as a consequence: while they were only three in 2004, this figure increased to 60 in 2007 (COM, 2008).

The first years of operation were understandably difficult. As all the experiences around the world clearly show, the functioning of a stakeholder advisory body requires several years to build confidence between the industry and the administrations, as well as to 'get the expectations right'. This is exactly what happened in the EU with the RACs.

One can wonder if one of the turning points in this evolution was the trip organised by the Commission with selected RAC representatives to the United States, where they visited some US Advisory Councils, generally regarded with certain envy by parts of the European industry as the ultimate model of a powerful, industry-dominated advisory body. Upon return from that trip, a distinguished member of the Pelagic RAC declared that after that visit, he appreciated the European RACs more than ever before.

But the value of the RACs can go beyond the issuing of recommendations on management measures. In certain cases, the RACs have demonstrated to be a privileged forum where the industry itself can develop a dialogue that may contribute to solve, or at least to reduce, long-standing conflicts between different interests. The best case in hand is that of the management of the anchovy fishery in the Bay of

Biscay. In this case, the discussion in the South Western Waters RAC between the Spanish and French industries around the ideas of the Commission on a management plan for that stock represented important progress in finding common solutions to a hitherto very conflicting situation.

The experience of the RACs, extremely positive as it is, required a number of years and a somewhat difficult 'learning curve' for both the RACs themselves and the Commission. Indeed, a number of factors made the first years of operation of the RACs difficult and sometimes frustrating. Perhaps two main questions were of particular relevance: getting the expectations right and building confidence, among the members themselves and between the RACs and the Commission.

A fundamental question was that of expectations. A number of stakeholders would have liked the RACs to have the power to provide binding advice. The realisation that their role was advisory, made evident when the Commission did not accept all their recommendations, was frustrating for some. Certain participants questioned the interest of investing in the process 'only to be later ignored by the Commission'. This issue of the expectations was in fact aggravated by misleading messages from the Commission itself to the effect that if recommendations were adopted by unanimity by the RAC (crucially, agreed by the fishing industry and NGOs) then they would be automatically accepted. Realising that this was not necessarily the case, produced considerable frustration in some RACs.

This issue of the expectations gave rise, in the first years to a commonly held view that became extremely popular among industry representatives: 'the Commission hears but doesn't listen'. This was the result of a number of cases where certain RAC recommendations were not followed by the Commission and, even in certain cases the reasons for the rejection were in their view not sufficiently well explained by the Commission services. This situation, however, has changed quite substantially in recent years. Today, the level of attendance by Commission officials and the quality of the replies from the Commission to the RAC advice, together with the increasing rate of acceptance of their recommendations, has dissipated considerably some of these original problems.

A second important question was that of the lack of trust among the actors. And this in itself has two components:

1 In some cases, there is no trust among fishermen of different Member States in a given RAC, as a result of long-standing disputes over the legitimacy of being there in the first place;
2 In most cases, the level of trust between fishermen and environmental NGOs is extremely low, as fishermen still see NGOs as an enemy.

As a consequence, debates are often very tactical and many participants hesitate to put all the data and problems on the table for open discussion, as they would not say certain things 'in front of certain others'. The personal experience of the author is that the most interesting things one can learn in some RAC meetings are often heard informally in coffee breaks.

Thirdly, the RACs do not have exclusive rights to be heard and stakeholders also feel free to utilise other platforms to convey their views. In some instances, the stakeholders themselves undermined the RAC advice by declaring that agreeing to recommendations in RACs would not prevent them from recommending something different to their ministers. In other cases, the Commission may have undermined the RACs by accepting meetings with more specific stakeholders, thus giving the impression that the RACs themselves are not the ideal platform to be heard by the Commission.

The lesson learned from that is that, if RACs (and their successors, the ACs) are to play a crucial role in policy making, they should really be privileged stakeholders, so all their members feel it is worth investing in stakeholders' advice in those fora.

RAC composition

Perhaps the most controversial question regarding the RACs is their composition. In fact, Decision 2004/585 establishes that in the general assembly and executive committee of the RACs, two thirds of the seats should be allotted to representatives of the fisheries sector and one third to representatives of other groups interested in the CFP. Among these other interests, environmental NGOs play a particularly important and contentious role.

This composition has been controversial from the beginning. The Mediterranean RAC took longer than any other to establish because for quite a long time the Mediterranean fishing industry claimed that their RAC should have a different composition with a lower representation from other interests. In other RACs NGOs were more readily accepted but the tension between fishermen and environmental NGOs has always been significant.

In fact, this phenomenon can be considered a paradox: as the number of actors in the consultation process increases, the influence of the traditional actors diminishes. So a governance mechanism set up to increase the buy-in by those affected more directly by the decisions (the fishermen) may actually produce, to some extent, the opposite effect (Suarez de Vivero et al., 2008). In fact, many fishermen interests have long considered that they can exert a more efficient influence outside the RACs than inside, notably by lobbying their Ministers prior to the relevant Councils. This is even more the case for certain NGOs, which find other ways to exert influence on CFP decisions outside the RACs: in some Member States, they have the upper hand in influencing governmental positions. But, especially in recent years under co-decision, they have proven very efficient in influencing political positions in the European Parliament.

In the recent reform of the CFP, the non-industry interests have been given a higher proportion (40%) of the seats in the new Advisory Councils, in recognition to the fact that fisheries management has clearly evolved in recent years in the sense of becoming a question of public interest, where more societal interests beyond the fishing industry have an influence in decision making.

Are RACs a success?

The RACs have clearly gone through a 'learning curve' from their establishment in 2004. This learning curve has applied also to the Commission, which has had to adapt and learn how to ensure that the RACs function properly and how to address the initial shortcomings. Overall and despite initial adjustments, the RACs are considered today as one of the most important improvements ever made to the CFP.

Since their establishment, their participation and active role has been increasing steadily. And, already in 2008 a first evaluation by the Commission showed a general satisfaction with their contribution to the CFP, although a number of areas for improvement were also identified (COM, 2008).

In certain occasions, the RACs have been instrumental in resolving certain long-standing issues that national administrations had trouble to sort out. By bringing together the industry from different Member States, the RAC provided for a forum where industry disputes could be argued without the direct intervention of administrations. The most important success in this regard is the agreement for a management regime for anchovy in the Bay of Biscay developed by the South Western Waters RAC in response to the difficulties surrounding the re-opening of the fishery of this stock after the collapse of 2005.

There were failures also, at least initially. Perhaps the clearest case is that of the Mediterranean, which took much more time than the remaining RACs simply to agree on its constitution. One can think that this reveals a lack of organisation among the Mediterranean fisheries associations. However, this would probably be simplistic.

Experience shows a very high level of organisation in many associative bodies of fishermen in the Mediterranean. But these organisations, which in some cases have developed very sophisticated systems of governance and self-regulation, traditionally tended to be local. And still today, when the main instrument for fisheries management in this area is the national and sub-regional management plans, the cooperation among different fishermen for the purpose of managing fisheries also remains local.

This results in lack of tradition of cooperation on a large scale in the Mediterranean basin, lack of tradition of discussing possible solutions together and little tradition of constituting large organisations capable of engaging in such kind of dialogue. This is probably why, in the absence of such tradition, it proved so difficult for the Med RAC to achieve the same degree of cooperation and organisation achieved by the Atlantic and North Sea RACs.

Overall and despite shortcomings, the main success of the RACs has been that they have turned into important actors in the CFP whose continuity and reinforcement is agreed by everyone. They definitely are there to stay.

Nevertheless, some fishermen have been critical of the RACs for a number of years. These have repeatedly voiced strong criticism of the RACs on account of the alleged follow up of their recommendations by the Commission. However, these critical remarks have been gradually reduced and have become relatively minor in recent years, as the 'learning curve' for them as well as for the Commission have reached levels of high mutual trust.

Scientific literature indicates that fishermen are critical with stakeholder consultation bodies notably on account of the lack of knowledge of what happens with their advice (Pita et al., 2010). In some cases, the Commission's meagre explanations of the consideration given to RAC advice fuelled this kind of reaction. However, considerable progress has been made in recent years and today all RAC recommendations are followed by the Commission with detailed explanations of its reaction to the advice.

In the context of the new policy adopted in 2013 and, particularly in the context of the implementation of the ban on discarding and with the open opportunity of regionalisation, the potential contribution of these stakeholder bodies will be substantially enhanced. But their challenges will increase consequently. And perhaps the main challenge is whether the newly established regionalisation process will imply a more important involvement of the stakeholders, or whether Member States will implement regionalisation through political negotiation among capitals, thus ignoring stakeholders advice. In that regard, it appears that the first agreements under regionalisation, the discard plans prepared by Member States in 2014, have generally been the result of political negotiation among capitals while, in certain cases, the advice of stakeholders has to some extent been ignored (see Chapter 17).

Non-governmental organisations

In the CFP the role of non-environmental organisations (NGOs), particularly those with an environmental interest, is crucial, and it is still growing. In today's CFP, environmental NGOs represent a major player, notably through the influence they exert on the legislators: both national governments and many members of the European Parliament do indeed take NGOs' views as a very important input in the way they define their political position on CFP issues.

The precedents

The best-known case of how environmental NGOs can have a crucial influence on fisheries policy is the dolphin-safe case initiated in the tuna fishery of the eastern Pacific Ocean. This has been described in detail by the late James Joseph (1994). In essence, the fishery for tuna in this area of the world was based on the use of dolphins as indicators of the presence of tuna schools. The association of the two kinds of species was such, that whenever dolphins were spotted in the surface, there was a very high probability that there would be a large school of tuna just below. Fishing for the tuna would, inevitably, imply a certain level of mortality of the dolphins.

This fact went unnoticed until an environmental NGO, the Earth Island Institute, produced a film showing dolphins being taken and killed by the large nets of tuna purse-seines. The impact of these images on the US public opinion was such that it prompted the US government to pass legislation banning this type of fishing and requiring dolphin-safe labelling for the tuna to be sold in the US market. This forced the US tuna fleet to stop this fishery and move to the western Pacific, where the association dolphin-tuna does not exist and where they remain to this day.

Other fleets, primarily the Mexican one, develop methods to fish over dolphins in a way that dolphin mortality would be very substantially reduced. These methods are today enshrined in the Agreement on the International Dolphin Protection Programme (AIDCP),[1] and allow fishing over dolphins under extremely strict conditions, including 100% observer coverage on all large purse-seiners and a system of transparent sanctioning that has no equivalent in the rest of the world.

This episode somehow set the tone for what would be, in the years to come, a crucial role of environmental NGOs in the development of fisheries policy including the CFP.

The evolution of the role of NGOs over the last years in the CFP has evolved considerably, at least in two particular aspects:

1 The gradual change in focus from iconic non-commercial species to mainstream commercial fisheries. While NGOs used to focus their activity on the protection of iconic species such as marine mammals, sea turtles and so on, their focus has evolved considerably and in the last 10 years they have been very active in other fishery-related fields, such as mainstream fisheries management (focused on specially endangered stocks such as cod, or Bluefin tuna) as well as on Illegal, Unregulated and Unreported fisheries;
2 The gradual evolution of certain NGOs (not all of them, though) from campaigners to formally recognised stakeholders. Their incorporation to ACFA and the RACs has marked this evolution.

Today, environmental NGOs have become a fundamental stakeholder in the CFP. The fact that some of them hire fisheries experts (some of them with a previous experience on fisheries management either in a national administration or in the Commission) is the best indicator that they are there to stay as fisheries stakeholders and that they want to exert their influence on the basis of a technically sound set of arguments.

From iconic species to mainstream fisheries

In Europe, the interest of environmental NGOs in fisheries issues has been growing steadily. Only 20 years ago, the interest of NGOs on maritime matters was clearly directed towards such concerns as the protection of marine mammals (whales and other cetaceans such as dolphins) and other non-commercial species such as seabirds or sea turtles. But there was comparatively little attention to the normal management of commercial fish.

This has dramatically changed in the last decade, with NGOs clearly increasing their interest and action on other fisheries issues, such as the management of endangered commercial fish species. Bluefin tuna in the Mediterranean and cod in the North Sea have been subject to important campaigning by the NGOs. Other species have also received considerable attention in recent times, such as sharks.

A more recent development is the interest of NGOs on issues related to illegal, unregulated and unreported fishing. In this regard, for example, NGOs have become a crucially important source of information on illegal fishing activities that the Commission currently uses in the context of the implementation of the IUU regulation.

Finally, there is a global trend towards the consideration of wider issues related to fishing, in particular the question of working conditions aboard certain vessels, with NGOs (such as the Environmental Justice Foundation, very active in western Africa) leading the international fight against slavery on board certain fishing vessels.

Widening the scope: from campaigners to formal stakeholders

Where the role of NGOs has become of notorious importance, is in transforming fisheries debates into issues of interest for the general public. In doing so, they place considerable influence on the political debate, making it much more prone to consider the positions of the public opinion at large.

From this point of view too, the role of NGOs has also evolved considerably. While their influence used to be informal, they have now become official 'stakeholders' on the same footing as the fishing industry. In the last decade they have gained a formal status as a stakeholder in the Union's consultation bodies, such as ACFA or the RACs, where they enjoy the status of statutory members. In these stakeholder bodies, NGOs act as counterbalance to the fishing industry.

This evolution, however, is not shared by all NGOs. Some of them still prefer to stay outside the system and to exert their influence through campaigning.

The incorporation of NGOs in the process of stakeholder consultation is not universally recognised as a success. There are several elements that cast a doubt into its effectiveness:

- The still important mutual reticence between NGOs and the fishing industry, which reduces the possibility of finding common ground and common recommendations within RACs;
- The alleged difficulties in following all the discussions in RACs and to make positive contributions to them. The level of attendance of RAC meetings by NGOs is lower than that of stakeholders from the industry. This may be due in some cases to budgetary limitations. But in other cases it may reflect the view that attendance is not always worth the investment;
- The well-proven capacity of NGOs to have an influence on other stages of the decision-making process leads certain NGOs to reconsider what are the channels to exert influence that have a better return for their investment. And RACs are not necessarily the best choice in that respect;
- Those NGOs who do not participate in the RACs find other ways to influence the political debate through access to the institutions and public administrations.

For these reasons, there is a tendency by those NGOs having engaged in the formal process of stakeholder consultation to wonder whether their efforts are more effective when deployed in different fora.

It is true that in recent years, while some NGOs invested in an active and cooperative participation in ACFA and the RACs, other have stayed out, preferring their traditional approach as public campaigners. In addition, NGOs have developed a new, very powerful methodology: lobbying the European Parliament. In this context and for those NGOs with limited resources, the question is how to be more influential, either through ACs or through public campaigning and lobbying.

Perhaps these considerations may be re-evaluated in the context of regionalisation. Here, the opinion of the new ACs may have a direct bearing on what Member States present as their proposal under regionalisation in the future CFP. In this process, whatever comes out will be handled by the Commission through a delegated act – without co-decision. This will make the debate within ACs more crucial and closer to the final decision. Perhaps in those circumstances environmental NGOs will decide to invest as much as they can in that process.

Other consultations

Beyond those held with stakeholders, the CFP includes other types of consultations. These can be classified in three broad categories: the additional consultation with public institutions in the European framework, the social dialogue and the consultations with the public at large.

Institutional consultations

The CFP has very little tradition of consulting other institutional bodies of the EU, such as the Economic and Social Committee or the Committee of the Regions. In fact, only on the occasion of very large new projects, such as the 10-year revision of the policy, have these bodies been consulted. It is important to explain why.

Concerning the Committee of the Regions, we must establish a clear distinction between the term 'regional' as understood by the Committee of the Regions as an EU institution (that is, regions understood as legal entities within Member States, some of them with a very high degree of self-rule) and the term 'regional' incorporated into the CFP in the 2013 reform, which refers to geographical spaces, normally coinciding with a sea basins or parts of them and thus having a supra-national nature.

Despite the fact that many regions do have a very significant stake in the CFP, the highly integrated nature of the CFP has traditionally made this policy hardly amenable to consideration at regional level. This is why the Committee of the Regions has never played a significant role in the development of this policy.

In addition, the traditional difficulties in the CFP tend to represent different national interest, rather than interest from maritime regions against the rest. Consequently, there has traditionally been relatively little room for agreed solutions to the CFP problems clearly agreed by a majority of the European fishing regions.

As for the Economic and Social Committee of the Union, its influence has also been limited in the CFP.

This is due, firstly, because the CFP has no basis and no tradition of regulating social policy. Beyond that, it is also true that, in general, the fishing sector and very particularly the small-scale subsector have not been strongly unionised anywhere in Europe and still today the notion of social dialogue and social bargaining is not nearly as developed in the fisheries sector as it is in many other industrial sectors in Europe.

Social dialogue

The social partners, however, have a certain tradition of dialogue at European level. The Sectorial Social Dialogue Committee for Sea fisheries (SSDC) was established in 1999. It is managed by DG Employment, Social Affairs and Equal opportunities – DG Maritime Affairs and Fisheries is invited to participate in meetings when Maritime Affairs and Fisheries issues are discussed. Employers are represented by the Association of National Organisations of Fishing Enterprises in the EU (Europêche) and the General Committee for Agricultural Cooperation in the EU (COGECA). The European Transport Workers' Federation (ETF) represents the employees. Both employers and employees had a representative in ACFA. The input of the SSRC focuses mainly on health and security on board, negotiations in ILO, fishermen training or social clauses in Fishery Partnership Agreements.

This framework on social dialogue is perhaps of particular relevance in the case of long-distance fishing, exerted by large vessels where workers are much more unionised than in the coastal sectors. In addition, this is a sector that is particularly sensitive in terms of social policy given the comparison between social legislation in European countries as compared to that of other states whose vessels operate and compete with European-flagged companies in the same fishing grounds.

Consultations with the public at large

This is a consultation instrument used in recent years, with the objective of testing out the public opinion on general issues of fisheries management. Modern technologies allow these consultations through the Internet, allowing for a level of public penetration and public input absolutely unheard of just a few years before.

The Commission has increasingly utilised this kind of consultations in a number of cases and today it is used systematically. For example, in the preparation of proposals such as the reform of the CFP in 2009, the aquaculture guidelines, or the proposal on the amendment to the shark fining regulation in 2010 were all preceded by a public consultation. In these cases, the Commission summarised the input from society at large, either as part of the proposal itself (COM, 2011), or in a specific document (SEC, 2010).

There are some lessons to be learned from that experience. One important one is that the interest of the European public opinion as regards fisheries issues seems to be more related to iconic species than to mainstream considerations on fisheries management modalities. The best example is that of shark-finning: the public consultation of this file drew more than a thousand contributions from the public, while the (more substantial one) on CFP reform barely drew around 400.

Another one is that these consultations tend to be dominated by well-organised lobbies, who can afford to prepare well-documented, high-quality contributions that are likely to be more influential than those of the general public.

In any case, as fisheries issues become better known for the European public and given the public nature of the fishery resources and the environment where they live, the need for this kind of consultation can only grow in the future as a basic requirement of modern, transparent and democratic societies.

Note

1 www.iattc.org/DolphinSafeENG.htm.

References

COM (1999). Communication from the Commission to the Council and the European Parliament. Review of the second series of regional fisheries workshops (1998/1999) and future outlook. Brussels, 7.1.2000. COM (1999) 747 final.

COM (2008). Communication from the Commission to the Council and the European Parliament. Review of the functioning of the Regional Advisory Councils. COM (2008) 364 final.

COM (2011). Proposal for a Regulation of the European Parliament and of the Council Amending Regulation (EC) No 1185/2003 on the removal of fins of sharks on board vessels Brussels, 21.11.2011 COM(2011) 798 final.

COWI (2008). Intermediate Evaluation of the Advisory Committee for Fisheries and Aquaculture (ACFA), August 2008, COWI / Nautilus Consultants / Framian.

EC (2004). Council Decision of 19 July 2004 establishing Regional Advisory Councils under the Common fisheries Policy (2004/585/EC). *OJEU* L 256, pp. 17–21.

EEC (1971). Décision de la Commission, du 25 février 1971, portant création d'un comité consultatif de la pêche. *OJ* No L 68, 22.3.71, p. 18.

EEC (1973). Commission Decision of 31 October 1973 on the Advisory Committee on Fisheries. (73/429/EEC). *OJ.* No L 355 of 24.12.73, p. 61–62.

EEC (1997). Commission Decision of 4 April 1997 on setting up a special aquaculture section of the Advisory Committee on fisheries. (97/247/EEC). *OJ.* No L 97 of 12.4.97, p. 28–29.

EEC (1999). Commission Decision of 14 July 1999 renewing the Advisory Committee for Fisheries and Aquaculture. (1999/478/EEC). *OJ.* No L 187 of 20.7.1999, p. 70–72.

Gray, T. and J. Hatchard (2003). The 2002 reform of the Common Fisheries Policy's system of governance – rhetoric or reality? *Marine Policy* **27**(2003): 545–554.

Joseph, J. (1994). The tuna-dolphin controversy in the Eastern Pacific Ocean: biological, economic and political impacts. *Ocean Development and International Law*, **25**: 1–30.

Lequesne, C. (2001). *L'Europe Bleue: A quoi sert une politique communautaire de la pêche?* Presses de la Fondation Nationale des Sciences Politiques, 239 pp.

Mikalse, K.H. and S. Jentoft (2008). Participatory practices across Europe: making stakeholders more responsible. *Marine Policy* **32**(2008): 169–177.

Pita, C., G.J. Pierce and I. Theodossiou (2010). Stakeholder's participation in fisheries management decision-making process: Fisher's perception of participation. *Marine Policy* **34**(2010): 1093–1102.

SEC (2010). Commission Staff Working Document. Synthesis of the Consultation on the Reform of the Common Fisheries Policy. Brussels, 16.4.2010 SEC(2010) 428 final.

Suarez de Vivero, J.L., J.C. Rodriguez Mateos and D. Florido del Corral (2008). The paradox of public participation in fisheries governance. The rising number of actors and the devolution process. *Marine Policy* **32**(2008): 319–325.

CHAPTER 13
The governance of the CFP

The CFP: a 'central command' policy?

The 'micromanagement from Brussels'

As the CFP is a policy of exclusive competence, it is implemented through regulations directly applicable in Member States, rather than by directives that have to be incorporated in national law. This fact is essential to understand the 'micromanagement from Brussels' that has characterised this policy from its inception. But this is not all.

As we saw in Chapter 2, the birth of the CFP was so complicated that it was necessary to negotiate all technical details for the policy to finally take shape. And this was possible because of the low number of Member States at the time. This 'central command' policy was probably the only possible option in 1983, in view of the political difficulties to set it up, and also because at that time the involvement of stakeholders and the administrative decentralisation were not yet an issue in EU policy.

However, this micro-managed CFP, that was both necessary and possible in 1983, has turned into a fully-fledged culture of micromanagement which, when applied to 28 Member States and several sea basins becomes increasingly difficult to manage.

The direct applicability of regulations and the very technical nature of many of these regulations, perhaps combined with this consideration of fisheries as the last frontier of national sovereignty among European States has indeed created a tradition where all details of the policy are considered 'political' and are decided by ministers, generally after long negotiations.

As a result, the CFP is arguably the European policy where Ministers have decided on more detailed, technical issues. This, in turn, resulted in 'technical' decisions being taken, all too often, on more political grounds. And this does very little to ensure that the policy takes into account the realities on the fishing grounds.

This type of decision making has long been identified as one of the weaknesses of the policy and it is certainly one of the reasons why the CFP has had a bad reputation of lack of 'buy-in' by stakeholders. Addressing and resolving this issue is a permanent challenge for the CFP.

The 'culture' of the December Council

There is no better example of this kind of policy making than the 'tradition' of the meeting of the Council of Fisheries Ministers that is held every December to decide on TAC levels for the following year.

There, the tradition is that the Commission makes proposals that inevitably imply certain reductions in fishing levels, in order to ensure the sustainability of the fish stocks and then Ministers try to obtain higher TACs than those proposed by the Commission, in what constitutes one of the best examples of 'nationalising the success and Europeanising the failure' that is so typical of national politics in many, if not all Member States on EU policy.

The Common Fisheries Policy: The Quest for Sustainability, First Edition. Ernesto Penas Lado.

This 'culture' is so deeply-rooted that it has made these Fisheries Councils famous for it. The negotiation has to go through the night (in at least two cases, 2002 and 2004, the work of the Commission services and the Council secretariat went on for three days and two nights without a minute of sleep). And Ministers often present themselves as having fought all night long to bring home slightly higher quotas than those proposed by the Commission. In some circumstances, some Ministers have even made a strong case of claiming that their national quota of a given stock had to be 903 tonnes instead of 902.

For many years all attempts to end this practice have been unsuccessful. In December 2005, a newly appointed female Minister, in her first ever Fisheries Council, declared her intention to stop such a 'macho nonsense'... only to end up in a night long negotiation all the same.

Another characteristic of this 'culture' of the December Council is the role of Coreper. Normally, in the decision-making process of the Union, Coreper acts as the 'Sherpa' to the summit, by deciding on less political issues so as to leave for ministers only the really political points.

However, in the run-up to the December Fisheries Council, the long-standing tradition was that Coreper did not decide anything: the tradition was that Coreper only served to identify 'three priorities per Member State', for the Council. This theoretical limitation to three issues was easily circumvented by Member States asking, typically, as one of their three priorities to have 'a higher TAC level for the following stocks:' and then following a long list of stocks for which TAC levels as proposed by the Commission were too low for them.

This was perhaps the ultimate expression of the 'culture' of the December Council: that even Coreper gave up its natural role to decide on the less political issues prior to the Council.

The recent introduction in Council of web-streaming, whereby the public can follow part of the debates through Internet, has provided very useful transparency into the process. However, its effects on decision making are not necessarily straightforward:

- On the one hand, Ministers have to present their case in a way that they are seen by their fisheries constituency as strongly defending its interests and this may not always be conducive towards compromise.
- On the other hand, they may fear negative reactions from other societal interests if they don't defend a righteous position.

Overall, this transparency probably only enhances the effect of stakeholders in their political position, whoever these stakeholders are and whatever their relative weight in each different Member State.

One can also wonder whether, beyond the public positions taken under web streaming, Ministers cannot negotiate more pragmatic solutions in the corridors, away from the transparent Council sessions. This is still true but experience shows that ministers who defend righteous principles in public and then try to negotiate solutions away from those principles tend to be increasingly discredited in the negotiations.

The psychology of decision making

The psychology of this decision making is largely determined by the short-term nature of the decisions traditionally taken at the December Council on TACs and quotas. To understand this we can look at Figure 13.1, where we compare the image of a long-term model showing the recovery of a fish stock, where the short-term sacrifices are compensated by the mid and long-term gains, with the image of an annual decision, where the short-term sacrifice is completely isolated from other considerations, giving rise to the psychological tendency to 'avoid bad news'.

In the first case (graph A), the decision to reduce a TAC, in accordance with the scientific advice, is an annual decision whose negative effects are evident (less economic activity, political criticism from stakeholders) but whose possible positive elements are not identified, or even discussed. In graph B, Ministers are also confronted with a difficult decision for the following year but that is done as part of a multiannual strategy whose rewards are spelled out by scientists in terms of stock recovery. In the first case, the decision is just a short term sacrifice, in the second case it is an investment with a long term reward.

It is also true that for that 'investment' to be fully acceptable to stakeholders (and subsequently to

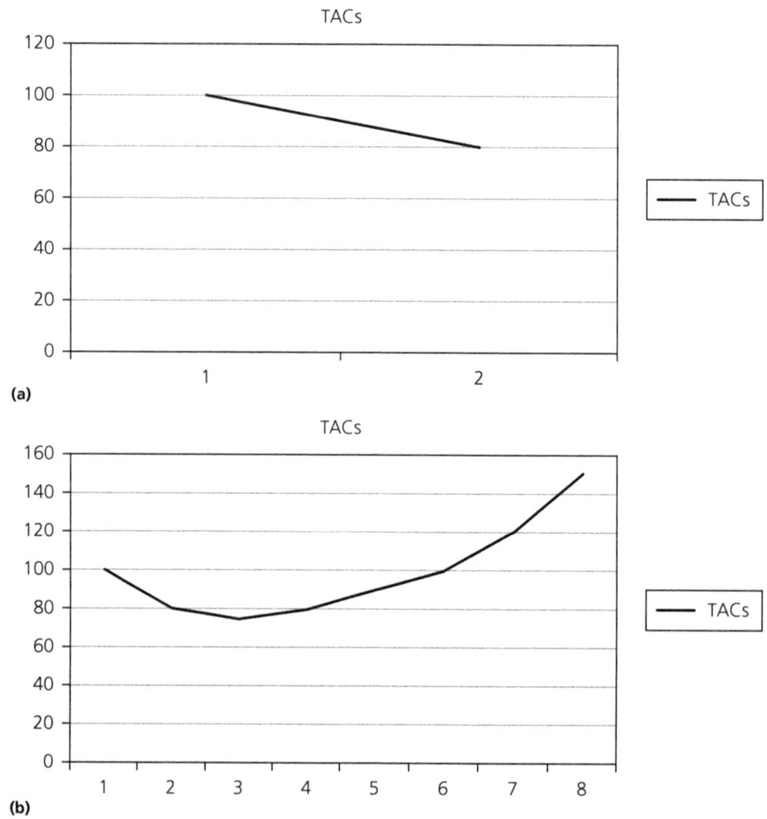

Figure 13.1 Graph (a) shows an annual decision on TACs for an overexploited stock. Graph (b) shows the TACs over the years for an overexploited stock as according to a long-term recovery plan.

ministers) one key condition is that the long-term strategy be gradual, that is, that the strategy should ensure that TACs are not dramatically reduced from one year to the next. The promise of stability in the levels of TAC of successive years is the single most important condition for the long-term approaches to be acceptable and to replace the short-term decision making.

In the 2000s, as long term plans developed and as the RACs increased their influence in decision making, the RAC recommendation of limiting annual TAC variations by plus or minus 15% in the long term plans has become practically a standard clause and a number of current plans actually include this condition as the ultimate guarantee that the annual 'sacrifices' will be moderate, which in turn is the condition *sine qua non* for the long term plans to be acceptable.

Horse trading

The negotiations of the December Council on TACs and quotas have often been described, with a rather derogatory tone, as 'horse trading'. The question is: can we consider this horse trading something illegitimate? This deserves some attention. Indeed, the nature of decision making in Council, under qualified majority, is such that negotiations among Member States to obtain support to their interest are a logical consequence. It is difficult to consider this practice illegitimate, or leading to bad decisions. The real question is not so much the fact that there are negotiations but rather the nature of some of these negotiations and, in particular, when it is the majority itself and not the substance behind it that matters at the end.

The really negative aspect of this horse trading was that decisions were made on TAC levels to accommodate the need for a Member State to vote for the deal, too often diluting policy objectives. Although the Commission traditionally established 'red lines' that concessions to Member States should never trespass, the reality of the negotiation was that these red lines were often ignored. And more importantly, the need to bring specific Member States into the majority of Council made it necessary to make concessions to that Member State on a given stock, whereas for other stocks in the same conservation status no such concession was necessary. The obvious result was that TAC levels for species under similar conservation status ended up having completely inconsistent TAC levels.

This was very difficult to explain to the industry, it gave the impression that the Union had double standards when deciding and all without transparency. It is this lack of consistency and transparency that really created the negative image of the horse trading.

Another effect of the horse-trading practice is what a group of colleagues came to know as the 'me too' approach.[1] The way this came about is a very telling example of how certain decisions used to be made (and are still today?) in the Council. The case came about with the introduction of a new research technique to evaluate the abundance of the Norway lobster (*Nephrops norvegicus*) stocks.

Abundance estimates based on fishery-dependent data, such as estimations of catch-per-unit-effort (CPUE) have many shortcomings which often cast serious doubts about such estimates. A new system was used in the North Sea: underwater cameras recording the number of burrows (where the Norway lobsters live) in the sea bottom. This technique was soon considered more reliable than classical CPUE estimates.

Ministers from the North Sea argued that based on the new camera estimates, the abundance of the North Sea stock was higher than previously estimated and the TAC was considerably increased as a consequence during the negotiation. The same technique was used in the Irish Sea but it did not result in a higher biomass estimate than with the previous methodology. However, the TAC was also increased because the new estimates removed the uncertainty of the CPUE data. And then the Bay of Biscay came into play. There, the new technique had not been used but if the TAC in the North Sea and the Irish Sea had been increased, then it was politically unacceptable not to increase it in this other area as well.

This is an example, among countless others, where certain decisions, even if well-based on evidence, are often seen as 'political favours' thus justifying the same kind of 'favour' to other Member States even if the science-based justification is not there.

The solution to this necessarily had to come through a systematisation of the decision-making process, as we saw in Chapter 4 (the 'policy statement') concerning the categories of stocks according to stock status.

'Paper fish' and governance

Another element that makes these negotiations difficult is the notion of 'fish to swap'. In many cases, Member States increase their bid for fish they are not interested in because under relative stability, this fish can have a real value as it can be traded through the system of 'quota swaps' with other Member States in exchange for the quotas of fish they are really interested in. This often makes Member States argue for higher TACs of stocks they do not fish, simply to have higher 'currency' to swap for their real target species.

It is also important to refer to one of the side effects of this culture: the so-called 'paper fish'. This phenomenon corresponds to the adoption of quotas for certain fish in certain areas that actually cannot be fully fished simply because the fish is not there, at least in the quantities allowed by the TAC levels. Why is this the case?

The paper fish phenomenon is partially a result of the horse trading, in which concessions to certain Member States are seen as requiring similar concessions to others for the sake of non-discrimination. In some cases, Member States want to have quotas for certain species not significantly found in a given management area, just in case the stock appears one day. In other cases Member States are interested in quotas as 'currency' to be exchanged with other Member States for other species of their interest, in the context of the annual quota swaps exercise.

Paper fish has been progressively 'cleaned up' in the annual TAC and quota regulation, and today there are considerably fewer than it here used to be. For a number of years the Commission has been trying to reduce these TAC levels, at least for those cases where the quota consumption for all Member States is consistently low. Progress, however, has been limited, as Member States are reluctant to reduce quotas that they may perhaps need one day, either to fish or to swap.

Allocating national quotas: a Member State prerogative

From its inception, the CFP established a clear-cut division of responsibilities on the allocation of fishing rights: total allowable catches are allocated at EU level (by Council decision only) but the allocation of national quotas among the fishermen of each Member State is an exclusive competence of the Member State.

This is important because it is one of the main mechanisms whereby each Member State can apply its own socioeconomic model to the management of fisheries. While the decision on TACs and the corresponding balancing out of biological, economic and social considerations is a European decision where different and contradictory socioeconomic and political considerations come into play, the way Member States distribute their national quotas to their industry allows them to apply their own preferences on the balance between economic rationality and social engineering.

Different Member States use extremely diverse allocation mechanisms. As we have seen in Chapter 5, some use tradable fishing rights and leave the question of allocation to market forces (although with a series of safeguards) to all or part of their quotas. Others, on the contrary, distribute national quotas amongst Producer Organisations and leave them to allocate their quotas internally to their members as they see fit. Others practice social engineering through centralised systems of allocation with the declared purpose of ensuring that the wealth generated by fishing be distributed among all fishermen, thus preventing the accumulation of rights in a few hands.

It is remarkable that these national allocation systems are in many cases very stable regardless of the government in place. When deciding on the national allocation systems, national traditions (liberal or protectionist) generally prevail over the ideology of the party in power, in a demonstration of the transversal character of political positions on fisheries management in many, if not most Member States.

Not surprisingly, Member States defend zealously this prerogative that they see as one of the few questions left in the CFP where they can decide on the preferred socioeconomic model for their fishing industry. In the context of the last reform of the CFP (see Chapter 16), some political groups of the European Parliament proposed the introduction of rules on how Member States should allocate their national quotas. Their idea was that Member States should be obliged by EU law to allocate the quotas in a way that would favour small scale fleets (allegedly squeezed out from quota allocation in some Member States) or more selective types of fishing (mostly static gear). The idea was completely rejected by Council, as all Member States saw the allocation of national quotas as a fundamental national competence that they did not wish to give up.

Improving decision making

All these elements have led to the December Councils being generally considered as a nightmare. The number of both political and technical issues piling up in one Council has been such, that this author has worked in several occasions three days and two nights straight without sleep. This has been particularly the case in December 2002 when the TAC Council coincided with the 2002 CFP reform or in December 2004 when the enlargement of that year significantly increased the number of Member States, the number of issues and the number of stocks to manage.

These experiences led the Commission to propose a number of changes in the process that would result in Ministers in December being able to concentrate in fewer, more political issues.

Front loading

The first of such mechanisms was the advancement of the discussion (and possible decision) on technical issues in the earlier stages of the process of preparation of the December Council.

This was the case, for example, with the extremely complex table of conditions and derogations for the implementation of the effort management of the cod recovery plan, where the complexity rendered extremely difficult to have meaningful discussions at ministerial level during the December Council.

These discussions, however, generally did not serve to resolve the issues but simply to understand them better, to extract the political elements from them so they could be decided by ministers.

A second mechanism was to give COREPER a more important role in deciding on a number of TACs prior to the Council, so as to allow ministers to concentrate on fewer, more political cases. This has been done since the mid-2000s and with Coreper having changed its tradition of simply 'identifying three priorities for the Council' as we described previously and has actually having taken decisions on easier cases. However, by and large, Coreper still plays a relatively minor role on the annual TAC decisions. The tradition of the negotiations at EU level that 'nothing is agreed until everything is upheld agreed', very in the CFP and this has always limited progress by Coreper.

Today, front loading is practised to its full extent but the result is to advance and improve the discussion of technically difficult issues, not to advance political decision.

The effects of the Policy Statement

In Chapter 4 we saw the development of the Policy Statement as an instrument to increase transparency and rationalise decisions on annual TACs. That document also represented a breakthrough in the way some of these decisions were taken, particularly through the introduction of categories of stocks according to their conservation status and proposed approaches for each category.

The merit of the categories of the policy statement was that Member States could no longer claim *ad hoc* increases in TAC levels. Their arguments had to change. Rather than proposing politically-led increases in TACs, they claimed instead that on the basis of alternative or updated scientific data, the stocks should actually belong in a different category. This was an important breakthrough, in that it turned the arguments from purely political into scientifically based and systematic. For that reason, the suppression of the detailed stock categories from the Policy Statement in recent years represented a step back in that process.

The Policy Statement also has another advantage: the possibility that, for a number of stocks, the TAC levels could actually go undisputed during the whole negotiation in Council. In fact, over the years, while the Council discussion does not necessarily always become simpler, the discussion on TAC levels themselves has been very considerably simplified and rationalised. The complexity of the December Councils has increased only because of new instruments (effort management) and new Member States having come into play but in terms of the decisions on TACs, the situation is considerably more rational than it was in the past.

In the context of the new CFP agreed in 2013, the Policy Statement can make another important contribution to decision making: the tradition of including a description of the evolution of the stocks allows this document to show the gradual progress toward the achievement of the long term goals. As the policy now has a much more precise objective, the achievement of fishing mortality at levels of maximum sustainable yield between 2015 and 2020, the annual reporting on progress towards this objective should become a powerful instrument to ensure that decisions go in the direction necessary to achieve it. Any decision against the trend toward achieving the objective would be shown up to the general public as not corresponding to the legally binding goal of the policy.

Streamlining TAC decisions

A further development of the TAC and quota approach has been the gradual tendency to avoid an accumulation of decision in one single Council. This accumulation was the result of certain factors, notably the date of delivery of the scientific advice for many stocks. On that basis, the possibility to frontload certain discussions prior to the scientific advice becoming available in October was obviously limited, beyond debating the general approaches presented in the policy statement. To allow for an effective streamlining of the process, it was necessary to advance the delivery of the scientific advice by ICES. And only then the different TAC decisions could be split between more than one Council.

The advance of the scientific advice was worked out in 2008 and 2009. For certain stocks (i.e. Baltic stocks), this was easy: the advice on these stocks was traditionally provided in June already. However, this was much more difficult for the Atlantic, since for a number of stocks, the final advice took into account fish surveys traditionally carried out in September. Finally, ICES found the formula to advance the delivery of the advice in a number of cases, facilitating the preparation of the proposal.

Since 2009 the TACs for the Baltic are decided in early September in a dedicated Council. The TACs for deep-sea stocks, adopted every second year, are dealt with also in a separate Council. In December 2012, the Commission even proposed to split from the main TACs those autonomously decided by the Union from those agreed with third parties in two separate regulations. But this initiative was later reversed, as Member States did not see any advantage in this approach.

Enlargement and decision making by Council

The decision making on TACs (by the Council) has also evolved considerably over time taking into account the successive enlargements of the Union. The number of Member States has grown from 6 to 28 in 30 years and this has obvious consequences in the way the Council conducts its business.

Nowadays, the Council has to streamline procedures to ensure that a qualified majority can be reached among 28 countries, of which 22 are coastal states and have fisheries interests. This means, for example, that a simple *tour de table* to test Member States' views takes more than an hour. It also results in less time being available to organise bilateral meetings with all Member States to try to find specific solutions for their claims. It also means that the possible contradictions among diverging Member States' claims can multiply exponentially.

As a consequence, it becomes much more difficult to achieve consensus solutions and this gives rise to an increased need to resort to qualified majority. This has gradually introduced important changes in the dynamics of the negotiations: there is much less time to listen to every individual Member State's claims and results in further concentration of the discussion on fewer, highly political points and more reckless search for qualified majority by the presidencies, as consensus in such a complex setting appears increasingly difficult to achieve.

This should in principle help to achieve the policy objectives. In the past, very often the quest for consensus meant that concessions had to be made to certain Member States that triggered requests for further concessions to others, creating a knock-on effect that resulted in a substantial watering down of the proposals. As qualified majority becomes the rule, it is easier to avoid this type of concessions and Member States increasingly understand that if their claims are not reasonable, the risk of being voted down in Council becomes a real one.

The CFP, a complex policy

The CFP has a well-earned reputation of being a complex policy. This complexity can be expressed in different ways and these are often combined in the same regulation:

- The technical complexity: the policy establishes very detailed rules and even small technical details are regulated. These often include specific rules to cater for special circumstances raised by Member States or the industry.
- Unclear rules. In a number of cases, the difficulty of the negotiations to agree on the final legislation results in unclear rules, as the need for political compromise leads to vague formulas that different Member States can interpret in their own way.
- The ever-increasing number of regulations and the tendency to resolve problems and legal gaps by adding new layers of legislation, not always

replacing the old ones. Part of this is inevitable: as the Union grows, new geographical areas and activities come into the CFP requiring regulatory measures.

- In some cases, the duplication of legal instruments to achieve similar goals.

This complexity is very often criticised. It has been argued that the complexity in the regulation of the CFP could increase the likelihood of misunderstanding and suboptimal decisions (Pastoors, 2014). However, in many cases the policy has also been criticised for not being clear enough, for example in defining its own objectives, thus suggesting additional text would be necessary. The Commission itself has contributed to this (COM, 2009).

More detailed legislation clarifies the rules but removes room for manoeuvre to adapt to special or unforeseen conditions. The debate on policy complexity is linked to the debate on what is expected from the legislation: to establish clear rules in every possible circumstance to ensure a 'level playing field' even at the cost of excessive rigidity and a high risk of inconsistency, or to establish more flexible rules to make legislation adaptable to changing circumstances.

That debate has traditionally been resolved in favour of the first option. Despite that, the complexity of the legal basis of the CFP is largely seen as a problem. However, reversing the trend is extremely difficult. Let us analyse why.

The ever increasing complexity

There are several reasons to explain the complexity:

- The tradition established in 1983 that in order to strike a difficult political deal it was necessary to consider everything, even technical details, as part of the negotiated deal. And, as we have indicated, this was still not a bad solution in 1983 with only 10 Member States. But such a tradition in 2013 with 28 Member States has an entirely different meaning and consequences.
- The tendency by Member States in Council (and now by MEPs in Parliament) to bring specific problems of their constituencies that require special derogations or conditions. For this reason, it is almost always the case that the final regulations adopted are more complex than the original Commission proposals.
- The lack of trust among Member States, which renders any proposal based on general objectives difficult to accept, as Member States do not trust each other implementation.
- As stated in previous pages, the relative easiness to include new policy layers, combined with the difficulty to eliminate the previous ones, resulting in a net accumulation of policy layers over time.

The regulations under the CFP tend to be longer and longer over time. An interesting study has shown that the word-count of the basic regulations of the CFP and their number of pages have been growing exponentially. The basic regulation of 1983 (Regulation 170/83) had 16 articles and 2771 words. That of 1992 (Regulation 3760/92) contained 21 articles and 4081 words. In 2002, the new basic regulation 2371/2002 had 36 articles and 9713 words. The current one, Regulation 1380/2013, has 59 articles and 21 986 words.

There are of course good reasons for this increase. The inclusion of more policy considerations, the further clarification of objectives and instruments, the more complete cross-references among policy pillars to ensure consistency, these are all good reasons why the length of the text needs to increase.

Perhaps the clearest case of the ever increasing complexity of the CFP is the technical conservation measures. A recent study the STECF showed that the number of technical measures regulations has been steadily growing since 1980 (Figure 13.2). Although some of these derogate old ones, the net effect is that of a net accumulation of technical rules over time, to a point where just keeping track of existing rules becomes a difficult task, not just for fishermen or inspectors, but for Commission officials as well.

However, the trend towards simplification of legislation has started already. Some simplification took place in the 2013 CFP reform. The best example is the new market regulation (EU, 2013a) that reduced to only one intervention mechanism the 6 mechanisms considered in the previous regulation (EC, 2000). More of this will be done in the coming years as part of the REFIT Commission initiative (COM, 2014) where the control Regulation (1224/2009)

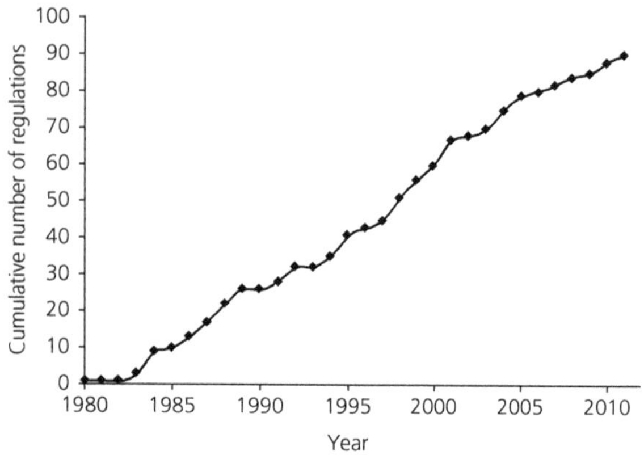

Figure 13.2 Number of regulations on technical measures in the CFP throughout the years. Source: STECF 2012.

will be subject to an evaluation with two aims of simplifying it and making it more 'fit for purpose'.

Despite these initiatives, the legislative simplification of the CFP remains an elusive objective. Everybody agrees with it but when it comes to deliver such simplification, most decision makers and the stakeholders behind them show a preference for the preservation of specific rules for their – often small – constituencies.

Everyone has contributed to that, including of course the Commission. Although the latter has certainly tried many times to propose simpler rules (that subsequently become much more complex as a result of the political negotiation) the Commission has had a tendency to require very frequent and detailed reporting by Member States on the results of the legislation. While reporting is an essential element for the implementation of certain regulations (the best example is the catch reporting system whereby Member States report on a monthly basis on the consumption of their national quotas under the TAC and quota regime) in many other occasions the requirement to report concerns annual information on the results of certain rules. Many of these reporting obligations, in fact, are followed by Member States in an extremely incomplete way (often not at all) and some of these reports, in turn, are also followed up by the Commission in a vary patchy manner too.

The Commission has taken some initiatives towards simplification of the CFP. Fisheries has been one of the areas where, by virtue of the very specific and detailed legislation, there was traditionally more room for manoeuvre for simplification. Although some simplification has certainly taken place, this has clearly not been enough to curb the ever-increasing number of rules in the CFP. It is worth referring to the proposal for simplification of Regulation 850/98 on technical measures presented by the Commission in 2008 (COM, 2008). Here the Commission introduced an innovative approach on the legislation of mesh size and included a bold de-centralisation of regionally-specific measures. However, the very innovative nature of this approach met considerable difficulties in Council, where the lack of trust among Member States and between these and the Commission prevented progress. The proposal was never adopted. Only after the 2013 CFP reform will such an approach be possible.

More progress has been done on the reporting side, where the tendency over the last years has been to reduce the reporting obligations to the really necessary and to turn access to information into a 'pull' exercise (Member States gather and store the information, the Commission 'pulls' it only if and when necessary) rather than the traditional 'push' one (where Member States must 'push' the information to the Commission on a routine basis). In any case, the evolution of the reporting obligations will largely depend on the European parliament's stance on the issue. It is clear that in the context of recent regulations, already adopted under co-decision, Council has generally tried to reduce compulsory reporting (in order to reduce the administrative burden of national administrations of Member States) whereas the European Parliament has favoured more reporting obligations, in order to increase transparency in the way the policy is run and monitored.

The number of regulations

The number of regulations in the CFP is ever increasing. This is to some extend inevitable: as the Union is enlarged, new areas and new fisheries come into play and new issues are regulated. This tendency, however, has been aggravated by the tendency of the policy to be more at ease with the addition of new instruments than with the elimination of existing ones.

The high legislative output has become a major feature of the CFP. And the tendency is not necessarily easy to reverse. While the number of new policy instruments tends to be more limited in recent times, the requirements of the policy still justify a high production of legal acts. For example, for 2015 the work programme of DG Mare included around 50 acts, among proposals for co-decision regulations, Commission implementing or delegated acts, Commission decisions, proposals for Council decisions for Council regulations and Commission communications. The latter are not legislative in nature but often they imply legislative proposals at a later stage.

Duplication of management instruments

Another characteristic of the CFP, over the years, has been its tendency to improve and innovate through the addition of new layers of regulation but these layers have not necessarily removed the previous ones, thus resulting on an accumulation of regulatory layers and, in a number of cases, on an increased complexity.

In some cases, this is the result of the need to adopt new measures while the old ones remain necessary. But in certain occasions double legislation has even been the result of the need for political compromise.

Two clear examples of this are the effort management of the recovery plans and the management of deep-sea stocks as of 2004.

Effort and recovery plans

As we saw in Chapter 4, the introduction of effort management, in addition to the existing management on the basis of TACs and quotas was introduced following recommendations from scientists that the later were not effectively controlling fishing mortality.

When considering the introduction of the new instrument, the Commission even considered the option of replacing the TACs by effort management, largely on account of the multi-species nature of most cod fisheries. However, Member States turned down this possibility immediately: as TACs and quotas were the 'carrier' of relative stability, they did not trust that the allocation keys of cod and other associated species could be easily translated into effort quotas. So the only way to accept to limit effort was to consider it an additional instrument to the catch limitations. As a result, the management plans for the most depleted stocks created duplicated management instruments and thus significantly increased policy complexity.

Deep sea stocks

Stocks of deep-sea fish were unexploited until relatively recently. Extremely varied, many species living in deep waters have in common their adaptation to a very constant environment, through long life spans and low reproduction rates. This makes some of these species extremely vulnerable to overexploitation.

The development of commercial fisheries for these stocks started in the mid-1980s. At the beginning of the 2000s the evidence of the fragility of some of these species started to be evident and the need to consider possible regulatory measures at EU level became an issue for the CFP.

The discussions on the management regime applicable to these species soon showed a marked split between two groups of Member States:

- Those who believed that the stocks needed to be managed through TACs and quotas, like any other stock in the continental shelf, and
- Those who believed that individual TACs were not well adapted to this multi-species fishery and that effort management was more adequate.

This discussion, although expressed in technical terms, was influenced by a question of national interest: those who preferred TACs were those Member States who had developed the fishery first and who thought that a system based on TACs would necessarily have to be based on historical catches to determine quota allocation. On the contrary, those who favoured effort management were those arriving more recently to the fishery, who thought that by establishing effort limitations as the management instrument, no relative stability would be fixed and the effort adjustments would have to take as a starting point the current levels of effort deployed.

The difficulties of the negotiation were such, that at the end the political compromise was to regulate the fishery both by catch and effort limitations and two parallel regulations were adopted simultaneously to manage these stocks: one establishing (biannual) TACs, started in 2002, and another one establishing effort restrictions (see Chapter 4). This duplication of instruments has survived to this day and one can certainly argue whether just one of the two regulations may suffice to properly regulate the fisheries for these vulnerable stocks.

The number of Member States in Council

The number of Member States in Council also has an influence in the level of complexity of the policy. As we saw in Chapter 2, the precedent of 1983 among ten Member States is that of establishing very detailed rules where all involved had an opportunity to introduce specific conditions or derogations to cater for specific problems. The number of these possible special conditions or derogations was relatively limited.

The problem comes when the same philosophy, by tradition, is applied to a Council with 28 Member States. The possible number of special conditions to be considered increases in proportion to the number of Member States. This happens, one way or another, in all EU policy areas but in the CFP this phenomenon is aggravated by the tradition of micro-management.

The tendency, however, is that the increase in complexity does not correspond exactly with the number of Member States. There is a clear reason for this: the end of the *de facto* consensus represented by the 'Luxembourg compromise' (see Chapter 2). As more regulations were adopted by qualified majority, a number of Member States have been voted down in Council when trying to introduce special conditions or derogations for their national interest. As a consequence, certain of these conditions have not been incorporated into EU law.

This tendency became very obvious in the reform of the CFP in 2002, the first one adopted by qualified majority. Subsequently, as the number of Member States increased through successive enlargements, the tendency to use qualified majority increases, thus is leaving a number of national requests for special conditions or derogations outside the final decisions. This implies that, although the number of Member States tends to increase the complexity of regulations, arguably the increase in complexity is somehow less than it could be if the traditional consensus seeking would continue.

The Treaty of Lisbon and co-decision

Co-decision and simplification

The Treaty of Lisbon introduced a fundamental change in the decision-making process of the CFP. While the ordinary legislative procedure had been so far the Council Regulation with a non-binding opinion from the European Parliament, the new Treaty established co-decision between the two institutions as the ordinary legislative procedure for the CFP. The experience of using this procedure in the CFP is still short but already very intense and meaningful.

Co-decision in principle increases the complexity of the decision-making process. However, the increased role of the European Parliament also means an increased democratic legitimacy of the decisions as well as the further consideration of European-wide political considerations beyond national interests. These elements can certainly enrich the decision making process.

A more difficult question is, however, how co-decision can cope with the amount of legislation that the CFP produces and revises every year.

As we saw previously, the CFP produces an impressive number of legislative proposals per year. The annual 'rolling programme' of DG Mare for 2015 contained, by April 2015, 15 acts adopted, 40 planned, 28 pending final green light and 16 in preparation. These acts are proposals for regulations, amendments to existing ones, Commission regulations and Commission decisions and communications. Although the number of proposals for co-decision acts is only a fraction of the previous, the co-legislators also intervene in the Commission's delegated acts, through a scrutiny period to make comments on the Commission's respect of its mandate, and ensure follow up to the Commission's communications through statements and positions. This results in a high demand for legislative and political discussion by the two co-legislators.

The question is: in a policy already considered as a 'central command' policy, far away from the realities of the fishing grounds, how will the co-decision address this problem. In general terms, Parliaments decide on laws, not on regulations. But as co-decision applies to the CFP, the Parliament in co-decision has inherited a legislative system where the political and technical decisions are taken together, in the same piece of legislation. This produces as an effect the need for the European Parliament to be involved in this mixture of political and technical issues all in one package.

The increasing complexity of decision making under co-decision makes this procedure arguably ill-suited for very frequent detailed technical decisions. One could expect, then, that co-legislators will eventually tend to concentrate legislation on the important political issues, while delegating technical rules to another level, be it the Commission or Member States under regionalisation. However, the on-going experience seems to indicate that this will take some time to materialise.

A good example is the experience of the provisional technical measures. This regulation was a 'left-over' of the annual TAC and quota regulation. As we saw in Chapter 4, the TAC and quota regulation was a framework that allowed relatively quick implementation of new technical rules (sometimes requested by Member States themselves). For many years an annex of the annual regulation on TACs and quotas contained a growing number of technical measures, for example the protection of certain Natura 2000 areas and so on.

Upon entry into force of the Treaty of Lisbon, this practice was no longer allowed. The Council then agreed to delete these provisions from subsequent TAC regulations but in order to ensure their continuity it adopted in 2009 a provisional, stand-alone regulation containing the said provisions (EC, 2009). As an interim solution, this regulation had a specific expiry date: 30 June 2011. The idea was that during this provisional period, these measures would be fully incorporated into the general technical measures framework, through an amendment to Regulation 850/98 and this required a co-decision act.

In the context of the ongoing CFP reform, it was considered that these provisional measures should be prolonged until 31 December 2012 but not any longer, given the character of 'interim solution' of these regulations. This required an amendment to the 2009 regulation, adopted in 2011 (EU, 2011). By the end of 2012, the provisional technical measures would need to be incorporated in a permanent regulation adopted by co-decision, as the provisional measures would expire.

The Commission made a proposal in this regard to incorporate all these measures into Regulation 850/98. This had to be done by the co-legislators before 1 January of 2013. The challenge was clear: these measures should be adopted quickly by the co-legislators in order to avoid a legal vacuum as of 1 January 2013. If co-legislators failed to agree, there would be Natura 2000 areas unprotected from fishing. In that context and in order to ensure an easy, non-contentious and quick adoption, the Commission resisted all attempts to include new, possible contentious elements in this proposal.

Yet, once that proposal was on the table, some Member States and some MEPs insisted on the inclusion of new provisions, some of them extremely detailed. The discussion on these new provisions proved contentious, particularly within the European Parliament. As a consequence, the adoption of the amendment to Regulation 850/98 was delayed until March 2013 (EU, 2013b) and, even if this went largely unnoticed, some Natura 2000 areas were completely unprotected from fishing for more than 2 months during 2013.

This delay in the adoption may have also been influenced by the on-going disagreement between the two co-legislators on the issue of the long-term plans. But whatever the case, this experience shows that the 'culture' of micro-management is so deeply rooted that it can go as far as delaying the implementation of certain legislation whose continuity is so essential for the credibility of the CFP.

The recently adopted approach on regionalisation represents an opportunity that could eventually relieve the co-decision process of a number of decisions that now may be taken under regionalisation.

Co-decision and societal interests

One fundamental change brought about by co-decision is that fisheries decisions become more amenable to considerations from societal interests other than those of the fishing stakeholders. The structure of the European Parliament is such, that the MEPs with fishery-dependent constituencies are concentrated in the PECH Committee. But these have a very small weight as compared with the number of MEPs who vote in the plenary. When the PECH committee presents a proposal for a Parliament's position, there is no guarantee at all that their proposal will be accepted by the plenary. And then MEPs in plenary are often likely to vote on the basis of a much wider scope of societal interests, beyond fisheries.

If the Council of ministers is also increasingly considering interests wider than those of the fishing industry, the introduction of co-decision in the CFP accelerates considerably the consideration of the CFP as a policy where the fishing industry is no longer the only stakeholder.

As an example, in 2013, when the European Parliament voted on its consent to a mandate to the Commission to negotiate a new fisheries agreement with Morocco, the members of the Fisheries Committee voted favourably to the mandate, on account of the interest of some of their constituencies to ensure the continuity of their economic activity. However, the Plenary of the EP voted against on this same procedure. The reason was the absence of references in the draft mandate to the relationship between the agreement and the population of the Western Sahara. Clearly, the Fisheries Committee voted primarily on the basis of fisheries considerations and interests. The Plenary, con the contrary, voted on the basis of much wider political considerations.

At the end, the Plenary gave its consent the signature of the Agreement but only after a solution was found on their initial request. This example illustrates that the fishing industry, which used to present their views to their fisheries departments only, now have a much wider audience to convince in terms of the future shape of the CFP and this represents a major challenge for their future.

Council versus Parliament: Articles 43(2) and 43(3)

There is an exception to co-decision as the ordinary legislative procedure in the CFP: the adoption of catch limitations remains an exclusive competence of the Council under Article 43(3).

Although it is not always easy to know the intention of the legislator when establishing the exception of Article 43(3), it is certainly sure that the annual nature of these decisions and the very limited time available every year to take such decisions, make the co-decision procedure unfit for that particular purpose. However, what may have been a rather practical arrangement, has somehow turned into a substantial institutional issue.

The problem came in the context of long-term management plans. The very objective of these plans has been, as we saw in Chapter 4, to provide a legally-binding long-term objective for the management of fisheries, so annual TAC decisions would not be taken on short-term considerations but as part of a multiannual strategy to achieve a long-term goal. Management plans established this framework through the so-called harvest control rules (HCR).

These rules are variable but in general they consist of a long-term goal (expressed in fishing mortality), sometimes intermediate objectives, sometimes also a time-frame for achieving objectives and maximum levels of inter-annual TAC variation. But in all cases, the HCR practically determine the way in which the annual TACs should be calculated, on the basis of the most recent scientific advice. And this is where the problem arose between Council and Parliament.

On the one hand, these HCR encapsulate like no other instrument the choice among the different objectives of Article 39 TFEU and, from this perspective, there is hardly anything more essential for the CFP than these rules. This militates in favour of basing such HCRs and, by extension, the management plans, on 43(2) and thus on co-decision.

On the other hand, if annual TAC decisions are entirely framed by the HCR in management plans, then these annual decisions are taken without any room for manoeuvre by Council and then Article 43(3) would be devoid of content for stocks under management plans.

The problem proved extremely difficult to resolve, to the point that, during the negotiations for the 2013 CFP reform, the issue represented a serious risk of blocking the negotiation between Council and Parliament on all the other issues. Consequently, it was agreed to use neutral language in the basic regulation ('neutral' means that the HCRs are referred to as being adopted 'under the rules of the Treaty' without indicating who does what) and to establish a Task Force between Council, Parliament and Commission to try to find a practical solution outside the text of the reformed CFP.

This task force met regularly as of September 2013, in the form of a kind of 'informal trilogue'. It met 10 times and discussed reflection papers prepared by the Commission and finally agreed on a text in April 2014, where the content of the multi-annual was described, with an indication of the parts of them that would be decided by co-decision, by Council alone, or by the newly established mechanism of regionalisation. Crucially, the task force agreed that the Fmsy objectives in management plans would be expressed in Fmsy ranges.

However, this agreement was not formalised by any of the two institutions. This means that only when specific proposals for long-term plans are presented by the Commission and discussed in future trilogues it will be possible to see if such agreement has provided a practical solution.

Co-decision and regionalisation

In Chapters 16 and 17 we will refer to the new regionalisation mechanism introduced in the CFP reform of 2013. This mechanism is voluntary but it carries a considerable potential to become an effective way in which decision making, on technical aspects, can be taken away from the complexities of co-decision and decided closer to the fishing grounds.

Indeed, regionalisation could be a very important mechanism whereby the legislative basis of the CFP could be fundamentally transformed, notably by replacing complex, technical rules decided by co-legislators into more flexible rules to be prepared by Member States and turned in EU law by the Commission, through a much faster procedure.

But for that to happen there has to be quite a substantial change in the way the legislation is conceived: if the co-legislators stick to the Council tradition of legislating the technical details, then they may leave relatively little room for regionalisation. If regionalisation is going to have a real content, both Council and Parliament will have to agree that they will only legislate on the big questions, leaving technical details to a different instance. This, however, when compared with the tradition of micro-management, may appear to co-legislators as giving-up decisional powers on certain questions that, although obviously technical and detailed, may still be perceived as highly political, at least for certain constituencies.

Whether the co-legislators will agree to leave ample room for regionalised decisions, as opposed by co-decision, only future practice will tell. In any case, given the complexity of the co-decision process, and the amount of legislative proposals required by an exclusive competence policy like the CFP, the tendency towards more regionalised decisions on technical matters may actually become a necessity, rather than a choice, for the co-legislators to be able to cope with the CFP's high legislative production.

Non-legislative elements for decision making

Beyond the legal questions that condition decision making, there are a number of non-legislative factors that have a growing influence in the way decisions are taken within the CFP. Perhaps the clearer ones since the inception of the CFP are the influence of public opinion and that of consumers.

The growing influence of the European public opinion in the decision-making process of the CFP is evident. In the 1980s and the 1990s, the main decisions on the CFP were taken by the Council alone and Ministers took decisions taking into account mainly the opinion of their national fisheries sectors. In that context, the public opinion was largely unaware of these decisions, or at least only fisheries constituencies were aware.

Things have changed substantially in recent years. The European public opinion has become much more interested than ever before on mainstream fisheries management issues. This is probably the result of several factors, including *inter alia* the work of NGOs in raising such issues with the public at large, the contribution of the Commission through awareness campaigns, or the decision by the Council of Ministers to make discussions public through web-streaming.

The changes are so dramatic that in some cases Ministers have taken political positions in Fisheries Councils that respond much more to the views of their public opinion than to that of their fishing sectors. As an example, in the discussions on the future discard ban in the last CFP reform, during a Fisheries Council in 2012 a Minister of one Member State took a very radical view in favour of the discard ban without any flexibility, despite the fact that the fishing sector of the Member State concerned had one of the most intractable problems of discarding. The reason for that position was simple: on the basis of initiatives from the public, their national parliament had adopted a resolution inviting the government to remove all discarding, whatever the consequences for the fishing sector.

Another important influence in decision making is the increasing weight of consumers and their growing demand for sustainable fish. As the demands from consumers for fish caught sustainably increase, so does the pressure from large supermarket chains to demand certified sustainable fish to their providers, sometimes for fear of boycotts by NGOs. In certain countries of northern Europe, the tendency is for certain supermarkets to sell only fish certified as being sustainable. We have referred to the Marine Stewardship Council in Chapter 4 as a world leading certification system that already covers 7% of total catches. This certification is already identified by an increasing number of consumers, allowing them to make informed choices about the level of sustainability of the fish they buy. This only underscores the potential for consumers to be leading actors in the CFP of the future.

This obviously transfers the pressure to the catching sectors, which increasingly see these certifications and the policies that justify them, as commercially attractive. While this phenomenon is already clear in certain parts of Europe, in other Member States the consumers are still in general not interested on issues of sustainability. However, it is probably just a matter of time that a similar trend will eventually appear in all European markets.

These changes have intensified in recent years. As we will see in Chapter 16, the last reform of the CFP brought about many more changes to the policy than in any of the previous reforms. These changes can largely be explained as a consequence of these trends.

Is co-management an option for the CFP?

By co-management one can understand a governance system where decisions on management are taken though some kind of cooperative process between administrations and stakeholders.

This is an issue of considerable interest that one way or another has long been in the public debate about the CFP. Indeed, co-management systems exist in many areas of the world, including Member States. And the debate will only increase in the context of the new CFP established in 2013, which sets out to simplify decision making and legislation in the CFP, notably through the development of 'results-oriented management', which can be considered a form of co-management, where fisheries managers establish the objectives and fishermen have room for manoeuvre to seek the specific solutions to achieve such objectives.

Pure co-management at European level is not an option under the CFP, since the decision-making procedures for this policy are well established in the Treaty, including the prerogatives for decision making of the Council of Ministers and the European Parliament. In addition, one of the main features of co-management is the relatively small scale in which it should work. Co-management in a large, complex scenario is not workable. In this context, the closest thing to co-management (beyond initiatives at national level) can be a system where the co-legislators and the Commission in its delegated acts, would legislate differently, not through detailed prescriptive rules but rather by setting objectives, that the industry would subsequently have room for manoeuvre to deliver in different ways.

This is an attractive possibility but whether it will materialise it will depend on a number of factors and notably on the key question of confidence building. Results-based management requires trust among different fleet segments, different Member States and Member States and the Commission. Trust that the industry can engage responsibly in this exercise. Trust that the different options to reach the same objective are similar, trust that no Member State, or no fleet segment, will use this methodology to take a competitive advantage against other and trust, ultimately, that the Commission, as guardian of the Treaty, will be convinced that the objectives as decided by the co-legislators of by the Commission itself through delegated acts are effectively being met.

This looks like a daunting task but is one that could at least be started at pilot level in areas where the relative simplicity of the fisheries, the reduced number of Member States and the good cooperation between them could make such projects possible.

As we saw in Chapter 4, certain systems of co-management have been used in recent years: the 'conservation credits' system in Scotland, where fishermen implement real-time closures when they identify areas of high abundance of small cod, is a well-known example.[2] The Scottish government considers this system a success, particularly on the basis of the 'buy-in' by the industry concerned. An independent evaluation of the contribution of this kind of scheme to the management of the fishery resources concerned would cast a very interesting light into the real potential of this type of management approach as part of the CFP.

Notes

1 Originally, *moi aussi* in French.

2 www.scotland.gov.uk/Topics/marine/Sea-Fisheries/19213/restrictions/closures.

References

COM (2008). Proposal for a Council Regulation concerning the conservation of fisheries resources through technical measures. COM(2008)0324 final.

COM (2009). Green Paper. Reform of the Common Fisheries Policy. Brussels, 22.4.2009. COM (2009) 163 final.

COM (2014). Communication from the Commission to the European Parliament, the council, the European Economic and social committee and the Committee of the Regions. Regulatory fitness and performance programme (REFIT): state of play and outlook. Brussels, 18.6.2014 COM(2014) 368 final.

EC (2000). Council Regulation (EC) No 104/2000 if 17 December 1999 on the common organisation of the markets in fishery and aquaculture products.

EC (2009). Council Regulation (EC) No 1288/2009 of 27 November 2009 establishing transitional technical measures from 1 January 2010 to 30 June 2011. *OJ* L347, 24.12.2009, p. 6.

EU (2011). Regulation (EU) No 579/2011 of the European Parliament and of the Council of 8 June 2011 amending Council Regulation (EC) No 850/98 for the conservation of fishery resources technical measures for the protection of juveniles of marine organisms and Council Regulation (EC) No 1288/2009 establishing transitional technical measures from 1 January 2010 to 30 June 2011. *OJ* L 165, 24.6.2011, p. 1.

EU (2013a). Regulation (EU) No 1379/2013 of the European Parliament and of the Council of 11 December 2013 on the common organisation of the markets in fishery and aquaculture products, amending Council Regulations (EC) No 1184/2006 and (EC) No 1224/2009 and repealing Council Regulation (EC) No 104/2000.

EU (2013b). Council Regulation (EU) No 227/2013 of the European Parliament and the Council of 13 March 2013 amending Council Regulation (EC) No 850/98 for the conservation of fishery resources through technical measures for the protection of juveniles of marine organisms and Council regulation (EC) No 1434/98 specifying conditions under which herring may be landed for industrial purposes other than direct human consumption. *OJ.* L 78, 20.3.2013, p. 1.

Pastoors, M. (2014). Exponential growth in the number of words used for the European Fisheries Policy (CFP): does better management require more text? *Marine Policy* **46**(2014): 101–104.

CHAPTER 14
Aquaculture

Is aquaculture part of the CFP?

A unique part of the CFP

Aquaculture is a unique part of the CFP. Indeed, the Union does not have exclusive competence on aquaculture as it does for capture fisheries. Aquaculture has obvious linkages with the CFP, notably in that it deals (largely but not exclusively) with the same marine species. However, aquaculture is much wider in scope, notably in relation with freshwater aquaculture species, where the CFP promotes financially their development while the management of their wild equivalents are entirely outside the scope of the CFP.

The aquaculture sector is somehow attached to the CFP for a number of aspects but it is not directly regulated under the terms of the CFP. At the same time, however, many EU policies beyond the CFP itself have an important bearing on the management and development of this sector.

In any case, aquaculture remains a very important (and promising) economic sector in the Union, particularly in certain coastal areas where this activity is part of a wider economic complex together with capture fisheries, providing vital economic development and jobs.

The relationship between aquaculture and capture fisheries has often been controversial, for a number of reasons:

- Some in the capture fisheries sectors still see aquaculture as a competitor in their markets;
- Fish aquaculture depends still largely from wild fish protein as the main source of food. This raises the issue of the potential effects of the further development of fish farming on the conservation of the wild fodder fish species;
- The large imports of aquaculture products to the EU market are resented as a threat to both European capture fisheries and aquaculture.

The particular characteristics of aquaculture has made this sector a very special case in terms of how the Union deals with it: largely outside the scope of the CFP in terms of management, largely managed at national level (and in some cases, at regional level in Member States), it is still a sector strongly affected by EU regulations and policies beyond the CFP. That makes aquaculture a particularly difficult sector to handle as part of the fisheries policy.

In the framework of the CFP, aquaculture has always been part of the structural policy. But until 2013 it had never been mentioned in the basic regulation of the CFP; the Union supported it economically but could not basically regulate it. And this financial support has traditionally been very unconditional and not always conducive to good, healthy development of sustainable aquaculture.

Some facts and figures on aquaculture in the EU

Aquaculture is one of the fastest growing food-producing sectors in the world, with an annual growth rate between 6 and 8% in recent years.

The Common Fisheries Policy: The Quest for Sustainability, First Edition. Ernesto Penas Lado.

Global production, in excess of 50 million tonnes a year since the turn of the century, has increased by one third in the new millennium, thanks notably to spectacular growth in Asia and South America (FAO, 2008).

The EU-28 aquaculture reached 1.27 million tonnes of production (worth €3.51 bn) in 2011 (JRC, 2013). This production is concentrated in 5 Member States: France, Spain, Greece, Italy and the UK, which together account for 77% in volume and 76% in value of the total production.

The sector is profitable. It recovered from the difficulties of 2008 and 2009 and in 2011 it showed a return of investment and profit margin of about 10%, higher than that of the average fishing fleet. Employment levels are in excess of 80 000 people, although much of the labour is part-time: in full-time equivalents total employment in Europe is estimated at 30 000. It is also worth mentioning that the sector comprised 15 000 companies, most of them (over 90%) micro-enterprises.

European aquaculture is diversified, comprising mainly three sub-sectors:

1 Marine fish farming, including salmon, turbot, sea bass and sea bream as the main species;
2 Mollusc aquaculture, dominated by oysters and mussels but with other species (clams etc.) also as an important subsector that overlaps between aquaculture and capture fisheries;
3 Fresh water fish farming, with species such as trout and carp dominating total production, without forgetting eels and others.

In addition, many diversified sectors exist notably to farm seaweeds and many less common animal species.

In terms of production, the European production has been relatively stable in recent years: from 1995 to 2009 it only increased by 10% from 1.2 to 1.3 million tonnes. And, in fact, this increase took place from 1995 to 2000 and has been stable, even decreasing slightly ever since. In comparison, aquaculture production in Norway from 1995 to 2009 grew by 246% (EUROSTAT, 2011).

The picture, in any case, is also very variable among Member States: the general stability in production at European level is in fact the combination of certain Member States increasing considerably their production of farmed fish (notably Greece, the UK, Spain and Ireland) while other Member States (Denmark, Germany, France) have reduced considerably their production in the same period. This shows that, unlike capture fisheries, where the common resources evolve in shared sea-basins for the benefit of all involved Member States, aquaculture is a domain where production still depends fundamentally on national policies.

The evolution of the EU policy in promoting aquaculture development

We can distinguish several phases in the way aquaculture development has been supported and managed at EU level:

- A first phase, roughly 1970 to 1992, where aquaculture was part of the structural policy but without any attempt to plan the activity over and beyond the provision of economic support for new aquaculture production;
- The 1992–2013 approach, where aquaculture was incorporated into the mainstream objectives of the CFP (including a specific mention of this sector in the basic regulation of the CFP). In this phase, the financial assistance to the sector (directly through the structural policy and indirectly through aquaculture research) an important legislative package (the 1991 package, see next) on water quality, sanitary conditions and so on was set up, turning aquaculture into a sector strongly conditioned by these regulations;
- The third phase, following the CFP reform of 2013, where aquaculture is considered a key sector into the new CFP and the emerging integrated maritime policy and where new additional instruments are put into place, notably to resolve the long-standing obstacles that have led European aquaculture to stagnation.

In the new phase, after the 2013 reform, aquaculture is considered a strategic sector whose development deserves action at very different levels, beyond the existing financial support and the sanitary and environmental legislation. In particular, the inclusion of a specific obligation for Member States to address their national management of this activity to remove barriers and red tape and the adoption of new EU instruments, such as maritime spatial planning, constitute an unprecedented attempt to provide a complete framework where the obstacles preventing European aquaculture to take off could be addressed.

EU legislation and aquaculture

Although aquaculture is not managed under the CFP, it is certainly influenced by a number of EU policies, notably beyond the CFP itself. This situates aquaculture in a somehow uncomfortable position: many EU policies and legislation affect aquaculture but very few of these policies and regulations have been designed for this sector. Very often, aquaculture is in the receiving end of the effects of EU policies designed with other more important objectives in mind. And the legislation applicable to aquaculture is not regrouped in a few legal acts; on the contrary, it is dispersed across a wide range of policy areas and regulations, many of them having aquaculture as a relatively minor interest. In any case, aquaculture has not been 'forgotten' by EU law; on the contrary, many policies have an impact on the sector, either directly or indirectly.

Aquaculture as part of the structural policy of the CFP

Aquaculture did not feature in the first structural regulations for the fisheries sectors, starting with the 1970 regulation (see Chapter 6). It was in 1986, in the wake of the accession of Spain and Portugal, that aquaculture was fully incorporated into the objectives of the structural policy: Regulation 4028/86 actually refers to fisheries and aquaculture in its very title (EEC, 1996).

In recent times, aquaculture investments have represented a very significant part of the total investments under the structural policy of the CFP. For example in the European Fisheries Fund between 2007–2013, aquaculture was represented the third most important sector in terms of total investment from this fund, with €389 million of EFF money invested until May 2013 (representing over €1 bn of total investment), only after permanent cessation and fish processing).

In the context of the last reform of the CFP, as we will see in Chapter 16, aquaculture received a very broad support and, for the first time, this was not just financial. The availability of EU funds for aquaculture has been made conditional on Member States' willingness to resolve the bottlenecks that prevent aquaculture development at national level, through the establishment of a national strategic plan where such bottlenecks would be addressed.

Aquaculture and environmental policy

The relationship aquaculture/environment has always been controversial. This is the result of the fact that aquaculture needs pristine environmental conditions, particularly in terms of water quality and these conditions are usually met in areas of high ecological value, which in turn tend to be a priority for nature conservation. Aquaculture is commonly seen by environmental interests as an activity of high potential for environmental degradation. This conflict has represented traditionally and still represents today a major challenge.

Aquaculture development in Europe has been significantly influenced by the EU environmental legislation. This influence has often been resented by many aquaculture producers as a negative one, notably because a number of aquaculture initiatives have been stopped as being incompatible with EU environmental law and, in general terms, some aquaculture producers consider that the high standards of EU environmental legislation stifle the development of the activity. Indeed, the reality is much more complex and two-sided.

On the one hand, it is true that in a number of cases, some of them very well publicised, certain aquaculture investments have not been authorised (by national authorities) following allegations of non-respect of EU law, in particular in the case of investments in Natura 2000 areas. On the other hand, however, environmental law has contributed to improve the quality of Europe's waters and clean water is aquaculture's most valuable asset. From that angle, environmental law has helped aquaculture to flourish. Overall, both policies should not be seen as contradictory: aquaculture needs clean water and a well-preserved environment and the aquaculture activity must ensure that it does not spoil the very environment it depends on.

Let us see the main environmental legislation and how it affects European aquaculture.

The directives on Environmental Impact Assessment

Aquaculture has been, from the mid-1980s, subject to EU legislation on environmental impact assessment (EEC, 1985). Recently revamped (EU, 2012), this legislation requires an environmental impact assessment of all major projects to be installed in coastal areas. This means that large aquaculture projects have been subject to this requirement.

This requirement has given rise to many (real or alleged) difficulties for aquaculture developments. In some Member States, in particular, discrepancies over the implementation of that requirement have blocked a number of projects and still today part of the aquaculture industry see this legislation as a major obstacle to aquaculture development.

The Habitats Directive and Natura 2000

The Habitats Directive (EEC, 1992) established the obligation by Member States to: 'contribute towards ensuring bio-diversity through the conservation of natural habitats and of wild fauna and flora in the European territory...'.

While the Directive refers to the European territory, it is understood that it applies also to the wild species and habitats in the marine waters under the jurisdiction of Member States. Beyond the conservation of wild species, the marine habitats to be protected on account of their high biodiversity constitute a network called 'Natura 2000', together with the areas to be protected in the context of the 'birds directive', adopted already in 1979 and revamped in 2009 (EC, 2010).

The problem has been that, in a number of cases, aquaculture projects in or close to Natura 2000 areas have been denied licenses on account of their alleged impact on the preservation of these areas. In certain cases, a large fish farming plant was denied a license in a European region for that reason, only to be allowed in another Member State ... also in a Natura 2000 area. This shows that it is not just EU law than can have an impact, it is also the way Member States implement EU law.

The legislative framework has not provided enough certainty to potential investors as to what aquaculture investments are and what are not compatible with the objectives of Natura 2000. And lack of certainty opens up the door to potential conflict and discourages investment. For this reason, the Commission decided to elaborate guidelines on how aquaculture can develop within the context of the implementation of the Natura 2000 network (European Commission, 2012).

These guidelines have a clear objective: providing more certainty to investors and national authorities alike as to what investments can or cannot be compatible with Natura 2000 areas. Similar guidelines had been prepared by the Commission for other economic sectors, generally with very positive results.

The Water Framework Directive

The Water Framework Directive (Directive 2000/60/EC) (EC, 2000) is a fundamental instrument of the environmental policy with a crucial effect on aquaculture. This effect is two-fold:

- On the one hand, the high standards of environmental protection represent in some cases a deterrent factor to the development of aquaculture. These high standards of water quality imply the need for the aquaculture industry to develop the technology to ensure that its own development does not endanger Europe's water quality.
- On the other hand, clean water is one of aquaculture's most precious assets. Ensuring high standards of water quality in Europe is a fundamental contribution to ensure that European aquaculture has areas of high water quality for its development.

Overall, this directive represents an opportunity and a challenge for the European aquaculture industry.

This directive sets out to achieve very high standards of quality in all European waters, marine and continental, including all those where aquaculture takes place. As high-quality, unpolluted water is a fundamental condition for good aquaculture development, this directive represents a vital part of the EU framework to develop sustainable aquaculture.

The directive puts together a number of previous pieces of legislation of importance for aquaculture. In this regard, it is of particular relevance the incorporation of the new framework of the former specific legislation on the quality of water for mollusc aquaculture (EEC, 1979), which established since 1979 specific rules to ensure that mollusc aquaculture could develop only in areas where a minimum standard of bacteriological quality could be ensured. The new framework is intended to ensure that the standards of the old legislation are maintained.

The Marine Strategy Framework Directive

As we saw in Chapter 11, the Marine Strategy Framework Directive (MSFD) sets out to achieve Good Environmental Status (GES) of European maritime waters by 2020. The notion of GES is implemented through 11 descriptors. Out of these descriptors, some are of particular relevance for marine aquaculture. In particular, descriptor 2 on the introduction of non-indigenous species by human activities and descriptor 8 on levels of contaminants in seawater and 9 on levels of contaminants in seafood have an obvious bearing on aquaculture practice. But other descriptors, such as the avoidance of eutrophication, or the preservation of seafloor integrity are also indirectly related to aquaculture.

In any case, as usual with environmental legislation, the MSFD represents at the same time a constraint and an opportunity for aquaculture. A constraint, because it imposes standards of environmental conservation that will require the aquaculture industry to invest in environmentally-friendly practice but an opportunity because it will ensure the quality of the marine environment on which aquaculture is crucially dependent.

Aquaculture and sanitary policy

This area of legislation was developed over the years although, as we saw previously, the main impetus came with the 1991 package.

Since 1979 the EU had legislation on the quality of seawater required for mollusc aquaculture (EEC, 1979). This legislation, inspired by the need to guarantee water quality in particular for the large oyster industry in France, provided a fundamental basis to establish control systems to guarantee that mollusc aquaculture will only take place in areas of high water quality. This responded to the need to reassure consumers that their shellfish arrived to the market with the highest possible quality and guarantees. As such, this legislation imposed controls in the way Member States managed their mollusc aquaculture to guarantee their quality. This legislation has been more recently replaced by Directive 2006/113/EC (EC, 2006a).

As mentioned earlier, the 1991 package was a fundamental step in establishing a legal framework where aquaculture could develop under conditions of animal health and consumer protection. Often criticised for the burden they represent for national administrations and the industry alike, they indeed established a framework where aquaculture would develop in a way to ensure consumer confidence in this sector. The elements of the package were:

- The Directive on the health conditions for the production of bivalve molluscs (EC, 1991a).
- The Directive on health conditions for the production and placing on the market of fishery products (EC, 1991b).
- The Directive on the animal health conditions governing the placing of the market of aquaculture animals and products (EC, 1991c).
- And, last but not least, the directive concerning urban waste management (EEC, 1991).

The question of animal health has been particularly contentious for aquaculture.

Today, this issue is regulated through the directive on animal health requirements for aquaculture animals (Directive 2006/88/EC) (EC, 2006b). This directive establishes a number of requirements for aquaculture facilities to be authorised by national authorities.

This legislation has also been considered at some stage as burdensome. However, it is important to bear in mind that, for at least two reasons, it is necessary to set very high standards for aquaculture production:

- The high farming densities that are necessary to ensure economic performance make the possible occurrence and extension of diseases an extremely high risk for aquaculture facilities.
- The fact that the sea (and river basin) has no boundaries makes it also necessary to have commonly high standard, so as to avoid that malpractice in one Member State could result in the extension of a disease to a neighbouring Member State.

Aquaculture and research policy

Aquaculture has always benefited from the European funds for research. As an illustration, the Sixth Framework Programme for Research (2002–2006) invested €98 million in 75 programmes for aquaculture, of which 29 projects corresponded to small and medium enterprises. These projects covered a variety of issues related to aquaculture, notably sustainable development, food quality and safety and policy support.[1] This support is maintained in the current EU research programme, Horizon 2020.

Alien species in aquaculture

This legislation, introduced in 2007, sets out to regulate the introduction of alien species in the European seas for their use in aquaculture (EC, 2007).

This regulation is justified in the light of certain experiences in the past, where the introduction of certain alien species in European waters, on account of their quality and/or fast growth rates, led to the unintended introduction of diseases associated to them that ended up extending to local species. There is certainly no better example of this than the unintended introduction of the parasite *Bonamia ostreae* in Western Europe in the 1980s, carried by imported oysters and that has ended up decimating the local populations of European flat oyster (*Ostrea edulis*) in most of its original areas of distribution.

This legislation requires that any introduction of an alien species for aquaculture be subject to a specific permit, accompanied to an impact assessment. It also defines whether certain species, initially alien but already well-established in European waters, can actually be considered alien in terms of their potential risks. As in other cases, the legislation carries a certain burden in terms of the requirements to introduce alien species for aquaculture production. These may be resented in the short term as yet another obstacle to aquaculture development. But as any European flat oyster producer would agree, this burden is worth assuming if it can serve to prevent another *Bonamia* case.

Stakeholders

Despite the absence of a legal basis to manage aquaculture, the stakeholder-consultation bodies established under the CFP umbrella have traditionally included aquaculture. As we saw in Chapter 12, the Advisory Council for Fisheries and Aquaculture, as its very name indicates, considered that consultation with aquaculture stakeholders was a fundamental part of its mandate. That underlines the importance that the Union has granted to aquaculture, over and beyond the limited regulatory competence.

This policy is, in fact, continued in the 2013 policy where, in the context of the suppression of ACFA (see Chapter 12) a new Advisory Council on aquaculture was set up.

This new body should play an important role in the new phase of aquaculture development and could, in particular, play a key role in improving the mutual understanding between the aquaculture industry and environmental organisations, which is a fundamental condition to allow for the development of sustainable aquaculture.

The aquaculture strategies

As we saw previously, many of the regulations and policies affecting aquaculture have not been designed with aquaculture as the main objective. In addition, these different instruments have not been developed in a way that would ensure a coherent package of policy instruments. This has led the Commission to try to put together all the relevant policies having an impact on aquaculture with the purpose of developing a coherent policy that would inspire future actions, from the point of view of an integrated, forward-looking approach. This has taken the form of aquaculture strategies.

2002: The first strategy

This first strategy (COM, 2002) intended to provide the elements to boost aquaculture production in Europe that, at that time already, was identified as having enormous potential for future growth. The strategy set out to achieve an annual growth rate of 4% in production and to create between 8000 and 10 000 jobs in the period 2003–2008.

The strategy rightly identified the areas where further progress was still necessary to boost aquaculture production. These included four areas of interest:

1 Encouraging economic viability. This included the need to address issues such as aquaculture feeds, the competition for space, marketing issues and so on;
2 Guaranteeing food safety animal health and welfare;
3 Addressing environmental effects and
4 Stimulating research.

On that basis, the strategy proposed an action plan to be implemented in the following two years with the aim of unlocking the growth potential of this sector in Europe.

The strategy, however, achieved relatively little progress and fundamentally failed to provide the elements necessary to allow for growth in production. Several reasons can explain that:

- The action plan was overambitious in terms of deliveries within two years. Achieving the goals clearly needed more time. In fact, a number of the actions could not be achieved within that time frame. The growth of aquaculture production before the turn of the century gave the Commission a false optimism concerning the growth prospects in the following years.
- There was a shortage of effective mechanisms to implement it, notably in relation with financial aid: the strategy was set up in the middle of a financial programme of the structural funds (FIFG 2000–2006) and it was not possible to incorporate the priorities of the action plan into the ongoing financial package.
- Since a number of bottlenecks were dependent upon Member State competence, the strategy could not, at the time, establish effective mechanisms to oblige or encourage Member States' administrations to implement the action plan, even though the strategy correctly identified the bottlenecks to resolve. This was, in particular, the case of the competition for space.
- At the time, the need for aquaculture development was not high enough in the political agenda and many of the actions did not get sufficient priority for their development.

In any case, the strategy was a very important new step, in that it put together all the issues affecting aquaculture, hitherto dispersed among many different EU and national policy strands in an integrated and coherent way. In doing so, the strategy succeeded to raise awareness about the problems being faced by this sector and about the need to address them in a coordinated manner.

2009: The second strategy

The second strategy, adopted in 2009, resulted from the evidence that the growth of aquaculture projected in the first strategy was not being realised. While aquaculture production was sky-rocketing around the world, EU aquaculture remained basically stagnant. As a result, the Commission understood that the whole strategy would need to be revised (COM, 2009). On this basis, the new strategy tried to build on the previous one but addressing the reasons why the first objective of the 2002 strategy (the increase in production) was not realised. In doing so, it also built on the new impetus provided by the emerging Integrated Maritime Policy (IMP) of the Union, in particular to inscribe aquaculture into the wider framework of IMP where this activity would benefit from a new focus on the development of the maritime economy of the Union.

In particular, the second strategy focused on the identification of the bottlenecks preventing aquaculture development. If EU aquaculture had an excellent research basis, a good market and generous public funding, why was it not growing? Doing the right diagnosis was here essential to find the solution. And it was clear that the diagnosis should cover all the phases and aspects of aquaculture development: not just what was done in Brussels about it but also, crucially, on what was done in Member States.

The strategy established three large objectives: (1) promoting competitiveness, (2) establishing conditions for sustainable growth and (3) improving the sector's image and governance.

Obvious objectives such as resolving the problem of the competition for space, or reducing the administrative burden, were given special emphasis. The issue of the bad environmental image of aquaculture and its consequences in terms of licensing were also particularly underlined and so was the question of the governance. However, the strategy carefully avoided any quantitative objective in terms of production.

But the crucial question was whether this second strategy would have both the political priority for its development and whether it could have the right instruments to make them applicable in practice, notably at Member State level. Actually, the opportunity to translate that strategy into more effective instruments came with the CFP reform of 2013.

2013: The strategic guidelines

As part of the process of CFP reform, the Commission presented its strategic guidelines, with the objective of mobilising national administrations to address the bottlenecks preventing aquaculture production, in an effort to work together at EU and national level in de-blocking these bottlenecks. As working method, the Commission proposed an 'open method of coordination', a voluntary process of cooperation on the basis of the guidelines and including joint efforts, exchange of best practice.

The guidelines built in the previous strategies and identified four areas of work for the Commission and Member States:

1. Simplifying administrative procedures. The Commission underlined that the time to grant an aquaculture license in EU Member States was often between 2 and 3 years, while in Norway, with the recent introduction of the 'single contact point' it has come down to 6 months;
2. Coordinated coastal spatial planning. This is the key element to resolve the competition for space and, at the same time, it is also crucial to reduce licensing time;
3. Enhancing the competitiveness of the sector. Full use of the support mechanisms (research under Horizon 2020, funds from EMFF etc.) to improve competitiveness;
4. Level playing field, by exploiting the competitive advantages of EU aquaculture. This includes, crucially, the exploitation of the high standards of environmental and consumer protection of EU products in the open market, as an important asset for consumers.

These guidelines constitute the basis for what we can consider a voluntary policy to promote aquaculture by all those interested.

Aquaculture strategy and CFP reform

Unlike the two previous cases, the 2013 guidelines were not an isolated effort but an integral part of an important revamping of the whole of the CFP. In this context (see Chapter 16), aquaculture featured for the first time in the history of the CFP as an important policy pillar in the basic regulation. But more importantly, a new instrument was established: the ex-ante conditionality in the EMFF, to the effect that access to aquaculture funds under the new structural instrument of the CFP would be made conditional upon presentation by the Member State of a strategic plan where the Member State concerned would endeavour to resolve the issues of excessive administrative burden and competition for space.

Furthermore and as established in the 2009 strategy, the guidelines on aquaculture and Natura 2000 were adopted by the Commission. These guidelines should clarify the conditions under which aquaculture development can take place in connection with Natura 2000 sites. These guidelines should remove the high level of uncertainty that exists today in this regard that in fact has considerably stifled aquaculture investment in a number of Member States.

At this point in time it is not possible to anticipate if this new instrument will be successful in unlocking the potential for growth in European aquaculture. But at least one can say that both the instruments and the strong political will to boost European aquaculture have been put into place as never before.

What future for European aquaculture?

European aquaculture: a unique mixture of strengths and weaknesses

The future of aquaculture in the context of the global fish supply is generally considered extremely promising. A study of the World Bank (2013) concluded, through modelling projections of current trends, that:

> Aquaculture will continue to fill the growing supply-demand gap in the face of rapidly expanding global fish demand and relatively stable capture fisheries. While total fish supply will likely be equally split between capture and aquaculture by 2030, the model predicts that 62% of food fish will be produced by aquaculture by 2030.

However, the same study considered that the growth of aquaculture worldwide will happen notably in areas such as Latin America or North Africa/Middle East, while production in Europe will remain stable. This is obviously based on trend analysis, thus reflecting the stagnation of European aquaculture in the last decade. Introducing the necessary changes to reverse this trend and prove these projections wrong is the big challenge for European aquaculture.

A different study on future trends by FAO is not optimistic about European aquaculture either. From 2007 to 2030, FAO estimates, the European consumption of fish will increase from 22 to 24 kg *per capita* and year, meaning a need for additional 1.6 million tonnes a year of seafood. According to this study, this additional supply will have to come from imports (FAO, 2007).

The future of EU aquaculture is, at the same time, promising and difficult. This apparent paradox is the result of a unique mixture of strengths and weaknesses. The future is promising because of the many strengths but these will not be unlocked unless the weaknesses are effectively addressed.

The strengths are very clear:

- A world-class research base for innovation (largely funded by the Union research policy);
- Extensive experience and expertise and know-how in the industry;
- A large, lucrative market for aquaculture products;
- Public funding available;
- A high political priority for development in many regions.

However, Europe still has a number of weaknesses that need to be fully addressed:

- A tradition of very heavy bureaucracy to obtain licenses for aquaculture in many Member States;
- The high population density of the European coasts and the very high and difficult competition for the access to physical space with other potential users;
- An image of certain aquaculture species (typically, those also available as wild species from capture fisheries) as a second-class product;
- An industry with few big players, dominated by small and medium enterprises;
- The frequent conflicts between aquaculture development and nature conservation.

The future of EU aquaculture lies in Europe's ability to resolve the weaknesses to unlock the full potential of this industry.

The challenges

In Chapter 16 we will see the elements of the 'new deal' on aquaculture as part of the CFP reform of 2013. But the challenges of European aquaculture are a mix of global issues for all this activity (such as that of the availability of fish meal) and specific European ones. Let's look at these challenges.

Fish feed: Aquaculture and wild fish

One of the most challenging issues for aquaculture development is that of the relationship between the farming of carnivorous fish and the pressure that this will exert on the exploitation of the main fish used for fish feed. This is indeed a global problem, not just a European one, since both aquaculture expansion and the exploitation and trade in fish meal and fish oil are global issues. Europe, despite important production of fodder fish (sandeel, Norway pout and others) is a net importer of fish meal and fish oil, particularly from the world's main provider of these commodities: the Peruvian anchoveta.

The problem is also made complicated by the fact that a variable part of fish meal and oil from fodder fish has also been used to feed terrestrial animals, so it is very difficult to establish a simple relationship between the growth of fish farming and the pressure on the wild fodder fish. There is a level of competition for fish meal between farmed fish and certain terrestrial animals that seems to be evolving in favour of aquaculture: while in 1960 fishmeal was used exclusively to feed pigs and poultry, the use of fish meal for aquaculture grew to 10% in 1980 and to 73% in 2010 and its consumption by pigs and poultry has declined as a consequence (Shepherd, 2012). This trend is not surprising since fish is more efficient in converting fish meal into human food than poultry or pigs. As fishmeal supply becomes more limiting, its use tends to concentrate in the sector that better converts it into food for humans.

In addition, many countries around the world have very ambitious plans to develop fish farming; if all these plans are realised, the competition for the – limited – sources of wild fish protein will intensify. This challenge is therefore two-fold: one of competition among countries for the limited supply of fish protein and oil and one of competition between fish farming and other livestock fed with fish meal. Whatever this competition may lead to, it is essential that it does not lead to overexploitation of Peruvian anchoveta or European sandeel, among other species.

In any case, two trends appear to solve this challenge: the ever increasing efficiency of conversion of fish meal into fish protein and, specially, the development of feeds alternative to fishmeal from wild fish, notably from vegetable origin, from fish trimmings and so on.

In addition, there is a growing debate as to whether the forage fish species could be better used for direct human consumption. Not surprisingly, countries like Peru are investing in the development of anchoveta as a species for direct human consumption, with a considerable increase in added value. As this idea unfolds, the availability of anchoveta for fish meal and oil will only shrink in the future.

Overall, the prospects are promising: forage fish can be exploited sustainably, regardless of their end use, and in fact the world's production of these fish has remained stable over the past 20 years despite a dramatic increase in the production of farmed fish. The fishmeal and fish oil industry depends one third from forage fish and another quarter from scraps produced when fish are processed for human consumption. In addition, the research on alternative feeds is well advanced, and oils from plants, waste processing, and even seaweed are used already to produce fish feed, to the point that today it is already possible to feed salmon on a fish-free diet.[2]

In any case, two trends seem clear in this debate: the improvement in the ratio of conversion of fish feed into fish biomass and the trend to develop alternative feeds to those based on wild fish meal.

The challenge is complex but it means something clear: the need to develop feeds from alternative origin, notably vegetable. In this regard, abundant research has been financed by the Union and indeed the trend observed in recent years is that of a gradual reduction of the dependence from wild fish oil and protein.

Although it is impossible to foretell the future, it seems clear that as the competition for forage fish increases and as its supply decreases (if a market for human consumption develops) those having developed alternative sources of fish feed would be in a better competitive position.

Aquaculture and marine spatial planning

The role of maritime and coastal planning to promote healthy development of aquaculture has been underlined by many authors (Penas, 1997). Unlike important aquaculture producers in the world, such as Norway or Chile, that have thousands of km of productive coast with a low population density, most of the European coast is densely populated and subject to many other economic activities. In these circumstances, competition for space is fierce, and finding a way to combine the different uses of the coastal and marine space in an orderly and rational way becomes paramount.

The recent adoption of a Directive on maritime spatial planning (EU, 2014) represents a potential major step in the direction of providing certainty about the future uses of the maritime space. This is of particular relevance for the possible development of offshore aquaculture, whose potential conflict with other uses of the maritime space will inevitably require the clarification, by Member States, of the use of such space under their jurisdiction.

The experience of coastal aquaculture, where competition for space has seriously contributed to stifle aquaculture investment, may not necessarily be repeated if Member States implement the directive successfully, so that aquaculture, if considered a promising development by each Member State, could enjoy areas where its development would be allowed without hindrance.

Aquaculture and administrative burden

In all analysis of the problems of European aquaculture this question stands out as an important deterrent to investment: in some cases, it has been reported by the industry that obtaining a license for aquaculture requires permission from 13 different authorities. This is not easy to resolve: the difficulties associated with the management of overcrowded coastal zones, and the number of possible interactions between aquaculture and other uses of that space require complex consultations with different departments representing different interests. So this is not just a problem of administrative complexity but rather the result of (largely unresolved) conflicting interactions among different uses (real or potential) of the coastal space.

Resolving this is not easy. But at least there are known solutions, already applicable in some Member states or regions:

- The idea of a 'one stop shop' where all the permits and requirements for an aquaculture investment could be centralised, thus providing for administrative simplification.
- The establishment, through integrated coastal planning, of areas identified for aquaculture development, where the licensing requirements could be simplified.

Many other solutions may exist, and since these solutions are a national (often regional) competence, the CFP cannot legislate on them. Instead, the Commission has identified the need to exchange best practice among national administrations and the industry, to identify good examples that could be used in other Member States or regions.

Aquaculture and environmental legislation

Environmental legislation has been considered, by part of the industry, as the main culprit for the stagnation of European aquaculture in recent years. This is short-sighted. With a long-term perspective, European aquaculture can only benefit from the need to develop under demanding environmental conditions. Only if it can meet these high environmental standards will aquaculture be able to take an important place in the intensive competition for physical space in the European coasts. And only if it capitalises on these high standards, can European aquaculture become more and more competitive with other alternative foods, aquatic or otherwise, in a Europe where the demand from consumers of high-quality, environmentally-friendly food will only increase in the years to come.

Improving the image of aquaculture

There are two areas where the image of aquaculture and its products must be improved: the environmental impact of the activity and the quality of its products as compared to the wild species from capture fisheries.

It is important to underline that, out of the three main aquaculture sectors in Europe, at least two can be counted for as contributing to achieve environmental goals. Mollusc aquaculture needs clean waters and this sector is second to none in terms of advocating high standards of water quality in Europe. As regards inland aquaculture in ponds, this activity has saved from agricultural use many wetlands from agricultural use in Europe, contributing indirectly to the preservation of biodiversity, for example by providing nesting areas for many birds.

In terms of product quality, many aquaculture species are increasingly difficult to distinguish by consumers from their wild counterparts. And indeed aquaculture species, produced locally, can also benefit from the campaigns of such initiatives as 'slow food'.[3]

The work of the Advisory Council on Aquaculture, where the industry and NGOs will sit together to make policy recommendations looks ideally placed to iron out the image problems that this activity still suffers from today.

Notes

1 Synopsis of Fisheries and Aquaculture research projects in the 6th Framework Programme. European Commission, 2008.
2 www.nmfs.noaa.gov/aquaculture/faqs/faq_feeds.html.
3 http://slowfood.com/slowfish/pagine/eng/pagina.lasso?-id_pg=141.

References

COM (2002). Communication from the Commission to the Council and the European Parliament – A strategy for the sustainable development of European aquaculture. COM/2002/0511 final.

COM (2009). Communication from the Commission to the European Parliament and the Council – Building a sustainable future for aquaculture – A new impetus for the Strategy for the Sustainable Development of European Aquaculture {SEC(2009) 453} {SEC(2009) 454} COM/2009/0162 final.

EC (1991a). Council Directive of 15 July 1991 laying down health conditions for the production and placing on the market of live bivalve molluscs (91/492/EC). *OJ.* l 268, 24.9.1991, p. 1.

EC (1991b). Council Directive of 22 July 1991 laying down the health conditions for the production and the placing on the market of fishery products (91/493/EC). *OJ* L 268, 24.9.1991, p. 15.

EC (1991c). Council Directive of 28 January 1991 concerning the animal health conditions governing the placing on the market of aquaculture animals and products (91/67/EEC). *OJ* L 46, 19.2.1991, p. 1.

EC (2000). Directive 2000/60/EC of the European Parliament and the Council of 23 October 2000 establishing a framework for Community action in the field of water policy. *OJ* L 327, 22.12.2000, p. 1.

EC (2006a). Directive 2006/113/EC of the European parliament and of the Council of 12 December 2006 on the quality required in shellfish waters. *OJ.* L 376, 27.12.2006, p. 14.

EC (2006b). Council Directive 2006/88/EC of 24 October 2006 on animal health requirements for aquaculture animals and products thereof and on the prevention of control of certain diseases in aquatic animals. *OJ* L 328, 24.11.2006, p. 14.

EC (2007). Council Regulation (EC) No 708/2007 of 11 June 2007 concerning use of alien and locally absent species in aquaculture. *OJ* L 168, 28.6.2007, p. 1.

EC (2010). Directive 2009/147/EC of the European Parliament and of the Council of 30 November 2009 on the conservation of wild birds. *OJ* L 20, 26.01.2010, p. 7–25.

EEC (1979). Council Directive 79/923/EEC of 30 October 1979 on the quality required of shellfish waters. *OJ* L 281, 10.11.1979, p. 47–52.

EEC (1985). Council Directive of 27 June 1985 on the assessment of the effects of certain public and private projects on the environment (85/337/EEC). *OJ* 175, 5.7.1983, p. 40.

EEC (1986). Council Regulation (EEC) No 4028/86 of 18 December 1986 on Community measures to improve and adapt structures in the fisheries and aquaculture sector. *OJ* L 376, 31.12.1986, p. 7–24.

EEC (1991). Council Directive of 21 May 1991 concerning urban waste water treatment (91/271/EEC). *OJ* L 135, 30.5.1991, p. 40.

EEC (1992). Council Directive 92/43/EEC of 21 May 1992 on the conservation of natural habitats and of wild fauna and flora. *OJ* L 206, 22.7.1992, p. 7.

EU (2012). Directive 2011/92/EU of the European Parliament and of the Council of 13 December 2011 on the assessment of the effects of certain public and private projects on the environment. *OJ* L 26, 28.1.2.12, p. 1.

EU (2014). Directive 2014/89/EU of the European Parliament and of the Council of 23 July 2014 establishing a framework for maritime spatial planning. *OJ* No L 257, 28.8.2014, p. 135.

European Commission (2012). Guidance on Aquaculture and Natura 2000.Sustainable aquaculture activities in the context of the Natura 2000 network. 89 pp.

EUROSTAT (2011). *Agriculture and Fisheries Statistics.* Eurostat, 2011 Edition.

FAO (2007). Future prospects for fish and fishery products. 4. Fish consumption in the European Union in 2015 and 2030. FAO Fisheries Circular No. 972/4, Part 1. Rome, 2007.

FAO (2008). FAO: the state of world aquaculture 2008 (data excluding aquatic plants).

JRC (2013). Summary of the 2013 Economic Performance report on EU Aquaculture Sector (STECF 13-XX). Scientific, Technical and Economic Committee for Fisheries. *JRC Scientific and Policy Reports*, 2013.

Penas E. (1997). *Elementos para una ordenación integral de los usos del medio litoral de Galicia. Xunta de Galicia*. Santiago de Compostela, 368 pp.

Shepherd, J. (2012). Aquaculture : Are the criticisms justified ? Feeding fish to fish. *World Agriculture* **3**(2): 11–18.

World Bank (2013.) Fish to 2030. Prospects for Fisheries and Aquaculture. World Bank Report number 83177-GLB. December 2013.

CHAPTER 15

Reforming the CFP: 1992 and 2002

Reforming the CFP

The CFP is reformed every 10 years approximately. Since its inception in 1983, the policy has therefore been reformed three times: in 1992, 2002 and 2013. The history of these reforms is a clear testimony of the policy's strengths and weaknesses, of the policy elements that resist the test of time and of those that do not.

In principle, the need for a revision every 10 years only applied, legally speaking, to a very limited number of provisions that, by their own nature, must be revised after a certain period. But the provisions under compulsory revision were very few. It has, however, always been the case that the Commission has used the occasion of their revision to propose a wider and more far-reaching review of the policy.

In 1992, Regulation 170/83 required the preparation by the Commission, in 1991, of a report on the application of the 12-mile regime. In addition, another report was required, also in 1991, on the effects of the implementation of the so-called 'Shetland Box'. So the legal requirements to revise the policy were very limited. And indeed, as we will see next, the 1992 reform was limited in its ambition and scope.

For the 2002 reform 2002, Regulation 3760/92 had extended the 12-mile regime for an additional 10 years. The implication was that, unless this regime was specifically continued after 2002, it would cease to exist. Beyond that, no other specific revision clause was mentioned. Again, the legal requirements for revision were very limited: only the 12-mile regime. However, the Commission took advantage of this review to propose a much wider and far-reaching reform.

Indeed, the 2002 reform had the foresight to propose an important revision after 3 years. In fact, Regulation 2371/2002 (EC, 2002a) established that: 'The Commission shall report to the European Parliament and the Council on the operation of the Common Fisheries Policy with respect to Chapters II [conservation and sustainability] and III [adjustment of fishing capacity] before the end of 2012'.

In addition, as it was already customary, this regulation extended the 12-mile regime for another 10 years and required a report by the Commission on its implementation before the end of 2012. This means that in 2002, for the first time, the CFP submitted itself to a review of two of its most fundamental pillars in 10 years' time.

In all cases, the reform process was preceded by a report by the Commission on the performance of the policy, followed by a public consultation, in a remarkable exercise of transparency and self-criticism. Few other fishery policies in the world are subject to that kind of critical scrutiny by everybody.

The Common Fisheries Policy: The Quest for Sustainability, First Edition. Ernesto Penas Lado.

The first reform: 1992

The first reform took place in 1992, 9 years after the birth of the Blue Europe in 1983. This reform was foreseen in Regulation 170/83 although, as explained previously, the legal requirement affected only a few provisions. But in addition to this requirement, this reform was made against the background of the transitional period for Spain and Portugal and the need to revise its conditions in 1995.

The context

After almost a decade of the CFP, the political environment for the first reform was officially conditioned by four main factors: (1) the withdrawal of Greenland from the EC, (2) the accession of Spain and Portugal, (3) the reunification of Germany and (4) the achievement of the internal market. However, a fifth factor was in fact the dominating one: the difficulties of agreeing to the policy between 1976 and 1983, still relatively fresh in the memories, left relatively little political appetite for a major reshuffling of the policy, at least in a majority of Member States.

As a result, the first reform of the CFP was not an ambitious one. There was, however, awareness that the policy would require some significant improvement. The evidence of significant overexploitation of some key resources was already there. And so was the evidence of the need to substantially reduce fleets (the Gulland report referred to in Chapter 5 was published in 1990), as well as that of improving control.

But the overall political climate at the time was that the basic deal of 1983 and the deal with Spain on accession (including the transitional period) should continue and that the policy should not fundamentally change but only be subject to certain adjustments.

To this effect, the Commission published in 1991 a report on the implementation of the first years of the CFP to gather opinions from stakeholders and Member States (SEC, 1991). In that paper, the Commission made a comprehensive analysis of the application of the first years of the CFP. A number of points can be underlined:

- The first warnings were expressed as regards the status of fishery resources. Although the report was not alarmist, it already claimed the need to reduce fishing pressure on certain stocks. It also warned about excessive catches of juvenile fish and proposed the improvement of selectivity through technical measures;
- The need to control fleet capacity as a mean to achieve sustainability of the fisheries appeared clearly as an objective for the first time;
- At the same time, however, it underlined the need to take economic factors into account and insisted on the need for a social policy to accompany the CFP, including aid to those leaving the profession and training;
- Interestingly, the problem of discards was already identified and even quantified, although no specific proposals were made to address it;
- It also addressed for the first time the need to consider multi-species approaches, based on a critical analysis of the annual TAC decisions;
- The need to improve the control of the CFP was also mentioned, although the report carefully avoided criticising Member States. However, a number of ideas to reinforce the control of the CFP were already presented. Overall, even if the language of the Commission was very careful and apparently bland, the identification of control as a central problem of the CFP was very clear;
- The report proposed to extend the regime of access, including the 12-mile regime of 1983 and the transitional period for Spain and Portugal until 2002, although it warned about the end of the 'Irish Box' in 1996.

It is also worth noting what the document did not propose: as in 1983, the report did not mention the Mediterranean at all. In the annex regarding the status of stocks, only Atlantic fisheries were referred to and no specific suggestion was made on whether, or how, the CFP should apply to the Mediterranean basin, even if, by the time, there were already four Member States with a Mediterranean coast.

Indeed, the 1991 report identified, albeit in a very soft language, the main problems of the CFP, that have continued to be the CFP's main weaknesses and challenges. It is to be noted that in subsequent analysis, carried out in 2001 and 2009 for the subsequent reforms, the language has changed but the problems identified have remained remarkably persistent.

The new basic regulation of 1992

The new basic regulation for the CFP (EEC, 1992), in fact consolidated a number of elements of the policy:

- It prolonged the regime applicable to the 12-mile territorial sea of Member States, including the special arrangements between the 6 and 12 miles;
- It also consolidated relative stability, with a specific mention to the maintenance of The Hague Preferences;
- It consolidated the 'Shetland box' area;
- It did not terminate the transitional regime applicable to Spain and Portugal, for which the end remained at its original date of 2002;
- It did not modify the basic rules for fisheries management that remained subject to the same combination of TACs and technical measures;
- It did not establish new rules or objectives regarding fleet policy. Although such policy started to evolve at the time. It is remarkable that this regulation did not mention at all the fleet policy as a pillar of the basic undertaking of the CFP.

Among the few novelties it is worth mentioning the following:

- It established the basis for the conservation policy to move, as relevant, to multi-species and multi-annual approaches. This idea, however, did take much more time to develop in practice. As we saw in Chapter 4, the CFP had to wait until the early 2000s to start developing the idea in practice;
- It re-established a scientific committee (the STECF) that could advise also on economic questions. Again, the practical implementation of this took considerably longer.

Finally, this new regulation did not actually address certain points:

1 The question of discards, underlined by the Commission as an issue, was not subject to any specific measure in the final regulation. Although the existence of the problem was recognised, it did not yet become a strong reputational issue for the CFP as it did years, after and addressing it was not high in it agenda of the time.
2 The Mediterranean, as in 1983, was completely absent from the basic regulation. However, despite the lack of specific provisions for this sea basin, this reform also identified the complete lack of a Mediterranean policy as a problem to be addressed, albeit in a way that would not interfere with the discussion of the Atlantic policy. The technical measures regulation of 1994 is the result of this awareness.
3 It is to be noted that, in 1992 already, there were evidences of the linkages among the different policy pillars. For example, how the fleet policy influenced the conservation policy. There was also several years' worth of experience in the application of the structural and the market policy. Likewise, the external policy was already well developed. Yet, the 1992 basic regulation did not refer at all to these issues. The policy was, still at the time, the conjunction of different policy pillars, based on different legal texts, without any legal linkage among them. And, in the absence of legal linkages, policy linkages were almost inexistent. This feature, in fact, was not really addressed until relatively recently in the CFP.

Overall, this second basic regulation of the CFP was rather conservative: although the problems with the CFP were already identified, the general preference for the preservation of the *status quo* prevailed over the need to start addressing the already obvious weaknesses of the policy.

Implementation of the first reform

As we will see, the adoption of a new legal basis does not necessarily mean that all the new objectives are achieved and, especially, that all the new possibilities are in fact developed. It is important therefore to refer to the practical implementation of this first reform to better judge its contribution to the CFP.

In this regard, it is very significant that some of the new elements introduced in the new 1992 basic regulation were barely implemented. Let us see some examples:

1 The idea of evolving the management of stocks from single-stock annual decisions to mixed fishery-multiannual approaches did not materialise. It was not until 10 years later, in the context of the discussion on the multiannual plan to save the cod stocks that the multi-annual plans started to be considered, as we saw in Chapter 4. The

main reason for this failure was Member States' reticence to lose room for manoeuvre for annual TAC decisions. It is also true that multiannual plans would require a significant change in the way the scientific advice was provided. In the absence of a clear demand, the scientific community did not provide long-term advice. In turn, the lack of long-term advice did not facilitate the introduction of long-term plans. This vicious circle had to be broken with the need to save the cod, as we saw in Chapter 4;

2 The incorporation of economic advice into management did not fully materialise either. In this case, the reason is probably that the first economic advice needed several years of building historical data series to start providing useful insight into these questions. It was not until 2000 that the Commission put together a system to ensure the systematic collection of economic data. This means that the information necessary to implement that commitment was not really available until the middle 2000s;

3 As we saw in Chapter 8, the implementation of the first Mediterranean regulation was considered deficient by the Commission in its Green Paper for the 2002 reform (see next).

Overall, this reform did not represent a substantial change either in the basic texts or indeed in the level of implementation of the policy.

The second reform: 2002

The context

The reform in 2002 took place in a context where the new evidence was emerging.

On the one hand, the strong impression that the structural policy and in particular the aid to the construction of new vessels, instead of addressing the problem of overcapacity, was in fact aggravating it. It is also to be underlined that this question took place in a broader context where the system of subsidies to the European fishing fleets was becoming a high-profile issue in the context of international trade negotiations.

A second important question was that, in the context of the discussion on the recovery of the cod, perhaps the first case in which the collapse of a fishery resource in the Union became a hot political issue, made it more clear than ever before that the CFP needed to have a system to consult stakeholders in a structured way; the difficulties to impose important sacrifices to recover overexploited resources made it more clear that a level of implication and 'buy-in' by the industry was becoming more necessary than ever.

Importantly, the evidence of the need to sustain efforts over time to recover the cod stocks also made it more clear that individual decisions on an annual basis would not suffice to ensure the recovery of these stocks: a more long-term perspective was needed and this opened the door to the acceptability by Member States of the long-term approaches to management proposed by the Commission.

Finally, other factors were also becoming obvious for the policy: the need to further integrate environmental concerns in the policy, as required by the Treaty, was also identified as a priority.

Over and beyond these objective elements, the reform process in 2002 was sufficiently far away from the political deal of 1983 and involved a number of Member States that were not part of that deal, so the political environment was more conducive to substantial reform than in 1992. In addition, the CFP had become at that time a favourite subject of criticism from almost all walks of life, including academia (Symes, 1997), and expectations for reform were higher than ever before.

The Commission prepared a 'Green Paper' in March 2001 to allow for a consultation period with all stakeholders and societal interests (COM, 2001). In this paper, the evidence of the shortcomings of the CFP was openly exposed, beyond the more careful language of the 1992 report. Although relative stability was questioned, the Commission concluded that such a principle was a guarantee of a stable allocation of fishing rights and that the problems of fleet overcapacity, which by that time were already very obvious (see Chapter 5), could potentially undermine such a stable allocation. Consequently, the Commission concluded that only if and when the overcapacity was resolved, could the principle of relative stability be revised. This was a strong motivation for the proposal to eliminate vessel construction.

The consultation process showed an important support to the establishment of stakeholder bodies. However, many contributions proposed that such bodies should have executive powers, not merely consultative ones. This would eventually lead to a certain degree of frustration when the Regional Advisory Councils were established.

On the long term plans, the consultation was generally positive but with nuances and reserves, as the lack of familiarity with the idea led to many opinions being wary about such plans tying up the hands of decision makers on annual decisions, a reticence that was not easy to overcome in the later negotiation.

But by far the most contentious issue was the response to the idea of suppressing aid to new constructions. This issue divided Member States in two camps: (1) those proposing the continuation of these aids, which named themselves 'friends of fisheries' in allusion to their alleged social sensitivity towards the fishing industry and (2) those favourable to the suppression, who by contrast named themselves 'friends of fish' in reference to their alleged higher sensitivity on conservation issues.

This division in two blocks was extremely solid and remained throughout the whole negotiation in Council.

The 2002 reform achieved three important headline objectives: suppressing funding for new vessel construction, establishing stakeholder consultation bodies (the Regional Advisory Councils) and establishing the basis for fishery management through long-term plans.

But beyond these headlines, 2002 was a year of intense production by the Commission of policy papers that short of being legally binding did actually have a bearing on the gradual development of new policy approaches. As we have seen in different chapters, the Commission adopted in 2002 important communications on the integration of environmental concerns in the CFP, on the need to address discarding, on the development of aquaculture and so on.

Vessel construction

The debate was influenced by both internal and external factors: the evidence on the result of the application of this instrument as regards fleet overcapacity and the wider context of the EU trade policy in the context of the World Trade Organisation (WTO).

Internally, the evidence that the conditions associated to new vessel constructions were such (in terms of the withdrawal of old vessels) that it was very difficult to dispute that this mechanism was indeed a contribution to increasing fleet capacity de facto. As we saw in Chapter 6, the consideration of vessel construction as a contributor to overcapacity is not self-evident, as new constructions were theoretically associated to the withdrawal of capacity of older vessels. But it was the way in which these withdrawals were legislated and how they were implemented that ended up discrediting the use of the instrument itself.

This is another example of a common phenomenon in fisheries management. Sometimes, the effects of the way in which an instrument is used is taken for the merits of the instrument itself, thus overlooking the fact that the constructions of new vessels, if and when applied with an effective high ratio of capacity withdrawal, could indeed be an effective instrument for capacity reduction.

Externally, the political discussion was influenced by the parallel discussions about subsidies in the context of the WTO, where the Union had long been under pressure on its fleet subsidies.

After the conclusion of the Uruguay Round in 1994, the issue of fisheries subsidies was discussed in WTO's Committee on Trade and Environment. The work of this committee led to the inclusion of the question of fishery subsidies in the next round of trade talks: the Doha Round. At their meeting in Qatar in November 2001, Trade Ministers agreed to: 'clarify and improve WTO disciplines on fisheries subsidies'.

Clearly, the Union had to take action to ban vessel construction in the context of WTO to ensure the consistency of its approach in that framework. The idea, developed already at that time, was to distinguish between 'bad' subsidies that would contribute to overfishing and 'good' subsidies that would contribute to ecological sustainability. In doing so, the Union could in fact preserve certain public interventions in the context of the structural policy.

The discussion was extremely difficult to the very last minute. In fact, vessel construction was not banned altogether. A phasing-out period of 2 years was agreed for vessels of less than 400 GT but beyond that, aid to vessel construction was discontinued through an amendment to the existing structural regulation (EC, 2002b).

Long-term plans

The decisions on the long-term plans were strongly influenced by the parallel discussions on the cod recovery plan. For this reason, the emphasis of the negotiation was to agree on principles applicable to heavily overfished stocks and how to drive them towards safe biological limits. But little discussion took place on the stocks already at safe biological limits and, in particular, what would be the long-term objective for those stocks. In fact, Article 6 of Regulation 2371/2002 establishes that the objective of these plans would be to: 'maintain stocks within safe biological limits for fisheries exploiting stocks at/ or within safe biological limits'.

As we saw in Chapter 4, the crisis of the cod stocks, which unfolded in parallel, was the key that determined Member States' support to the idea of long-term management.

But the real difficulty in the negotiation was the role to be played in long-term plans by the management of effort as an instrument. As we saw, the scientific community strongly recommended the use of effort management in cases of stock recovery, since they considered that TACs were not limiting fishing mortality. In that vein, the Commission wanted to include effort management as an integral part of recovery plans.

In the political discussion, however, it was clear that the use of effort was not supported by too many Member States. While its use could be admitted in cases of extremely overexploited stocks, it could not be so in cases of stocks that were inside biological limits. This gave rise to two articles on long-term plans: one about recovery plans for stocks outside safe biological limits (Article 5 of Regulation 2371/2002), which included effort management as an instrument and another one on management plans for stocks inside safe biological limits (Article 6), which excluded effort management.

This seemed a good political solution but in fact was one which would soon show its limitations. In fact, a number of stocks were around safe biological limits (SBL). In these cases, the stocks could well be inside SBL in one year and outside SBL the next year. Or the other way round. In such cases, what is the legal basis of the corresponding plan? Other cases are those where the plan regulates two closely associated species. If one is inside SBL and the other is outside, which is the legal basis for the plan? Article 5 or Article 6?

These cases occurred in practice, as we saw in Chapter 4. And the solution had to be that the plans were not based on Articles 5 or 6 but rather on the general Article of the CFP (Article 37 at the time).

This illustrates why legislation should never ignore natural fluctuations and variability.

Regional Advisory Councils

This part of the reform was, in principle, less contentious. All Member States signed up to the principle that the policy should increase the buy-in from the industry. However, it was by no means easy. Perhaps the biggest difficulty was that some Member States saw the consultation bodies as a precedent of what, they feared, could eventually lead to a process of re-nationalisation of the CFP. It is interesting to note that these fears have been expressed again in connection with the 2013 CFP reform and the process of regionalisation.

Certainly, the fact that some Member States floated the idea of these stakeholder bodies as having executive powers and not just advisory ones did little to dissipate these fears. And indeed a number of stakeholders used the opportunity to propose the idea of the future Regional advisory Councils as bodies with real executive powers.

It was obvious, however, that there was no legal basis in the Treaty for any decision making to be done by stakeholder bodies. The advisory nature of the bodies had to be accepted by some as the result of legal constraints but this did not prevent a level of disappointment. This frustration, interestingly, was seen again in the CFP reform of 2013 in connection to the regionalisation process.

The debate moved then onto the question of the composition, with two models being discussed: for some, the stakeholders would need to be primarily, or exclusively, the fishing industry, whose economic future depends on the decisions made under the CFP. For others, the marine resources belong to the whole of society, so the stakeholders should be a fair representation of all legitimate societal interests.

At the end, the solution was an intermediate one: to give representation to all legitimate stakeholders but to give the fishing industry a higher overall weight (2/3 of the members of the Executive committee).

The definition of the RACs already introduced the notion of 'regionalisation'. In fact, a number of them they were organised according to regional approaches, an anticipation of the regionalisation that would materialise a decade later. Others, on the contrary, because of the nature of the fisheries they covered, were not regional.

It is also interesting to note that the process of regionalisation gave rise to a demand, by a number of stakeholders, for the establishment of relatively small, sub-regional RACs. Requests to constitute an Irish Sea RAC, or a RAC for the outermost regions of Spain and Portugal, were voiced. This was a fundamental question: how large or small should RACs be? Should they concentrate in small areas at the risk of having many cases overlapping two or more RACs? Or should they be very large at the cost of losing a real regional dimension?

At the end, the final balance of establishing seven RACs, of which five of a strict regional nature (Baltic, North sea, North western waters, South western waters and Mediterranean) and two of a larger scope (pelagic and distant fishing) was a reasonable balance that fundamentally has never been seriously questioned.

Other issues

The new basic regulation emerging from the negotiation was Council regulation 2371/2002 (EC, 2002a). This new text enshrined the establishment of the RACs and that of the multiannual plans, while the suppression of aid to new construction was done through an amendment to Regulation 1263/99 establishing the second FIFG (see Chapter 6). The new regulation attempted, for the first time, to integrate the different elements of the CFP into an integrated set of rules. It was much more explicit in defining policy objectives than either the 1983 or the 1992 texts.

Policy objectives

Although the objectives of the CFP are spelled out in Article 39 TFEU, it has always been difficult to translate these objectives into operational guidance when taking decisions. This question was raised in this reform for the first time. As a result, the policy objectives were more clearly spelled out, placing the term sustainability (ecological, economic and social) as the fundamental objective of the CFP.

This was an important step, in that it identified the element that the CFP should aim at: the sustainable character of the activity. However, the question of how to achieve the three objectives at the time and how to combine them in the short term, this remained undefined.

Scope

Contrary to previous texts, the new basic regulation clarified its scope, not only in terms of the geographical areas covered (the Mediterranean was included as part of the CFP in a basic regulation for the first time!) but also in terms of the marine waters covered and the organisms living in them. In this context, the organisms that have different life stages in marine and fresh waters (anadromous like salmon and catadromous like eels) were only partially covered by the new regulation, as these organisms were considered part of the CFP only 'during their marine life'.

This would have consequences when it came to adopt regulations on these species: both salmon (Chapter 4) and eels (Chapter 11) had to be managed through a complex combination of European and national measures.

New fleet policy

As we explained in Chapter 5, the CFP reform in 2002 suppressed the Multiannual Guidance Programmes (MAGPs) replacing them by a new policy: the obligation for Member States to: 'put in place measures to adjust the fishing capacity of their fleets in order to achieve a stable and enduring balance between such fishing capacity and their fishing opportunities'. This new obligation was enshrined, for the first time, in the body of the basic regulation through a dedicated Chapter. The loss of a specific instrument of fleet policy was replaced by a generic obligation.

This generic obligation was complemented by others: to submit an annual report on how such objective was being achieved and to ensure the functioning of the 'entry/exit' regime whereby the capacity withdrawn with public aid cannot be recovered, so the overall ceiling of the Member State has to be adjusted downward accordingly.

It is noteworthy the lack of political debate on this point. Rather than analysing whether the fleet objectives were the right ones, Council unanimously preferred to scrap the old system of MAGPs, and replace it by another one without compulsory fleet reductions. This was not necessarily the result an evidence of having achieved the objectives of fleet adjustment, but rather the result of the high unpopularity of the compulsory reductions of fleet, among industry and national administrations alike.

Control

We have referred in Chapter 9 to the difficulties of discussing control matters within the CFP. In 2002, the new basic regulation consolidated certain principles on fisheries control through a dedicated Chapter but it did not establish significant new elements in the control policy.

However, the idea of building a mechanism to enhance coordination and improve efficiency in the control activity by Member States gradually made headway, eventually leading up to the decision, taken in 2004 to establish the European Fisheries Control Agency. If not legally, at least politically EFCA was also a result of the 2002 CFP reform (see Chapter 9).

A new scientific basis

The 2002 reform introduced some new elements (long-term plans, regional advisory Councils) that substantially changed the needs for scientific advice. These new needs were the need to provide for advice on long-term approaches, including effort management and the need to engage scientists in a dialogue with the stakeholders (Penas, 2007). In addition, the new basic regulation enshrined the STECF as the formal provider of scientific advice. This implied a net increase in the number of specific requests for advice to this body, even at the cost of duplicating the work: in many occasions, even if the advice had already been provided by ICES, STECF had to formally provide it again, as this was a legal requirement. This produced duplication of efforts and, sometimes, contradictory advice between the two bodies, as we saw in Chapter 10.

What the 2002 reform missed: Maximum Sustainable Yield

It is also important to refer to the important question that was NOT part of the 2002 reform: the introduction of the notion of Maximum Sustainable Yield (MSY) as a long term objective of the policy. The reason of this was the timing of the international commitment to pursue this objective in international fisheries.

In effect, although the notion of MSY was well known as a principle of fisheries management internationally and even certain international organisations of fisheries management, such as ICCAT, had established MSY-like objectives in their founding convention as far back as 1964, the notion was not widely consolidated as an objective for fisheries management.

In 2002, however, the notion of MSY as a global objective for the management of fisheries was enshrined at the World Summit on Sustainable Development, held in Johannesburg that year. Among other objectives, the summit agreed on eight measures to achieve sustainable fisheries.[1] Among them, the first one was: 'Maintain or restore stocks to levels that can produce the maximum sustainable yield with the aim of achieving these goals for depleted stocks on an urgent basis and where possible not later than 2015'.

Beyond this international political commitment, the notion of MSY also represented an opportunity, at EU level, to resolve one of the oldest conundrums of the CFP: how to combine the different (and often incompatible in the short term) objectives of the policy as laid down in the Treaty (today's Article 39 TFEU).

The CFP traditionally lacked a clear operational objective, beyond the combination of general objectives of the Treaty. And this lack of a more specific long-term objective for fisheries management resulted in a focus on short-term measures. In this context, the notion of MSY could be considered as the point around which fisheries would find, in the mid and long term, the best combination of the objectives of Article 39 TFEU (which, as we saw in Chapter 1, have been the same since 1957).

In sum, the combination of the political commitment from Johannesburg, together with the internal need to give the CFP an operational long-term objective for fisheries management, represented a very important opportunity to include the MSY notion as a fundamental objective of the new CFP.

This idea, however, did not materialise in the new, reformed CFP because of the close coincidence on timing. The whole notion needed more substantial discussion before the political commitment in Johannesburg could be translated into hard law. The MSY notion was well known by scientists but many fisheries managers and the industry did not fully grasp its consequences and opportunities and as a result it was extremely difficult to engage in a political debate around that concept. The basic regulation on the CFP adopted in December 2002 did not include an objective based on MSY.

As a result, the text agreed for the long-term plans (Article 6 of Regulation 2371/2002) had a major flaw: it did not specify what would be the long-term objective for those stocks already inside safe biological limits.

In these circumstances, the Commission decided not to wait until the next reform in 10-year time but to bring forward the notion of MSY as the long-term objective, both in the on-going discussions on long-term plans as well as in the context of annual TAC discussions. This was a very slow and difficult process, in which many difficulties had to be overcome:

- The policy required advice on F_{msy} levels that was not readily available from the scientific community. For this reason, some of the earlier proposals were based on MSY-like estimates of long-term fishing mortality levels. This made the policy somehow confusing;
- The lack of a clear legal basis (as in Articles 5 and 6 of Regulation 2371/2002 referred to previously) seriously hampered this approach, as sometimes Member States considered the Commission's approach to reach F_{msy} levels as unwarranted in legal terms;
- Certain stocks were so far away from the MSY objective, that the acceptance of such objective, even as a long-term one, raised resistance as a consequence of the strong reduction of TACs in the short term that it would require.

These initiatives materialised finally in the presentation of a document laying down the concept and its application as a guiding principle of the conservation policy (COM, 2006). This was obviously short of establishing a legal basis, since the idea of proposing an amendment to the basic regulation was quickly turned down as non-viable. This communication, however, although short of legal consequences, had a very positive effect in the sense that it made the concept better known, better understood and more readily acceptable in practice.

The communication had to deal with considerable scepticism at the beginning. The shortage of MSY projections by the scientific community at the time was perhaps the biggest one. Stakeholders and administrations were reluctant to engage into a long term commitment to what they saw as a very elusive target. In addition, the question of how to handle MSY in mixed fisheries was already an issue at the time.

This justified, in some cases, a two-step approach. Even if for certain stocks we did not know the estimation of the F_{msy} as a possible long-term objective, this was not an immediate problem for the most severely overfished stocks. In these, long-term plans must focus on recovering to sustainable levels in a first phase. Only when this is achieved, the long-term objective around a well-evaluated MSY could be decided. Unfortunately, for many stocks the conservation status was so poor, that the MSY objective remained really very far off, so the emphasis should be put in ensuring that the stock status moved in the right direction, even before we could determine exactly what the long-term objective could be.

Since 2006 and until the notion of MSY as a long-term objective was enshrined in the 2013 reform, this objective was gradually incorporated into the successive proposals for long-term plans. As we saw in Chapter 4, the different long-term plans gradually incorporated MSY as their long-term objective, even in the absence of a legal obligation in the basic regulation to do so.

In any case, the 2002 reform was a lost opportunity to enshrine this notion in the CFP. As we saw in Chapter 4, this required a considerable effort to implement, without a legal basis, until its consolidation in the CFP reform of 2013, as we will see in the next Chapter.

The implementation of the 2002 reform

Again, the real measure of the success of the 2002 reform must be seen in terms of its practical implementation in the following years.

Long-term plans

The introduction of Long-Term Management plans (LTMPs) constitutes a considerable success. Despite initial reticence, today these instruments are widely recognised by everybody as the essential element of the conservation system under the CFP. It is probably not by chance that the implementation of LTMPs coincides with the real reduction of fishing mortality and the start of the recovery of fish stocks in the Atlantic, as we saw in Chapter 4.

There can be several reasons for a successful implementation of long-term plans. One is, beyond doubt, the need to urgently address the recovery of certain key stocks in the brink of collapse, such as North Sea cod, as we saw in Chapter 4. Another one may be, as in other cases mentioned and in line with Mike Holden's remarks, the availability in the Commission services of officials and other staff with in-depth knowledge of the fisheries and the challenges at stake.

As regards the implementation of effort management in long-term plans, we analysed implementation in Chapter 4. This implementation was very patchy and it is not easy to draw clear conclusions. In any case, it is clear that, whatever the objective merits of the effort management, Member States and the industry have increasingly questioned its continuity. Arguments about excessive burden (since effort limitations are applied in addition to catch limitations) and lack of added value (at least when implementation is poor) have gradually undermined the image of this instrument. It is also true that, when effort management has been properly implemented, it did produce some effects, particularly forcing fleet reductions in certain fisheries. For this reason, too, the use of effort management is considered by many Member States as undesirable.

The adoption of long-term plans has slowed down in recent years. The introduction of Impact Assessments prior to the presentation of these plans together with a more important involvement of the RACs has certainly improved the quality of the plans but it has slowed down the pace at which they are prepared and adopted. And, as the Treaty of Lisbon entered into force, the lack of agreement on the application of Articles 43(2) and 43(3) TFEU has stopped the adoption of new plans.

The only negative aspect is that the coverage of long-term plans is still limited. Many important stocks are still outside these plans and have to be managed on the basis of annual decisions.

However, this problem is mitigated considerably by the adoption of the MSY as the objective of the policy in the 2013 reform. This means that even in the absence of long-term plans, there is a clearer framework now within which annual TAC decisions must be taken: whatever the considerations, TACs now have to be decided in a way that is conducive to the achievement of the MSY objective between 2015 and 2020.

Implementation of the RACs

As we saw in Chapter 12, the establishment of the RACs has been another success, beyond doubt. Again, despite initial difficulties and frustrations, RACs have become an essential element in the CFP and the growing importance of their role is recognised by everybody.

Fleet capacity and the suspension of aid to new constructions

The suppression of aid to new vessel construction was a breakthrough in political terms. However, its effective implementation has fallen short of what was necessary to address the problem of overcapacity, as we analysed in Chapter 5. As can be observed in Figure 5.3, there is hardly a break in the evolution of fleet capacity before and after aid to construction of new vessels was discontinued.

It could certainly be argued that without that measure, the problem of overcapacity in Europe may have been worse. Obviously, this will remain pure speculation.

As we saw in Chapter 5, the implementation of the new fleet policy, by leaving Member States the freedom to evaluate (or to fail to do so) their overcapacity cannot be considered a success. The new policy, based on fleet ceilings that are formally complied with but that have little meaning, have not necessarily provided better results than the previous policy based on MAGPs, as we saw in Chapter 5.

As regards the suspension of the aid to new constructions, the implementation was correct from the point of view of compliance. A different question is whether this suspension contributed to address the problem of fleet capacity. In this regard, too, the evolution of European fleets over the years does not seem to show any major discontinuity as of 2004, when this type of aid was suspended. This fact is intriguing and casts doubt about the traditional wisdom that aid for new constructions is the worst contribution to overcapacity. On the positive side, however, the elimination of this type of aid has improved the stance of the Union in international fora, such as the World Trade Organisation, where subsidies to the construction of vessels have been seriously questioned.

Other issues

The commitment to revise and improve the situation of fisheries management in the Mediterranean was also relatively well implemented at least in terms of legislative development. The new Mediterranean regulation adopted in 2006, together with the reinforcement of the GFCM in recent years has made a considerable difference: for the first time, the CFP has a fisheries management policy in this basin. However, the practical application of the new regulation has advanced relatively slowly and still today the results of this policy are not evident.

The important elements of the improvement of the control regime of the CFP have also materialised, although it took the Union several years to do so: the new control regulation, including a number of new instruments, was adopted in 2009 (see Chapter 9), 7 years after the adoption of the reform itself.

The question of the scope proved very difficult: on the basis of the new legal basis, the Commission took the first initiatives to manage the European eel and prepared a plan for the management of Baltic salmon. As these resources were part of the CFP 'during their marine life', the effective regulation of these stocks had to be subject to a unique combination of EU fisheries measures together with national measures in their river basin, in a combination of CFP and national rules that still has to prove its effectiveness.

Note

1 www.unmillenniumproject.org/documents/131302_wssd_report_reissued.pdf.

References

COM (2001). Green Paper on the future of the Common Fisheries Policy. Brussels, 20.3.2001. COM (2001) 135 final.

COM (2006). Communication from the Commission to the Council and the European Parliament implementing sustainability in EU fisheries through maximum sustainable yield. COM(2006) 360 final. Brussels, 4.7.2006 term sustainability. *ICES Journal of Marine Science Advance Access* 2007: p. 1–8.

EC (2002a). Council Regulation (EC). No 2371/2002 of 20 December 2002 on the conservation and sustainable exploitation of fisheries resources under the Common Fisheries Policy. *OJ* L 358, 31.12.2002, p. 59.

EC (2002b). Council Regulation (EC.) No 2369/2002 of 20 December 2002 amending Regulation (EC) No 2792/1999 laying down the detailed rules and arrangements regarding

Community structural assistance in the fisheries sector. *OJ* L 358, 31.12.2002, p. 49–56.

EEC (1992). Council Regulation (EEC) No 3760/92 of 20 December 1992 establishing a Community system for fisheries and aquaculture). *OJEC* No L 389 of 31.12.1992, pp. 1–14.

Penas, E. (2007). The fishery conservation policy of the European Union after 2002: towards long-term sustainability. *ICES Journal of Marine Science Advance Access* 2007: 1–8.

SEC (1991). Commission des Communautés Européennes. Rapport 1991 de la Commission au Conseil et au Parlement sur la Politique commune de la Pêche. SEC (591) 2288. Bruxelles, le 4 Décembre 1991.

Symes, D. (1997). The European Community's Common Fisheries Policy. *Ocean and Coastal Management* **35**(2–3): 137–155.

CHAPTER 16
The CFP reform of 2013

The context for reform and the Green Paper

The reform in 2012: the *status quo* is not an option

As we saw in the previous Chapter, the 2002 Regulation actually foresaw a more ambitious policy review in 2012 than ever before. But as we have seen, the real driving force for CFP reform is not just the legal requirements but the evidence that changes are necessary. In that regard, the reform foreseen for 2012 started on a more ambitious note than any of the other reforms. Why? There were a number of reasons for this:

- The growing interest of environmental NGOs in the mainstream fishery management issues and consequently the growing interest by the European public opinion on these issues
- The example of the CFP as a policy excessively centralised 'in Brussels' and the reputational issue that it represented for the Union as a whole;
- The existence of certain issues that increasingly made headlines and drew the interest of the general public. The existence of important levels of discards in European fisheries was the best example;
- The publication in 2007 of a report by the European Court of Auditors on the control of the CFP[1] had a considerable impact in the Commission, which realised that a policy with such a poor level of compliance would need a considerable revamp.

These points can be summarised as one: the fact that the CFP and some of its associated problems were becoming a political issue over and beyond the limited field of interest of the fisheries stakeholders.

Against this background, the exposure of the CFP to the European public opinion and particularly the use of this policy by Eurosceptics in their campaigning against the EU and who never hesitated to consider this policy a failure beyond possible reform, made it also an important point of principle to demonstrate that a much-criticised policy could indeed be substantially reformed from inside and that calls for abandoning this policy on account of the 'impossibility of reform' were unjustified.

The Green Paper

The Commission presented its 'Green Paper' on CFP reform in 2009 (COM, 2009). In view of this context, this paper was a rather exceptional exercise of self-criticism on the policy. In fact, many sectors of the industry reacted negatively to this document by indicating that the Green Paper presented a catastrophic view of the policy that was not justified, in yet another example of the 'we don't like it but don't change it' effect that we saw for the 2002 reform.

The Green Paper expressed the Commission's diagnostic of the shortcomings of the CFP and then proposed a number of solutions. The paper identified five structural deficiencies of the CFP as follows:

1 There is no clear long-term objective for the policy. The different objectives of Article 39 TFEU are not worked out into clear operational objectives that would provide clear policy guidance;

The Common Fisheries Policy: The Quest for Sustainability, First Edition. Ernesto Penas Lado.

2 The decision-making system encourages the prevalence of short-term considerations over long-term objectives;
3 The policy is too much based on 'micro-management from Brussels' and does not give sufficient self-responsibility to the industry;
4 Overcapacity has never really been addressed efficiently and remains a problem for the CFP;
5 There is little will to effectively enforce the policy, both at the level of national administrations and the industry.

The Green Paper had a generally favourable reception, particularly on grounds of its very open recognition of the need to revamp the CFP. It was presented as an example of how the Union can be able to be self-critical and courageous to propose policy change.

From the Green Paper to the actual proposals on CFP reform presented by the Commission there was a long and intense consultation process. More than 400 submissions were presented to the Green Paper, from different types of organisations and individuals: 75 from registered organisations, 86 from public authorities (European, national and subnational), 146 from unregistered organisations and 110 from individual citizens.

In general terms, the consultation process showed a very high degree of agreement with the Commission's diagnostic but a lot less on the specific Commission's remedies. Again, the 'don't like it but don't change it' effect was still there, although less marked than in previous occasions.

This process of consultation largely explains many of the features of the Commission proposal. This consultation was summarised by the Commission in a specific communication (SEC, 2010). In particular, a number of questions emerged as majority opinions:

- The support to the existing regime applicable to the 12 nautical miles was almost unanimous;
- The support to relative stability was clear, although with significant exceptions;
- The issue of the individual transferable rights was very divisive, with many opinions favourable and many others fiercely opposed.

Overall, the public consultation did have an important bearing on the Commission's proposals. In particular, the clear messages in favour of preserving the 12-mile regime and the support to relative stability were a confirmation of the lack of appetite for change in these two areas.

The proposals

The impact assessment

This was the first ever CFP reform done after an Impact Assessment. This impact assessment compared four main options, corresponding to the different weight of biological conservation vis-à-vis socio-economic considerations. This comparison is the fundamental one for the CFP: it is this balance between biology and socioeconomics that can be construed as a combination of the objectives of Article 39 TFEU. On this basis, the impact assessment compared, against the baseline scenario (the *status quo*), four main options with two sub-options (SEC, 2011):

1 In a first option, achieving environmental sustainability with a flexible time horizon while limiting short-term negative social impacts. This included *inter alia* tradable fishing rights and no fleet subsidies.
2 In a second option, achieving environmental sustainability without any flexibility regarding the time horizon. This included tradable rights only at Member State level, no structural policy and no third country agreements.
3 In a third option, achieving environmental sustainability within a flexible time horizon while minimising negative short-term social impact. It included a structural policy focused on social issues.
4 A fourth option, achieving environmental sustainability within a flexible time horizon while limiting short-term negative economic and social impacts but without EU-led tradable quotas.

The four options considered different priorities to solve the essential conundrum of the CFP: whether biological considerations should prevail over socio-economic considerations or the other way round. Option one was the most balanced, option 2 the most environmentally led and options 3 and 4 the most socially led. The three options considered that achieving environmental sustainability would be in any case a pre-condition for the other economic and social objectives to be achieved. The options were tested against five objectives: environmental, social, economic, governance and others and were tested at European level as well as in four European regions where fisheries are of particular importance: Brittany, Scotland, Galicia and Sicily.

The assessment looked at effect in the mid-term (2017) and long term (2022). The main conclusions of the assessment were the following:

- The *status quo* option had the lowest performance of all, showing the need for change.
- Option 1 had the best overall performance under most criteria. The sub-option including individual transferable rights performed substantially better than the basic option 1, thus showing the interest of that instrument.
- Options 2 and 3 performed better than the baseline option 1 but significantly worse than option 1 with ITRs.
- Option 4 performed only better than option 3 in the mid-term, but was the worst option in the long term.

The impact assessment was based on modelled projections that can always be questioned. However, they represent the best 'educated guesses' of what the different options would mean in practice, through sensitivity analysis to test the robustness of the different assumptions. In that sense, the message was clear: the best option for the CFP would be to target ecological sustainability in a flexible timing, with limitations (but not excessive ones) to reduce short-term economic impacts and including ITRs to address overcapacity. But beyond this clear preference, the IA also drew a fundamental conclusion: the *status quo* for the CFP was simply not an option. This also contributed to the momentum about radical change in the CFP.

The package

On the basis of these consultations, the Commission presented during 2011 three legislative proposals and three other accompanying documents. The legislative proposals were:

- A proposal for a new basic regulation, to replace Regulation 2371/2002 (COM, 2011a);
- A proposal for a new market regime for fishery products (COM, 2011b);
- A proposal for a new financial instrument: the European Maritime and Fisheries Fund (EMFF) to replace the European fisheries Fund (COM, 2011c).

The non-legislative ones were:

- A Communication on the implementation of the 2002 reform as regards the conservation and fleet policy, as well as on the possible continuation of the 12-mile regime. This was a requirement of Regulation 2371/2002 (COM, 2011d). The Commission proposed here the continuation of the non-contentious 12-mile regime.
- A communication on the future external fishery policy. Although not legally required, this communication was an opportunity to make a package with the other CFP reform proposals (COM, 2011e). This would be subject to a specific discussion by both Parliament and Council (see next).
- An overarching communication to explain how the different elements of the reform constituted a consistent policy package. This document had a pedagogic objective: to explain how the different elements of the policy would interact in a consistent manner. This was the first time the CFP had the opportunity to look at all its policy pillars as a package (COM, 2011f).

A particularly interesting circumstance was that, contrary to previous reforms, the timing of the reform coincided with that of the renewal of the financial instrument, the EFF and that of the market policy instrument. This provided a unique opportunity to propose a reform that would constitute a package with a series of structural and market measures to make it possible.

The discussion of the new basic regulation quickly concentrated on the so-called 'big four', the most substantial and contentious issues: (1) the MSY objectives, (2) the discard ban, (3) regionalisation and (4) transferable fishing concessions and fleet policy. Let us analyse these big four and the remaining issues.

The discard ban

Why a discard ban?

The problem of discards has been identified by the Commission as a problem of the CFP as early as in 1992 at least. And, in the 2000s, different initiatives were taken. However, it is fair to say that they had a rather limited effect in terms of policy development. In addition, the Commission, in its Green Paper of 2009 (see previously) did not include discarding among the five 'structural deficiencies' of the CFP.

It is true, however, that large levels of discarding are to a certain extent the result of some of such structural deficiencies. Why then did the discard ban become the most visible and even iconic element of the CFP reform of 2013?

The answer is not straightforward but without a doubt, the international context and the nature of discards as a major 'reputational issue' for the CFP played a key role.

Discarding is very significant in world fisheries. A FAO study in 2005 estimated the total level of discards in world fisheries to be around 7.3 million tonnes, corresponding to 8% of the global commercial catch in weight. Of that amount, approximately 1.3 million tonnes (corresponding to 23% of the total catches) were discarded in the Northeast Atlantic (FAO, 2005).

In 2006, a specific study gave a more concrete picture of the problem in European waters (SEC, 2006): For example:

- In the Baltic, cod discards were less than 10% but herring discard in demersal fisheries reached levels of 100%.
- In the North Sea, demersal trawls and seines discarded 20% of catches on average, trammel nets 40% of the pelagic fish caught, while pelagic trawls had a low discard rate. Beam trawls on the other hand had 40–60% discard rates of target and non-target species.
- In the North West Atlantic, pelagic trawls had low discarding, while beam trawls had even more than 100% discards of herring, 40% of hake and 80% of whiting.

These discard levels were due to a wide variety of reasons: from economic ones (maximising quotas by retaining on board only the most valuable fish) to legal ones (obligation to discard undersized and over-quota fish) including side-effects of national quota allocation systems (many vessels did not have quotas of stocks they caught even if their Member States quotas were not exhausted). But whatever the reasons, this report clearly showed that the problem of discards was a real one, that Europe had to deal with.

These very high levels of discards were becoming, in recent years, a major source of embarrassment for the CFP, both in Europe and abroad. In Europe, the high level of discarding of fish was used frequently by Eurosceptics to denounce the absurdness of European policy. The increasing awareness of European public opinion, particularly in certain Member States, created a high societal demand for a clear policy to address that problem.

Overall, the example of discards demonstrated a tendency that we already referred to in Chapter 13: the increasing influence of public opinion on policy issues of the CFP. Clearly, discarding was not at all the fishing industry's greatest preoccupation. But it became the European public opinion's greatest issue.

These high levels of discards are well known to be a combination of both economic factors (discarding the low value fish) as well as legislative ones (the TAC and quota system and technical measures do establish an obligation to discard over-quota and undersized fish).

The legislative solution of discarding can be addressed partially: technical measures can be changed so as to remove the obligation to discard. However, removing discarding due to the TAC and quota system would have posed a major political problem, since this system, under relative stability, is very largely supported by a majority Member States and stakeholders alike. In addition, the economic discarding, by definition, cannot be easily handled in legislative terms.

In these circumstances and following the example of a number of countries around the world, political momentum gathered around the idea of a radical approach to address the problem: a ban on discarding.

Why a top-down approach?

The approach proposed for the discard ban was criticised by much of the industry as being too prescriptive and 'top-down' and thus contrary to the very philosophy the Commission itself promoted in the Green Paper. That proposal, at that juncture, can only be understood in a larger, more historical context.

In fact, the Commission started to raise discards as a problem for the CFP already in 1992, as we have seen. And the question was reiterated by the Commission in parallel to the CFP reform of 2002. At that time, the Commission published a communication calling upon Member States and the industry to tackle and resolve this problem in a gradual, cooperative manner (COM, 2002). In this communication, the Commission proposed a gradual approach to the problem, undertaking pilot projects and incorporating discard avoidance measures in the different multiannual plans that would follow, in consultation with stakeholders.

It is noteworthy that this communication was adopted shortly before the adoption by Council of the CFP reform of December 2002. Yet, it had no impact on it, as the new basic regulation adopted at the time did not establish specific provisions on discard avoidance. Likewise, the recovery and management plans adopted after the CFP reform of 2002 practically ignored the issue. Clearly, the debate was not mature at the time on the need and opportunity to address the problem of discards in the CFP.

In 2005, FAO published its report (mentioned previously) on discarding. In that report, FAO advocated a gradual, cooperative approach to tackle the problem, through: 'the development of more robust methods of estimating discards, allowance for discards in fishery management plans, development of by-catch management plans and promotion of best practices for by-catch reduction and mitigation of incidental catches' (FAO, 2005).

The Commission insisted in the need to tackle discarding some time later and published in 2007 a new communication on the subject (COM, 2007). At that time, consultations with stakeholders clearly indicated that, rather than imposing a discard ban as a top-down political decision, it was preferable to try to move gradually towards a reduction of discards, under an approach based on cooperation with the stakeholders and bottom-up approaches. On that basis, the Commission expressed its intention to: 'progressively eliminate discards and reduce unwanted catches in European fisheries'.

On that basis, it undertook to discuss the way forward with Member States and stakeholders and prepare an action plan for the elimination of discards. That action plan, however, never materialised. There were a number of extremely interesting projects to eliminate or substantially reduce discarding but these were all isolated efforts.

Beyond those positive examples, by and large the consultative, bottom-up approach proposed in 2007 did not provide real results. In fact, there was no specific follow-up from that call for action. Either the industry and Member States did not see the need and opportunity of embarking in that objective, or the Commission failed to find the right incentives for that to happen.

The only result of these initiatives was the adoption, in the mid-2000s, of a ban on high-grading (the practice of discarding legal fish of relatively lesser value to ensure a maximum economic output from available quotas). This ban has been in force ever since but its actual implementation remains very questionable.

In any case and other than a few very positive initiatives, it has to be admitted that the bottom-up approach favoured by the Commission at the time did not provide tangible results.

This is no surprise. Although the problem of discards was already identified by certain quarters of the European public opinion as an important issue and notably a reputational one for the CFP, the industry did not see any particular advantage in moving in that direction. A non-discard policy would imply changes that may affect direct fishermen's revenues without a clear perspective of possible trade-offs.

The lesson learned from this was that, if the problem of discards was to be addressed, this would require a top-down political decision to do so. Only by setting political objectives by co-legislators it would be possible to mobilise both the industry and national administrations in the search for practical solutions to this problem. Rather than proposing a policy on the basis of what seemed feasible, the Commission proposed an ambitious policy goal in order to mobilise everybody to find the technical solutions to make that policy possible.

This was the approach proposed by the Commission and, by and large, this basic postulate was not fundamentally questioned in the political process for the adoption of the regulation. The discussion took place on the way to achieve the objective, including the timing and, particularly on the need for certain derogations or flexibility mechanisms to take into account the different practical difficulties to apply a discard ban in many European fisheries.

The top-down approach is criticised by many in the industry and the scientific community. It has been argued that the public awareness about the discard problem focused managers' attention onto the symptom of fisheries management rather than on the underlying causes (Borges, 2013). Likewise, the long experience of Nordic countries applying the discard ban was not fully reflected: systems cannot be designed once and for all, and Nordic countries developed their discard ban schemes gradually over many years and learning from accumulated experience (Johnsen and Eliasen, 2011).

The Union's failure to address this problem on a gradual basis, since the first warnings were given in 1992 and repeated years after, led to a situation where the public awareness about the problem reduced the room for manoeuvre to resolve the problem gradually: the political pressure was there for quick solutions, and the experience of other countries on gradual approaches had to be ignored. This is yet another demonstration that, when a problem is identified, and bottom-up gradual solutions are proposed, it is essential that that these be taken seriously. Otherwise, at the end the pressure to solve the problem almost inevitably results in top-down approaches being imposed.

The practical difficulties and the need for flexibility

A number of difficulties had to be addressed in this context. It is important to underline that the basic objective of removing the problem of discarding in European fisheries was shared by practically everyone in both Council and Parliament. The main difficulties expressed were related to the applicability of the discard ban in the fishing grounds. In particular:

- Some delegations (Spain and to a lesser extent Belgium) indicated that, given their very unbalanced relative stability (in certain areas they have quotas for only a fraction of all the regulated species in the catch) they could not implement the discard ban (they could not retain on board species for which their quota was zero). This problem was recognised by the Commission, which undertook to find practical solutions;
- Others referred to the difficulties of implementing the prohibition to market undersized fish for human consumption. Particularly in the Mediterranean, it was underlined that it is very difficult to find alternative marketing possibilities to the human consumption markets. In fact, up to the last moment, the idea of excluding small fish from the discard ban in the Mediterranean was proposed by a number of Member States and MEPs.

As a consequence of these difficulties, the discussion focused considerably into the flexibility mechanisms that would allow Member States cope with these problems. At the end, a combination of such mechanisms was adopted. Over and beyond the use of quota swaps (that remained voluntary), the flexibility mechanisms established in Article 15 of Regulation 1380/2013 provide for possible solutions to most of the problems. The discussion of these mechanisms was the most difficult part of the negotiation, in that it opposed two conflicting views:

1. Those who wanted the discard ban to be radical and were reticent to accept too much flexibility for fear that such flexibility would allow the continuation of current practices;
2. Those who thought that excessive rigidity in the objectives would fail to take into account the real difficulties in the fishing grounds, even by those doing their best to implement the ban.

The result was the adoption of three types of flexibility mechanisms:

1. The exemption for survival. This was the least contentious. Fishery products that survive to discarding can, for obvious reasons, be exempted. The identification of such species was left for decision under discard plans for regionalisation.
2. The *de minimis* exemption. This was by far the most difficult issue. The motivation for this was largely the realisation that in some fisheries it was almost impossible to be able to retain all fish on board, either for lack of quotas or for insufficient space in the holds or because certain fish deteriorate on

board in long fishing trips. Although the principle was agreed relatively early, the maximum percentage of the *de minimis* level was subject to intensive negotiation both in terms of the maximum percentages and in terms of their application (by trip, by year, by species, by fishery and so on).

3 The cross-reporting of retained fish to another TAC. This was a specific flexibility to cater for the limitations imposed by relative stability. If a Member State did not have quota for a certain stock caught as by-catch and could not manage to obtain some quota through swapping, then the solution would be to count the catches of the species concerned against the quota for their target species.

All along the negotiation, the maximum percentages for the application of these flexibility mechanisms were discussed from two opposing angles: those who privileged a discard ban that would not give the impression of having too many ways around it (that would allegedly discredit the policy) and those who privileged an approach that would be workable in practice.

It is difficult to assess, however, whether the most difficult case as, those associated with very unbalanced quota allocations, can adequately be solved through a combination of quota swaps and the implementation of these flexibility mechanisms.

It is also important to underline that the basic rules established in Article 15 must be supplemented by more detailed provisions by fishery. Actually, the very description of which fisheries fall under the different obligations or under different time scales, remains undefined in the basic regulation. Likewise, the decision on specific *de minimis* rules, or the definition of species/gears for which survival from discarding is high (which would remain outside the discard ban) must be specified somewhere. The idea was that the future long-term management plans should provide the framework for these decisions to be made. But it became obvious that many of these plans may take a relatively long time to develop. Many of them will certainly not be in force by the time the discard ban enters into force. This made it necessary to create the notion of the 'discard plan' where the detailed rules referred to previously would be established provisionally, for a period of 3 years, until the relevant management plans are in force.

Behind this solution there was a basic mistrust by Parliament, who feared that if discard plans were not limited in time, then Council may hide behind this fact to delay adoption of the long-term plans and thus continue to decide on annual TACs outside any multiannual framework agreed with Parliament.

Discard ban and TAC levels

This question featured as a very important one, particularly in the Council debate prior to the trilogue. In principle, the question is simple: for a constant fishing mortality rate, if all the fish must be landed then TAC levels should be higher, the difference being the amount of fish that was previously discarded and not counted against the quota.

But the debate was more complex, as certain positions suggested that higher TACs should actually work as a political compensation for having to implement the discard ban. In practice, calls for higher TACs on account of the discard ban have been made even before the discard ban itself enters into force.

This issue featured already at the first important Council where the Commission proposal was discussed: June 2012, under the Danish presidency. At the meeting some delegations were open to the possibility of higher TACs as a kind of political reward for accepting the discard ban. This idea, however, could entail a risk of not meeting the objectives of MSY and, on that basis, Council agreed on a declaration to the effect that the quota top-ups should in any case not endanger the achievement of the MSY objectives.

At the end, the legislators clearly established that the implementation of the discard ban should not undermine the achievement of the MSY objectives. This clearly implies that, when establishing quota top-ups, these top-ups should be of a magnitude that does not increase fishing mortality. In other words, that it will correspond to the part of the catch that took place already but was discarded, so the higher TACs will correspond to more fish being landed but not to more fish being caught.

Regionalisation

Regionalising the CFP: easier said than done

Initially considered a key element to de-centralise a policy long considered as 'micromanaged from Brussels', the regionalisation had to face two major challenges: the limits imposed by the Treaty and the reticence by some to move into an unknown territory.

The Treaty does not have any legal basis for any regional decision making in the Union. From this point of departure, it was clear that a process of regional decision-making should be rather informal and end up in one of the normal decision-making mechanisms of the Union, when it comes to turning the regional agreements into EU law. This fact produced initially a considerable degree of frustration among those having very high expectations about this process but as the discussion advanced and the Legal Services of the three institutions clarified the limits of regionalisation in the Treaty, the negotiation focused on finding ways to allow for a high degree of regionalisation while respecting all the decision-making mechanisms of the Union.

Although the idea of de-centralising the policy was not openly opposed by anybody, in practice a number of national administrations were less than enthusiastic in engaging in this process. The reasons for this were very varied:

- For some Member States, regionalisation was seen with mistrust as a possible open door to the renationalisation of the policy;
- Others feared that a process based on a heavier involvement of national administrations would play against the interest of small Member States, whose limited human resources could eventually put them in a disadvantage vis-à-vis larger (and wealthier) administrations;
- Others considered that the regionalisation would imply more work for national administrations, something they could not do in view of their own national staff policies. This applied even to some large, wealthy Member States;
- Others expressed profound mistrust in some of their neighbours. As regionalisation would need to rely more on national efforts to deliver common goals, the lack of trust on each other was also seen by some as an obstacle;
- Last but not least, as it became obvious that Member States could not run the regionalised CFP through national legislation and that the best solution was that the Commission would translate regional agreements into EU law, many Member States also mistrusted the Commission playing that role, or wanted to give the Commission a 'rubberstamping' role that would be against the Commission's right of initiative.

All these arguments had something in common: they were all expressed in private; no Member State would recognise these fears in public. This in fact constituted an obstacle to the process and eventually led to a system that still relied considerably on well-known EU decision-making procedures.

Many of these points boiled down to one theme: mistrust. Whatever the legal basis established for regionalisation, it would only work if trust is built among the actors, both among Member States and between these and the Commission.

Regionalisation and national law: the need for delegated acts

This question was also very important. One obvious way to ensure that regionalisation was a flexible mechanism would have been to allow Member States, after having agreed on a set of measures, to implement them through national legislation, with a mechanism to ensure that all Member States incorporated the agreed measures in the same way.

This option, however, had a flaw: by relying on national legislation, Member States would have problems to enforce the agreed measures on each other's vessels: the inspection services of a Member State cannot enforce national law of another Member State.

This was a key factor in the debate: the option of implementing regionalised approaches through identical national legislation was soon abandoned as impractical. In the end, it was obvious that the only way of ensuring an effective way to implement the measures agreed among Member States was to adopt them as Commission acts. To that effect, the co-legislators accepted to empower the Commission to adopt such acts.

This decision, however, proved also very delicate. Beyond the general reticence by Council to give empowerment to the Commission under delegated acts (an institutional question that goes far beyond fisheries), this solution posed a problem: if Member States present a proposal agreed among them, should the Commission have any room for manoeuvre to change it, on account of possible contradiction with EU principles?

This was a difficult political and legal question, in which the Member States more interested in regionalisation wanted the mechanism to be one where 'Brussels' would not really have a say. On the other hand, for the Commission, a mere 'rubberstamping' role of Member States' decisions was unacceptable, as contrary to its exclusive right of initiative.

At the end, a practical solution was found: where regionalisation applies, Member States propose measures through unanimous agreement (including consultation with stakeholders) and then submit such proposal to the Commission. If the Commission does not have any difficulty with the proposal, it turns it into EU law through a delegated act. If the Commission considers that the proposal by Member States violates EU law or is contrary to the objectives of the CFP, it can refuse to implement it. In that case, it makes a proposal for its adoption under the normal co-decision procedure. But in no case the Commission can 'cherry-pick' the elements of the proposal by Member States that it would adopt or reject: either the Commission accepts the proposal and implements it, or it makes a proposal for co-decision. This solution allows for ample room for manoeuvre for Member States to agree on common measures while respecting the Commission's right of initiative.

Maximum sustainable yield

Accepting the principle

The 2013 CFP reform successfully consolidated the notion of Maximum Sustainable Yield (MSY) as the objective of fisheries management. As we saw in Chapter 15, this is not a new policy objective. But, by including it as a legal obligation, there is now a powerful instrument to ensure that both long-term plans and annual TACs are decided on the basis of achieving that objective. In fact, this should result in a quicker advance to this objective than has been the case in the past.

It is important to underline that nobody questioned the objective of establishing the notion of MSY as the objective for the management of fish stocks in EU waters. Important differences, however, existed on certain questions:

- The target date to deliver the objective of having stocks at MSY levels;
- The use of fishing mortality levels (F_{msy}) or biomass levels (B_{msy}), as the leading parameter or, at least, as a non-binding reference;
- How to handle stocks where the scientific information does not allow the identification of a clear MSY objective;
- How to handle mixed fisheries where the individual MSY levels for all the species associated may not be possible to reach at the same time.

At the end, the solution was to establish the F_{msy} as the parameter on which the management would be based, with the notion of B_{msy} as an aspirational objective, devoid of legally binding character.

Is MSY a balanced objective?

As we saw in Chapter 4, the MSY objective was de facto used by the Commission in its proposals after 2002 and, after our communication on this matter in 2003, even more so. But the MSY approach is more far-reaching. In its diagnostic of the shortcomings of the CFP, as summarised in the Green Paper, the Commission clearly identified the lack of a clear-cut long-term objective as one of the shortcomings of the CFP. Then the question was what would be the point around which the different objectives of Article 39 TFEU can all be reasonably reached in a balanced manner?

In the 2002 reform this problem was resolved by establishing that the objective of the CFP was to achieve 'biological, economic and social sustainability'. The problem is that this objective does not provide guidance as to which one of the three aspects of sustainability should be privileged when taking short-term decisions. So a definition of an objective was necessary that would provide a better reply to the basic question. The answer is: somewhere around MSY levels.

Of course this is easier said than done (or justified). There are many questions around this. Perhaps the main one is: even assuming the MSY is the best proxy for a balanced application of Article 39 TFEU, how do we get there? What happens in the transition? The reply here has two main elements: the need to be gradual and the biological sustainability as a pre-condition for the other two.

First, the need for a gradual approach bringing fishery resources to an MSY level by a given date may imply very dramatic reductions of fishing possibilities in the short term. Providing a certain flexibility to achieve the necessary reductions in order to minimise economic disruption was therefore a fundamental element.

A second question was, in the transition towards MSY levels, which elements should prevail the biological or the economic? Here, it is important to understand that the biological sustainability is a precondition for the economic and social sustainability to be achievable. Unless the biological resource is sustainable, there is no possibility of ensuring the sustainability of the economic operations or the sustainability of the employments. For this reason, the balanced achievement of the objectives of Article 39 TFEU requires that the biological sustainability be assured as the guiding principle of fisheries management.

MSY by when?

As we saw in the previous chapter, the political agreement on the international implementation of the MSY objective referred to 2015 as a target date, 'where possible'. On the other hand, other political and legal commitments already assumed that the objective of MSY should in any case be reached by 2020. This limit of 2020 was not chosen by chance. One key element was the fact that the objective of MSY in 2020 was already consolidated in the context of the Marine Strategy Framework Directive (in particular the descriptor 3 of fisheries, see Chapter 11). By choosing 2020 as the final delivery date, the two pieces of EU legislation are perfectly aligned. The question was then how to qualify the cases where reaching MSY in 2015 would not be possible. One question was clarified quickly: the lack of scientific advice should not be a reason to delay the achievement of MSY to 2020. The solution to lack of clear scientific recommendation is handled next through the approach on MSY proxies. The reason is very simple: if the lack of a clear scientific advice is a reason to delay the implementation of the MSY objective, this may work as an incentive not to have good scientific advice.

The co-legislators understood that the regulation should not give a bonus to those not fully cooperating in the collection of data to allow for good scientific advice. Although the language does not reflect this point, this was clearly identified by legislators as a clear principle. However, the final text does not specify any criterion to establish whether reaching MSY in 2015 will be possible or not. Determining the cases where MSY will be delayed to later years, up to 2020 will be a matter for the practical implementation.

Which MSY: based on fishing mortality or biomass?

This was another difficult issue. MSY is a notion that can be implemented on the basis of two parameters: the fishing mortality corresponding to MSY (F_{msy}) or the biomass corresponding to MSY (B_{msy}). This is important because, depending upon the selected approach, there can be different consequences on the policy.

A policy based on F_{msy} implies that the fishing activity will take a part of what nature produces, whatever that amount is (that is outside our control). But that implies that B_{msy} objectives may or may not be reached, depending upon nature's contribution to the abundance of the stock.

On the contrary, a strategy based on B_{msy} would imply that, even under very unfavourable natural conditions for the development of the stock, the policy should aim to the achievement of these objectives in biomass, which in certain occasions could imply very strong reductions of fishing or even closures of fisheries to achieve an objective that would in any case remain elusive.

This was a politically charged discussion. Beyond scientific aspects, it is clear that a policy that targets B_{msy} would result in much more precautionary measures. In some cases, where environmental conditions are unfavourable to the reproduction of the stock, even the closure of fisheries may not suffice to recover the stock to the estimated B_{msy} levels.

The example of cod recovery (see Chapter 4) was very relevant at that juncture. As we saw, the initial objective of that plan was expressed in biomass. However, no amount of fishery restrictions has so far been able to rebuild the biomass of some cod stocks to those target levels. The scientific community has understood that some of these levels are simply not possible to achieve under current climatic and ecological conditions, even under a complete closure of the fishery. For that reason, the objective was changed later on to one expressed in fishing mortality.

At the end, the co-legislators understood that the B_{msy} objective, largely outside the control of fisheries managers, should remain only an aspirational one.

Why not Maximum Economic Yield?

The notion of Maximum Economic Yield (MEY) is often suggested as a better alternative to MSY as a fishery management objective. The notion of MEY implies that, if we compare total sustainable yield curves against the total aggregated cost of fishing, the point corresponding to the maximum benefit (the higher difference between yields and total costs) is generally found at lower levels of fishing than for MSY. That implies that, beyond the economic consequences as such (a more profitable and rational fishing industry), exploiting fisheries at MEY level would normally imply more conservationist strategies than MSY. If that is the case, why should not MEY be the objective of the CFP?

The difference between the two concepts can be seen in Figure 16.1. The yield reaches a maximum at the MSY level and subsequently decreases, as increased effort leads to resource to collapse. However, the costs associated to the fishing activity increase linearly as the fishing effort increases. The MEY would correspond to the level where the difference between yields and costs is highest, that is, where the

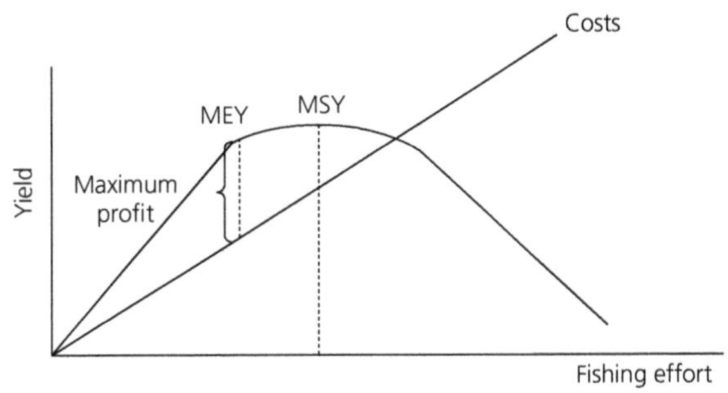

Figure 16.1 Comparison between Maximum Sustainable Yield (MSY) and Maximum Economic Yield (MEY) expressed as the point of highest difference between yield and costs.

profit margin of the fishing activity is highest, and this tends to be found at levels of fishing effort lower than those corresponding to MSY. From that point of view, MEY would seem to correspond both to economic rationality (high profitability) and to a more conservationist (lower level of fishing effort and mortality) approach for the resource. Little wonder this option is generally preferred by many in the environmental community. Why then was this option not chosen?

The reason for this must be found in the deeply rooted diversity of political and socio-economic objectives that Member States have for fisheries management. While some Member States consider that their fishing sector should be economically self-sustainable, for others the main political objective of fisheries management is to preserve current employment levels and to protect the social fabric of fishery-dependent coastal areas. While the former could be more sympathetic to the notion of MEY, the latter would consider such notion as contrary to their objective: MEY would represent a lower level of activity; fewer vessels and lower employment, an unacceptable scenario for all those whose main political objective is to maintain the current levels of economic activity in fisheries areas.

These differences of approach are legitimate in democratic societies and, as we saw earlier, the CFP does not try to impose upon Member States any specific socioeconomic model on the way they should exploit their fishing rights. For this reason, the notion of MEY did not make headway in the negotiation leading up to the adoption of the reform.

MSY for mixed fisheries

This issue was also difficult for one simple and fundamental reason: in mixed fisheries, it may not be possible to ensure that all species caught together can be exploited at their individual F_{msy} levels. In some cases, it is to be expected that certain stock in the mix may have a worse conservation status that the others. Then there would be two options:

1 To adapt the exploitation of the mixture of the stocks to the levels required to ensure that the F_{msy} of the most endangered stock is not exceeded. This would mean, in practice, that the rate of exploitation of the healthier stocks would be lower than that corresponding to F_{msy} and then healthier stocks would be underexploited to protect the most endangered one;

2 To adapt the exploitation to that of the most economically important or healthier stocks, at the cost of exerting fishing mortality levels in excess of F_{msy} for the most endangered stock. This would imply taking full advantage of the sustainable yield of healthier stock, at the cost of overexploitation of the weaker stock.

The solution to this conundrum by the co-legislators was to require that, when dealing with mixed fisheries, special measures would be taken to address the problem of certain species.

This means that the plans may have to include specific measures (unspecified) to provide extra protection to the most vulnerable stock, so as to allow at the same time full exploitation at MSY level of the other associated stocks in the fishery. These specific measures could for example take the form of closed areas, whereby allowing the full exploitation of the healthy stocks while bringing extra protection to those under threat.

Transferable fishing concessions and fleet policy

Why a system of transferable rights?

In Chapter 5 we referred to the limited success of the fleet policy both with and without Multiannual guidance programmes (MAGPs). We also referred to the critical report of the Court of Auditors on the CFP's fleet policy. In addition, we have shown how the current fleet ceilings are such that they do not limit fleet size in practice.

These elements clearly pointed at the need to revise the current state of affairs on fleet policy: the *status quo* was not an option. Going back to the old MAGPs, suppressed in 2002, was not an option either. A fresh approach was necessary. This approach was also considered in a context where the continuation of a high level of structural aid could still not be guaranteed (the decision on the Multiannual Financial Framework, establishing the financial resources for all EU policies in the period 2014–2020 took place at a later stage), so the solution had to consider the possibility of having to deal with the problem of overcapacity without the same level of funds available for scrapping.

The idea of proposing transferable rights was not new. In the preparation of the 2002 reform, the Commission already hinted at the possibility of revising relative stability as an instrument that was no longer in conformity with the reality of the fishing grounds. However, the Commission considered that for that to happen without causing a negative impact it was necessary to resolve the problem of overcapacity first. Only when the overcapacity was resolved would a rights-based system be applicable (COM, 2002). In that context, the rights-based system was thought of as a way to preserve a certain balance once the main imbalance between fleet and resources was resolved.

In contrast, the approach proposed in this reform was the opposite: that a rights-based system was the main instrument to balance fleets and resources. The main reason leading to this proposal was the double evidence that the fleet policy of the CFP was not satisfactory and the only instrument for real capacity reduction, the scrapping aid, was strongly criticised by the Court of Auditors.

Rights-based management (or, their more common expression, the Individual transferable quotas, ITQs) are used in a number of countries around the world, such as New Zealand, Australia, Norway, Iceland and, in certain fisheries, the USA and Canada. This type of management is also applied in a number of Member States, where all or part of their fleets are managed according to this system. In the particular, the Netherlands and Denmark have been applying ITQs for some time and other Member States (Estonia, Sweden, Spain, etc.) have applied them to part of their fleets.

There is no unanimity when judging these systems, although practically all countries having used them consider them positive in general terms.

It is important to note that, somehow, the evolution towards ITQs has not always been a bold, innovative move. In a number of cases, as soon as the entry to fisheries is limited, a market is established *de facto* on the use of such fishing rights. In a study on the implementation of rights-based management in Europe, the authors underline that: 'Management systems tend to evolve unavoidably into the direction of individual property rights regimes from the moment that a limitation has changed into a right and vested interests have arisen' (Davidse, 1997).

The evidence of implementation of these systems is that the fleet capacity is generally reduced and better aligned with available resources. This is notably due to the effects of ITQs in stimulating management behaviour in fishermen.

For ITQs to work, however, there must be a number of preconditions. One key condition is that the enforcement is effective, since: 'Under a regime of individual property rights, effective protection of the rights leads to proper protection of the fish stock' (Davidse, 1997).

There are, however, a number of dissenting voices against the benefits of rights-based management. Among these, the idea that these systems represent a privatisation of a common good, or that they work to the detriment of small-scale fishermen, or against those applying the most ecologically friendly (but not necessarily the most profitable) fishing techniques. Typically, those against rights-based management are those who fear their fishing rights could be bought by other interests, from different countries, different fleet segments and even different regions from the same country.

Why did TFCs fail?

The fleet policy is without doubt the biggest disappointment of the 2013 policy reform.

Conscious of the limited effectiveness of the fleet policy in force over the years and aware of the political difficulty to make acceptable a policy based on ambitious objectives of fleet capacity adjustments, the Commission tried for the first time the use of a market mechanism to handle the question of capacity: the Transferable Fishing Concessions (TFCs). These TFCs were modelled after the many examples of Individual Transferable Quotas (ITQs), today applicable – generally successfully – in many countries in the world including a number of EU Member States.

The specific name proposed by the Commission was carefully crafted to avoid concerns expressed by a number of stakeholders during the consultations. For some, 'quotas' would be too easily mixed with national quotas under relative stability. For others, such rights could not be privatised: in some Member States the fishery resources are a public property that cannot be privatised. For these reasons, the Commission invented a new name that would get away from confusing expressions and that would underline the fact that the quotas allocated would belong to the Member States as a public good: the word 'concessions' was introduced as a clear indication that, under any national law, concessions are by definition a public property that is lent to private operators to use and exploit during a limited time for the benefit of society.

The idea of TFCs transferable among Member States found very strong opposition in many Member States that feared that their fishing industries would sell their fishing rights to foreign operators, notably from Spain and the Netherlands, the two Member States whose industries have invested more significantly abroad. Again, the fears expressed in Chapter 1 became evident. It is also true that much of the industry, especially the small scale, was strongly opposed to any idea of a tradable system of fishing rights. Last but not least, European NGOs were not favourable to this instrument that they feared would favour large-scale interests rather than the small-scale fleets they favour.

The Commission, in order to preserve the a minimum of support to TFCs as an instrument acceptable by Member States, decided to propose transferability restricted to within Member States. The Commission also included other safeguards to address the concerns expressed during the consultations, such as the need to protect the small scale fishing from this system and the need to prevent speculators to enter the market of fishing rights. These concerns were adequately addressed by the Commission proposal, by excluding vessels of less than 12 m from the compulsory obligation to use this system and by requiring the holders of fishing concessions to be owners of active fishing vessels (in order to prevent speculators from buying fishing rights), among other safeguards.

In addition, the proposal included the elimination of the system of fleet capacity ceilings (which as we saw in Chapter 5, is largely useless) for all those fleets segments managed under the TFC system. This would complement the use of the instrument with administrative simplification.

Even under these conditions, the TFCs were not accepted by a majority of Council and Parliament and even if it was clear that they did not necessarily have a significantly better alternative. A number of reasons explain this:

- The use of a market mechanism in a world characterised by a long tradition of paternalistic management was resented by many as too revolutionary and not adaptable to the culture of traditional fishermen. Certain stakeholders qualified TFCs as a 'capitalistic drift' in the policy. And certain opinions even indicated that if a market of fishing rights was created, then too many fishermen would actually sell these rights to larger operators (!), in yet another demonstration that, for a part of the industry and a fair share of national administrations, the fishing sector is not a normal economic sector where the principles of the free market economy can apply.
- The impression that although the system proposed excluded international transfers of rights and provided for the exclusion of small-scale fleets, the evolution of the system, once in place, to include both elements would be legally unstoppable through litigation. In other words, the limitations and safeguards in the Commission proposal were considered by some as relatively easy to overcome in the courts of justice.
- For others, the pure market mechanisms would fail to achieve their real goal: not a quantitative reduction of fleet capacity but rather a qualitative reduction of the most undesirable types of fishing. For certain stakeholders (typically NGOs) the problem of overcapacity was more qualitative (the need to reduce vessels using 'bad' gear such as trawlers), rather than quantitative (just reducing overall capacity).

The proposal on TFCs was supported only by a few Member States and clearly rejected by the European Parliament and was not adopted, remaining just as a voluntary instrument for Member States. It is also noteworthy that the reticence against this instrument by a number of MEPs in Parliament was such, the even the possibility of funding the establishment of such mechanisms through the EMFF (see next) was extended to all management mechanisms, to avoid any specific mention of TFCs in that regulation.

The new fleet policy

In the absence of support to the proposal on TFCs, it became obvious that there was no clear alternative instrument for fleet management. Almost inevitably, the fall back was a system which is fundamentally the same as that applied since 2002, although with some positive changes:

- A reinforced role for the Commission in establishing guidelines. In fact, the regulation establishes the obligation by Member States to follow the guidelines provided for by the Commission. The Commission has published guidelines in 2013 (EU, 2013a).
- A clear linkage with the new structural instrument, the EMFF (see next). The availability of funds for scrapping was made dependent upon the preparation of a serious evaluation of the existence of fleet overcapacity in certain segments. By making scrapping under EMFF possible only for segments identified as having overcapacity, the new fleet policy will force Member States to be much clearer in the identification of overcapacity problems. The fact that this mechanism is part of the ex-ante conditionality mechanisms (see also next) gives the Commission a much more important role in demanding more serious and selective evaluations of overcapacity in national fleets than ever before.

- The clear establishment of an entry/exit regime requiring new entries in the fleet to be compensated by withdrawals of the same capacity, even without public aid, is a guarantee against a rebuilding of fleet capacity even if there is room for manoeuvre in terms of the overall capacity ceilings.

Overall, the new regime is largely based on the old one, but with certain instruments that, if used properly, should result in a better evaluation of overcapacity and a more selective system to address and resolve it.

Other issues in the basic regulation

Over and beyond the discussion of the 'big four', other issues also featured high in the different discussions and, particularly, in the *trilogue* between Council and Parliament.

The policy objectives

As we saw previously, the Green Paper had identified the lack of clarity on the long term CFP objectives as one of the five structural deficiencies of the policy. Indeed, the objectives of 2002, the biological, economic and social sustainability did provide guidance on what the ultimate objective would be but not at all on how the transition should be organised.

While maintaining that the three 'sustainabilities' (biological, social and economic) should stay as long term objectives on the same footing, it was, however, understood that one of the three, the biological sustainability, was actually a precondition for the other two to be achieved. If the resources are not biologically sustainable, then economic and social sustainability could not be ensured.

This led to the conclusion that the long term objective of the policy can only be achieved if the biological sustainability could be achieved, albeit in a reasonable period of time allowing for a gradual transition. This philosophy ultimately inspired the establishment of the MSY objective as the leading parameter to guide policy achievement in the long term: by 2020.

Long-term management plans

That the long-term management plans (LTMPs) had become a cornerstone of the CFP was beyond doubt. This was in fact one of the foregone conclusions of the political debate. However, these instruments, already introduced in the 2002 reform, were consolidated as a fundamental instrument for the future CFP, in particular they were reinforced in two aspects: their multi-specific dimension and the integration of environmental considerations.

The traditional EU system of single-stock TACs and quotas is poorly adapted to manage multi-species fisheries. These pose a number of problems that are very difficult to address. In particular, TAC levels for different species closely associated in the same fishery are established on the basis of single-species considerations and this often results in catch limitations that are exhausted at very different times of the year. That means that when the fishery for a stock is closed but the fishery for an associated species can continue, catches on the former de facto continue but with two further negative effects: (1) discarding and (2) unaccounted fishing mortality. For these reasons, there was wide consensus that the future multi-annual plans should, where relevant, be multi-species.

In addition, these long-term plans were recognised by both co-legislators as the main management instrument. It is noteworthy that, when it was made evident that these plans would take time to develop and that the specific rules to implement the discard ban could not wait until all fisheries concerned would be covered by long-term plans, the Parliament in particular insisted that the discard plans would be of a limited duration, 3 years, so as to ensure that in the end all the discard ban technical rules would be fully integrated in these management plans. The idea was that the discard plans would be just a transitional instrument until such time as all the important long-term management plans would be adopted.

It is also important to underline that the co-legislators considered such plans as the ideal 'carrier' of measures with an environmental objective, such as those to protect marine mammals or seabirds. The idea was that, instead of horizontal regulations for that purpose (such as Regulation 812/2004, see

Chapter 11) the environmental measures would be part of the mainstream fisheries management rules as part of the management plans. This will represent an important cultural change, and as such it will probably require some time to implement.

The composition of Advisory Councils

The Advisory Councils (ACs) are the successors of the RACs of the 2002 reform. The issue of their composition was important notably in the context of regionalisation. If the initiative on management measures was to come from regional approaches, where the stakeholders should play a key role, what would be then the adequate composition of these bodies to ensure that all legitimate interests are represented in a balanced way?

The debate here centred around two possible models:

1. One where the fishermen, as the most directly affected stakeholder, would have a reinforced capacity to influence recommendations;
2. One where the weight of the other stakeholders would be increased. As ACs were called to play a key role in the future CFP, the idea was that these ACs would have a representation that would reflect, in a more balanced way, the weight of other societal interests beyond those of the fishing industry.

As a compromise, the co-legislators decided that the fishing industry should represent 60% of the seats and other stakeholders would occupy 40%. This represents an evolution, since the RACs of the 2002 reform provided for a two-thirds representation of fishermen and a one-third representation of other interests. This evolution should reflect the fact that the new Advisory Councils, as compared to the old RACs, will have an important extra role: actively participating in shaping up proposals for management measures under co-decisions. In that regard, a more balanced composition between the fishing industry and other interests would correspond better to the new CFP where, as we have seen, societal interest other than the industry have an increasing role to play.

The co-legislators also agreed that the details of the functioning of the new ACs would be decided by the Commission through a delegated act. This was adopted by the Commission in 2015 (EU, 2015).

Integration of environmental concerns

In a new CFP with a significant environmental component, the issue of the integration of the environmental concerns into the fisheries policy was also an important question. This had two components: (1) the general consideration of environmental objectives in the mainstream of the CFP and (2) the introduction of a mechanism to ensure that the implementation of environmental legislation (notably the habitats directive) within the CFP would be done rapidly and efficiently.

Environmental objectives as part of mainstream fishery management measures

As we saw in Chapter 11, some of the instruments of the CFP dealing with environmental objectives (the best example is the regulation to protect marine mammals) are a stand-alone instrument, disconnected from the mainstream fisheries management measures and in some cases enjoying a relatively low priority for implementation and compliance.

The co-legislators considered that, in order to improve on this situation, the solution was to ensure the full integration of these instruments into the mainstream fishery management instruments, notably the long-term management plans. This means that, in the future, management plans should contain, if and when relevant, specific measures to ensure the protection of marine mammals and to do so according to the same philosophy that inspires the new management approach: sea-basin specific, de-centralised and results-based. This will certainly represent an important cultural change, which will likely require a gradual implementation.

Implementation of environmental legislation

This was another important contribution of the new basic regulation. As we saw in Chapter 11, the CFP after the Treaty of Lisbon lacked a specific fast-track mechanism to implement fisheries measures under the CFP to contribute to protect maritime Natura 2000 designated by Member States. This had to be addressed by the reform.

This was not easy: from the beginning it became obvious that Natura 2000 and the future environmental measures under the marine strategy framework directive could not possibly be treated under co-decision. This would create a contradiction: Member States being obliged to take environmental protection measures under environmental legislation and, when doing so, they could see their initiatives blocked in the context of the CFP if the co-legislators so decided. This anomaly required a fast-track solution in which the CFP would not decide whether or not the CFP will contribute to protect the areas concerned but only how that protection would be formulated under the CFP. And, for this purpose, the only possibility was a Commission delegated act.

And indeed the co-legislators established that the fisheries measures under the CFP necessary to contribute to the implementation of Regulations 2008/56/EC (the marine strategy directive: EC, 2008), 2009/147/EC (the birds directive: EC, 2010) and 9243/EEC (the habitats directive: EEC, 1992) would be done through Commission delegated act. Such delegated act will intervene only as a result of the initiative of the concerned Member States under regionalisation.

The regulation also clarified the case of measures where only vessels from one Member State would be affected. This is typically the case of Natura 2000 sites within 6 miles (in the 6–12-mile coastal band, vessels from other Member States can operate). In this case, the Member State can take these measures through national legislation. Finally, the regulation also allowed the Commission to take urgent measures to protect Natura 2000 sites in case of evident danger to the conservation of these sites in the absence of Member States' initiatives.

Closed areas

This question had at least two components:

1 The discussion on the possible continuation of two closed areas that resulted from the basic deal of the CFP in 1983 (the Shetland Box) and the one that replaced the Irish Box of the accession of Spain in 1986; the biologically sensitive zone of 2003.
2 The wider issue of the role of fishery protected areas (FPAs) as a mainstream instrument for fisheries management, an instrument advocated by a number of stakeholders and scientists.

The Commission proposal, for the first time since 1983, did not propose the continuation of the 'Shetland Box' (see Chapter 2), which had featured in all basic CFP regulations since 1983. On previous occasions, the UK delegation had insisted on the continuation of this measure. However, the obsolescence of this measure was such (the scientific evaluation of this area did not show any measurable benefit) (SEC, 2007) that even the Member State benefiting from it as part of the 1983 CFP deal did not consider it necessary to continue its implementation.

On the biologically sensitive zone the Commission decided not to raise the issue at all. Since this measure was not part of the basic regulation 2371/2002 but part of the 'Western Waters' Regulation (EC, 2003) it was not necessary to raise the issue. On the contrary, the Commission considered that given the political difficulties around the adoption of the Western Waters Regulation, it was preferable that the issue of the biologically sensitive zone did not pollute the difficult discussion on CFP reform.

As we saw in Chapter 4, the use of closed areas as an instrument for fisheries management has been debated for quite a long time. Although the possibility of using such areas is part of the instruments always available in the CFP, there has never been a real push for this kind of instrument. Beyond the reasons already explained in Chapter 4, fisheries protected zones as a main instrument for fisheries management in the CFP have a major political flaw: they can interfere with relative stability.

In this case, on the initiative of the European Parliament, a new provision was adopted requiring Member States to present, through regionalisation, fish stock recovery areas, with a particular emphasis on areas of high concentration of juvenile fish, to be totally or partially protected from specific types of fishing. The real objective of this proposal, however, was not to turn these areas into a mainstream management instrument but to use them, combined with funding from EMFF, as a way to develop a culture of management by small-scale fishers. However, these closed areas would be established on the initiative of Member States and for a variety of reasons. Overall, there was no clear idea of how these areas would fit into the mainstream management measures (TACs and technical measures).

The concerned Article also establishes that: 'The Union shall continue to provide additional protection to existing biologically sensitive areas'. This appears as an obvious reference to the continuation of the 'biologically sensitive zone' around Ireland in the context of the Western Waters Regulation.

This article, together with current obligations such as those related to Natura 2000 and current technical measures establish an ample legal basis for the use of closed areas as one of the management instruments of the CFP. However, calls from a number of stakeholders to consider Fish Protection Areas as the main management instrument were not considered, on account of the clear political prevalence of the TAC and quota system as the main instrument for the CFP.

Aquaculture

One clear breakthrough in the new CFP is that, for the first time, the policy sets specifically the objective of developing sustainable aquaculture in Europe, as part of our overall implementation of Article 39 TFEU.

True, aquaculture as such has never been alien to the CFP but traditionally the CFP limited itself to the funding of aquaculture investment in the framework of the successive programmes of structural policy for the fisheries sector. However, despite all the elements at hand, European aquaculture has been stagnant for the last decade, while the overall global production of aquaculture products has been increasing dramatically (see Chapter 14). And this, despite the availability of the many fundamental strengths at its disposal: excellent research, good infrastructures, good funding and so on.

In a strategic communication of the future of aquaculture (COM/SEC, 2009), the Commission analysed the reasons for this stagnation and concluded that the two key factors for this stagnation were:

1 the difficulty and shortage of physical space in European coasts, and
2 the extremely long and difficult licensing systems in Member States.

In addition, it was also recognised that the relationship between aquaculture and environmental law and particularly Natura 2000, was not clear enough and this resulted in frequent conflict between the two that discouraged investment.

On that basis, the idea was that providing further financing to aquaculture development under the structural funds would not solve the problem, unless Member States, under their own competence, did the necessary to correct these two main shortcomings. In addition, the Commission published guidelines on how aquaculture interacts with Natura 2000, in order to create a clearer framework that would remove uncertainty and allow sustainable development (EC, 2012). The preparation of a national plan specifying how each Member State, under its national competence, would resolve the problem of availability of space and how they will simplify licensing, was established as a pre-condition for Member States to have access to funding for aquaculture. This was a real breakthrough.

Control

The 2013 reform did not modify the control instruments of the CFP. The reason is simple: a major reshuffling of the control rules had taken place in the previous years, when Council adopted Regulation 1224/2009 on a new control regime for the CFP. With a new regulation so recently adopted, it was considered that a further reform of the control regime was not necessary.

However, the question of the control of the CFP was an issue. On the initiative of the European Parliament, the idea was accepted of establishing a working group where the performance of Member States would be examined in a fully transparent manner. This idea was accepted because in the light of the future implementation of the EMFF, which included significant elements of conditionality (i.e. that the reimbursement of funds could be withheld or even suspended on the basis of bad performance by Member States on their control obligations) it will become increasingly necessary to ensure that any possible application of this type of instrument be done in a fully fair and transparent manner.

The working group was subsequently set up by the Commission (COM, 2014a). However, because of the terms and conditions of the inter-institutional agreement between the Council, the European Parliament and the Commission, the Parliament cannot be directly involved in policy implementation; That means that the compliance group will be able to examine questions related to how the control policy is working and what changes may be necessary but it will not be able to discuss possible 'naming and shaming' of underperforming Member States, for example.

Scientific advice and data collection

This new Chapter of the basic regulation is also innovative. The obligation by Member States to provide data for scientific advice was also for the first time incorporated into the basic obligations under the CFP.

Also importantly, the regulation diversified the possible source of scientific advice. In the 2002 regulation, the STECF was considered as the only provider of such advice. As a consequence, all advice had to go through this body, thus resulting in duplicated work at best and contradictory advice at worst, though only in certain occasions. By admitting scientific advice from different bodies (and not exclusively from STECF), the CFP can diversify the sources of advice, so as to avoid duplications and allowing a division of the advice between different providers, notably ICES and STECF.

On data collection, the regulation also provided an important breakthrough: the provision of data to prepare the scientific advice that underpins the policy was made into an obligation for Member States. An obligation, in any case, accompanied with abundant financial support (see next on the EMFF). This is intended to remove the classical argument about the difficulty to take measures because the scientific advice is missing. In the future CFP, there should be fundamentally no important advice missing.

Small-scale fishing

This issue also featured as an important one all the way through the negotiation. Initially, the Commission proposed a division between small and large-scale fishing only for one specific issue: the exclusion of small-scale fishing from the compulsory use of transferable fishing concessions. As the TFCs were not supported by the co-legislators already at an early stage, the question remained one of how to define small-scale fishing. This was politically important even if no specific provision in the basic regulation actually applied differently to small-scale fishing. However, the discussion had a political value as a precedent for the discussion on the EMFF, where a different funding regime was to be made available in principle for the small-scale sector.

Delegated and implementing acts

It is also important to look at this important question: what were the entitlements for the Commission to establish detailed rules through delegated and/or implementing acts. Although it looks bureaucratic, this issue is very important for the future of the CFP and particularly as regards the problems related to the perception of the policy as a complex, central command policy. It is obvious that the complex procedure of co-decision between Council and Parliament cannot possibly deal with all the detailed EU rules concerning such a complex, exclusive competence policy as the CFP. It is therefore necessary that the co-legislators delegate to the Commission the adoption of a series of technical rules.

It must be underlined that this discussion took place in a context of a very unsettled picture in terms of the interpretation of the Treaty of Lisbon and also against the background of a relatively low level of trust by Council in the Commission. The origin of this mistrust is varied, but in many cases is related to the Council's view of the Commission as 'too conservationist' and therefore too prone to take politically unpalatable positions when given a mandate to adopt technical rules. This produced a paradox: while Council and Parliament recognised that decision-making is heavy and that fast-track procedures were necessary, the lack of trust in the Commission limited considerably the number of cases where the co-legislators were ready to give the Commission the mandate to do delegated acts.

At the end, however, an important number of delegated acts were foreseen, but on condition that the mandate given to the Commission by the co-legislators was narrowly defined so as to avoid significant room for manoeuvre for the Commission in adopting the rules.

The external dimension

The CFP reform of 2013 included, for the first time, specific provisions on this pillar in the basic regulation of the CFP (see Communication previously). This was innovative: never before the CFP had established the principle in EU law that the external action of the Union on fisheries management should be inspired by the same principles and objectives that lead the policy in EU waters. This fact was fundamentally undisputed in the negotiation. With the establishment of a specific Chapter on the basic regulation, the dichotomy between the domestic and the external policy will disappear.

However, the external sector was still subject to a specific debate. Over and beyond the provisions included in the context of the new basic regulation, Council and Parliament also adopted their respective resolutions, where they laid down the political orientations that will inspire the development of this policy pillar in the years to come. The Commission proposed that the external action be focused on three main axes:

Contributing to long-term sustainability worldwide

This objective means the recognition that the Union, as the world's largest market for fish, also has to play a leading role in promoting the values of sustainable fishing wordwide. This meant in particular:

- Fighting against illegal fishing and tackling the problem of fleet capacity worldwide.
- Completing the institutional architecture (notably through the United Nations) of multilateral governance of fisheries.
- Reinforcing and completing the network of Regional Fisheries Management Organisations.

These questions were not new and the Union had already been active on all these areas. However, the proposal was to make this not just a matter of policy in practice but to turn this into a legally binding objective of the new policy.

Towards sustainable fisheries agreements

The Commission here was very clear to indicate its intention to continue developing bilateral fisheries relationships but to do it under clear new principles that would put the accent on sustainability and cooperation with the third countries concerned. This included:

- Placing a new emphasis on the need for these agreements to ensure sustainability. This means also that the agreements should be compatible with the activity of local fishermen.
- Ensuring a better scientific basis to achieve sustainability.
- Reinforcing the governance of the agreements, including a new proposal on the management of fishing authorisations, conditions on reflagging and so on, to ensure that no abusive practices can take place under these agreements.
- Improvement of governance in the third countries concerned.

All these proposals were accompanied with a proposed change of name, from Fisheries Partnership Agreements to Sustainable Fisheries Partnership Agreements, as a clear political signal that sustainability was the new fundamental dimension of these agreements.

Coherence with other EU policies

This objective was also a new one: in recognition that bilateral agreements had been largely stand-alone efforts on fisheries relationships with relative little connection with other EU policies, the Commission proposed that the synergies between bilateral fisheries agreements and other EU external policies be reinforced and made a compulsory condition for the future development of these agreements.

The ideas proposed by the Commission in its communication were by and large endorsed by Council (CEU, 2012) and Parliament (European Parliament, 2012).

The new orientation of the external policy represents a fundamental change as compared with the previous policy, both in terms of the full integration of these objectives into the mainstream objectives of the CFP and in terms of the considerable clarification of the objectives of that policy pillar.

It is important to refer to what these principles represent. Part of the European industry had misgivings about these new principles, notably on grounds that, by making our own policy so demanding, notably in terms of the conditions for bilateral agreements, the Union will become 'non-competitive' to sign agreements with third countries in the face of other possible partners ready to sign bilateral fisheries agreements on a much less scrupulous basis. In addition, the emphasis on the private contribution of EU fishermen to pay for access is also seen by some sectors as abandoning the traditional scheme and turning to a 'privatised' system where vessel owners and not taxpayers, should be paying for access.

Both misgivings should be looked at from a strategic perspective. On the private fees for access, it is important to bear in mind that the bilateral fisheries agreements have been discussed in the World Trade Organisation (WTO) where many voices claim these are unacceptable subsidies. Ensuring that access is paid by the vessel owners, while the agreements with public money will pay for development and governance is the best way to protect the agreements from possible allegations, in WTO or elsewhere, that they constitute undue subsidies to the continuation of the activity of the EU fleets.

On the other hand, the strong criticism expressed against the EU bilateral fisheries agreements since they instruments were instituted (criticism largely coming from stakeholders from inside the EU) has made these instruments increasingly difficult to get accepted. In addition, international pressure is mounting against abusive practices. Overall, if the fisheries agreements are to have a future, they will have to address the criticism.

The new orientations of the external policy may appear as difficult to implement, or to give rise to short term difficulties, particularly when we compare the external action of the EU with that of other countries that engage in bilateral and multilateral fisheries relationships with much fewer constraints. However, in strategic terms, the new approach represents the best way to preserve these relationships (and the role of the legitimate European interests in them) in a world that inevitably has to evolve towards more controlled and more responsible fishing practices.

The new market regulation

The new market regulation represented another opportunity to revisit this policy pillar, existing as part of the policy since 1970, in the light of the accompanying proposals to revamp the CFP. In particular, this proposal coincided with the increased awareness of European consumers in sustainable fish. This provided a very good opportunity to give the fishing industry a higher responsibility in managing supply and demand, through empowerment of producer organisations (POs) and a lower contribution from public funding on market intervention mechanisms. In addition, it was also an opportunity to give consumers an increased role in the policy by inviting them to make choices about the fish they would buy.

Importantly, in parallel to this proposal the Commission launched a public campaign in favour of the consumption of sustainable fish, trying to raise the awareness of European consumers of the role they could play with their selective choices in the fish markets. The obvious complement to this campaign was to provide European consumers with the right labelling so they would be sufficiently informed to make the right choices. Therefore, new, more ambitious labelling requirements for fish and fish products were an essential element of this new regulation.

These labelling requirements, however, did not include compulsory information about the sustainability of the fish. The reason is that the Union, still today, does not have a publicly defined labelling on fish sustainability. Earlier attempts by the Commission services to establish such a system were rejected on grounds of insufficient need to legislate at European level. Consequently, the new labelling requirements focused on the nature of the fish, the area of catch, the date of landing and so on. Considerations on sustainability had to be left as voluntary.

The Commission proposal set out to achieve several objectives:

- To empower the industry, notably through producer organisations to take their own responsibility when managing supply and demand of fish, through a completely new instrument: the

production and marketing plans. This is crucially important to improve the share by the fishing industry of the added value generated in the production chain. But also crucially, to empower POs to play a key role in handling some of the consequences of the reform: the increasing amount of unwanted fish arriving to ports as a consequence of the discard ban;
- To revamp the market mechanisms to give consumers a better opportunity to make choices when buying their fish. This was crucial to involve the society at large in the CFP. The key to give consumers a say in their choices is to provide them with better information: better labelling was one of the key objectives;
- To reduce the number of intervention mechanisms: from six mechanisms under the previous regulation of 2000 (EC, 2000), to only one: storage aid. This represented an important simplification effort but also a change in philosophy: from a strong emphasis on market support mechanisms, to a focus on the capacity of producer organisations to regulate markets on their own, with the storage aid as a mechanism to assist the industry in difficult, unforeseen circumstances and, crucially, to contribute to find solutions to handle the unwanted fish to be landed as a result of the discard ban;
- To develop the market intelligence by means of the market observatory (known as EUMOFA) with the objective of providing market information for operators to maximise the economic return of their activity and also to help find solutions to the problem of unwanted landings.

The negotiation led to the adoption of Regulation 1379/2013 (EU, 2013b). This new regulation fully accepted the Commission's ideas on the essential role to be played by the POs through their Production and Marketing Plans, which represent a fundamental change in the market policy after 40 years: emphasis on the empowerment of POs instead of intervention mechanisms with public money.

One of the most difficult issues in the negotiation was that of the minimum compulsory labelling requirements, where the interests of certain parts of the industry to have more complete information on the origin and nature of the products had to be compounded by the interests of part of the processing industry who feared that the diversified origin of their raw materials would become a competitive disadvantage in the European markets.

Another difficult question was that of the date of catch. Initially, the Commission had proposed that the fresh fish should indicate the date of catch, in a way to favour small-scale coastal fishing, whose short fishing trips result in the product being marketed very shortly after their catch. However, this proposal would potentially be very negative for long-distance fishing, which often spends 3 weeks at sea and land the fish 1 or 2 weeks after the catch, yet their fish is of impeccable quality. To avoid an unjustified bad image to the fish coming from long-distance vessels, the final date reflected is that of landing.

Overall, the new market regulation represents the biggest historical departure from the traditional market policy for fishery products. As we saw in Chapter 2, this policy dates back to 1970 and from then it had remained remarkably stable. The new approach represents the departure from the old policy, based on many diverse and horizontal intervention mechanisms, to another one, considerably simplified, where the private operators (notably through producer organisations) and the consumers in general (through improved labelling) will play a much more important role than ever before.

It is also to be noted that the budget for this policy was transferred from direct to shared management, in another example of further de-centralisation of the CFP. In fact, the EMFF (see next) allocated nearly €45 million to implement the new market measures, for the period 2014–2020.

The new structural instrument: the European Maritime and Fisheries Fund (EMFF)

As an essential part of the package of CFP reform of 2013 we must count the new structural instrument for fisheries, the European Maritime and Fisheries Fund (EMFF), adopted in early 2014 (EU, 2014a).

As we saw previously, the almost parallel timing of this proposal with that of the basic regulation was a very fortunate circumstance. Never before could the CFP and its financial instrument be discussed one just after the other: this provided a unique opportunity to discuss the EMFF as the instrument to help the implementation of the newly devised CFP and to maximise consistency between these policy pillars. This opportunity, however, was mitigated by the tendency, as we have seen throughout this book, for the CFP and its instruments to have a considerable inertia and which experience difficulty in breaking with the past. Overall, the discussion on the EMFF was a mixture between the need for an innovative instrument to accompany the changes introduced, a few months before, in the CFP and the interest by many to preserve several of the traditional instruments of the structural policy in fisheries.

The proposal

The possibility to discuss the CFP reform and the financial instrument practically in parallel allowed an unprecedented level of consistency between the new policy objectives for the CFP and the financial means to support them. For the first time ever, the financial instrument, negotiated barely a few months later than the basic regulation, could be much better tailor-made to fit the needs of implementing the new policy.

But the new financial instrument was also influenced by another important factor: the general policy objective of maximising the synergies among the different pillars of the structural policy of the Union and the objective of ensuring that all funds would contribute to the same objective: the implementation of the Europe 2020 strategy. This led to a new approach on the structural funds. Renamed as ESI funds (European Structural and Investment funds) the five EU structural funds were regrouped under a more coherent approach, consisting of a new common provisions regulation (EU, 2013c), which pooled together all common rules for all funds and a common political approach in their preparation, notably through the Strategic Plans by Member State. On that basis, the five funds developed their own proposals.

In that context, the new financial instrument was strongly conditioned by a number of factors:

- The new financial framework imposed by the situation of economic crisis in Europe, which clearly underscored the need to be much more selective with the use of the structural funds in general;
- The need to be more strategic in the spending of European public funds. The Commission published in 2010 a communication on a strategy for smart, sustainable and inclusive growth, known as the 'Europe 2020 Strategy' (COM, 2010) that established three clear axes for the European policy action for the next years in all policy areas: (1) innovation, (2) green growth and (3) socially inclusive growth. These three criteria should inspire all policy areas, including naturally the CFP;
- The need to inscribe the financial support policy for the CFP in the political package of CFP reform, therefore considering the new fund as an integral part of a coherent package of delivery of a new CFP;
- The need also to consider the linkages between fisheries areas and the wider maritime economy at large and thus the need to incorporate in the fund a specific instrument to favour the integral development of the maritime economy;
- Finally, the need to rationalise and streamline the expenditure of public funds for the CFP. This called for the decision to coordinate all the structural funds into one single architecture, in order to develop synergies and reduce the administrative burden in Member States. But it also had a consequence for the Commission services by transferring a number of elements hitherto financed directly by the Commission to the system of shared management with Member States that applies to the fisheries structural funds.

On this basis, the Commission presented a proposal (COM, 2013) that would become an instrument to help Member States achieve the new goals of the CFP. For this purpose, the proposal included a number of new elements as compared with the previous fund, the EFF:

- Conditionality. The proposal made a special emphasis on rendering the access to public funds conditional to the fulfilment of certain conditions, particularly in terms of administrative capacity in areas such as control and data collection.

- Suppression of aid to the fleet, notably scrapping and temporary laying-ups the proposal to suppress scrapping aid was a consequence of the proposal for the use of a market instrument (the TFCs, see previously) to handle fleet policy.
- Inclusion of existing and new funding schemes into a single instrument, including schemes hitherto managed under central management by the Commission. The specific support schemes to outermost regions was also included. In fact, this proposal regrouped all funding of the CFP, with the only exception of the funds for bilateral fisheries agreements.
- The widening of the scope of the instrument, notably by adding the funding of activities beyond fisheries and aquaculture but related with the implementation of integrated maritime policy. This financing, however, was limited as the main bulk of investment in the maritime economy would be realised under the other ESI funds, notably the regional funds and the Horizon 2020 research programme.
- Closer linkages to the delivery of CFP objectives, notably through the introduction of performance indicators.
- Ring-fencing of the funding of certain core responsibilities for Member States in running the policy beyond the conditionality referred to previously, minimum percentages of the total budget were ring-fenced to finance control and data collection, to ensure that Member States could not argue lack of funds to fail to fulfil their obligations under the CFP.
- A continued emphasis on the community-led local development (the continuation of the Axis 4 of EFF), as an instrument with a demonstrated ability to promote growth and jobs in fishery-dependent communities.
- A strong emphasis on aquaculture, albeit subject to the presentation of a national strategy whereby the Member States interested in such funds should address and resolve the bottlenecks (see Chapter 14) that have impeded aquaculture development in recent years.

But the proposal was also characterised by what it did not propose:

- Scrapping aid this was consistent with the proposal on TFCs. If the question of fleet capacity was to be handled through a market mechanism, then no public money should be spent on scrapping. However, the Commission always considered that if the TFCs were not adopted, then the inclusion of some – limited – scrapping would be inevitable.
- Aid to temporary laying-ups. Here the Commission considered that this aid had no significant contribution to address fleet overcapacity. On the contrary, as we saw in Chapter 6, this kind of aid, when provided on a regular basis, can delay or prevent the resolution of fleet capacity problems.
- Engine replacement, as it was considered that this type of measure was conducive to increasing capacity.

The proposal was presented when the basic regulation was still being negotiated and therefore it was made consistent with the ongoing trend in the negotiation of the former. However, as the basic regulation was adopted in May 2013, prior to the start of the *trilogue* on the EMFF, this could be discussed in the light of the decisions taken already by the co-legislators in the basic regulation. This provided an unprecedented opportunity to shape up the structural instrument in a way that would be consistent with the newly agreed CFP.

The negotiation and the final regulation

In the *trilogue*, the negotiation took place under the political pressure of finalising the agreement as soon as possible.

To prepare the *trilogue*, Council decided relatively soon a common approach in July 2013. In this approach, it is noteworthy that the old division between 'friends of fisheries' and 'friends of fish' that characterised the 2002 reform was not nearly as evident. There are two reasons for this:

1 Firstly, the number and complexity of Member States was much higher than in 2002: 27 instead of 15, including a new block of land-locked Member States with specific interests, often very different from those of coastal Member States.

2 Secondly, the old 'friends of fishing' were not necessarily asking for money for new constructions of new fishing vessels. Although there were still some voices asking for this possibility to be reconsidered, these did not come necessarily from the old 'friends of fisheries'.

This resulted in a much more complex but perhaps less confrontational negotiation, as there were no longer two mutually antagonistic blocks. Finally, a deal was struck on 28 January 2014, the EMFF being the last of the five structural funds to be agreed, despite its low economic importance compared to the other four. Again, the political sensitivity of fisheries issues, over and beyond its economic value, came into play.

The following issues were of particular relevance in the negotiation:

- The nature and conditions of the fleet measures;
- The allocation of funds among different policy pillars and among shared management with Member States and direct management by the Commission. In particular, the amount of money that Member States should dedicate to control, data collection and integrated maritime policy;
- The innovative vis-à-vis traditional actions eligible for funding;
- The conditions on ex-ante conditionalities (that is, the mechanisms to link access to EMFF funds to compliance with key aspects of the policy);
- The conditions attached to the funding of aquaculture and, particularly, the practical implementation of the notion of 'sustainable aquaculture'.

At the end, the new regulation included a wide palette of measures at the disposal of Member States: fleet, social, environmental, scientific, aquaculture, markets, processing, maritime policy and so on, thus constituting an unprecedented variety of measures at the disposal of Member States. Let us look at these measures.

Fleet measures

This whole Chapter was discussed under a very different political background, especially as compared to the 1992 and 2002 CFP reforms. At that time, fleet issues were very high in the agenda. As we saw, in 2002 the suppression of aid to construction of new vessels was the most important and contentious political issue. In 2013, however, none of the co-legislators considered that the problem of overcapacity was so significant that should justify new or dramatically different measures. As a consequence, the fleet measures under EMFF were considered by both much more in the light of how to modernise the aging European fleets than from the angle of how to tackle overcapacity.

Since the co-legislators agreed to reject the Commission's proposal on a rights-based management system to deal with overcapacity, they had no option but to accept the most traditional fleet management instruments: scrapping and temporary laying-ups. However, the aid to new construction (suggested by a certain, though limited, number of Member States in Council) was not accepted: it would have meant a return to the past that co-legislators decided not to take. Likewise, aid to the modernisation of vessels (also advocated by some Member States) was not considered, as being part of the 'old system' arguably conducive to fleet overcapacity. On the contrary, investments on vessels were accepted for the purpose of increasing health and safety conditions, or to favour the adoption of more selective fishing, or to facilitate the implementation of the discard ban.

In addition, even the restricted fleet measures were considered a 'necessary evil' and, consequently, over and beyond the limitations in eligibility of the scrapping funds, a total financial cap was imposed on the fleet measures, corresponding to 15% of the total investment by Member States or €6 million, whichever the highest. This 15% compares dramatically with the percentages allocated to fleet policy in earlier instruments, which reached 50% of overall investment in FIFG1 and over 30% in FIFG2 (see Chapter 6).

Scrapping

The scrapping aid was finally agreed, as co-legislators understood that, in the absence of the market mechanism proposed by the Commission for fleet policy, the system should still have the possibility of providing for public funds to facilitate fleet adjustment through scrapping. However, for the first time and largely drawing on the report of the Court of

Auditors on the fleet policy,[2] already referred to in these pages, it was decided that the scrapping would be limited in time and budget, thus representing for the first time ever a phasing out of this instrument on 31 December 2017:

- Firstly, the funds available for scrapping were part of the said overall ceiling of 15% for fleet measures.
- Secondly, the funds are available only until December 2017, that is, scrapping must be planned at the beginning of the programming period, to underline its nature as a provisional instrument to address overcapacity.
- Finally, the scrapping funds are available to Member States only after the presentation of a scrapping plan where the Member State analyses the fleet segments having overcapacity, according to Commission guidelines, so as to ensure that the scrapping is not granted on a 'first-come-first-serve' basis but according to a plan to target the overcapacity in the fleet segments where it is well identified. With these provisions, the legislators wanted to address the allegations of the Court of Auditors referred to previously.

Overall, the absence of TFCs and the limited time and budget available for scrapping and all together with the continuation of the system of fleet ceilings that is no longer limiting capacity (see Chapter 5) means that the CFP will have very limited instruments to adjust fleet capacity in practice, even if Member States would evaluate their own overcapacity as substantial.

Temporary laying-ups

The temporary laying-ups were also subject to a difficult debate, not so much on the need and opportunity to consider them (where both co-legislators generally agreed) but rather on the level of conditions attached. The idea, in any case, was that such an instrument should not become a permanent aid to fishermen but rather constitute an instrument to tackle unforeseen situations. The Commission had not proposed this measure, reflecting the experience that this instrument, when used in cases that are not real urgencies, can indeed contribute to delay or prevent structural adjustments in certain fleet segments.

At the end, the corresponding article (Article 33) agreed by the co-legislators referred to this instrument as being eligible in case of emergency measures 'or of conservation measures referred to in Article 7 (of the basic Regulation 1380/2013)'. This means that in practice, temporary laying-ups will be eligible in the event of any conservation measure, including TACs, technical measures and so on. This represents a substantial widening of the scope of this instrument, as compared to previous structural regulations.

Whether or not this was the intention of the co-legislators, the result is that the conditions attached to temporary laying-ups are less strict than ever before. However, the impact of this kind of measure will be limited by the overall ceiling applicable to fleet measures as well as by the limitation to 6 months of the period that any vessel can obtain funding during the whole 2014–2020 period.

Engine replacement

New vessel construction having been turned down by co-legislators, a very intense debate took place on the need and opportunity to consider the financing of vessel modernisation and, notably, engine replacement.

In favour of this possibility were arguments about vessel safety and climate change: replacing old, fuel-guzzling engines by new, cleaner and effective ones. Against this possibility, changing engines represent a very marginal gain in fuel consumption (for a very costly investment of taxpayers' money) and entail a high risk of *de facto* creating new overcapacity. In addition, the fuel crisis of 2008 showed that fuel consumption can be substantially reduced without necessarily changing engines, through certain changes in fishing behaviour.

The trilogue discussed at length what kind of limitation and safeguard should be established for engine replacement to be eligible. Of particular difficulty was the decision on the fleet segments that may have access to this type of aid, with the European Parliament advocating access only to small-scale vessels and excluding trawlers and Council advocating also larger vessels including trawlers. At the end, the compromise was to grant this type of aid but:

- Admitting engine replacement but without increase in engine power or depending on vessel size, requiring that the new engine should have lower power than the replaced one. On this condition, the aid was made available beyond the small-scale, to vessels up to 24 m.

- Trawlers, after very difficult negotiation, were also admitted as eligible, as Council insisted that in certain Member States trawlers constituted the bulk of their fleet.
- In addition, the vessels eligible should belong to a fleet segment defined as not having overcapacity, in the context of the implementation of the national reports on fleet policy (see previously).

It should be noted that this measure was included, among others, such as the funding of energy efficiency audits, as part of a general measure aimed at mitigating climate change, in a clear indication by the co-legislators that the real motivation for admitting this instrument was the reduction of the production of carbon dioxide by the fishing fleets.

Aquaculture

The regulation established a wide range of measures to promote sustainable aquaculture. While this emphasis encompassed investments on production and not just on innovation, the key question was to ensure that the aquaculture promoted by the EMFF would be environmentally friendly. The discussion focused on the balance between aquaculture development and nature conservation and how to relate the respect to environmental legislation to the eligibility of individual projects. Here, the question was that the respect of environmental objectives could not always be clearly related to an individual aquaculture project, so it was difficult for the legislator to establish that a given project could not be eligible for funding on account of lack of respect of environmental legislation by the Member State concerned. It was also agreed that the aquaculture projects using genetically modified organisms would not be eligible for funding under the EMFF.

The new funding possibilities for aquaculture are extremely varied but they are also conditional. In effect, the experience of previous structural instruments, where abundant public funding has not broken the situation of stagnation of European aquaculture production, shows that funding alone will not allow European aquaculture to take off. Thus, the EMFF included an important novelty: the funding is conditional to Member States presenting a strategic plan addressing the bottlenecks that stifle investment, such as excessive red-tape and – unresolved – competition for space. Although the legal requirements for this plan are included in the basic regulation (see previously), the EMFF established the possibilities for funding on the basis of that condition.

Aquaculture funding comprises, beyond productive investment, innovation, advisory services for aquaculture farms, promotion of human capital and networking, the promotion of sustainable practices, the promotion of organic aquaculture, animal and human health measures, aquaculture as provider of environmental services and aquaculture stock insurance.

This extremely wide range of measures shows the high political priority accorded by co-legislators (responding to the Commission's initiative) to the sustainable development of this activity, not just in terms of increasing production but fundamentally by helping the sector in the search for solutions to the many challenges of the activity. It is noteworthy for example, the support to aquaculture of aquaculture as a provider of ecosystem services. This reflects the fact that, in particular fish farming in ponds represents much more than just farming: the activity allows the preservation of wetlands in continental Europe, thus providing for nesting places for birds among other environmental services.

Marketing

The provisions on marketing were discussed after the adoption of Regulation 1379/2013 on the common market organisation for fisheries products (see previously) so it was possible to ensure that the EMFF funds would be tailored to cater for the objectives of the new market framework.

Financial aid was provided to help Producer Organisations prepare their production and marketing plans, considered a key element in empowering the fishing sector to take its own responsibility in the management and marketing of the fishery resources, in particular through planning of the offer to adapt it to consumer demand. In this context, the EMFF included financial aid also to help measures to improve the marketing of products, including for example the marketing of species of low value (important to facilitate the discard ban) or the marketing approaches that would give producers a higher stake in the establishment of the final prices to the consumer.

In that vein, the most difficult discussion was that of whether the only intervention mechanism left, the storage aid, would benefit from EMFF support during all the duration of this financial instrument (until 2020) or whether it should be phased out. This debate, as ever, was about liberal versus interventionist market approaches. The co-legislators decided to make this aid available only until the end of 2018.

Outermost regions

The EMFF did not change the nature of the existing structural measures for the French, Portuguese and Spanish outermost regions. However, these measures were incorporated into the main structural instrument under shared management, as part of the structural package of the Member states concerned under the CFP.

The financial allocations were increased as compared to the previous period. This was only partially due to the increase in the number of these regions: the French archipelago of Mayotte in the Indian Ocean became a *Département d'outre mer* and since 2014 an outermost region of the EU and was consequently added to the list of territories benefiting from special treatment under Article 349 TFEU. But beyond this change, it was generally recognised that the specific problems of some of these territories would deserve an increase in the existing allocations.

Control and data collection

The level and conditions of funding of these two fundamental instruments dragged on until the end as one of the most difficult issues to resolve in the final negotiation. It should be noted that the Commission proposal to de-centralise the bulk of the aid to control means to shared management (instead of direct management as was the case till 2013) was welcomed as a step in the direction of decentralising the management of EU funds.

As we saw previously, the need to finance the control activities by Member States was agreed by everyone. However, there were strong discrepancies between Council and Parliament concerning the level of financing: while the Parliament advocated a much higher level of spending on control matters, Council advocated a lower figure, on account of the very low absorption rate of the funds managed by the Commission in previous years. Council argued, if funds are not fully used today, increasing the funds will only increase the amount of unused money that could be used elsewhere. As we saw in Chapter 9, in previous years Member States did not manage to spend all the funds available for the purchase of fisheries control means, despite the fact that the Commission had offered up to 90% of Commission's co-financing.

But for Parliament the improvement of control remained a fundamental priority. The solution was to consider as a compromise an intermediate figure but one that in any case represents a considerable increase in the funds dedicated to control (€580 mn). To compensate the problem of low absorption, the co-financing rate was established at to 90%.

In addition, very favourable conditions were established in terms of the eligibility of actions: Member States could use these funds for a wide variety of actions to improve their national control systems. This was of crucial importance in the context of the ex-ante conditionality on control: if Member States have to carry out an action plan to address their administrative capacity before 2016 in order to have access to the EMFF financing, it is essential that the EMFF contemplates as eligible action all those needed by Member States to address their administrative capacity weaknesses.

As indicated previously, the collection of data became an obligation for Member States under the CFP. The discussion, very much like in the case of control, was between two approaches: that of the Parliament, advocating larger amounts for data collection and that of Council, advocating an important but lower amount of funds for this purpose, notably on the basis of relatively weak absorption (though higher than for control) of the funds under the data collection regulation.

At the end, the solution was to increase the budget considerably (up to €520 mn) while increasing the co-financing rate from the former 50% (and from the Commission's proposed 65%) to 80%. Overall, this represents an important increase in the total figure dedicated to data collection.

Another important feature was the consideration of *ex-ante* conditionalities for both control and data collection, among other conditionality mechanisms (see next).

Processing industry

The fish processing industry has been benefitting from the funding of the structural policy for many years. And the Commission has always been favourable to the continuation of such funding. However, this does not necessarily means funding it under EMFF. Indeed, funding this industry has been possible under other structural funds for a long time. So for the Commission the question was: Let us reserve the more limited EMFF funds for other objectives more directly linked to the implementation of the reformed CFP, and let the processing industry be financed through other funds, notably regional funds.

At the end, however, the co-legislators preferred to have a provision for the funding of processing of fishery and aquaculture products in the EMFF, although such provision limited the grant aid mechanism to small and medium enterprises, while large ones would have to be supported through financial engineering mechanisms. Behind this decision lied probably the fear amongst national fisheries administrations that when deciding on the expenditure of the large ESI funds (notably Regional Funds), the processing industry would risk being squeezed out by other economic sectors. It was important therefore to have the possibility of funding the processing industry under EMFF at least as a safety net.

The debate also focused on whether these measures should be limited to small and medium enterprises (SMEs) or also to larger companies. The idea behind the debate was the earlier criticism to the structural policy for financing projects by large multinationals that should not require taxpayers' money to invest. Furthermore, it was argued that even if such aid was desirable in some cases, large companies should not be allowed to squeeze out smaller companies from the access to – limited – funds. However, some Member States in Council insisted that investments by large companies were extremely important to create jobs and that public aid was often a key factor when deciding such investments. To resolve this situation, the co-legislators decided to include the financing of processing industry through grant aids only for SMEs. Larger companies could benefit from financial aid only in the form of financial instruments, as laid down in the common provision regulation for all ESI funds (EU, 2013c).

Social measures

The wide range of social measures and the very favourable conditions established for a number of them are one of the most remarkable characteristics of the EMFF, in yet another demonstration of the importance of the social angle of the CFP.

Start-up support for young fishermen

This measure, fundamentally advocated by the European Parliament, provides aid to young (under 40) fishermen to facilitate their access to the profession by financing them up to 25% of the acquisition of a vessel (up to a maximum of €75 000). The measure addresses the alleged difficulty of young fishermen to enter the profession. In the absence of aid to new constructions, this measure and its attached conditions became a difficult issue.

Promotion of human capital and social dialogue

Under this section, EMFF provides for life-long learning as well as training on a variety of topics. It also includes the promotion of employment for women, as well as the support to the social dialogue between bosses and maritime workers. It should be noted that this section also applies to partners of self-employed fishermen, in order to widen the scope of these measures to include not just to professional fishermen but a more extensive population dependent on the fishing activity in Europe's coastal regions.

Mutual funds for climatic events and environmental incidents

This measure, despite the title, is in reality a social one, to compensate fishermen for their economic losses as a result of unforeseen environmental incidents. In case of such events as oil spillages, these provisions allow for the granting of compensation to all those affected by the stoppage of their activity.

Health and safety

This was, probably, the less contentious part of the regulation; one where everyone agreed on the need to provide continuous support to the improvement of health and safety conditions on board fishing vessels, in yet another demonstration of the point made in Chapter 6: that the social angle of the CFP will continue to be more important than extended wisdom seems to suggest.

Diversification of income

This was a very interesting debate, around a fundamental question: should the EMFF funds be used to invite fishermen to leave the profession? While the co-legislators agreed on the need to provide fishermen with complementary sources of income (while remaining fishermen), important disagreements appeared around the question of diversification measures to encourage fishermen to leave the activity. In particular, a number of MEPs were adamantly opposed to any policy that would be interpreted as 'inviting fishermen to leave the profession'. Therefore, measures for diversification were considered mainly as complementary, not as alternative to the fishing activity.

As we saw in Chapter 6, the structural policy has considered in the past instruments to accompany the retirement of fishermen, in an attempt to accompany the fleet adjustment measures with concomitant social measures for the redundant fishermen. In the new EMFF, very much in line with the new fleet policy that did not consider any kind of compulsory fleet reduction, the emphasis was not to invite fishermen to quit and reduce overall fishing pressure but rather to ensure that fishermen remain fishermen, albeit if necessary as a part time profession, complemented by other economic activities. This is all consistent with the general consideration (see previously) of fleet overcapacity as a secondary problem to the CFP.

Environmental measures

These measures constitute a major evolution of the CFP: never before the structural policy of the CFP contained so many and so varied instruments to deal with environmental questions, thus reflecting the orientation of Europe 2020 explained previously.

Limitation of the impact of fishing on the environment

This measure is intended to contribute to adapt the European fleet to the new obligations of the CFP, notably the discard ban, by financing the adaptation of vessel structure, equipment and gear to fishing methods that will increase selectivity and reduce the amounts of unwanted catches.

This measure is essential in the context of the necessary changes in fishing practice that will be inevitable as a result of the entry into force of the discard ban. These changes will not necessarily be easy in all cases, and the difficulties to tackle are not only technical, as we will see in Chapter 17 but at least everything concerning the technical changes to adapt to more selective fishing (without increasing the overall ability to catch fish) will have a financial support until 2020.

Protection and restoration of marine biodiversity

This measure provides for the financing of a wide number of measures destined to improve the conservation status, or the awareness about it, of marine areas of high diversity. The potential beneficiaries of this kind of measure are extremely wide: from fishermen to governments and even NGOs.

Energy efficiency and mitigation of climate change

As indicated previously this measure contemplates investments with the aim of reducing fuel consumption. In addition to the contentious issue of engine replacement (see previously) this section included investments and energy audits that contribute to improve fuel efficiency. This is a direct way in which EMFF contributes to the overall objectives of reducing CO_2 emissions.

In the negotiation, the promoters of engine replacement and even new vessel construction argued that such measures were necessary to fight climate change, by replacing old, fuel-guzzling engines and old fashioned vessel designs by new ones with a lower environmental impact. However, the evidence of the 2008 fuel crisis, where the European fleets showed a remarkable capacity to reduce fuel consumption without changing engines showed the considerable potential of alternative measures to reduce carbon emissions. Those alternative measures are exactly the subject of this provision.

Inland fishing

This instrument is the best example of the increasing influence of land-locked Member States in the CFP. Never before, the structural policy considered such a wide range of instruments to promote, from all points of view, land-locked activities, ranging from inland fishing to the preservation of aquatic Natura 2000 areas. The measures even allowed for the funding of engine replacement for inland fishing vessels, even though their activity is outside the scope of the management measures under the CFP.

Other measures

Innovation

This provision reflects the initial emphasis of the Commission to dedicate an important part of the EMFF budget to innovation, under the philosophy that the EMFF should be an instrument to promote change, not to preserve the *status quo* and corresponding to the philosophy of the Europe 2020 initiative, which identified innovation as one of the three main pillars of all future development in the Union. Although the adoption of this instrument was not contentious, the addition of a number of new instruments during the trilogue had the effect of diluting the initial dimension of innovation in the original Commission proposal.

In any case, whether or not the innovation angle will be an important part of the implementation of the EMFF, it will fundamentally depend on Member States' choices when preparing their operational programmes. As we saw in Chapter 6, while some Member States want innovation to be the main focus of their structural policy under the EMFF, others are more likely to favour initiatives in the direction of preserving the current order of things. This will fundamentally remain a national choice, only within the overall limits of expenditure by pillar explained previously.

Exploitation of unwanted catches

As we saw previously, the discard ban was one of the most fundamental changes introduced in the CFP in 2013. Obviously, one of the practical problems to resolve in the implementation of this new policy is the use of unwanted catches, landed under the new landing obligation but otherwise discarded on account of their low commercial value. This provision foresees the funding of activities both on board the vessels and on land with the purpose of increasing the added value or opening new market possibilities for the otherwise unwanted fish. It obviously constitutes one of the best examples of how the EMFF was designed to contribute to achieve the goals of the new, reformed CFP.

Development of systems to allocate fishing opportunities

This measure was intended in principle as a reaction to the opposition of the co-legislators to the idea of rights-based management. If such systems were not compulsory, the idea was then to leave them voluntary but to provide aid to those wishing to develop them. However, the opposition to this instrument, both by a number of Member States and a number of MEPs was so radical, that at the end the measure was extended to all systems of allocation fishing rights, so as to avoid any specific mention to the TFCs.

Partnerships industry/science

As we saw in Chapter 10, the integration of fishermen's views into the scientific advice, beyond anecdotal information, requires a more systematic approach whereby fishermen's information can be reliably considered as scientifically valuable data. There are very good examples of how this systematisation can be established through a fishermen/scientists partnership. The EMFF considered the funding of such partnerships as important.

Fishing ports

This measure, in principle, looks like a continuation of traditional structural support to port infrastructure. However, there is an important novelty as compared with earlier structural instrument: that the aid is limited to the improvement of existing infrastructure. This is due to the fact that, in earlier periods, the availability of abundant funds for port infrastructure led, in a number of cases, to a certain 'inflation' of such structures as auction halls and others, many of which are today underutilised. For this reason, the provision specifies that the funding shall not cover the construction of new ports, new landing sites or new auction halls.

Community-led local development

Community-led local development (represented by the Axis 4 of EFF, see Chapter 6) was generally considered as a very promising part of the structural policy and therefore enjoyed ample support from both co-legislators. The only contentious issue was the question of the composition of the Fisheries Local Action Groups (FLAGs). Here, two schools of thought developed: those in favour of fishermen-dominated FLAGs and those in favour of a much wider composition, including non-fisheries interests. The former argued that the instrument was for fishing communities and that a large composition may end up by other interests squeezing fishermen out of the projects. The latter argued that, by enlarging the composition of the FLAGs, these would increase their capacity to tackle a wider range of innovative projects.

At the end, a compromise formula was agreed, to the effect that the FLAGs would 'ensure a significant representation of the fisheries and/or aquaculture sectors', thus leaving the choice to Member States but in a way that the FLAGs would largely correspond to the sectorial interests of the CFP.

Small-scale fishing

This was one of the favourite and most political topics in the negotiation. In particular, the European Parliament made itself a strong advocate of small-scale fishing and took as a priority to ensure that the small-scale sector would not be squeezed out of the access to the EMFF funds, by more powerful and better organised industrial fleets. Council, on the contrary, generally took the view that the problems of the sector are not limited to the small scale and that larger vessels also need access to a number of instruments.

The debate included also a very difficult question on how to define small-scale fishing. The Commission had examined that question and considered a variety of possible definitions. At the end, the Commission concluded that the traditional criterion (vessels of less than 12 m of length overall, excluding trawlers) was the less contentious, in terms of simplicity and applicability. The debate in the trilogue reproduced these considerations and after strong criticism to the Commission for not considering various other alternatives, the co-legislators came to the same conclusion: that any alternative definition would have more problems than advantages.

In addition, a very divisive issue was the question of trawling. Considered by the Parliament as a fishing technique that should be excluded from certain instruments, the Council generally saw no reason to exclude trawling which, in certain Member States, is practised by the bulk of their fleet.

At the end, the compromise was to include larger vessels and trawlers but in exchange of special conditions to ensure that the small-scale fishing would not be squeezed out from the funding under EMFF, as was arguably the case in some Member States under EFF. To do so, three mechanisms were agreed fundamentally upon insistence of the European Parliament:

1. The obligation to include in the operational programme a specific plan for the small-scale fishing sector (applicable only to Member States having more than 1000 of these vessels, to avoid the unnecessary administrative burden for those Member States that have no small-scale sector);
2. The obligation to give priority for 60% of the funds available for engine replacement to the small-scale vessels (applicable to Member States with such fleet segments);
3. The higher rate of co-financing for projects involving small-scale vessels.

Overall, it can certainly be said that the new EMFF allows for a very favourable treatment to small-scale fishing for Member States with significant small-scale sectors.

Integrated maritime policy

In line with the Commission's new emphasis on considering the maritime economy as a whole, requiring specific integrated instruments, the EMFF also included the funding of maritime activities, beyond fisheries and aquaculture. This did not come without controversy, though. In particular, fisheries constituencies were reticent to accept that part of 'their' structural money would be allocated to other beneficiaries, to support non-fisheries activities.

However, Member States in Council accepted this principle relatively easily, although with the idea of using this money to fund two specific questions: the implementation of the Marine Strategy Framework Directive and the development of integrated maritime surveillance.

Conditionality

This question was an important novelty in the fisheries structural fund. As we saw in Chapter 6, one of the problems of the structural policy was that it has always been unconditional: if the expenditure was eligible, it was accepted, regardless of its contribution to achieve CFP goals. This had to change: the coincidence in time between the new policy and the new financial instrument presented a unique opportunity to use the EMMF funds as an instrument for Member States to address and resolve their traditional weaknesses in implementing the CFP.

The question of conditionality was established for the whole of the five structural funds in the context of the 'common provisions regulation'. This took the form of the so-called '*ex-ante* conditionality', which means that the structural funds are available to Member States only insofar as these have the administrative capacity to spend the funds properly and along with the EU priorities.

In addition, the EMFF added two other fund-specific instruments: a specific ex-ante conditionality on certain areas where administrative capacity is key to ensure Member States can fulfil their basic obligations under the CFP and the possibility of interrupting or suspending payments to the Member State on account of their non-compliance with the objectives of the policy.

Last but not least, the EMFF also included provisions on the admissibility of beneficiaries on the basis of their performance record with the rules of the CFP. These three mechanisms make the EMFF the most advanced fund in terms of the conditionality of the public aid.

Ex-ante conditionality

The idea of the *ex-ante* conditionality is intended to work as an incentive for Member States to solve their problems of administrative capacity to handle the obligations of the CFP. In line with the general *ex-ante* conditionality of the Common provisions regulation, the EMFF-specific *ex-ante* conditionality applies to control and data collection and concerns the administrative capacity to exert these basic functions of the CFP.

In the negotiation it was argued that suspending the access to EU funds would certainly not help the Member States concerned to improve these weaknesses, so the mechanism was considered by some as counter-productive. However, the ultimate objective of the instrument is not to suspend funds but to use it as leverage to force Member States to address their weaknesses, by giving them time (until 2016) to do so, with an important contribution from the structural funds themselves.

For control and data collection, Member States should demonstrate that they possess the necessary administrative capacity to exert their obligations in these two areas. In case they cannot demonstrate such capacity at the moment of adopting their national operational programme under the EMFF, they should accept an action plan to acquire such capacity before the end of 2016. In doing so, they can use the – very generous – EMFF funds allocated for that purpose.

It is important to underline that the objective of such conditionality is not to close access to EMFF funds to those member States having difficulties (e.g. related to the financial crisis). Certain stakeholders expressed misgivings about this: if Member States have financial problems to do control and data collection, closing access to EMFF funds will aggravate the problem, not resolve it. However, the philosophy is the opposite: to have access to the funds the Member States concerned should be forced to embark on a plan to address their weaknesses using the EMFF funds to do so.

On the fleet report, the conditionality consisted of obliging Member states to conduct a serious evaluation of their overcapacity problems if they want to benefit from scrapping funds, as a way to ensure that such scrapping is well targeted to the fleet segments where it is really needed.

As regards aquaculture the conditionality was related to the need to ensure that Member States would 'do their homework' in terms of facilitating licensing, reducing bureaucracy and carrying out integrated coastal management as a precondition to ensure that the EMFF funds earmarked for this activity would result in a real increase in sustainable production, unlike in the previous funds.

Conditionality on policy performance

Another important new mechanism is the possibility to interrupt or suspend EMFF funding to a Member State in case of a serious non-compliance with the rules of the CFP. Unlike the *ex-ante* conditionality, this mechanism may be triggered at any time and is not related to administrative capacity but to non-compliance for whatever the reason, giving rise to serious damage to the marine biological resources. This mechanism is established in addition to other existing mechanisms to address poor policy performance with the CFP, in the context of Control Regulation 1224/2009. Together with the latter, it constitutes a very powerful set of instruments at the disposal of the Commission to ensure an adequate compliance level with the future CFP.

Conditionality for individual beneficiaries

But the conditionality has another important angle: that private operators identified as serious offenders to the rules of the CFP should not benefit from the taxpayer's money. This principle was agreed by everyone but its practical implementation required a specific legal instrument, to be adopted by the Commission as a delegated act.

This delegated act was a very difficult exercise. The basic text required that beneficiaries with one infringement would be excluded. Given the high frequency of certain infringements, this would result in a very high percentage of beneficiaries being excluded for funding at least for a year. This seemed out of proportion, as the intention of the legislation was to exclude the big offenders, and not those committing minor faults. Consequently the delegated act had to combine two contradicting objectives: to ensure that the Commission did not establish rules outside of its mandate but to establish rules that would be proportional, and this was done by interpreting that certain minor infringements would only trigger non-eligibility for aid in case of a certain number of accumulated faults.

Investments by the Commission under direct management

Finally, the EMFF also considered an important part of the budget for a number of initiatives that are not amenable to shared management with Member States due to their horizontal nature. These activities therefore remained under the responsibility of the Commission. A total of €647 million was allocated for that purpose. The activities included:

- *Scientific advice*. This budget allows the Commission, *inter alia*, to fund the STECF, to associate the Joint Research Centre in the provision of scientific advice and, crucially, to pay for the services of ICES in providing the advice. In its proposal, the Commission considered also de possibility of having funds to promote directly the enlargement of the human base for fisheries research in Europe, currently rather limited (see Chapter 10). However, the negotiation on budget allocations among different policy priorities between the co-legislators at the end gave priority to control and data collection.
- *Control*. In addition to the €580 million allocated for control under shared management, the Commission will dispose of €123 million for the purchase or renting of control means to be used jointly by Member States as part of joint control programmes. This mechanism is very important as a way to promote sea-basin cooperation among Member States in control. As such, this instrument can contribute directly to enhance regional cooperation and, indirectly, can also help EFCA

by providing funds to help Member States' participation in the Joint Deployment Plans (see Chapter 9).

- *Governance*. This section includes the funding of the new Advisory Councils that replace the Regional Advisory Councils of the previous reform. The funding remained stable as compared with the existing RACs: a maximum of €250 000 per annum. However, in the specific conditions for the functioning of the ACs, in a Commission delegated act (COM, 2014b), the ACs are given the possibility to seek extra funding, beyond the Commission's annual grant.
- *Voluntary contribution to international organisations.* This provision foresees the extra-budgetary contributions to RFMOs, in order to reinforce their activity beyond the limitations imposed in the budgets of these organisations. Typically, these funds are used to pay for research programmes that are crucial to ensure the management of key resources by these RFMOs.
- *Integrated maritime policy.* In addition to the funds allocated under shared management, this provision allows the Commission to finance joint actions, difficult to undertake jointly by Member States, to develop maritime policy.
- *Market intelligence.* This is the funding for the EUMOFA referred to previously.

Overall, the financial package adopted under the EMFF represents, more than ever before, a tailor-made instrument to adapt the European fishing sector to the future CFP.

The allocation of funds

This highly political issue has two components: (1) the allocation of the total envelope of funds for the EMFF among Member States and (2) the allocation of funds among the policy pillars. The two issues had a very different political process in the way they were negotiated. In particular, the first issue was exclusively a matter for Council and the Commission, with the European Parliament remaining neutral. The second, however, was subject to a very difficult negotiation between the two co-legislators, since the Parliament made it a question of principle that certain actions (in particular the reinforcement of control and data collection) be given a higher level of priority in the allocation of funds.

On the allocation of funds among Member States

This exercise, as ever, was politically delicate. The criteria for allocation of the funds as laid down in Article 17 of the regulation were a favourite item for Ministers in the various Council discussions. The main political issue was that, unlike EFF where the allocation was based on cohesion criteria (i.e. the allocation of funds was largely influenced by the level of wealth of the Member States) the new EMFF was proposed as a sectorial fund and consequently the allocation of the financial envelope should be more proportional to the size of the fishing and aquaculture sectors in Member States.

This new development produced significant changes in the allocations: under EFF certain Member States with a low *per capita* income had received large financial packages, in some cases not proportionate with their relatively small fishing sectors. On the contrary, other Member States with high income levels and large fishing sectors had received low financial packages. The new allocation should then rebalance these cases.

This was done by the Commission through a very careful process of discreet negotiations with Member States, particularly with those that would 'lose' funds as compared with EFF on account of their small fishing and aquaculture sectors. At all times, the Commission considered that the Council discussions on allocation criteria were not conducive to a solution, since any agreement to be agreed would still have to be applied in practice through measuring such criteria. This could consume much time and political effort in Council for nothing.

In any case, the political discussion in Council on the criteria of Article 17 continued more as a question of principle than on the basis of their real effect on the allocation of EMMF funds. Interestingly, certain Member States were against the use of the level of consumption of existing funds as a criterion, fearing that this would be a precedent for other funds, where insufficient budget execution (a problem in particular for certain new Member States) could lead to lower allocation of structural funds in the future. Others (typically net contributors to the EU budget) had the opposite position: Member States that cannot manage to execute their funds should not benefit from structural aid at these high levels.

At the end, a practical solution was found while avoiding a hopeless negotiation on allocation criteria. The final allocation of funds (EU, 2014b) was not politically contested by any Member State.

On the allocation of funds among policy pillars

This was one of the most difficult questions in the negotiation. The main discrepancy between the co-legislators was the amount of money to be dedicated to control and data collection. The European Parliament made it a question of principle that a good control was essential for the success of the CFP and, to that effect, more money should be dedicated to support Member States' expenditure on control means. In addition, the improvement of the scientific advice was also a clear priority for the Parliament and they advocated a much higher budget allocation to the collection of scientific data.

Council did not disagree with this position in principle but considered that on the basis of the relatively low absorption of the budget for these two questions in the period 2007–2013 such increase would simply not be spent by Member States.

At the end, an agreement was possible by increasing the amount on control and data collection (though less than initially requested by the EP) but combining this increase with an increase in the co-financing rate, to increase absorption by Member States: data collection would be 80% co-financed by the Commission (only 20% by the Member State) and control 90% (Member States only 10%). Overall, the total expenditure in data collection was increased to €520 million and in control to €580 million. In addition, around €123 million were allocated for control purposes under direct management by the Commission, to finance joint control means shared by Member States on a regional basis (e.g. to buy or lease a patrol vessel to be jointly used to control an fishery or a sea basin).

The regulation distributed the total EMFF budget of €5 749 331 660 for shared management (i.e. for funds to be spent by Member States under their national Operational programmes) among the budgetary Chapters allocated to the different pillars, as follows:

- €4 340 800 000 (75.5%) to the sustainable development of fisheries and aquaculture and fisheries areas, to marketing and processing and to technical assistance;
- €580 000 000 (10.1%)to control and enforcement;
- €520 000 000 (9%) to data collection;
- €192 500 000 (3.3%) to outermost regions;
- €44 976 000 (0.8%) to storage aid; and
- €71 055 600 (1.2%) to integrated maritime policy.

Importantly, these amounts have different meaning: in the case of control and data collection, they are ring-fenced. That means that the EMFF can spend more on these two questions but not less. In any case, transfer between the two remains possible.

The remaining €647 275 400 for direct management by the Commission was allocated (on an indicative basis) as follows:

- Governance of integrated maritime policy, 5%
- Cross-sectorial activities (integrated maritime policy), 33%
- Innovation and new technologies, 2%
- Protection of the marine environment (implementation of the Marine Strategy Framework Directive), 5%
- Support to the scientific advice (STECF, ICES), 11%
- Control and enforced means (shared means among Member States within a fishery or sea basin), 19%
- Voluntary contributions to international organisations (the contribution to the regular budget of RFMOs, like the financial compensation for bilateral fisheries agreements are paid by a different budget), 10%
- Funding of Advisory Councils and communication, 9%
- Market intelligence (EUMOFA), 6%.

The Commission has a room for manoeuvre of 5% between these indicative allocations and may be authorised by co-legislators to change these percentages if and when necessary.

The implementation of this regulation requires a number of delegated and implementing acts that are being developed in 2014 and 2015.

Notes

1 Special report no. 7/2007 on the control, inspection and sanction systems relating to the rules on conservation of Community fisheries resources together with Commission replies.

2 Court of Auditors Special Report 12 – Have EU measures contributed to adapting the capacity of the fishing fleets to available fishing opportunities?

References

Borges, L. (2013). The evolution of a discard policy in Europe. *Fish and Fisheries* Doi: 10.1111/faf.12062.

CEU (2012). Council Conclusions on a Communication from the Commission on the Externaldimension of the Common Fisheries Policy. Council of the European Union. Brussels 1 March 2012 7086/12 PECHE 66.

CMO (2007). Communication from the Commission to the Council and the European Parliament. A policy to reduce unwanted by-catches and eliminate discards in European Fisheries. COM(2007) 136 final. Brussels, 28.3.2007.

COM (2001). Green Paper on the future of the Common Fisheries Policy. Brussels, 20.3.2001, COM(2001) 135 final.

COM (2002). Communication from the Commission to the Council and the European Parliament. On a Community Action Plan to reduce discards of fish. Brussels, 26.11.2002. COM(2002) 656 final.

COM (2009). Green Paper. Reform of the Common Fisheries Policy. Brussels, 22.4.2009. COM (2009) 163 final.

COM (2010). Communication from the Commission. EUROPE 2020. A strategy for smart, sustainable.

COM (2011a). Proposal for a Regulation of the European Parliament and of the Council on the Common Fisheries Policy. Brussels, 13.7.2011, COM(2011) 425 final.

COM (2011b). Proposal for a Regulation of the European Parliament and the Council on the common organisation of the markets in fishery and aquaculture products. Brussels, 13.7.2011, COM(2011) 416 final.

COM (2011c). Proposal for a regulation of the European Parliament and of the Council on the European Maritime and Fisheries fund (repealing Council Regulation (EC) No 1198/2006 and Council regulation (EC) No 861/2006 and Council Regulation XXX/2011 on integrated maritime policy. Brussels, 2.12.2011, COM(2011) 804 final.

COM (2011d). Report from the Commission to the European Parliament, the Council, the European Economic and social Committee and the Committee of the Regions on reporting obligations under Council regulation (EC) No 2371/2002 of 20 December 2002 on the conservation and sustainable exploitation of fishery resources under the Common Fisheries Policy. Brussels, 13.7.2011, COM(2011) 418 final.

COM (2011e). Communication from the Commission to the European Parliament, the Council, the European Economic and social Committee and the Committee of the Regions on the external dimension of the common fisheries policy. Brussels, 13.7.2011, COM(2011) 424 final.

COM (2011f). Communication from the Commission to the European Parliament, the Council, the European Economic and Social Committee and the Committee of the Regions. Reform of the Common Fisheries Policy. Brussels, 13.7.2011, COM(2011) 417 final.

COM (2013). Proposal for a Regulation of the European Parliament and of the Council on the European Maritime and Fisheries Fund [repealing Council Regulation (EC) No 1198/2006 and Council Regulation(EC) No 861/2006 and Council Regulation No XXX/2011 on integrated maritime policy. COM(2013) 245 final and inclusive growth. Brussels, 3.3.2010. COM(2010) 2020 final.

COM (2014a). COM decision of 16.12.2014 setting up the expert group on compliance with the obligations under the Union fisheries control system. Brussels, 16.12.2014 COM(2014) 9643 final.

COM (2014b). Commission Delegated Regulation of 9.10.2014 laying down detailed rules on the functioning of the Advisory Councils under the Common Fisheries Policy – COM(2014) 7012 final.

COM/SEC (2009). Communication from the Commission to the European Parliament and the Council - Building a sustainable future for aquaculture - A new impetus for the Strategy for the Sustainable Development of European Aquaculture {SEC(2009) 453} {SEC(2009) 454} COM/2009/0162 final.

Davidse, W.P., coordinator (1997). Property rights in fishing. Effects on the industry and effectiveness for fishery management. Dienst Landbouwkindig Onderzoek. Landbouw-Economisch Instituut. *Onderzoekverslag* **159**: 329 pp.

EC (2000). Council Regulation (EC) No 104/2000 of 17 December 1999 on the common organisation of the markets in fishery and aquaculture products. *OJ* L 017, 21.1.2000, p. 22.

EC (2003). Council Regulation (EC) No. 1954/2003 of 4 November 2003 on the management of the fishing effort relating to certain Community fishing areas and resources and modifying Regulation (EC) No 2847/93 and repealing Regulations (EC) No 685/95 and (EC) No 2027/95. OJEU L 289 of 7.11.2003, pp. 1–7.

EC (2008). Council Directive 2008/56/EC of the European Parliament and of the Council of 17 June 2008 establishing a framework for community action in the field of marine environmental policy (Marine Strategy Framework Directive). *OJ* L 164, 25.6.2008, p. 19–40.

EC (2009). Council Directive 2009/147/EC of the European Parliament and of the Council of 30 November 2009 on the conservation of wild birds. *OJ* L 20, 26.1.2010, p. 7–25.

EC (2012). Guidance on Aquaculture and Natura 2000. Sustainable aquaculture in the context of the Natura 2000 Network. European Commission. *Environment* 2012: 89.

EEC (1992). Council Directive 92/43/EEC of 21 May 1992 on the conservation of natural habitats and of wild fauna and flora. *OJ* L 206, 22.7.1992, p. 7–50.

EU (2013a). Communication from the Commission to the European Parliament and the Council – Guidelines for the analysis of the balance between fishing capacity and fishing opportunities according to Art 22 of Regulation (EU) No 1380/2013 of the European Parliament and the Council on the Common Fisheries Policy.

EU (2013b). Regulation (EU) No 1379/2013 of the European Parliament and of the Council of 11 December 2013 on the common organisation of the markets in fishery and aquaculture products, amending Council Regulations (EC) No 1184/2006 and (EC) No 1224/2009 and repealing Council Regulation (EC) No 104/2000. *OJ* L 354, 28.12.2013, p. 1.

EU (2013c). Regulation (EU) No 1303/2013 of the European Parliament and of the Council of 17 December 2013 laying down common provisions on the European Regional Development Fund, the European Social Fund, the Cohesion Fund, the European Agricultural Fund for Rural Development and the European Maritime and Fisheries Fund and laying down general provisions on the European Regional Development Fund, the European Social Fund, the Cohesion Fund and the European Maritime and Fisheries Fund and repealing Council Regulation (EC) No 1083/2006. *OJ* No L 347, 20.12.2013, p. 320.

EU (2014a). Regulation (EU) No 508/2014 of the European Parliament and of the Council of 15 May 2014 on the European Maritime and Fisheries Fund and repealing Council Regulations (EC) No 2328/2003, (EC) No 861/2006, (EC) No 1198/2006 and (EC) No 791/2007 and Regulation (EU) No 1255/2011 of the European Parliament and of the Council. *OJ* L 149, 20.5.2014, p. 1.

EU (2014b). Commission Implementing Decision of 11 June 2014 setting the annual breakdown by Member State of the global resources of the European Maritime and Fisheries Fund available in the framework of shared management for the period 2014–2020. (2014/372/EU). *OJ* L 180, 20.6.2014, p. 18.

EU (2015). Commission Delegated Regulation (EU) 2015/242 of 9 October 2014 laying down detailed rules on the functioning of the Advisory Councils under the Common Fisheries Policy. *OJ* L 41, 17.2.2015, p. 1.

European Parliament (2012). European Parliament resolution of 22 November 2012 on the external dimension of the Common Fisheries Policy (2011/2318(INI).

FAO (2005). FAO Fisheries Technical Paper no. 470. FAO, Rome, 2005.

Johnsen, J.P. and S. Eliasen (2011). Solving complex fisheries management problems: what the EU can learn from the Nordic experiences of reduction of discards. *Marine Policy* **35**: 130–139.

SEC (2006). Discards from Community vessels. STECF Plenary meeting, Ispra, 6–10 November 2006. SEC (2006).

SEC (2007). Evaluation of closed area schemes. STECF Plenary meeting. Ispra, 5–9 November 2007. SEC(2007).

SEC (2010). Commission Staff Working Document. Synthesis of the Consultation on the Reform of the Common Fisheries Policy. Brussels, 16.4.2010 SEC(2010) 428 final.

SEC (2011). Commission Staff Working Paper. Impact Assessment accompanying Commission proposal for a regulation of the European Parliament and the Council o the Common Fisheries Policy (repealing Regulation (EC) No 2371/2002). Brussels, 13.07.2011, SEC(2011) 89.

CHAPTER 17

What's next?

Implementing the new CFP: a daily affair

The CFP, as probably any other policy, will not be judged on its theoretical basis, but rather on its implementation in practice. Once the new legal texts have been adopted, the real measure of the future success of the CFP will lay on its implementation. A new legal basis does not necessarily translate into real policy change particularly as regards the use of voluntary instruments, such as regionalised approaches. New rules established in the policy, such as the discard ban or the conservation policy geared to achieving MSY, will only be successful if properly implemented and this requires both political will and sufficient means from all involved.

There are two levels of implementation that are crucial for the success of the policy:

1 The effective implementation of the compulsory objectives of the new policy (e.g. the MSY objectives and the discard ban); and

2 The practical implementation of the voluntary mechanisms (e.g. the regionalisation, or the development of aquaculture).

The compulsory implementation of the specific obligations has a clear dimension of control and enforcement. But not only that: effective implementation must also be feasible in the real world and a number of factors may contribute to effective implementation, in the form of incentives, financial assistance and so on, to make obligations easier to comply with. In some cases, the objective itself to be achieved is defined in a rather loose way, thus requiring additional guidance as to how these objectives can be achieved and verified.

The implementation of voluntary instruments is a different issue altogether. The experience of previous reforms shows that voluntary mechanisms, or those that are very loosely defined at the end, tend not to be implemented effectively. As we have seen, certain ideas have had a legal basis in the CFP as a voluntary or a possible mechanism for a relatively long time but they have not been implemented in practice. The best example is probably that of the multi-species management plans. Here, the question is what can we do to favour the practical implementation of these instruments and what kind of incentive would be necessary to turn the possibility into a reality.

Finally, certain objectives have been achieved gradually even in the absence of a legal basis: the gradual implementation of the MSY objective, even without a specific legal basis in the 2002 reform is the best example. So the question here would be: what are the possible issues that, over and beyond the obligations (or open possibilities) of the basic regulation, are likely to become important issues for the next 10 years?

Let us examine these questions.

The Common Fisheries Policy: The Quest for Sustainability, First Edition. Ernesto Penas Lado.

Implementation of the MSY objectives

This fundamental objective of the new policy also poses a number of interesting challenges, both from the point of view of the delivery date and from that of the way to apply the objective in mixed fisheries and on the scientific advice to underpin the solutions.

MSY in 2015 or 2020?

The new basic regulation, 1380/2013 does not specify when it will 'not be possible' to deliver the MSY target in 2015, so such target could be delayed to 2020 at the latest. One thing is clear, however, in the intention of the legislator: the insufficiency of scientific advice should not be a reason to delay the delivery of the MSY target. If this was the case, then this would act as a negative incentive to ensure that the scientific information becomes available. In addition, the recent development of 'proxies' to F_{msy} (see next) removes the pretext that the objective cannot be delivered because it is not known.

One can then suggest that the degree to which the MSY target can be delayed from 2015 to later years, up to 2020 would be fundamentally the distance from the current levels of fishing mortality and the F_{msy} objective. In other words, the farther away from the F_{msy} target a stock is, the more it would be justified to provide extra time to reach the objective, in order to secure a gradual approach in reaching it. The Regulation, in any case, is silent on this point, so in practice it will be in the context of multiannual plans that this delivery date can be fixed, on the basis of the criteria established by the co-legislators.

MSY proxies

Another challenge for the implementation of the MSY objective is the case of data-limited stocks, in which the F_{msy} may be difficult to estimate. Although the situation has improved in recent years, still a significant number of European stocks don't have a well estimated F_{msy} level. This implies the need to develop proxies to this parameter. Here, the scientific community has indeed developed interesting alternatives that can be used.

The scientific community has dealt with this issue and has developed proxies to be used in case of data-limited stocks. ICES, in particular has recently reviewed methodologies for that purpose (ICES, 2014). In the USA, a similar development has also taken place (Newman et al., 2015).

A second question is that, in 2015, many stocks are still not be covered by a multiannual plan, where the long term objective (expressed in terms of a range of F_{msy} to be achieved by an end date) is spelled out. This means that in a number of stocks, the only instrument available to advance towards the MSY objective will be the annual TACs.

In this regard, the introduction by ICES of the advice on the 'MSY transition', whereby TAC projections are provided that would result in MSY being achieved in several years.

MSY in multi-species fisheries

A third question is the implementation of F_{msy} in mixed fisheries. As we saw in Chapter 15, the solution of the co-legislators is that the multi-annual pans will contain 'specific alternative measures' to deal with species in the mix that require special protection on the basis of their poor conservation status, at least as compared with the other species in the mix. This is also a very important challenge, particularly for those cases where the species to be protected is very difficult to separate from the others in the catch.

Certain avenues will have to be explored in this regard. Selective trawls, to develop further the species selectivity of the gear can be one solution. Closed areas, as far as the different species associated will have slightly different areas of occurrence, can also be used.

This will be a very difficult task, in particular since the specific measures to protect the most threatened stock while pursuing MSY-level fishing to the other stocks in the fishery will not necessarily be the same for different fleets, using different gear. The alternative of regulating different alternative fishing practices in detail would again increase detailed management and legislative complexity. This area, in fact, seems ideal to test out alternative approaches based on results-based management under regional approaches.

It must also bear in mind that multi-species approaches require a very high level of scientific knowledge on technological and ecological interactions among the main species associated in a given fishery and such knowledge is far from complete, even in the best studied areas such as the Baltic or the North Sea.

Implementing the discard ban

This is going to be one of the first and most important challenges for the implementation of the reformed CFP. The CFP has established a new obligation but has not necessarily removed all the sources of discarding in EU fisheries: relative stability is the best example. Good implementation will require a number of initiatives. Firstly, to modify the current legislation to make it compatible with the discard ban and secondly, to prepare the secondary legislation establishing flexibility rules.

The necessary changes in current legislation: the 'omnibus' proposal

Existing legislation had to be adapted to avoid conflicting or contradictory provisions regarding the obligation to land all catches. The clearest example is the treatment of fish under the minimum landing size under technical measures regulations. While previous legislation obliged fishermen to throw back to the sea all the fish under that minimum size, the landing obligation will oblige them to land that fish. The previous provisions on the minimum landing sizes must therefore be adapted to that effect.

Another example is the current structure of technical conservation measures, where mesh sizes are related to specific catch compositions: if the latter do not correspond to those admissible for the mesh size being used, some fish must be discarded. This also required a modification to ensure that no contradiction exists in legislation when the discard ban enters into force in a given fishery.

The Commission proposal to eliminate these inconsistencies was difficult to agree by the co-legislators, to the point that it could not be adopted on time: by 1 January 2015, when the discard ban entered into force for certain fisheries, this regulation had not yet been adopted, and the negotiation to iron out the last discrepancies between Council and Parliament was finalised in January 2015. That means that during the beginning of 2015, the discard ban had to be implemented on the basis of a 'common sense' approach on how to interpret the inconsistencies between the old and the new legislation.

The regulation was finally published in May 2015 (EU, 2015).

Improving selectivity: the next generation of technical measures

An obvious contribution to facilitate the discard ban is the revamping of technical conservation measures, to improve both size selectivity and inter-species selectivity. Indeed, the Commission has argued that the rules applicable to the discard ban should concern the 'residual' level of unwanted catches that would result from the reduction of discarding through changes in fishing practice. Everyone agrees to the principle that the best way to avoid discarding an unwanted fish is to avoid catching it in the first place.

The Commission intends to make a new proposal on technical measures, along the lines of the new, regionalised, results/oriented philosophy of the 2013 reform. The role of technical measures to reduce discarding is well demonstrated by science. However, the experience gathered also shows something very important: technical measures that increase selectivity can result in the short term in lower catch rates, and thus lower income. This constitutes a disincentive for their effective implementation. Therefore the important potential role of technical measures to reduce discard levels is not just a matter of adopting new legislation but fundamentally about finding the right incentives for fishermen to see an advantage in their application (Enever et al., 2009).

Technical measures are traditionally easy to circumvent by fishermen. There are countless ways of neutralising the effects of gear with a theoretically high selectivity. Ultimately, much of the further selectivity needed can be achieved through fishermen's behaviour and practice, something that by its own nature cannot be legislated. Finding the right incentives for fishermen to increase the level of selectivity of their fishing will remain a key factor to ensure effective implementation of the discard ban. In this respect, the EMFF will provide financial support for the development and acquisition of more selective fishing gear.

Preparing the future rules

The idea of the discard ban is that the catch of unwanted fish must be reduced to the maximum through selectivity. Then, the flexibility measures under the discard ban must be applied to the residual catches of unwanted fish, once the effort on the improvement of selectivity has been done. This implies a revamping of the technical measures, in a way that will also deliver other objectives of CFP reform: the new technical measures should increase selectivity, with a special emphasis of regionalised, results-based and flexible approaches.

Another fundamental question will be the implementation of the flexibility rules adopted in the 2013 reform and, in particular, the ways to apply the *de minimis* rule, the exemption of species with a high survival rate to discarding and so on.

The intention of the co-legislators was to ensure that the details about the implementation of the discard ban be included in the long-term management plans (LTMPs). However, given the difficulty to ensure that all fisheries be covered by such plans by the time the discard ban enters into force led to the consideration of the 'discard plans' with a duration of 3 years. These plans must contain a number of elements that may not always be necessarily simple:

- Definition of the specific fisheries covered by the discard ban. The basic regulation defines these fisheries by the 'leading species' and the area. But identifying what is, for example, a 'pelagic fishery' in cases of demersal fishing catching large amounts of pelagic fish, can be an example of the kind of problem to be clarified;
- The definition of the species that can survive to the discarding. This will be difficult notably because of incomplete or contradictory data, the considerable diversity of situations (making it very difficult to give a straight yes/no answer to the basic question) and so on;
- The definition of the level of *de minimis* allowance to discard within the limits established in the basic regulation. This will be of a particularly difficult nature.

The challenge is two-fold: providing the knowledge base for such decisions and ensuring the effective involvement of Member States and stakeholders in preparing these decisions under regionalisation.

The first generation of discard plans were adopted in December 2014, for the Baltic (EU, 2014a), another for certain pelagic fisheries in South Western waters (EU, 2014b), for certain small pelagic fisheries and fisheries for industrial purposes in the North Sea (EU, 2014c), for certain pelagic fisheries in North Western waters (EU, 2014b) and for small pelagic fisheries in the Mediterranean (EU, 2014b). Some lessons can be drawn from this experience:

The plans were developed on time. Member States wanted to demonstrate that the regionalisation process behind the preparation of the plans would work in practice and national administrations invested heavily in ensuring that the necessary agreements were struck. This required in some cases relatively high political intervention.

Overall, the discard plans showed already the difficulty of balancing out regionalisation and harmonisation: the *de minimis* rules are different rules for different plans. A pelagic vessel crossing the lines between zones affected by different discard plans may, towards the second half of the year and depending upon the availability of quotas for choke species, be subject to different rules on the amount of fish they can actually discard under the *de minimis* rules.

The problem of 'choke species'

This problem was actually raised by a few Member States in the political process leading up to the adoption of the reform. However, as we enter into the implementation phase, the problem is being raised by more Member States, albeit on a different name: what Spain considered in the negotiation a problem of 'relative stability' other Member States now call a problem of 'choke species'. Both questions respond to the same problem: how can a Member State cope with the discard ban in areas where their national quota of one stock is very low, or even zero, while their quotas of other species are significant? Will they have to stop fishing when the quotas of the 'choke' species are exhausted? The problem is two-fold:

1 In some cases the TAC levels in the area are relatively consistent with the catch composition but the relative stability is not. That means that the TACs of two or more species can be exhausted more or less simultaneously during the year. However, what applies to the TAC does not apply necessarily to the national quotas, which, in certain cases, are very unbalanced for a given Member State.
2 In some other cases the problem is not so much relative stability but very different TAC levels for stocks that are caught in similar amounts.

In the first case, the most obvious solution to this problem is quota swapping but on a very different basis as compared to the traditional practice since 1983: instead of exchanging target species, Member States would now need to exchange by-catch species for which they have zero or low quotas. Since this mechanism is voluntary, it remains to be seen whether the discard ban will produce an altogether different 'market' of fishing rights, particularly as a function of Member States' availability of large quotas of less interesting species to be used as 'currency' for the exchanges.

A question is whether the need for this 'currency' to exchange quotas may lead to a tendency in Council to 'push TACs upwards' against scientific advice. Clearly, this is another example where the implementation of the discard ban should not be made at the expense of the overall MSY objectives.

In addition to swapping, the challenge of choke species can be handled through the flexibility mechanisms considered under Article 15 of the regulation: the *de minimis*, the counting of some catches to the quotas of other species and so on. But only the experience will tell if these mechanisms are enough to resolve the problem.

In case these mechanisms are not sufficient in some cases, the political debate may lead in the future to consider the real need an opportunity to have certain species under the TAC and quota system at all. For example, certain secondary species having TACs (a typical example is dab and flounder in the North Sea) are secondary and do not determine fishermen's behaviour, which is a function of the main target species for the flatfish fishery. Is there really a conservation value in having a catch limitation for dab and flounder in the context of the discard ban? Other examples can be mentioned in other areas. This is a delicate matter since it touches upon relative stability. It is therefore for the Council to evaluate if it is worth having TACs for certain secondary species under TACs and relative stability without a real conservation value but with negative implications on the implementation of the discard ban.

On the second question (the different levels of TAC based on single-stock considerations) the solution may come from the multispecies approaches (including Fmsy ranges), where the fishing levels for different species will no longer be looked at from the perspective of the single stock but as a more coherent level of fishing to achieve fishery objectives.

Facilitating the discard ban

Implementing the new provisions on the discard ban will not be simple. A number of initiatives must be put into place to facilitate this implementation.

Another fundamental question to facilitate the implementation of the discard ban is the need, by Member States, to change the way they handle the traditional 'quota swaps'. While these have traditionally focused on obtaining more of the commercially interesting species for Member States, now the implementation of the discard ban will force them to consider swaps also to complete their respective 'basket of fishing rights' so as to minimise the inevitable discarding. The Commission must stay neutral, since these quota swaps are voluntary: they are done by Member States and communicated to the Commission. However, the Commission can play a key role by providing data on the catch composition of different fisheries so as to highlight to discrepancies between the fishing rights under relative stability and the actual catch composition of the fleets.

Another question to develop is the role to be played by the new financing instrument, the EMFF, to facilitate the necessary behavioural change that should necessarily accompany an effective implementation of the discard ban. Adapting to the discard ban will require in many cases adaptations to the storage facilities on board, for vessels to be able to cope with catches that are today discarded. But, crucially, it will involve the development of storage facilities in ports, for the fish that will not be sold for human consumption, as well as possible new markets for these fish.

It is clear that the 'nonsense' of a policy where large amounts of fish were discarded could not be changed for another possible nonsense: that of fish arriving to ports and rotting for lack of economic use. This is something that represents a huge challenge: what to do with the hitherto unwanted fish in European ports and very particular in small scale ports that are far away from industrial areas where there can be easier alternatives for the unwanted fish, such as fishmeal or pet food factories. Finding practical solutions to this problem is crucial to the image and credibility of the policy.

Among other solutions, one can find some inspiration in certain projects developed under the Axis 4 of the European Fisheries fund. One project, for example, has successfully developed the preparation of croquettes to feed aquaculture fish, made from broken fish usually not sold in auction markets. But many other ideas are also developed or being developed. Clearly, there is no one-size-fits-all answer and a variety of solutions will have to be developed in the coming years. As we saw in Chapter 16, the EMFF allows for the co-financing of these initiatives.

Controlling the discard ban

This is one of the most difficult challenges for the future implementation of the discard ban. The control of this kind of measure is widely recognised as difficult. In fact, the only theoretically effective way to check the implementation of this measure is through direct observation of the activity of the vessels. Other methods, such as the comparison between landings of a vessel with the landings of another vessel considered a 'reference', can be useful to provide insight into possible offenders but can hardly be a solid proof of lack of implementation.

If the best solution is direct observation, there are fundamentally two models available: the use of observers on board and the fully documented fishery through CCTV cameras.

The first option is expensive and only viable for relatively large vessels. It could be a solution for only a limited number of fisheries. The second option can be a solution for certain fisheries (where they are already applied) but it is difficult to see their extension to all fisheries, both for technical reasons (in very complex and diverse fisheries the system is not efficient, at least for now) and political ones (the CCTV cameras are considered by some Member States as being incompatible with constitutional rights of privacy).

In any case, given the nature of the decision on the discard ban, and its entry into force for a high number of species without the benefit of a gradual learning process, it seems clear that solutions will not be necessarily ready-made from day one. It has been argued that in the experience of Nordic countries, a strong and rigid control and enforcement regime can undermine the legitimacy of the system and be counterproductive in regard to reducing discard, and that participation of those most concerned by the rules is important for the legitimacy and functionality of the system (Johnsen and Eliasen, 2011).

Discard ban and conservation policy: a new paradigm

As we saw in Chapter 15, there is an obvious relationship between the implementation of the discard ban and the way in which new TACs will be fixed. Theoretically, under the discard ban TACs should no longer be 'landing quotas' but 'catch quotas' and this means that for stocks subject to the discard ban, it would be justified to adopt 'quota uplifts' to compensate for the obligation to land the fish hitherto discarded (and killed anyway). This opens up a new challenge: providing the necessary knowledge base for this new approach.

In theory, it is easy to anticipate the change: new TACs under the discard ban should be 'fishing mortality neutral'. That is, the new TACs should be fixed in a way that the estimated level of discards will be included in the total allowable catch, so the total amount of fish landed will *de facto* be higher but the total amount of fish killed by fishing will remain the same. This is, however, much easier said than done, for at least two reasons:

1 The knowledge of real discard levels is very incomplete and often unreliable;
2 The real level of discards is not necessarily constant: fishermen can change fishing practice in a way that will reduce discards. A TAC based on earlier estimations of discards may actually result in an increase in fishing mortality if fishermen fish in a different way than previously.

But beyond TAC fixing, there will be other important consequences for the conservation policy: its effects on the minimum conservation reference sizes (MCRS).

As these will determine whether the fish can be sold for human consumption or not, the establishment of these MCRS will inevitably have to take into account parameters other than the age of maturity of the fish and the selectivity of the gear.

In fact, the higher the MCRS, the higher the amount of fish that will not be sold for human consumption and the higher the difficulty to comply with the discard ban. Very ambitious MCRS will require very high selectivity (something that may be economically difficult to assume) and may make the discard ban more difficult to enforce. On the other hand, lower MRCS will allow for a better enforcement of the discard ban but then one can question whether this could lead to the further development of fisheries for small fish, which in certain markets can fetch good prices for human consumption.

The situation can be made more complex by the development of regionalisation, which contemplates the possibility of regionally decided MCRS. While regionalised MCRS can be a good opportunity to better adapt CFP rules to specific regional conditions, it also carries a risk of establishing very different rules in neighbouring areas, making it more difficult to understand the policy and possible creating the impression of lack of 'level playing field'. Striking a good balance between regionalisation and 'level playing field' will constitute a major challenge for the practical implementation of this policy in the years to come.

It is important, again, to ensure that the new policy does not lead to a possible negative impression in the public opinion: if and when the regionalisation process leads to lower (or no) MCRS, there is a risk that the public perception of the new policy will be that, under the old policy, nobody could catch and sell small fish; but now it is legal to do so. This could be a – superficial – impression by the general public that would be extremely difficult to counter. This will certainly require very careful explanation of the reasons why this situation would be possible.

Regionalisation

This question poses a particularly difficult angle: unlike the former issues, regionalisation is not compulsory. So the question here is, being voluntary for Member States, should the Commission promote it actively? Or should it leave it for Member States to do it just as much as they want?

The answer is not straightforward. In principle, if we consider regionalisation as the best way to avoid the 'micromanagement from Brussels' and to bring decision making on technical matters closer to the fishing grounds, the Commission should actively pursue this process. However, given the strong dependency of this process on good cooperation among Member States, it is this level of collaboration (very variable among different sea basins) that will probably indicate the extent to which the future policy can rely on regionally adopted measures.

This cooperation is not just a question of national administrations. In particular, if the new law making under regionalisation is to deliver its main objective (getting decision making closer to the stakeholders and to the fishing grounds) then a crucial aspect of cooperation will be that within the Advisory Councils. Whether the fishing industry and other interests are capable of working together to make proposals in a proactive manner, or whether they will sit back and wait to be consulted by national administrations at some stage, that will be the decisive in determining whether the CFP will effectively get closer to the fishing grounds, or whether it will simply replace the 'Brussels centralism' by a brand new centralism of the capitals.

Enhancing regional cooperation

The first challenge is for Member States to establish an adequate environment for cooperation. And this requires a level of trust that, in many cases, is not yet there. Cooperation is particularly good in cases where the fisheries take place in a limited-access regime where only coastal States fish in the area. For this reason, the cooperation in the Baltic is by far the most advanced as only Baltic Member States are fishing there. In addition, the Member States concerned have a long tradition of regional cooperation under the old IBSFC, before the 2004 enlargement. Last but not least, the role of third States, in this case Russia, is relatively minor, as corresponds to the relatively small EEZs of the Russian Federation in this area.

A similar case of good cooperation is the North Sea where, again, only coastal states fish. In this case, however, the intervention of the third State in question, Norway, is much more crucial since there is an old tradition of regulating the numerous shared stocks together. Combining the internal procedures for regionalisation with the necessary negotiation with Norway (which involves exchanges of fishing quotas outside the North Sea) is probably the biggest challenge in this area.

In other areas of Union waters the cooperation will be more difficult. The north-western waters and the Mediterranean are the two clearest cases. In the western waters area there is still an important tension between the coastal Member States and the other Member States (notably Spain and France) that have long-distance fleets operating in the area. This makes cooperation difficult at times.

In the Mediterranean, the cooperation is made very difficult by at least three factors: (1) the extraordinary diversity and complexity of this basin, (2) the importance of non-EU States and (3) the existence of important political conflict between a number of countries. To overcome these difficulties, it will probably be inevitable to go sub-regional. In other words, to start with the cooperation in sub-areas of the Mediterranean where the main coastal states are EU Member States and have no serious political conflict with third countries. The best example is the Adriatic, a sea largely covered by EU members or candidate countries and where no major maritime conflict exists. Another example can be the Gulf of Lyons, shared by three EU Member States without any significant conflict.

The first examples of regional cooperation have been very encouraging: the need to present discard plans before the end of 2014, to establish more specific rules for the implementation of the discard ban on pelagic fish as of 2015 have been successful: in all sea basins where discard problems exist, Member States have indeed been able to present to the Commission agreed proposals for discard plans. These were not always easy to negotiate but ultimately national administrations were able to find agreements and the Commission could subsequently adopt these plans as delegated acts.

An important factor to make regionalisation a success is, in any case, the establishment of an informal dialogue between Member States and the Commission concerning the preparation of proposals under regionalisation. As we saw in Chapter 16, the regionalisation process gives the Commission two choices: accept the proposal from Member States to adopt it as a delegated act, or reject it and make a proposal for co-decision, but the Commission cannot 'cherry-pick' the parts of the proposal that it agrees with and forget the rest.

This 'black or white' approach can appear too rigid and perhaps conducive to considerable frustration if and when the proposal from Member States is not fully consistent with EU law or with policy objectives. The solution to this could come from the previously mentioned informal dialogue: the Commission could indicate possible difficulties with the draft measures before they are formally adopted, so after this informal consultation Member States could re-submit proposals that they knew would find no obstacle from the Commission.

Regionalisation versus harmonisation

A second element that will influence the implementation of regionalisation is the fact that, although regional decisions only affect Member States directly involved, the tradition of the CFP makes Member States fear that whatever is accepted in one area may end up being imposed elsewhere.

The challenges here are already evident in the case of the initiative to advance the discard ban policy in the fisheries in Skagerrak. In this case, the Member States involved in the fishery (in particular Sweden and Denmark), together with Norway, took the initiative to advance the discard ban. In doing so, they anticipated the regionalisation of the policy.

When the Commission proposal following the recommendations prepared by the concerned Member States was presented in the Council, a number of Member States without any direct interest in the Skagerrak fishery opposed some of the proposed measures, either because they believe that the specific measures for the Skagerrak would break the 'level playing field' among fishermen of different Member States, or in some cases because they did not want to create precedents that may at the end be extended to other areas where they fish. The latter was clearly the case of the possible compulsory use of CCTV cameras to ensure 'fully documented fisheries'.

Beyond the question of the establishment of precedents, there is a traditional contradiction in the way stakeholders see regionalisation: they want it when it means that they can adapt the rules to their particular local conditions but when the same is done for neighbouring waters, those crossing the lines complain about lack of harmonisation. This tension between regionalisation and harmonisation is as old as the CFP itself (Lequesne, 2001), and the implementation of the discard plans in 2015 already showed that it still exists in the new policy scenario: the fact that the different regional discard plans had different rules in neighbouring areas was resented by the industry fishing in different areas as an unnecessary complexity.

In that vein, the most far reaching effect of regionalisation may be a possible breach of harmonised approaches on certain policy areas. The best example is perhaps that of the future development of minimum conservation reference sizes (MCRS).

In fact, these can be adapted as part of regional approaches to facilitate the discard ban. It is obvious that adapting MCRS to local conditions (either by establishing region-specific MCRS, or to suppress them altogether for certain species) may be a powerful instrument to facilitate the implementation of the discard ban.

However, this flexibility comes at a risk: that the regionalisation may lead to a multiplicity of different MCRS for the same species in different European regions. The risks associated with this scenario are varied:

- greater difficulties for controlling the transport and the markets, requiring very sophisticated traceability systems to judge whether a fish of a given size has been sold for human consumption legally or illegally;
- creating an image of the CFP as having 'different standards' in different areas, giving rise to allegations of lack of level playing field or even discrimination;
- making this policy impossible to understand by normal citizens that want to make intelligent choices on what fish to buy: it may become extremely difficult to ascertain if a given fish in a market or a restaurant is legal or illegal, depending upon the area where is caught;
- adding difficulties to the possibility of applying MCRS to imported fish. If a fish has many different MCRS (or even no MCRS in certain areas) in different European areas, which MCRS is applied to imported fish?

It can certainly be argued that the problem is not new: already in the pre-reform policy there were different minimum landing sizes for the same species in some areas, particularly in the Mediterranean as compared to the Atlantic. This can cause considerable difficulties already (see Chapter 4). However, the problem could eventually multiply under regionalisation.

Beyond these possible effects, the main challenge for regionalisation is the possible inertia of the current system. Regionalisation means that there may be different rules in different regions, and this may be resented as a lack of a 'level playing field', a principle that many Member States are adamant to preserve.

In addition, regionalisation also means a certain level of uncertainty as to how the final technical rules (to be agreed by Member States) will look like. The search for certainty, another traditional characteristic of the CFP, may play against a process that cannot always provide the certainty of the rules agreed in Brussels. Last, but not least, some of the fears expressed about regionalisation (see Chapter 16) in terms for example of large national administration dominating small ones may reappear in the decision making process, and may limit the scope and ambition of what Member States may be ready to leave for the regionalisation process.

The role of stakeholders

As the CFP moves towards regionalisation, a new challenge emerges: the role of stakeholders in this process. The basic regulation requires the future Advisory Councils to be consulted. This compulsory consultation, however, does not clarify how and what moment it will take place. Nor does it establish any obligation by Member States to take their advice into account.

In that context, one can consider two possibilities: (1) that Member States will try to build their regional cooperation on the ACs and try to associate them to the process from the beginning, or (2) that they will cooperate at the level of national administrations alone and then consult the ACs at the end of the process. This will be a crucial choice for national administrations, in particular to see if the regionalisation process will simply take decisions 'away from Brussels' or indeed 'closer to the fishing grounds'. These are two different concepts and the basic regulation does not provide guidance on the approach to take. So the choice is for Member States to change a policy traditionally seen as 'micromanagement from Brussels' to another one of 'micromanagement from capitals' or else to change this to a policy with a higher ownership by the stakeholders themselves.

For stakeholders too this will be a fundamental challenge. The new Advisory Councils replacing the RACs will still have to do their more traditional work of making contributions upstream of the decision-making process but they will also have to provide advice on the specific regional measures foreseen under the new basic regulation. This represents a major change in the way they work: from reactive to a Commission proposal, they will have to be proactive to make proposals. In addition, in some cases they may have to provide a substantial part of the confidence building that is a pre-condition for good regionalisation. As we saw in Chapter 12, certain RACs have contributed substantially to build trust and reduce conflict among fishermen of different Member States, in a more effective way then their national authorities. The confidence-building necessary to make regionalisation work can one day heavily depend on the ability of stakeholders to understand each other even beyond their own national administrations.

For decades, the CFP has been seen as a top-down policy, where the interests of stakeholders were not sufficiently taken into account, thus leading to a lack of 'buy-in' of the industry into the policy. This has been repeated so many times that has become accepted as a fact. Since 2002, the CFP has enjoyed a system a consultation with stakeholders through two types of bodies: ACFA and the RACs. I have referred to the experience of these over the last few years in Chapter 12.

Despite the important progress in this regard, thanks in particular to the work of the RACs, the accepted view remains. Now that the policy has gone a step forward, by allowing for decisions being originated by Member States under a regionalisation framework and crucially with stakeholder consultation, the question is, will this new bottom-up approach be fully taken up by the industry? And will national administrations be open to this?

The very recent experience of the first discard plans for pelagic stocks (see previously) gives us a first impression about these questions, and the picture that emerges is very mixed: in the case of the North Sea, the discard plan presented by Member States under regionalisation largely ignored the advice from the Pelagic Advisory Council, pointing to a model where regionalisation is led by national administrations. In contrast, the discard plan for the Mediterranean followed the recommendations of the Mediterranean Advisory Council, setting the tone for a stakeholder-led model.

Which one of the two will consolidate as the future orientation? This is difficult to foresee, particularly because the first discard plans were done under the pressure of time, and with different political expectations about the regionalisation process in different geographical areas, so these plans may not necessarily constitute a reliable indication of how the process will unfold in the future. In any case, as stated previously, the regionalisation was conceived as a way to get decision-making closer to the fishing grounds, and if this remains the focus of regionalisation, one can identify the AC-led approach as closer to that objective.

In that connection, it is necessary to reflect on whether the ACs will have the means to play this reinforced role in the new CFP. The grant aid given by the Commission for the functioning of the ACs (€250 000 *per annum*) is the same as for the old RACs. Yet, the new ACs have a far more ambitious mandate. One can certainly wonder if the continuation of the same level of funding is consistent with the much-enhanced role of the stakeholder advisory bodies in the new CFP.

Long-term multi-species management plans

This is also a very considerable challenge. Here, a first difficulty to resolve is to obtain the adequate scientific advice. The recent example of the Baltic Sea is an indication of the difficulties that this will represent. But in any case the availability of scientific advice will be a function of the management choices to be made in future years. These choices encompass a number of basic questions:

- A proactive, planned approach *vi-à-vis* a reactive one led by regionalised initiative;
- The type of multi-species relationship: technological (gear related) or ecological (predator-prey relationships);
- The scope (species and geographic areas) of the plans, including the relationship between regional and sub-regional approaches *vis-à-vis* the management of widely distributed species.

Let us look at these questions in some detail.

Planned versus bottom-up approach

It is obvious that the system cannot absorb too many plans to be developed at once. Neither the Commission nor the co-legislators can handle too many of these plans at the same time. Prioritising is thus essential.

The question is, then, what plans should come first? And how will the priorities be established? A question associated to that is: will the plans reflect a programmed approach, where the plans will be presented according to certain objective criteria? Or will they reflect the priorities established by the Member States and the Advisory Councils through bottom-up initiatives?

In principle the programmed approach seems preferable for a number of reasons. Firstly, because a bottom-up approach is likely to produce a piecemeal approach, where a number of individual stocks would be subject to plans as suggested from Member States or stakeholders, where others would not. This would produce an important gap in the way the discard ban is implemented: since the discard plans have a 3-year duration, the specific conditions applicable to the discard ban would in some cases be incorporated in the plans and consolidated, while in stocks outside the plans there would theoretically be no discard plans.

Furthermore, the reform also considers the multi-annual plans as the main 'carrier' of environmentally oriented measures. If the plans are patchy, ensuring only a partial coverage of stocks, then the implementation of these environmental measures would also be patchy. Overall, a systematic approach to cover all important stocks would seem preferable to ensure a more complete and consistent implementation of the policy.

What multi-species approach?

The multi-species advice has not evolved too much in recent times. Since the 2000s, ICES provides in its annual advice a table summarising the level of association of different stocks in the same fishery, for different areas, although the picture is much clearer for the best-studied area: the North Sea. But this table deals only with one aspect: the so-called technological relationships, that is, species that to some extent are caught by the same gear in the same fishery.

But another dimension of the multi-species management is the so-called ecological relationship, that is, notably the predator-prey relationships among species and how the abundance of a prey species influences the abundance of their predator and the other way round. Here, the evidence available is thinner and the development of this kind of approach will require a significant investment in new research, involving the study of stomach contents to assess these predator-prey relationships in a much more compete and reliable manner than today. Without this investment, the development of appropriate advice to tackle these relationships actually may never happen and the real development of the ecosystem approach to fisheries management may remain a theoretical possibility for the foreseeable future.

In any case, there seems to be no alternative to a gradual, step by step development of plans that would start by establishing objectives based on the individual estimates of MSY per stock. And, as the necessary scientific basis will unfold, incorporate gradually the real multi-species elements as further refinements.

The pace at which this development will take place is difficult to predict but at least in two aspects we can anticipate it will be slow:

1 The complexity of ecological relationships, which will make multi-species ecological models extremely difficult to develop in most areas. Even if this may seem feasible in the best-know areas (i.e. the North Sea), it is difficult to see how they could be developed in other areas where the level of scientific information is consistently weaker.
2 The high cost of the research necessary to underpin ecological models. In the context of limited human and financial resources for fisheries research, it is not easy to imagine that the new necessary research will find easy funding in the tears to come, despite the significant increase in the funds for data collection.

Overall, it seems clear that the first generation of new, multiannual plans will be in reality a collection of individual stock plans presented and packaged together, as is the case in the first of these proposals, regarding all the important stocks in the Baltic Sea, adopted by the Commission already in 2014 (COM, 2014a). The proposition of more ecologically integrated plans will have to be done gradually, as the necessary scientific advice develops.

Scope of the plans

The plans developed from 2002 onwards were fundamentally single species plans or, as we saw in Chapter 4 two-species plans at best. The evolution towards multi-species plans will inevitably raise a number of questions. The main one is that the distribution of species is very variable and their level of containment on one well-defined geographical zone, very variable. In many fishing areas, fishery resources are a mixture of stocks of a relatively local distribution, with others with a much wider distribution. And while regional or sub-regional plans can be adequate to manage the former, they are not adapted to manage the latter. Of course this phenomenon is also variable: while it is almost inexistent in the Baltic, due to its self-containment, it is higher in the North Sea and much higher in the western waters and the Mediterranean.

So the question is: what would be the ideal geographical scope of multiannual plans? The answer is relatively straightforward for the Baltic: practically all stocks (with the only exception of the western Baltic herring) are contained in the Baltic, which means that all the relevant stocks can be adequately managed through a plan corresponding to the basin.

For the North Sea the situation is less clear. While some stocks are self-contained with the North Sea (and even so, there are issues of stock boundary in the Channel) other stocks clearly overlap their distributions with the west of Scotland. Monkfish is a case in hand: a regional management plan for the North Sea would not entirely cover all areas of distribution of this stock, extended between ICES areas IV and VI.

The question is aggravated in the western waters. There, the case of certain areas such as the Celtic sea illustrate the problem: while a management plan for the species in this area could encompass many demersal species with stocks identified as 'Celtic sea stocks', other key species of the area, such as hake, have a much wider area of distribution and the hake stock could not possibly be managed on the basis of Celtic Sea considerations alone.

Of course, sea-basin integrated plans are very complex and the availability of the scientific advice will be a crucial factor to determine at what speed these plans can be presented and adopted.

There is also a very important inter-institutional challenge: the more stocks are regulated under management plans, the more the European Parliament can have a say in the management of EU stocks. This implies that if the management plans represent a piecemeal approach then the EP will be excluded from the management of the stocks not covered. Finally, the discard plans have a 3-year duration only. That means that within three years of the implementation of these plans, the conditions associated to the discard ban (*de minimis*, survivors etc.) will have to be incorporated into the management plans to ensure their continuity.

Other issues

Improving the scientific advice

The new CFP will be more dependent on the scientific advice than ever before. The reformed CFP will demand both more advice and new advice.

The new CFP will require new advice because of new objectives are being set. While all the traditional advice will still be necessary, new forms of advice will be required:

- The advice will now require specific estimates of MSY, or proxies and will also require specific estimates of discard levels to calculate quota top-ups as related to the discard ban;
- The advice on multi-species relationships will be essential to allow for the development of multi-species models;
- The general implementation of the ecosystem approach will also require abundant new evidence and advice;
- The improvement of the economic analysis will also require further and more specialised advice.

The new CFP will also require more advice: the extended number of 'clients' to the advice (the regional groupings of Member States under regionalisation, the Advisory Councils) and the increasing demand for transparency will put additional pressure on an overstretched scientific community. And with these new clients there will also be a growing demand for advice that will be easily understandable. In recent years ICES has started to produce advice in the form of 'popular advice' where the traditionally cryptic language and presentations, understandable only for scientists, is complemented by easier advice summaries for non-specialists.[1]

All these new requirements will have to be met by the European scientific community in a context of a human and financial basis for research that, at best, will not be expanding. This makes it essential that the fisheries research and advice in Europe should increase its cost effectiveness.

Probably, the best framework within which to achieve this is that offered by the data collection regulation. This will provide an ideal opportunity to refocus research and data collection on the real priorities, where the historical series of data will be of particular relevance. But is also be an opportunity to advance in the coordination of efforts among research institutions around Europe.

Today, there is still much duplication in the system, with several European institutes undertaking research on the same topic. The idea of enhancing coordination among them to avoid replication and obtain a more efficient use of the limited human and financial resources could be developed in this context. And so too could be that of broadening the scope of possible contributors to fisheries science: while this has been traditionally produced by national or regional marine research institutions, the potential of, for example, university departments to contribute to this end remains today limited to economic science and could perhaps be expanded to other areas of fishery research.

Natura 2000 sites

This is one of the cases where the Commission may be forced to take very difficult decisions. As we saw in Chapter 16, the new regulation foresees the possibility for the Commission to take urgent actions to protect Natura 2000 areas in case that there is a serious risk for a site and in the absence of Member States recommendations. This raises several problems:

- that the Commission will have to decide on (temporary) fisheries measures where Member States have not agreed. This would put the Commission in a very difficult position of favouring some Member States' interests and arguments over others;
- that these measures will protect the site from fishing but not from other uses. That means that an area may receive protection from fishing in the CFP but this would not prevent the development in these areas of other human activities, unrelated to fisheries, that may have a negative impact on the site;
- that the measures will reflect the 'urgency' of protecting a site that perhaps should have been protected long time ago only if Member States had taken the corresponding initiative.

It is difficult to predict how much use of this provision will be necessary in the future but for these reasons it would seem that it would actually only be a last resort in exceptional cases.

Fishery Protected Areas: a lost opportunity?

As we saw in the previous chapter, the CFP has a new article on the possible establishment of closed areas on the initiative of Member States. However, this article falls short of the expectations of a number of stakeholders and authors, who are strong advocates of Fishery Protected Areas (FPAs) as a mainstream management instrument, and not just as a secondary measure (Gell and Roberts, 2003).

The idea of using this instrument is actually very old. The French scientist Marcel Hérubel proposed them for demersal resources already a century ago (Roberts, 2007). Since then many experiences have been accumulated, with various degrees of success, including some in EU waters (see Chapter 4). One of the best examples of positive results is that of Georges Bank in Northeast America. There, the establishment of large – seasonal – closed areas from the mid-1990s has resulted in more effective conservation of numerous resource and non-resource species (Murawski, 2000). The question is: will these positive experiences motivate Member States to try to turn FPAs into a more fundamental management instrument than they are today?

Actually, the reply may be different for different areas. In the Atlantic, with large fishing grounds shared by several Member States, any alternative to the TACs and quotas as the main management instrument may be resisted on account of real or perceived effects on relative stability. FPAs are not necessarily 'neutral' in the way they affect national fleets, and depending upon their place and extension, they may be considered as a way to undermine relative stability: they may prevent one or more Member States to fully exploit their national quotas. Furthermore, the establishment of large FPAs poses the question of whether the fishing effort that would not take place in the areas concerned would be redeployed or reduced. If it is redeployed, then the question is whether the positive effects of the FPAs will offset the effects of increased effort elsewhere. And if the total effort is reduced, then we have to bear in mind the rejection by Member States of the effort management regimes in recent years. For these reasons, the perspective of spatial management as a mainstream instrument of the CFP looks rather unlikely so far as relative stability will remain.

In the Mediterranean, on the contrary, the use of FPAs (already considered a mainstream instrument in Regulation 1967/2006, see Chapter 8) could meet much less opposition since the TAC system is not applicable and many fisheries are not shared by two or more Member States. Although the implementation of these areas in the 2006 regulation seems to be less than impressive, as the Mediterranean moves to achieving the MSY objective by 2020, the use of FPAs may prove a more fundamental instrument than has ever been before.

The role of consumers

The awareness of European consumers of the role they can play to promote sustainable practices, through their choices in the shops and fish markets has increased dramatically in recent years. A good example of this is the evolution of the world's best-known – private – certification system: the Marine Stewardship Council (MSC).

By 31 March 2013 this certification had been given to 198 fisheries (for 88 species) around the world. These certifications apply to very different fisheries, from large-scale cod fishing in the waters of the Northeast Atlantic, to local razor-clam diver gatherers. The certification works in 106 countries and it covers 7% of the wild-caught seafood in the world, for a value of US$3.5 billion in the retail sector globally, where buyers' awareness of the system is estimated at 30%.[2]

The success of MSC (although the absolute figures here may appear unimpressive, it is the trend that matters) indicates that the consumers are becoming increasingly aware of the importance of their choice. The new labelling requirements established in the new market regulation (see Chapter 16) should contribute to increase consumer's awareness of what they buy.

The question is then whether private certifications like MSC will be the only answer to the consumers' increasing demand for sustainable fish, or whether new, public certification systems will be opportune and necessary. In this regard, the new market regulation foresees a feasibility study of a possible public, European system of certification of sustainable fish. Earlier initiatives to establish such a system did not materialise since Member States were very divided and the Commission considered that the case for EU legislation on this field was not sufficiently clear. After the CFP reform of 2013 and in the light of the important role that consumers have acquired in the new CFP, this question will have to be discussed again.

The future of management by fishing effort

Fishing effort remains, theoretically, a possibility under the new basic regulation. However, as we saw in Chapter 4, effort management has always been used, in the context of the CFP, as an add-on to the TAC and quota system. As a result, whatever its objective merits, effort schemes have been increasingly under pressure from Member States and stakeholders alike and the first new generation of multiannual plans after the reform (the proposal for multispecies plan for the Baltic (COM, 2014b)) excludes this instrument altogether. Although the Commission does not justify this decision, the political debate in the 2010s has been consistently critical of effort management as an unnecessary complexity. This may indicate the tendency to be followed in subsequent plans.

Effort management, however, has its advantages if applied as the main management instruments, although it is not free from downsides. The experience of the Faroe Islands, who introduced a days-at-sea system to manage its fisheries for cod, haddock and saithe in 1996 has very positive elements, particularly in achieving an appropriate balance in catch opportunities among fleet components and has minimised interference among gears. However, it did not achieve the objective of controlling the mortality of cod and saithe at the desired levels, partly because the number of days allocated was too high (Jákuppsstovu, et al., 2007).

The experience of effort management as a duplicate instrument is unlikely to be reproduced in a CFP where the objective of simplification has been clearly identified. The only possibility for effort management would be to replace catch limitations in certain cases. However, given the strong political attachment of most Member States and stakeholders to relative stability, there would have to be an extremely strong case for the use of effort management. The management of certain mixed demersal fisheries under the discard ban is probably the only scenario where the use of effort management could theoretically present certain advantages. But for that to become reality it would be necessary to ensure that the current relative stability as expressed by catch quotas could be credibly and reliably translated into effort quotas, something that so far has never been done.

The new CFP under co-decision

Co-decision and policy complexity

The adoption of the reform is relatively close in time to the incorporation of co-decision as the ordinary legislative procedure for the CFP. One important question in this regard is how the co-decision process is going to cope with the number and technical detail of the future CFP. Although the regionalisation procedure should result in a lower level of 'micro-management from Brussels' one can wonder how the policy is going to be handled if and when the regionalisation is not successful and the adoption of such questions as discard plans and implementation of environmental questions ends up in co-decision.

The co-decision mechanism is well adapted to policy fields where the legislative production is low and the issues are of real political relevance. But the CFP has a fundamental characteristic: its extraordinary legislative production and its technical complexity. In this regard, it is legitimate to ask the question: how can such a complex procedure as co-decision cope with the amount of technical detail in the CFP?

Probably the answer to this question is: it will depend on whether the regionalisation process really takes off and works efficiently if the technical discussions are left by the co-legislators to the regional level, the co-legislators will be able to concentrate on the main objectives and principles, which is the kind of work that normally corresponds to Ministers and Parliaments.

Adaptation of legislation

This issue has another very important angle. Co-decision being generally a slow process, how will the future CFP deal with all those cases where there is a need for a rather rapid adaptation of the existing legislation? There can be several cases of this kind.

An obvious one is the frequent need to change a plan because of a re-estimation by scientists of key parameters. The best case in hand is the F_{msy} estimates and ranges. If scientists change their estimation of the F_{msy} of a stock, how quickly can the CFP introduce this fundamental change in the corresponding LTMP?

Another example is what happened in the Baltic in 2008 regarding the so-called 'summer ban' for cod. That summer ban was fixed by Council regulation in specific dates. That year, however, the duration of the ice cover in the Eastern Baltic was longer than usual. And, consequently, the time available for effective cod fishing between the melting of the ice cover and the start of the summer ban was shorter than usual.

In these circumstances, stakeholders asked for a change in the dates of the summer ban. However, in the time available to do so it was simply not possible to bring about a change in a Council regulation to make the summer ban more flexible, because this required co-decision between Council and Parliament, a procedure that just cannot be done in a few weeks. This is just an example, among very many others, showing that a CFP where all technical details are decided by co-decision will simply not be responsive to real-time changes. If the dates of the summer ban are to be adapted to changing circumstances, such as those mentioned previously, it is clear that fast-track procedures are necessary, and that requires that the rules be adopted at a lower level than co-decision.

The role of the Commission

Finally, this new system also places an important responsibility on the Commission, in a number of key areas: (1) Regionalisation, (2) Enforcement, (3) Trust building and (4) Further initiatives.

On regionalisation, the Commission will have to strike a very difficult balance between the need to accommodate regional proposals, to provide a real content to this process, while at the same time ensuring a level playing field among fishermen across Europe and exerting its role as guardian of the Treaty by ensuring that regional decision making respects Union principles.

This balance is delicate. If regional recommendations are simply rubberstamped by the Commission, the risk is that different regional legislation may *de facto* tend to establish different standards for different fishermen, something that would in some cases be contrary to the principle of non-discrimination. On the occasion of the adoption of the first of these regionalised approaches, the discard plans adopted in December 2014, the Pelagic Advisory Council already underlined the lack of harmonisation between the discard plans between the North Sea and the west of Scotland, two areas subject to different recommendations that their vessels cross regularly, thus having to change rules in doing so.

On the contrary, if the Commission is over-zealous in preserving the level playing field, then those investing in regionalisation may have the feeling that their efforts are all for nothing. The notion that the Brussels machine crushes all regional creativity would easily produce deep frustration.

There is probably no alternative but to learn from experience. The precedent of the RACs clearly shows that at the beginning there can be some frustration when the Commission rejects certain proposals. However, over time they have learned to deal with this and to make recommendations that they know the Commission can accept because they are compatible with Union principles and policies. In all likelihood, this kind of learning curve will also take place in regionalisation. It is therefore important that we see this process as a gradual one that may not provide spectacular results overnight but rather as one that will unfold and gradually improve over the next years. Again, as always in fisheries management, having patience and determination at the same time may be the key for success.

On enforcement, this new CFP has an unprecedented number of instruments: from those of the new control regulation to the new provisions on conditionality of the EMFF. Using these instruments to their full potential can further improve compliance with the CFP.

Also as we saw before, certain processes will require a great deal of confidence building. The role that the Commission can play here is crucial. In particular, the process of adoption of the provisions on regionalisation clearly showed that there is no better alternative for the adoption of more technical rules than the Commission acts. But Member States do not always trust the Commission and can only agree to such delegations if it is under very controlled conditions.

Finally, although it is difficult at this stage to foresee what new developments will be required in the CFP beyond those elements identified and legislated in the new basic regulation, the Commission can play a fundamental role in ensuring that, even without an initial legal basis, the initiatives necessary to have the CFP updated to new developments are effectively taken.

All these questions require from the Commission an adequate combination of political determination to exert its role of guardian of the Treaty, and technical expertise to do its job properly. In this regard, the comment by Mike Holden (see Chapter 2) on the importance of technical expertise on fisheries matters in the Commission services is still pertinent today. As a representative of the fishing industry put it to this author, the Commission needs a critical mass of officials who can 'tell a sardine from a herring'. This is all the more important when the Commission interacts with the outside world and the other EU institutions. Stakeholders appreciate that Commission officials know and care about fisheries, and the fisheries expertise gives the Commission officials an extra added value when it comes to facilitate agreements in the co-decision process. Whatever the evolution of the policy, a minimum level of fisheries expertise in the Commission services will always be essential.

Will the new CFP prevent the 'tragedy of the commons'?

As we saw in Chapter 1, fisheries management represents the ultimate example of Garret Hardin's 'tragedy of the commons'. We also referred to the theories developed by Elinor Ostrom on how to organise the 'coercion' mechanisms that Garret Hardin himself identified as the solution to avoid the tragedy (Ostrom, 1990). So one can ask the question as to how the CFP, and particularly the CFP as re-designed by co-legislators after the 2013 reform, will fare in connection with Ostrom's suggested solutions to the tragedy.

In fact, Ostrom proposed 8 principles that should inspire the management systems to be successful in avoiding the tragedy of the commons. Although she developed such principles with the idea of 'local' communities in mind and not specifically for fisheries, most of these principles can be tested *mutatis mutandis* against the specific case of the CFP. Let us examine these principles:

- *Clearly defined boundaries (effective exclusion of external un-entitled parties)*

  In principle the boundaries of the people entitled to exert the fishing activity is clearly delineated in the CFP. There is, however, a remaining problem in this regard: the question of the perceived legitimacy (or lack thereof) of some of the entitled parties. While this is not at all a problem in certain areas (North Sea, Baltic, Mediterranean, Black Sea, where only coastal States are entitled to fish for quota species) in others, particularly the West of Scotland and the bay of Biscay, the entitlement to fish by certain operators from distant ports is still, if accepted legally, called into question in practice. This results in a lack of trust among operators that undermines the cooperation and weakens the links among peers.

- *Rules regarding the appropriation and provision of common resources that are adapted to local conditions*

  The adaptation of the rules to local conditions is a permanent struggle of the CFP and, despite some advances it is still largely not achieved. Traditionally, either the local conditions are ignored as part of the overall negotiating compromise to adopt the regulations, or in some cases the local conditions are included as part of the main body of the regulation, thus giving rise to a high level of technical detail, as we saw in Chapter 13.

  In this regard, the regionalisation adopted in 2013 provides for a unique opportunity to legislate in a way that the local conditions will be taken into account but not as part of the main body of the legislation. If the regionalisation is fully used to make more results-based legislation where local conditions can be easily incorporated, then this second principle could well develop in the years to come.

- *Collective choice arrangements that allow most resource appropriations to participate in the decision-making process*

  Undoubtedly, the future ACs should be the appropriate instrument to apply this third principle. But for that to be the case, it is necessary that at least two conditions be fulfilled: (1) that the ACs will indeed be considered by the stakeholders as really representative of their interests and (2) that the institutions really take their input into account.

  On the first condition, important challenges lay ahead. In particular, the representativeness of small scale fishing interests is very difficult. This is due to the very structure of small scale fishing: they tend to be associated in a number of small associations that only represent themselves and not necessarily other small scale fishermen from other regions. Since they cannot all be represented, small scale fishermen still have difficulties to present common views that can be considered as representative of their whole sector. In contrast, the most industrial sectors are normally associated in umbrella organisations where they work out easily common positions, thus having a much greater impact in the recommendations presented by the ACs.

  On the second condition, the key question is the extent to which national administrations will associate the ACs into the development of the regionalised approaches to decision making. Indeed, the ACs recognise that important progress has been made in terms of how the Commission follows their advice. But now it will be for national administrations to demonstrate that the policy so criticised for its 'central command from Brussels' will not be replaced by another with 'central command from the capitals'.

- *Effective monitoring by monitors who are part or accountable to the appropriators*

  This principle raises at least two questions: (1) whether the industry can actually be directly involved in the control of the activity and (2) whether the inspectors in Member States are really 'accountable' to the appropriations.

As we saw in Chapter 9, improving the control of the CFP is a very long-standing struggle and the fact that the diagnostic of the control system of the CFP has changed relatively little in 30 years does not contribute to an optimistic future. But the 2009 regulation and other developments (conditionality under EMFF) establish a much more powerful instrument to achieve good compliance.

The question of the industry's self-control is very difficult. For that to happen, a number of conditions will have to appear and it is doubtful they would ever exist in the CFP

- *A scale of graduated sanctions to resource appropriators who violate community rules*

In the CFP this principle should be considered as a system of graduated and harmonised sanctions. All Member States have a system of sanctions for fishery offences. Another question, however, is whether these systems are effectively dissuasive, whether they are harmonised among Member States and, crucially, how they are applied in practice.

The control regulation lays down a clear obligation for Member States to have such legislation. For the Commission, it is a high priority to ensure that these requirements are effectively implemented.

- *Mechanisms of conflict resolution that are cheap and easy to access*

In general terms, the way the CFP resolves conflicts is through the rulings of the European Court of Justice. Whether the access to the Court is cheap and easy, is probably a very arguable question. In any case, the CFP clearly has a system to resolve conflicts and, over and beyond the recourse to the Court (which in general terms is a last resource) there are plenty of other mechanisms (including informal mediation by the Commission services) that can come into play when a conflict appears.

- *Self-determination of the community recognised by high-level authorities*

The self-determination in the CFP must be considered in its particular legal context: there is no such thing as self-determination under the CFP. However, as self-determination we could consider the ability of part of the industry to adopt its own rules within a wider framework adopted by the co-legislators. In other words, the self-determination can be the mechanism whereby the industry, under regionalisation, could agree its own rules to deliver certain pre-determined policy objectives. However, this self-determination is difficult to deliver in practice, for at least two reasons:

✓ The considerable mistrust that has dominated the industry, that results in a call for fisheries managers to legislate to ensure a 'level playing field'. Only if the industry within the ACs can work out conditions to build trust among operators will this self-determination be possible.

✓ The resistance of certain administrations in Member States to lose the power of decision to stakeholders.

There is, however, an area where this self-regulation can develop into a much more powerful mechanism: the market policy, where the empowerment of Producer Organisations to regulate supply and adapt it to demand and their new role in finding solutions for the marketing of fish not allowed for human consumption under the landing obligation, could imply a substantially higher degree of self-rule by fishermen that could contribute to fulfil this condition.

- *In the case of larger common-pool resources, organisation in the form of multiple layers of nested enterprises, with small local common pool resources at the base level*

The European fishery resources are certainly a very large common pool of resources. The 'nested enterprises' can be considered as corresponding to the different levels of integration of fisheries interests, from producer organisations to larger umbrella organisations to the level of European-wide organisations, each intervening at different levels of governance.

Overall, the CFP has, by and large, most of the instruments identified by Ostrom to avoid the tragedy. From this perspective, the CFP can be considered as a policy with the potential to avoid the tragedy.

There is a downside, however: if not all, some of these instruments have existed from the beginning of the policy and yet they did not avoid the clear deterioration of the 'common' during the 1980s and 1990s.

That, again, leads us to a conclusion that is not at all new for the policy: with the existing instruments the policy can be successful; the condition for that to happen is that the existing mechanisms are used properly.

Fisheries in a wider maritime context: integrated maritime policy

When looking at the future of the CFP it is also important to remember that the CFP and the fishing activity do not exist in a bubble; they are part of a much wider context: that of the maritime economy. It is therefore pertinent to wonder what can be the effect on the CFP of the development, in recent years, of a specific policy framework to cater for the integrated management of the maritime activity: the integrated maritime policy (IMP).

In Chapter 11 we already saw the relationship of the CFP with the environmental pillar of IMP: the Marine Strategy Framework Directive. But IMP is much wider. In particular, other policy areas have already and will have even more in the future, a considerable effect on the way the European seas and its resources are handled.

Among the different sectors, the one that most clearly interacts with fishing is the energy sector. Beyond the traditional oil and gas drilling, the development of wind farms in marine areas will change considerably the uses, priorities and the governance of the maritime space; and this will undoubtedly have effects on the management of fisheries, notably by setting aside for exploitation areas of interest for the fishing industry.

Some of these new developments will pose serious challenges for the future of fishing. But many others will indeed be opportunities for the fishing industry and, specially, for the coastal communities dependent on fishing.

By and large, the most important of these elements is the so-called 'blue growth' initiative put forward by the Commission in recent years.

Blue growth

The blue growth initiative is a long term strategy to promote economic growth and jobs in the most promising areas of economic activity in the European oceans (COM, 2014c). In this context, five areas have been identified as particularly promising for further economic growth: (1) aquaculture, (2) coastal tourism, (3) marine biotechnology, (4) ocean energy and (5) seabed mining.

The development of these sectors is both a challenge and an opportunity for fishing communities. It is a challenge because it will obviously imply a future strong competition for the use of the maritime space: certain current fishing areas may be reclaimed for alternative uses and this may limit the marine surface available for the fishing activity. On the other hand, the development of these sectors may provide more jobs for the fishery-dependent areas of Europe. How to balance out the different interests involved will constitute a major challenge for both regulators and the civil society.

In any case, there is plenty of potential for the development of synergies among the different and apparently incompatible uses of the maritime space. For example, a wind farm will be incompatible with fishing with large scale vessels, for obvious reasons of safety. However, some experiments are already being developed of certain forms of aquaculture around wind farms. In addition, the areas closed for fishing around wind farms could become areas of reproduction of species, thus providing indirect benefits to the fishing industry.

The way in which different uses of the maritime space will relate to fisheries is a fascinating challenge for the future. Finding ways to combine different activities in the same marine space is a fundamental scientific and technical challenge. Studying the economic trade-offs among the different uses of that space, and in particular considering the 'externalities' of the different uses will also be a challenge for economic science. In particular, the evaluation of the so-called 'economic services', that is, the evaluation of the economic benefits deriving from the preservation of a functioning ecosystem, will be a fundamental part of the necessary knowledge basis for the blue growth idea to develop in a way that will be sustainable and compatible with the fishing activity.

The blue growth initiative will use as essential components certain instruments such as maritime spatial planning, marine knowledge and integrated maritime surveillance.

Maritime Spatial Planning

A particularly important development in this regard is the new Directive on Maritime Spatial Planning (EU, 2014d), which establishes the obligation for Member States to plan the development of maritime activities in waters under their jurisdiction according to general principles of spatial planning. The value of this initiative is obvious as an instrument to promote the 'certainty' in the future use of the maritime space, a key factor in promoting investment in the maritime economy.

But in terms of the effects on the CFP this directive can have an important value: ensuring that the future uses of the maritime space are decided on the basis of good planning, where all important actors (including the fishing industry) can have a say in the process on an equal footing with other stakeholders.

This is important because the tendency over the last few years is that other activities, particularly those related to the energy sectors, take initiatives with little regard for their effects on the fishing activity, using the political priority and financial muscle of the energy sectors. As we saw in Chapter 11, the consideration of fisheries in a wider economic context of maritime activities can be a good opportunity for the fisheries (and aquaculture) activities to be considered on an equal footing with other activities in the same space in terms of its ecological, economic and social value. The lack of such framework can only favour the 'law of the jungle' where fisheries can simply not compete with other activities with far stronger financial muscle and political influence.

Marine knowledge

In Chapter 10 we referred to the Data Collection Regulation and its importance for the CFP as a knowledge-based policy. The equivalent of the data collection at the level of integrated maritime activities is the initiative on the European Marine Observation and Data Network (EMODNET) (COM, 2010). Its declared three aims are: (1) improving the efficiency of public administration at all levels, (2) providing industry with a marine knowledge infrastructure that enables it to compete in the global economy and (3) deepening knowledge of the marine world and its relationship with human activity in order to support scientific research and reduce uncertainties in parameters such as sea level rise or biodiversity.

The third aim, in particular, can be of key relevance for the CFP. The fishing industry will have to face many challenges in the future. Global warming and the preservation of biodiversity, in particular, loom large as significant challenges for the sector. And these challenges will require in-depth knowledge of the oceanic processes and the role that the fishing activity plays in such processes.

A recent example can illustrate this. In December 2013 the European Parliament adopted a first reading resolution on the Commission's proposal to phase out bottom trawling in deep waters.[3] In its position, the Parliament requested four additional years and, crucially, in-depth knowledge of the effects of bottom trawling on benthic communities not yet protected under the Natura 2000 network.

In the coming years, the challenge of preserving biodiversity will only increase. Basing decisions in that regard on in-depth knowledge will certainly require a considerable investment in our knowledge of physical, chemical and biological parameters of the European oceans.

Some external challenges for the CFP

Implementing the reform agreed in 2013 is quite a challenge. But as experience shows, within a lapse of 10 years till the next CFP reform, many new challenges may come up that are not specifically contemplated in the text agreed in 2013. The best example is that of MSY, or discards, both not taken on board by the CFP reform of 2002 but which became very important policy issues in the meantime. So the question is, over and beyond what was agreed by the co-legislators in 2013, what challenges will the CFP have to face that are not part of this reform?

This is obviously a difficult question to answer, as many issues becoming important for the CFP were at their time very difficult to anticipate. But with all the necessary caveats, it is possible to identify certain issues that, one way or another may become an important challenge for the CFP in the years to come.

The effects of climate change

This is perhaps the most difficult challenge to face, because both the effects and the time scale of their occurrence are very difficult to predict. However, whatever the real dimension of the problem of global warming, it is bound to produce a number of effects on the CFP:

Changes in the distribution of the food organisms essential for the commercial species

Shifts in the abundance and composition of phytoplankton and zooplankton can have significant effects on the marine ecosystems as we know them today. There is increasing evidence of the effects of climate change on ocean productivity (Behrenfeld et al., 2006). Changes in sea water acidity (as a result of higher concentrations of CO_2 being dissolved in sea water) may also produce significant effects. In particular, many phytoplankton species have calcareous carapaces. The formation of these carapaces is inhibited by acidity and thus phytoplankton abundance may suffer. These changes will only add additional stress to marine populations, thus making them more sensitive to possible overexploitation.

At the same time, the question of the carrying capacity of the European seas will be in question. Will European oceans be able to produce the same amount of biomass, not only for fisheries production but also for other living resources?

Changes in the distribution of fish stocks

Scientists have already alerted about the significant changes in the distribution patters of commercial species. In the Baltic Sea, expected climate changes will make waters warmer and fresher. Marine-tolerant species will be disadvantaged and their distributions will partially contract from the Baltic, while habitats for fresh fish will likely expand. Fishing fleets presently targeting marine species (cod, herring etc.) will likely have to relocate to more marine areas or switch to other species which tolerate lower salinity (Mackenzie et al., 2007).

In the North Sea a similar phenomenon is observed. The distributions of exploited and non-exploited fishes have responded markedly to recent increases in sea temperature, with nearly two-thirds of the species shifting in mean latitude, depth or both in the last 25 years (Perry et al., 2005). In general terms, the evidence is that that the occurrence of species in temperate waters is moving northwards (in the northern hemisphere), through a phenomenon of species replacement that will change species composition, though not necessarily overall productivity.

These changes are very visible already: the abundance of red mullet and even anchovy in the North Sea in recent years are good examples. But in more general terms there seems to be a northward shift in the abundance of certain species as a consequence of the higher sea water temperatures.

The effects of this phenomenon can be significant in the future CFP. Even if total productivity does not change, the species composition of the different European seas will do. These changes can indeed have effects on the way fish resources and fishing rights match with each other under the (geographically rigid) relative stability. Whether this will lead to possible changes in allocation keys by area, or in the establishment of new or revised management areas, is difficult to tell given the difficulty to reopen any relative stability key. However, as these changes in species distribution become more apparent, it will be difficult to avoid that possibility.

These changes, perhaps combined with the implementation of the discard ban, are likely to increase, perhaps dramatically, the mismatch between what fishermen will find in their nets and what they (or their Member States) are entitled to fish in any given area. This will be a major test for relative stability as we know it today. Since this principle enjoys ample support from most Member States and many stakeholders, the question is not whether relative stability will survive but rather how it will have to be adapted, adjusted or even, in certain cases, reinvented to be able to face these challenges.

The deterioration of coastal infrastructure by rising sea levels

This may also represent a challenge in terms of the future structural policy that may have to place in future years a different focus in funding priorities.

The fishing industry may also have to contribute to general climate change mitigation policies. The contribution of the fishing industry to the mitigation policies would probably be small in absolute terms: fisheries are responsible for 1.2% of the global oil consumption. However, for the fishing industry the mitigation measures may have important consequences. The main one would probably be the reduction of CO_2 emissions. The EMFF includes a series of measures to mitigate climate change by the fishing fleets.

Possible increases in fuel prices

Any increase in fuel prices (either as a result of fluctuations in crude oil prices, or as a consequence of a change in the fiscal policy on fuels) may have a very important bearing on the future of the European industry.

The fishing industry is very fuel-thirsty. Although consumption is very variable, its total average is about 0.45 kg of fuel consumed per kg of fish landed (or 640 litres per ton of landed fish).[4] But such consumption levels can reach values as high as 4 kg of fuel per kg of fish landed, when using such gear as beam trawls. Overall, fuel represents on average 23% of the running costs of the European fishing industry, second only to labour costs with 32%.[5]

In addition, fuel for fishing vessels is not taxed in Europe (or indeed in many other countries around the world), so any increase in fuel prices cannot possibly be compensated by a reduction in the levels of taxation. These circumstances make the fishing industry particularly vulnerable to significant increases in fuel prices.

We have some recent experience in this regard: the oil crisis of 2008. We have referred to this experience already in Chapter 6 and it is of particular relevance here to refer to the conclusions of the most recent Commission report on the performance of the fishing fleets (see Chapter 6): that the so-called fuel package was not used by most Member States and that most of the adaptations to the rising fuel practices came about through changes in fishing practice.

Another question is, however, in the event of another oil crisis like the one of 1973, how would the fishing sector adapt? The fact that the fishing industry resisted the shock of 1973 does not guarantee the same possibility in the future: in the pre-1973 world the prices were so low that the sector had ample room for manoeuvre to absorb it. In 2008, with a much narrower room for manoeuvre to absorb significant increases in fuel prices, the European industry adapted surprisingly well, through a number of adaptations in fishing practice that even made the special structural action designed to help the industry largely unutilised. The question is, after the adaptations already made as a result of the 2008 crisis, how much room for manoeuvre to absorb further significant increases in fuel prices has the European fishing industry?

Possible changes in the ownership of the means of production

This is an issue that by and large remains hidden. True, the case of 'quota hoppers' (see Chapter 3) made headlines at some stage. However, the question of the property of the fishing fleets has long been a non-issue under the CFP.

The allocation of quotas under relative stability and the absence of a European-wide market of fishing rights may give the impression that the ownership of fleets and fishing rights is still largely national. However, behind the binome national quotas/national vessels there is a significant amount of foreign ownership of vessels, either by other EU Member States or by third countries. Although no systematic studies have been carried out in the past on the magnitude of this phenomenon,[6] the everyday evidence suggests that it may be very significant in a number of fleet segments, and certainly so in the aquaculture sector, where a number of fish farming companies are in the hands of Norwegian interests.

It is conceivable that if the property of the means of production change substantially from national to foreign ownership, or from local ownership to investors, the stakeholder input into the policy may also substantially change in the years to come, with possibly significant consequences for the CFP.

Possible changes in the world's fisheries governance

In Chapter 7 we referred to the possible evolution of the world's fisheries governance. It is clear that this would have a considerable impact on the way the CFP is shaped up today. In fact, beyond fisheries, there are a number of developments that clearly point at a possible evolution of the world's maritime governance and this will affect fisheries, one way or another. These will undoubtedly affect the Union as a key global player and the largest market of fish in the world.

Much has been written about the future of marine resources in a world where increasing population will inevitably push for higher levels of exploitation of natural resources. In particular, the scientific journals *Science* and *Nature* have published since the 1990s generally pessimistic projections of the world's fish production, including certain recent doomsday scenarios (Marra, 2005). These pessimistic views actually coincided to some extent with the pessimistic view of the Commission in its 2009 Green Paper on CFP reform (see Chapter 16).

But more recent data (see Chapter 4) show that the efforts to reduce fishing pressure in the last 15 years have started to show positive results in Europe. In fact, certain authors did not share the pessimistic views, indicating that current management is slowly bearing fruit and that rather than disregarding current management approaches as inefficient, we should expand and use the current instruments to achieve sustainability (Hilborn, 2007).

Whatever the projections, the future of the world's fishery resources will continue to draw much attention and initiatives. Possible areas where global fisheries governance may evolve in future years are *inter alia*:

- A new focus on the protection of marine areas, notably beyond zones of national jurisdiction (the BBNJ initiative, see Chapter 7). This may be accelerated by the tendency to extend oil and gas drilling to areas hitherto unexploited, such as Arctic waters following the thaw of the polar ice cap. The possibility that this initiative will result in significant areas of the high seas being protected from certain types of fishing cannot be overruled.
- An evolution of the Law of the Sea. Although this seems unlikely at this stage, the possibility of increased tension around the *status quo*, particularly if the multilateral management in the high seas does not work properly, could also provoke future changes through unilateral action by coastal States.
- A questioning of the legal *status quo* in the allocation of fishing rights, promoted by developing coastal States wishing to have a bigger share of the international ocean's resources. This is not only likely but a clearly observed trend already.
- New developments in the World Trade Organisation regarding subsidies to the fishing activity, at least for large-scale vessels.
- New initiatives to fight ocean pollution, such as those regarding the removal of plastics from the world's oceans.

These and many other possible changes would have implications in the CFP and not only on the external sector: the CFP in European waters would also be affected.

It is clear that the Union should play a leading role in these developments. With its global influence (notably its 'soft power') Europe can ensure that these developments correspond to the multilateral, cooperative approaches that the EU promotes in all policy areas, and not to unilateral action by coastal or other States. Ensuring healthy and stable management of the world's fishery resources through concerted multilateral action is not only the best solution for the world's fisheries, it is also the best way to defend the legitimate interests of European fishermen in that context.

Closing remarks

This book draws a certain number of conclusions and lessons learned. These are varied, but there is perhaps one that summarises them all: the key importance of the forward-looking policies as opposed to those of managing only the present and the near future. All the important improvements in the policy have required long-term vision and courage to confront the strong inertia of the system. If the formula for success is always elusive, the formula for failure is clear: to manage only the day-to-day business and to surrender to the short-term difficulties. This is of particular relevance today, when the development of the information society provides ample social repercussion to any short-term difficulty and when public debates based on short, immediate messaging often focus on superficial issues, not necessarily on substantial ones.

Whatever the challenges of the future implementation of the reformed CFP, this author will contend that the key for the success of the policy is to be always inspired by long-term goals and not just by short-term considerations. According to an old Greek proverb: 'A society grows great when old men plant trees whose shade they know they will never sit in'.

While this can certainly apply to many policy areas, fisheries management, and consequently the CFP represent the ultimate example of how a policy needs to be able to look at future goals even at the cost of being unpopular in the short-term.

From the previous pages, one can conclude that the best developments in the CFP came about when the long-term objectives prevailed, that is, when decision makers planted trees even if they knew they might never sit in their shade.

We saw in Chapter 4 (in particular in Figure 4.5) that the recovery of European fish stocks required taking difficult decisions whose results became apparent only several years later. This is by far the most important lesson to be drawn from more than 30 years of implementation of the CFP.

Notes

1 http://ices.dk/publications/our-publications/Pages/Popular-advice.aspx.
2 Marine Stewardship Council. Annual Report 2012/13.
3 European Parliament legislative resolution of 10 December 2013 on the proposal for a regulation of the European Parliament and of the Council establishing specific conditions to fishing for deep-sea stocks in the North Sea.
4 Data from the European Commission Services.
5 Profitability of the EU fishing fleet. European Parliament, 2013.
6 A study on this topic is currently being carried out for the Commission.

References

Behrenfeld, M.J., R.T. O'Malley, D.A. Siegel, C.R. McClain, J.L. Sarmiento, G.C. Feldman, A.J. Milligan, P.G. Falkowski, R.M. Letelier and E.M. Boss (2006). Climate-driven trends in contemporary ocean productivity. *Nature* **444**: 752–755.

COM (2010). Communication from the Commission to the European Parliament and the Council. Marine knowledge 2020 marine data and observation for smart and sustainable growth. Brussels, 8.9.2010. COM(2010) 461 final.

COM (2014a). Proposal for a Regulation of the European Parliament and of the Council establishing a multiannual plan for the stocks of cod, herring and sprat in the Baltic Sea and the fisheries exploiting those stocks, amending Council Regulation (EC) No 2187/2005 and repealing Council Regulation (EC) No 1098/2207 – COM(2014) 614 final.

COM (2014b). Proposal for a regulation of the European Parliament and of the Council establishing a multiannual plan for the stocks of cod, herring and sprat in the Baltic Sea and the fisheries exploiting those stocks, amending Council Regulation (EC) No 2187/2005 and repealing Council Regulation (EC) No 1098/2007. COM/2014/0614 final.

COM (2014c). Communication from the Commission: Innovation in the Blue Economy: realising the potential of our seas and oceans for jobs and growth – COM(2014) 254/2 (13/05/2014).

Enever, R., Revill, A.S. and A. Grant (2009). Discarding in the North Sea and on the historical efficacy of gear-based technical measures in reducing discards. *Fisheries Research* **95**: 40–46.

EU (2014a). Commission Delegated Regulation (EU) No 1396/2014 of 20 October 2014 establishing a discard plan in the Baltic Sea. *OJ* L 370, 30.12.2014, p. 40.

EU (2014b). Commission Delegated Regulation (EU) No 1394/2014 of 20 October 2014 establishing a discard plan for certain pelagic fisheries in south-western waters. *OJ* L 370, 30.12.2014, p. 21, 25, 31.

EU (2014c). Commission Delegated Regulation (EU) No 1395/2014 of 20 October 2014 establishing a discard plan for certain small pelagic fisheries and fisheries for industrial purposes in the North Sea. *OJ* L 370, 30.12.2014, p. 35.

EU (2014d). Directive 2014/89/EU of the European Parliament and of the Council of 23 July 2014 establishing a framework for maritime spatial planning. *OJ* No L 257, 28.8.2014, p. 135.

EU (2015). Regulation (EU) 2015/812 of the European Parliament and of the Council of 20 May 2015 amending Council Regulations (EC) No 850/98, (EC) No 2187/2005, (EC) No 1967/2006, (EC) No 1098/2007,

(EC) No 254/2002, (EC) No 2347/2002 and (EC) No 1224/2009, and Regulations (EU) No 1379/2013 and (EU) No 1380/2013 of the European Parliament and of the Council, as regards the landing obligation, and repealing Council Regulation (EC) No 1434/98. OJ L133, 29.5.2015, p.1.

Gell, F.R. and C.M. Roberts (2003). Benefits beyond boundaries: the fishery effects of marine reserves. *Trends in Ecology and Evolution* **18**(9): 118–455.

Hilborn, R. (2007). Reinterpreting the State of Fisheries and their Management. *Ecosystems* **10**: 1362–1369.

ICES (2014). Report of the workshop on the development of quantitative assessments methodologies based on life-story traits, exploitation characteristics and other relevant parameters for data-limited stocks. 27–31 October. Lisbon, Portugal. *ICES WKLIFE IV REPORT 2014*. ICES CM 2014/ACOM: 54.

Jákuppsstovu, S.H., L.R. Cruz, J.-J. Maguire and J. Reinert (2007). Effort regulation of the demersal fisheries at the Faroe Islands: a 10-year appraisal. *ICES Journal of Marine Science* **64**: 730–737.

Johnsen, J.P. and S. Eliasen (2011). Solving complex fisheries management problems: what the EU can learn from the Nordic experiences of reduction of discards. *Marine Policy* **35**: 130–139.

Lequesne, C. (2001). L'Europe Bleue: A quoi sert une politique communautaire de la pêche ? Presses de la Fondation Nationale des Sciences Politiques, 239 pp.

Mackenzie, B.R., H. Gislason, C. Möllmann and F.W. Köster (2007). Impact of 2&st century climate change on the Baltic Sea fish community and fisheries. *Global Change Biology* **13**(7): 1348–1367.

Marra, J. (2005) When will we tame the oceans? *Nature* **436**: 175–176.

Murawski, S.A., R. Brown, H.-L. Lai, P.J. Rago and L. Hendrickson (2000). Large-scale closed areas as a fishery-management tool in temperate marine systems: the Georges Bank experience. *Bulletin of Marine Science* **66**(3): 775–798.

Newman, D., J. Berkson and L. Suatoni (2015). Current methods for setting catch limits for data-limited fish stocks in the United States. *Fisheries Research* **164**(2015): 86–93.

Ostrom, E. (1990). *Governing the Commons: The Evolution of Institutions for Collective Action*. Cambridge: Cambridge University Press.

Perry, A.L., P.J. Low, J.R. Ellis and J.D. Reynolds (2005). Climate change and distribution shifts in marine fisheries. *Science* **308**(5730): 1912–1915.

Roberts, C. (2007). *The Unnatural History of the Sea*. Island Press/Shearwater Books. 435 pp.

Glossary

Advisory Committee for Fisheries and Aquaculture (ACFA). Stakeholder consultative body established in the CFP to advise the Commission on management measures. It operated from 1983 to 2013.

Advisory Councils (ACs). Stakeholder consultation bodies set up to advice decision-making on issues related to the implementation of the CFP. They replaced the Regional Advisory Councils (RACs) after the 2013 reform.

Anadromous fish. Fish that migrate from the sea up into fresh water to spawn, or stay entirely in fresh water and migrate upstream to spawn. An example is salmon.

B_{msy}. Level of biomass of a stock that can support the maximum sustainable yield from that stock.

Catadromous fish. Fish that migrate from fresh water down into the sea to spawn, or stay entirely in fresh water and migrate downstream to spawn. An example is European eel.

COREPER. ***COmité des REpresentants PERmanents*** (Committee of permanent representatives), body of the European Council of Ministers that prepares the meetings of the Council of Ministers by deciding on the less political questions, leaving for decision by ministers only the most political issues.

Council of Ministers of the European Union. One of the instritutions of the European Union. Constituted by Ministers from the Union's 28 Member States. The Council meets regularly by policy areas. Fisheries and Agriculture are dealt with by the AgriFish Council.

Co-legislators. Term referring to the two institutions (the Council of Ministers and the European Parliament) that jointly adopt EU legislation on fisheries since the entry into force of the Treaty of Lisbon.

Demersal fish. Fish that live and feed on or near the sea bottom.

DG Mare. Directorate General responsible for fisheries and maritime affairs in the European Commission. The name replaced the original of DG XIV and later on DG FISH. It turned into DG Mare when it incorporated the handling of integrated maritime policy in addition to fisheries policy.

Directive. Type of EU legislation aiming at harmonising national legislations. They do not apply directly to European citizens: they have to be transposed into national law.

Electroning Registration System (ERS). Also referred to as the **Electronic logbook**. Electronic system to record the quantities of fish caught and discarded under the CFP. It replaced the paper logbook.

European Fisheries Fund (EFF). Instrument of the structural policy of the CFP between 2007 and 2013.

European Maritime and Fisheries Fund (EMFF). New financial fund to support fisheries and aquaculture between 2014 and 2020.

European Commission. Usually referred to in the book as 'the Commission', is the executive body of the European Union.

European Community (EC). Upon the entry into force of the Maastricht Treaty in 1993, the EEC was renamed the *European Community* (*EC*) to reflect that it covered a wider range of policy. This was also when the three European Communities, including the EC, were collectively made to constitute the first of the three pillars of the European Union (EU), which the treaty also founded. The EC existed in this form until it was abolished by the 2009 Treaty of Lisbon, which merged the EU's former pillars and provided that the EU would 'replace and succeed the European Community'.

European Currency Unit (ECU). Accountancy unit of European currencies, precursor of the Euro, Single European Currency.

European Economic Community (EEC). The European Economic Community (EEC) was the international organisation created by the Treaty of Rome of 1957.

European Parliament (EP). One of the institutions of the European Union.

European Structural and Investment Funds (ESI funds). The five structural funds of the Union, which, as of 2014, are regrouped under certain common provisions and subject to coherent integrated programming. These are the European Regional Development Fund (ERDF),

The Common Fisheries Policy: The Quest for Sustainability, First Edition. Ernesto Penas Lado.

the European Social Fund (ESF), the Cohesion Fund (CF), the European Agricultural Fund for Rural Development (EAFRD) and the European Maritime & Fisheries Fund (EMFF).

European Union (EU). The European Union (EU) is a politico-economic union of 28 Member States that are located primarily in Europe. The EU operates through a system of supranational institutions and intergovernmental negotiated decisions by the Member States. The institutions are: the European Commission, the Council of the European Union, the European Council, the Court of Justice of the European Union, the European Central Bank, the Auditors and the European Parliament. The European Parliament is elected every five years by EU citizens.

Exclusive Economic Zone (EEZ). Sea zone prescribed by the United Nations Convention on the Law of the Sea over which a state has special rights regarding the exploration and use of marine resources, including energy production from water and wind.

Food and Agriculture Organisation (FAO). Agency of the United Nations that leads international efforts to defeat hunger. Serving both developed and developing countries. FAO is also a source of knowledge and information, and helps developing countries and countries in transition modernise and improve agriculture, forestry and fisheries practices.

FIFG: Financial Instrument for Fisheries Guidance. Instrument of the fisheries structural policy that applied from 2000 to 2006.

Fishing effort. The amount of fishing gear of a specific gear type used on the fishing grounds over a given unit of time. For example, hours trawled per day, number of hooks set per day or number of hauls of a beach seine per day.

Fishing mortality. Mathematical expression of the part of the total rate of deaths of fish due to fishing. Often expressed as a rate that indicates the percentage of the population caught in 1 year.

F_{msy}. Fishing mortality rate that, if applied constantly, would result in Maximum Sustainable Yield.

Fish stocks. Fish stocks are subpopulations of a particular species of fish, for which intrinsic parameters (growth, recruitment, mortality and fishing mortality) are traditionally regarded as the significant factors determining the stock's population dynamics.

General Fisheries Commission for the Mediterranean (GFCM). Fisheries management organisation that, under the umbrella of FAO, is responsible for multilateral management of fisheries in the Mediterranean and Black Sea basins.

Gross Tonnage (GT). It is a unitless index related to a ship's overall internal volume. Gross tonnage is different from gross register tonnage. Gross tonnage is calculated by measuring a ship's volume (from keel to funnel, to the outside of the hull framing) and applying a mathematical formula. In the CFP it is one of the two parameters to define fleet capacity.

Highly migratory fish. Fish species that undertake migrations of significant but variable distances across oceans for feeding, often on forage fish, or reproduction, and also have wide geographic distributions. Thus, these species are found both inside the 200 mile exclusive economic zones and in the high seas outside these zones.

Integrated Maritime Policy (IMP). New policy approach by the Commission whereby all aspects relating to the European maritime activities are looked at together in an integrated manner.

Inter-American Tropical Tuna Commission (IATTC). International commission that is responsible for the conservation and management of tuna and other marine resources in the eastern Pacific Ocean.

International Baltic Sea Fisheries Commission (IBSFC). Organisation responsible for the management of fishery resources in the Baltic Sea. Established in 1973, it was discontinued in 2005 following the accession to the EU of four Baltic States and replaced by a bilateral agreement between the EU and the Russian Federation for the purpose of jointly managing Baltic fish stocks.

International Commission for the Conservation of Atlantic Tuna (ICCAT). Organisation established in 1967 to manage highly migratory fish in the Atlantic and Mediterranean seas. The headquarters are in Madrid, Spain. The EU is a member since 1998.

International Council for the Exploration of the Sea (ICES). Scientific organisation that coordinates marine and fisheries science in the North Atlantic since 1902. It is the main provider of scientific advice to the CFP.

International Commission for North Atlantic Fisheries (ICNAF). International fishery management organisation for the Northwest Atlantic that preceded the Northwest Atlantic Fisheries Organisation (NAFO).

Indian Ocean Tuna Commission (IOTC). Commission responsible for the management of tuna and tuna-like species in the Indian Ocean.

Individual transferable Quotas (ITQs). System of allocation of fishing rights to individuals or companies, allowing them to trade them under certain conditions.

Illegal, Unreported and Unregulated fishing (IUU). Fishing activity that is exerted around the world by vessels that do not respect fishery rules, do not report their activity or operate under conditions of lack of regulatory activity.

Multiannual Guidance Programmes (MAGPs). Programmes established between 1983 and 2001 whereby each Member States had objectives to adjust the capacity of their fishing fleets to available resources. They were suppressed in 2002.

Maximum Economic Yield (MEY). The value of the largest (average) positive difference between total revenues and total costs of fishing with all inputs valued at their opportunity costs in a fishery under stable conditions.

Maximum Sustainable Yield (MSY). The highest theoretical equilibrium yield that can be continuously taken (on average) from a stock under existing (average) environmental conditions without affecting significantly the reproduction process.

MEPs. Members of the European Parliament.

'**Micromanagement.**' Term commonly used to refer to the traditional decision-making process of the CFP, where Ministers often decide on regulations that include very detailed technical rules.

Marine strategy Framework Directive (MSFD). EU environmental legislation establishing the objective of achieving a 'good environmental status' for the waters around Europe by 2020.

Northwest Atlantic Fishery Organisation (NAFO). Intergovernmental organisation with a mandate to provide scientific advice and management of fisheries in the north-western part of the Atlantic Ocean.

Natura 2000. Network of areas protected to preserve species and habitats of high ecological value. They apply to the EU territory but also to the marine areas under the jurisdiction of Member States and may include measures to protect these areas from possible negative impacts of fishing.

North East Atlantic Fishery Commission (NEAFC). Regional fisheries management organisation that maintains controls over fishing and fishing-related acts in the North East Atlantic Ocean.

Paper fish. Expression related to certain decisions to set catch quotas in excess of real available fishing possibilities, or in some cases in areas where the fish does not occur. So these fish exist only on paper, not in the fishing grounds.

Pelagic fish. Fish that occur in surface marine waters.

Regulation. Legal act establishing rules applicable by the Member States of the EU. They are the normal legislative act of the CFP. Until 2009, they were adopted only by the Council of Ministers. Upon entry into force of the Treaty of Lisbon, the regulations of the CFP are (in most cases) adopted jointly by the European Parliament and the Council of Ministers.

Relative stability. Principle adopted in 1983 whereby fishing possibilities are allocated to Member States according to allocation keys that are stable and not re-opened for discussion.

Regional Advisory Councils (RACs). Stakeholder consultation bodies established in the CFP reform of 2002 to advise the European Commission on matters related to the implementation of the CFP.

Regionalisation. Term used to refer to the delegation of certain decisions of the CFP to a group of Member States of the EU acting through collaboration, normally with respect to a given sea basin. It does not refer at all to regions within Member States.

Scrapping. Destruction of an active fishing vessel. Practice traditionally subsidised by the EU through public aid.

Scientific, Technical and Economic Committee for Fisheries (STECF). Scientific advisory body established by the European Commission to advice on the conservation pillar of the CFP. Established initially in 1983, it was re-established in its current form by adding fisheries economics in its remit in 1992.

Straddling stocks. Fish stocks distributed between zones of national jurisdiction and international waters.

Structural policy. Policy of the EU applicable to different economic sectors with the aim of improving the structures of production to ensure their long term competitiveness.

Technical Conservation Measures. Fishery rules whose objective is to ensure the selectivity of the fishing activity. They take the form of limitations in the size of nets, areas where fishing is restricted, minimum fish sizes and so on.

TFEU. see Treaty of Lisbon.

Total Allowable Catches (TACs). Annual limitations of catches by fish stock, for all fleets concerned. In the EU, TACs are divided among Member States in national quotas.

Temporary cessation. Also referred to as '**temporary laying-ups**'. Action of subsidising the temporary stoppage of activity of fishing vessels, for different purposes.

Transferable Fisheries Concessions (TFCs). Mechanism of transfer of quotas among vessel owners of Member States of the Union, proposed by the Commission but not accepted by the co-legislators in the reform of the CFP in 2013.

Trawl. A cone or funnel-shaped net that is towed through the water by one or more vessels. They can be towed in the water column (pelagic trawl) or in contact with the sea bottom (bottom trawl).

Treaty of Accession. The legal text that establishes the conditions for the accession of a new country to the European Union.

Treaty of Lisbon. The Treaty of Lisbon is an international agreement which amends the two treaties which form the constitutional basis of the European Union (EU). The Treaty of Lisbon was signed by the EU Member States on 13 December 2007 and entered into force on 1 December 2009. At Lisbon, the Treaty of Rome was renamed to the Treaty on the Functioning of the European Union **(TFEU)**.

Treaty of Rome. Initial Treaty establishing the European Community with six countries in 1958. New Treaties have been developed over the years and new countries have joined the Union, until today's treaty of Lisbon with 28 member countries.

Trilogue. Term referring to the tripartite negotiation to adopt regulations under the CFP, as the ordinary legislative procedure. Upon a proposal from the Commission, the Council of Ministers and the European Parliament negotiate the final terms of a regulation, with the participation of the Commission. The term is a neologism meaning 'dialogue among three'.

United Nations Convention on the Law of the Sea. The Law of the Sea Convention defines the rights and responsibilities of nations with respect to their use of the world's oceans, establishing guidelines for businesses, the environment, and the management of marine natural resources. Signed in 1982, it entered into force in 1994.

United Nations Fish Stock Agreement (UNFSA). The United Nations Agreement for the Implementation of the Provisions of the United Nations Convention on the Law of the Sea of 10 December 1982 relating to the Conservation and Management of Straddling Fish Stocks and Highly Migratory Fish Stocks (in force as from 11 December 2001).

Vessel Monitoring System (VMS). System of satellite tracking of vessels. Also commonly known as the 'blue box'.

Western Waters. Atlantic waters off the west coast of the European Union, from the west of Scotland to the Canary Islands, where a specific effort limitation regime applies.

Index

The Common Fisheries Policy: The Quest for Sustainability, First Edition. Ernesto Penas Lado.

www.ingramcontent.com/pod-product-compliance
Ingram Content Group UK Ltd.
Pitfield, Milton Keynes, MK11 3LW, UK
UKHW050842101026
15624UKWH00018B/444

9 781119 085645